Molecularly Imprinted Polymers as Artificial Antibodies for the Environmental Health

Santanu Patra · Mika Sillanpaa
Editors

Molecularly Imprinted Polymers as Artificial Antibodies for the Environmental Health

A Step Towards Achieving the Sustainable Development Goals

Springer

Editors
Santanu Patra
IDUN, Department of Health Technology
Technical University of Denmark
Lyngby, Denmark

Mika Sillanpaa
Department of Chemical Engineering
Technology
University of Johannesburg
Doornfontein, South Africa

ISBN 978-3-031-58994-2 ISBN 978-3-031-58995-9 (eBook)
https://doi.org/10.1007/978-3-031-58995-9

This Springer imprint is published by the registered company Springer Nature Switzerland AG
The registered company address is: Gewerbestrasse 11, 6330 Cham, Switzerland

Preface

Protecting environmental health has become increasingly crucial in the face of the expanding implications of urbanization, industrialization, and subsequent ecological concerns. Researchers are constantly trying to establish new, innovative, and efficient materials to identify and reduce environmental threats. Artificial antibodies, which are synthetic and aim to imitate the binding capabilities and selectivity of natural antibodies, are one of the most encouraging developments in this field. This book presents the idea of artificial antibodies and their manifestation in the form of molecularly imprinted polymers (MIPs). Their superlative capacity to recognize and bind specific target molecules with high affinity makes them exceptional in the detection and separation of environmental pollutants. In light of the growing interest in artificial antibodies, which are all centered on human health, the purpose of this book is to concentrate on the application of artificial antibodies to improve environmental health.

By delving into the complex realm of MIPs, this book investigates the transformational potential of these materials for protecting our ecosystems and, eventually, our health. By examining these artificial antibodies, the book discusses their crucial role in the fight against environmental pollutants, showing their importance as a scientific marvel and a beacon of hope for a healthy planet. It examines how MIPs can be integrated into analytical instrumentation and detection systems to enhance environmental analysis.

The book chapters cover the fundamental principles of MIPs and their synthesis methodologies, offering readers a solid understanding of these unique materials. It delves into the design and selection of template molecules for imprinting, as well as polymerization techniques and strategies for optimizing MIP performance. The chapters highlight the case studies and analytical applications of MIPs in environmental health—including the sensors, separation and purification techniques, sample preparation, and preconcentration methods. The final chapter also discusses the limitations and challenges in the practical implementation of MIPs.

This book is composed for a large readership, including university students and researchers from diverse backgrounds such as chemistry, materials science, polymers, and environmental science and technology. It can be used as a reference book for both undergraduate and graduate students along with the academicians, researchers, and industries in these fields. We hope that the chapters of this book will provide the readers with valuable insight into state-of-the-art molecularly imprinting technologies and their application in environmental health.

Lyngby, Denmark Santanu Patra
Doornfontein, South Africa Mika Sillanpaa
February 2024

Contents

Chapter 1
Overview of Molecular Recognition and the Concept of MIPs

Yeşeren Saylan, Özge Altıntaş, Özgecan Erdem, Fatih Inci, and Adil Denizli

1.1 Introduction

Molecular recognition is a fundamental phenomenon governing the interaction between molecules in the realm of chemistry and biology. It plays a central role in a diverse range of processes, spanning from the precise interaction of enzymes to their substrates in biological systems to the selective capture of target molecules in chemical sensors and separation technologies (Tashiro and Shionoya 2020). At its core, molecular recognition involves the ability of molecules to selectively recognize and interact with one another based on their unique structural and chemical characteristics (Gellman 1997). One intriguing aspect of molecular recognition is the development and utilization of molecularly imprinted polymers (MIPs). MIPs represent a remarkable approach in the field of materials science and nanotechnology, where synthetic polymers are designed to mimic the selective binding capabilities of natural receptors, such as antibodies and enzymes (Zhou et al. 2022). This concept has paved the way for the creation of tailor-made materials with the ability to bind and recognize to specific target molecules with high selectivity and affinity (Haupt et al. 2011). Molecular imprinting represents a unique process in which the target molecule or its modified version, serves as a template. This template guides the arrangement and copolymerization of interacting monomers, resulting in the creation of a solid shell resembling a cast. Initially, these monomers form complexes with the template, facilitated by non-covalent or covalent interactions (Murdaya et al. 2022). After the polymerization process and the subsequent template removal, the polymer develops

Y. Saylan · Ö. Altıntaş · A. Denizli (✉)
Department of Chemistry, Hacettepe University, Ankara, Turkey
e-mail: denizli@hacettepe.edu.tr

Ö. Erdem · F. Inci
UNAM-National Nanotechnology Research Center, Bilkent University, Ankara, Turkey

F. Inci
Institute of Materials Science and Nanotechnology, Bilkent University, Ankara, Turkey

S. Patra and M. Sillanpaa (eds.), *Molecularly Imprinted Polymers as Artificial Antibodies for the Environmental Health*, https://doi.org/10.1007/978-3-031-58995-9_1

binding sites. These binding sites possess a congruence in size, shape and functional group positioning with the original template and they are securely anchored within the cross-linked structure (Saylan and Denizli 2019). Essentially, the polymer retains a molecular memory that enables it to selectively rebind the template molecule. Consequently, molecularly imprinted polymers exhibit a crucial characteristic shared with biological receptors: the skill to specifically bind and recognize particular target molecules (Saylan et al. 2020a). Nonetheless, from a structural viewpoint, MIPs exhibit substantial differences compared to biological receptors. In proteins, individual monomers (amino acids) are meticulously arranged in a sequence, ultimately folding into well-defined secondary and tertiary structures (Zhang et al. 2022). In contrast, MIPs incorporate monomers randomly, lack control over chain length and exhibit a highly intricate, chaotic network due to extensive cross-linking. While the imprinting process creates specific binding sites, their population often remains diverse, influenced by equilibria governing template-monomer complex formation and dynamic polymer chain growth before complete polymerization (Refaat et al. 2019). The resulting varied pore size distribution in the end material, along with the confinement of binding sites within the bulk, can result in slow mass transfer. It's worth noting that MIPs may contain thousands or even millions of binding sites, a notable difference from biological receptors, which generally have few or just one (Rahman et al. 2022). Although not universally problematic, these characteristics can limit the substitution of MIPs for antibodies in certain applications (Refaat et al. 2019). Consequently, current research in the field of MIPs is directed toward identifying solutions or alternative approaches to address these potential limitations. One of the appealing aspects of molecular imprinting is its versatility, allowing it to be applied to a wide array of target molecules (Ali and Omer 2022). The technique has been successfully employed to imprint small organic molecules, such as sugars, steroids, nucleotides, peptides, amino acids, pesticides and pharmaceuticals (Bozal-Palabiyik et al. 2020). This technique is now considered a well-established and nearly routine practice. Additionally, MIPs can be synthesized in various physical forms such as quasi-soluble nanogels, nanocomposites, nanostructured films, thin films, nanowires, nanospheres and including porous microspheres (Rahman et al. 2022).

In this chapter, we will delve into the fascinating world of molecular recognition and explore the underlying principles of MIPs. We will discuss the key concepts, applications and the intricate molecular imprinting process that allows scientists and engineers to engineer materials capable of molecular recognition. Furthermore, we will explore the diverse range of applications where MIPs have found utility, from pharmaceuticals and environmental monitoring to medical and biotechnology. As we delve deeper into this topic, we will gain a better understanding of how molecular recognition and MIPs are shaping the landscape of modern science and technology.

1.2 Molecular Recognition

Molecular recognition is a fascinating and vital phenomenon that underpins numerous processes in chemistry, biology and materials science. Molecules can interact selectively with one another based on their unique structural and chemical properties (Cao et al. 2019). This selective interaction is akin to a molecular handshake, where molecules recognize each other and form specific, complementary bonds, enabling a wide array of essential functions and applications (Chen et al. 2020; Persch et al. 2015).

- The basic principles of molecular recognition are as follows:

  Lock-and-key principle: The fundamental concept of molecular recognition can be likened to a lock-and-key mechanism. In this analogy, molecules act as keys, and their binding sites or receptors serve as locks. For a successful interaction to occur, the key (molecule) must precisely fit into the lock (receptor), ensuring a secure and specific connection (Chatterji 2016).
  Complementary interactions: Molecular recognition is driven by various intermolecular forces, such as hydrophobic effects, electrostatic interactions, van der Waals forces and hydrogen bonding. These forces play a pivotal role in determining the strength and specificity of molecular interactions (Paleos et al. 2001).
  Specificity and affinity: Molecular recognition is highly specific, meaning that a molecule will typically bind selectively to its cognate receptor or target. This specificity arises from the precise geometric and chemical complementarity between the interacting molecules. Additionally, molecular interactions can exhibit varying levels of affinity, reflecting the strength of the bond formed between the molecules (Skerra 2007).

- The biological significance of molecular recognition is as follows:

  Enzyme–substrate binding: Enzymes recognize and bind to specific substrates, facilitating chemical reactions essential for metabolism and cellular functions.
  Antigen–antibody interactions: The immune system relies on molecular recognition to neutralize and identify foreign invaders, such as viruses and bacteria (Ciferri 2021).
  Hormone-receptor interactions: Hormones bind to receptors on target cells, triggering a cascade of cellular responses that regulate bodily functions (Behncken and Waters 1999).
  Cell–cell communication: Cells recognize and respond to signaling molecules, enabling coordination within tissues and organs (Krengel and Bousquet 2014).

- Some of the technological applications of molecular recognition are as follows:

  Chemical sensors: Molecularly selective materials are used in chemical sensors to detect specific analytes, ranging from pollutants in the environment to biomarkers in medical diagnostics (Pirondini and Dalcanale 2007).

Drug design: Understanding molecular recognition is crucial in designing pharmaceutical compounds that selectively interact with their target proteins, minimizing side effects and maximizing therapeutic efficacy (Babine and Bender 1997).
Separation technologies: Molecular recognition-based materials, such as Molecularly Imprinted Polymers (MIPs), are employed in chromatography and filtration for the separation and purification of molecules (Moyer et al. 2013).
Molecular electronics: Researchers are exploring molecular recognition for the development of molecular switches and devices, potentially revolutionizing the field of electronics (Heath and Ratner 2003).

1.3 Molecular Imprinting Method

The molecular imprinting method is a fascinating and powerful technique in the field of materials science and chemistry. It is designed to create highly specific recognition sites, often referred to as molecular memory, within a polymer matrix. These recognition sites are tailored to bind selectively with a particular target molecule, making molecular imprinting an essential tool for various applications, including sensors, drug delivery and chromatography (Chen et al. 2016). The process of molecular imprinting begins with the selection of a template, known as the target molecule. This template molecule serves as the blueprint for creating complementary binding sites within a polymer matrix (Tarannum et al. 2020). The template is mixed with a functional monomer, which can form reversible covalent bonds with the template. Cross-linking agents are also introduced to link the monomers together, creating a three-dimensional polymer network (Piletsky 2006; Whitcombe et al. 2014). Once the polymerization process is complete, the template molecule is removed, leaving behind cavities or imprints with specific sizes, shapes and chemical functionalities that precisely match the template molecule (Saylan and Denizli 2020). These imprints are essentially the molecular memory of the target molecule (Arabi et al. 2021; Kandimalla and Ju 2004). Usually, there are two primary approaches for creating MIPs, relying on interactions between the functional monomer and the template. Covalent imprinting, which operates in a stoichiometric manner, ensures that functional monomer residues are exclusively present within the imprinted cavities (Huang et al. 2021). This approach is traditional and frequently relies on reversible condensation reactions involving Schiff's base, ketals/acetals, and boronate esters. Nonetheless, covalent imprinting is seen as somewhat rigid, primarily due to the constraints imposed by reversible condensation reactions. Additionally, achieving thermodynamic equilibrium can be challenging because strong covalent interactions lead to slow binding and dissociation (Dong et al. 2021).

In contrast, noncovalent imprinting relies on p-p interactions, van der Waals forces, hydrogen bonding, and ionic interactions. Among these, hydrogen bonding, especially between methacrylic acid (MAA) groups and primary amines in nonpolar solvents, stands out as the most prevalent form of noncovalent interaction (Włoch and Datta 2019). Lately, noncovalent imprinting has garnered attention as a common synthesis approach, owing to its straightforward operation and the rapidity of binding and removal processes. Nonetheless, noncovalent imprinting is sensitive to even slight disruptions in the interactions that stabilize the complex, such as the presence of water, rendering it less robust (Ndunda 2020). To amalgamate the longevity associated with covalent imprinting and the swift target uptake observed in noncovalent imprinting, a novel technique referred to as semicovalent imprinting has surfaced (Chen et al. 2016). This approach offers an intermediate solution in which the template is covalently linked to the functional monomer, while the process of template rebinding relies on noncovalent interactions (Akgönüllü et al. 2023).

1.4 Molecularly Imprinted Polymers

MIPs stand as a captivating and groundbreaking category of materials, finding application in a diverse array of fields. These engineered polymers possess a distinctive ability: the capacity to selectively identify and attach to precise target molecules at the molecular level. Often referred to as plastic antibodies, MIPs mimic the role of their biological counterparts in molecular recognition (Piletsky et al. 2001). The primary advantage of MIPs lies in their capacity to selectively identify and attach to the template molecule or molecules that closely resemble its structure (Lusina and Cegłowski 2022). This selectivity emerges from the precise molecular interactions between the imprints within the polymer and the target molecule. These interactions can encompass electrostatic attractions, van der Waals forces and hydrogen bonding (Shimizu and Stephenson 2010; Wang et al. 2023c). This selectivity makes MIPs invaluable in various applications where precise molecular recognition is required (Mahony et al. 2005). MIPs are created through a process known as molecular imprinting. This process involves polymerizing a mixture of cross-linkers and monomers in the presence of a template molecule (Lusina and Cegłowski 2022). The template molecule is the specific molecule to which the resulting MIP will be designed to bind. After the polymerization process reaches its conclusion, the template molecule is removed, resulting in binding sites within the polymer network that match the size, shape and functionalities of the target compound (Aghoutane et al. 2020). These imprinted polymers are characterized by their stability, resilience and ability to withstand a broad spectrum of pH levels, solvents and temperature variations (Cormack and Elorza 2004). Therefore, MIPs emulates the selective retention of target molecules (similar to antibody-antigen interactions) without the associated stability constraints observed in natural receptors (Mujahid et al. 2023).

Moreover, the production of MIPs is notably uncomplicated and budget-friendly, presenting a practical substitute for utilizing natural receptors. There are three primary methodologies for MIP synthesis: covalent, noncovalent and semi-covalent methods. (Włoch and Datta 2019). In the covalent approach, reversible covalent bonds are established between the template molecule and monomers before polymerization. Following this, the template is removed from the polymer by breaking the corresponding covalent bonds, which are then reformed upon the rebinding of the target compound (Li et al. 2019). This technique yields a polymer with a comparatively consistent distribution of binding sites, thereby diminishing the presence of nonspecific sites owing to the robust stability of template-monomer interactions (Pandey et al. 2023). Nonetheless, the covalent approach comes with limitations because it can be challenging to design a template-monomer complex in which covalent bond formation and cleavage can readily take place under mild conditions (Rajpal et al. 2023). The semi-covalent approach provides an intermediate alternative. In this scenario, the template is covalently linked to a functional monomer, but the process of template rebinding relies exclusively on noncovalent interactions (Ensafi and Nasr-Esfahani 2021). The noncovalent approach, pioneered by Arshady and Mosbach, revolves around the establishment of relatively weak noncovalent interactions, including hydrogen bonds and ionic interactions, between the template molecule and chosen monomers before the polymerization process begins (Arshady and Mosbach 1981). This approach is the most adopted method for producing MIPs, primarily due to its simplicity and the extensive selection of available monomers (Ding et al. 2020; Saylan et al. 2023; Turiel and Martín-Esteban 2010).

1.4.1 Concept of MIPs

MIPs represent a pioneering concept in the realm of materials science and molecular recognition. They are a class of synthetic polymers engineered with a specific purpose: to selectively bind and recognize to target molecules with a high degree of precision (Wulff 2013). The foundational principle of MIPs revolves around the concept of molecular imprinting, a process that mirrors the natural lock-and-key mechanism of molecular recognition but within a synthetic framework (Zaidi 2020). At the heart of MIPs lies the concept of molecular imprinting. This intricate process begins with the selection of a target molecule, often referred to as the template. This template molecule possesses unique structural and chemical characteristics that are of interest for selective recognition (Muratsugu et al. 2020). The template is then embedded within a mixture of monomers and cross-linkers (Suzaei et al. 2023; Tse Sum Bui et al. 2023). During polymerization, these monomers and cross-linkers form a polymer network around the template molecule. This polymerization process imprints the template's molecular structure onto the polymer matrix, creating binding sites that are specifically complementary in terms of chemical functionality, shape and size to the template (Mayes and Mosbach 1997). A standard MIP synthesis protocol consists of several key components, including a solvent, a polymerization initiator, a cross-linker, a functional monomer and template (Refaat et al. 2019). To enhance the properties of MIPs, numerous endeavors have been undertaken, as the polymerization reaction is affected by a variety of factors. These factors include the monomer and cross-linker type and quantity, the selection of solvent and initiator, and the temperature and duration of the polymerization process (Madikizela et al. 2022). A summary of the various templates, monomers, various production techniques and applications of MIPs is given in Table 1.1.

Functional monomers: Functional monomers play a crucial role in the formation of templates with pre-polymerization complexes. These monomers possess specific functional groups that engage in interactions and binding with the template molecules (Zhang et al. 2010). The choice of an appropriate functional monomer is of paramount importance to establish a robust connection with the template, ultimately leading to the creation of well-defined antibody-antigen complexes or donor-receptor during the polymerization process (Chen et al. 2016). The selection of the most suitable functional monomer is a critical factor in tailoring specific cavities designed to mimic template molecules. It's worth noting that the quantity of functional monomers employed in the creation of MIPs is finite, which can impose limitations on their selectivity and potential applications (Hasanah et al. 2021).

To establish a strong and precise interaction with the template, it is imperative to design and incorporate a novel functional monomer characterized by a high degree of specificity and selectivity towards the target analyte (Elaine et al. 2022). Functional monomers typically feature two common domains: one for molecular recognition and another for polymerization. The interactions and mechanisms between the template molecule and the functional monomer take place during the pre-polymerization

Table 1.1 Various templates, monomers, fabrication techniques and applications of MIPs

Template	Monomers	Fabrication techniques	Applications
Proteins	Methacrylate derivatives, acrylic acid	Surface imprinting, covalent imprinting	Bio-sensing, drug targeting, biomimetic catalysis
Peptides	Methacrylates, styrene derivatives	Mini-emulsion polymerization, sol–gel polymerization	Biomolecular recognition, diagnostics
Nucleotides	Methacrylates, vinyl monomers	Electrochemical polymerization	DNA sensing, diagnostic devices
Amino acids	Methacrylates, acrylamide derivatives	Molecularly imprinted membranes	Enantioselective separation, chiral recognition
Drugs	Acrylates, methacrylates	Suspension polymerization	Controlled drug release, therapeutic delivery
Environmental pollutants	Methacrylates, acrylamide derivatives	Emulsion polymerization, bulk polymerization	Water purification, sensor development
Small molecules	Acrylamide, methacrylate derivatives	Bulk polymerization, precipitation polymerization	Drug delivery, sensing, separation

phase, and the outcome is heavily influenced by the quantity and quality of the recognition component within a MIP (Cowen et al. 2020). Functional monomers encompass various types, including carboxylic acids, sulfonic acids, heteroaromatics, anilines and pyrroles, among others (Pratama et al. 2020). Examples of carboxylic acid-based monomers comprise benzoic acid, acrylic acid and methacrylic acid (MAA) (Nishchaya et al. 2023). On the other hand, sulfonic acid monomers encompass compounds like acrylamide methylpropane sulfonic acid, while heteroaromatic base monomers encompass vinylimidazole and vinylpyridine (BelBruno 2018).

Templates: The choice of a template plays a crucial role in organizing functional groups within functional monomers. The functional group within a template, which may include ester, amide, hydroxyl, carboxyl or amino groups, significantly influences the performance of MIPs (Kadhem et al. 2021). Templates with highly polar groups can facilitate the creation of high-performance MIPs because they lead to the formation of more stable molecular complexes. Polymers that can establish hydrogen bonds with functional monomers have the potential to display a high degree of affinity and selectivity, thanks to the directional, saturating and strong nature of hydrogen bonds (Morsi et al. 2023). The goal of molecular imprinting is to synthesize MIPs that closely resemble biological receptors, with the potential to replace them in various applications (Wang et al. 2023a). An ideal template must meet three key criteria: first, its functional groups should not hinder the polymerization process; second, it should demonstrate exceptional chemical stability throughout polymerization; and third, its functional groups should be capable of forming complexes with functional monomers (Ensafi et al. 2021). Templates for specific targets find broad application in various fields, including clinical, industrial, chemical, pharmaceutical, biological and environmental domains (Haupt 2010). The creation of molecular imprints for organic molecules like tyrosine, tetracycline and atrazine, as well as ion-imprinted polymers tailored for toxic heavy metals such as aluminum ions, mercury ions and chromium ions, has been effectively accomplished. While similar approaches can be used to imprint biological macromolecules like proteins or viruses, the imprinting technique for these complex templates remains a challenge (Soufi et al. 2021). Macromolecules like proteins are exceptionally complex and have nonspecific recognition sites on their surfaces, often due to the presence of charged residues (Janiak and Kofinas 2007).

Initiators: An initiator acts as a substance that initiates the beginning of polymer chains and accelerates the polymerization process. The quantity of initiators used is considerably less than the number of functional monomers. The techniques utilized in MIP preparation include electropolymerization, photopolymerization, and free radical polymerization (FRP) (Chen et al. 2016). FRP can be initiated either through the application of heat (thermal) or exposure to light (photochemical). It is versatile and can be applied to a broad spectrum of functional groups and template structures. (Malik et al. 2019). Various types of initiators find use in the MIP preparation process, with common ones including potassium persulfate (KPS), dimethyl acetal of benzyl (BDK), benzoyl peroxide (BPO), azobisdimethylvaleronitrile (ADVN) and azobisisobutyronitrile (AIBN) (Pratama et al. 2020; Xu et al. 2020a). Photopolymerization, a form of FRP, commences through exposure to light. This approach offers several advantages, lower reaction temperatures, minimal waste generation, increased productivity and reduced energy consumption (Lee et al. 2001). However, the depth of light penetration is constrained and depends on the wavelength and spectral distribution. The photochemical initiation process involves subjecting relevant photoinitiators to UV irradiation (Paruli et al. 2021). Typically, organic photoinitiators are employed in photopolymerization. Organic polymerization photoinitiators are categorized into various classes based on their function (Fuchs et al. 2012). Often, a mixed type of photoinitiator is used to achieve the best overall curing speed and minimal color formation. Commonly used organic photoinitiators include cationic photoinitiators, thioxanthones, miscellaneous compounds, benzophenone, benzyl/benzoin compounds and acetophenone (Pratama et al. 2020). Electropolymerization is a commencement procedure that manages the quantity of cycles or amperage directed towards an electrode to produce a polymer film with a predetermined thickness (Zembrzuska et al. 2019). An electrosynthesized polymer can serve as the framework to anchor MIP particles and enable their interaction with the electronic surface. In this method, MIP particles are dispersed in a monomer solution, and then polymerization takes place. Electropolymerization proves to be a valuable approach for creating MIPs, especially when crafting electrochemical sensors (Peng and Su 2015; Roushani and Zalpour 2021).

Solvent: Porogens play a vital role in the polymerization process by serving as dispersing agents and agents for creating pores within the polymer matrix. These substances are responsible for dissolving all the components used in the synthesis of MIPs (Song et al. 2022). The polarity of a porogen can profoundly influence the relationships between functional monomers and the template, especially in non-covalent interaction systems. In non-covalent imprinting techniques, it is common to use a non-polar or less polar organic porogen, like chloroform, acetonitrile or toluene, to attain the highest level of imprinting efficiency (Esfandyari-Manesh et al. 2011). Ultimately, the selection of solvent or porogen is what dictates the adsorption and morphological attributes of the resulting polymer (Schmidt et al. 2005).

Crosslinkers: A crosslinker plays a crucial role in arranging the functional monomers around template molecules as the polymerization takes place. Integrating these elements into the production of MIPs leads to the formation of a polymer that is extensively interconnected and rigid once the template has been extracted (Yan and Row 2006). The specificity and binding capacity of an MIP are directly impacted by the specific crosslinking agents chosen and the amount used. Generally, an inadequate crosslinker can result in mechanical instability in the MIP (Du et al. 2014). Conversely, the increased use of crosslinker reduces the quantity of recognition sites within the MIP. In the preparation of MIPs, a combination of non-covalent and covalent interactions is employed (Bonatti et al. 2021). There is a distinct interaction that takes place between the functional monomer and the template when the analyte is present. In the context of covalent imprinting, the resulting imprinted cavity retains a portion of the functional monomer and upholds a stoichiometric ratio (Spivak 2005). Covalent imprinting relies on reversible covalent bonds and typically employs crosslinkers such as triallyl isocyanurate (TAIC) (Rahman et al. 2012), bis-(1-(tert-butylperoxy)-1-methylethyl)-benzene (BIPB) (Pratama et al. 2020) or dicumylperoxide (DCP) (Idil et al. 2021). In contrast, non-covalent imprinting relies on π–π interactions, Van der Waals forces, hydrogen bonds and ionic interactions (He et al. 2020). Of these, hydrogen bonds are the most frequently utilized. Non-covalent imprinting methods have gained wider recognition in MIP production and utilize crosslinkers like ethylene glycol dimethacrylate (EGDMA) (Iturralde et al. 2014), methylenediacrylamide (MBAA) (Lanza and Sellergren 2004), divinylbenzene (DVB) (Ma and Row 2018), diisopropenylbenzene (McCluskey et al. 2007), bisacryloylamidopyridine (Divya et al. 2023), N,O-bismethacryloyl ethanolamine (NOBE) (Sibrian-Vazquez and Spivak 2004), glycidyl methacrylate (GMA) (Wang et al. 2020) and trimethylene trimethylolpropane (TRIM) (Muhammad et al. 2012).

1.4.2 Commercializaton of MIPs

MIPs have emerged as a versatile and promising technology with a wide range of applications, driving their commercialization in various industries (Gkika et al. 2024) So, these synthetic polymers are designed to mimic the molecular recognition properties of natural receptors, such as antibodies and enzymes. The ability to selectively bind to specific target molecules makes MIPs valuable in numerous real-world contexts (Bräuer et al. 2021). Key benefits stem from the inherent stability, chemical inertness, reusability, and scalability of these imprinted materials (Chen et al. 2016). In recent years, there has been a growing interest in the commercialization of MIPs, driven by their unique properties and versatile applications (Wang et al. 2023a; Ayivi et al. 2023; Suzaei et al. 2023). One of the notable applications of MIPs is in sensing and diagnostics. These polymers can be tailored to recognize and capture specific analytes, such as toxins, pathogens, or biomarkers. In the medical field, MIP-based sensors are employed for the detection of diseases, offering rapid and accurate diagnostic results. The commercialization of MIP-based diagnostic tools is gaining traction, particularly in point-of-care testing and personalized medicine (Li et al. 2024) MIPs find extensive use in environmental monitoring for the detection and removal of pollutants (Rahman et al. 2022).They can be designed to selectively adsorb contaminants like heavy metals, pesticides, or hydrocarbons. The commercial applications of MIPs in water and air purification systems contribute to sustainable environmental practices, ensuring the safety of ecosystems and human health (Ostovan et al. 2022). In the pharmaceutical industry, MIPs are utilized for drug delivery systems and separation processes (Zaidi 2016). The tailored molecular recognition properties of MIPs enable controlled release of drugs, enhancing therapeutic efficacy while minimizing side effects. The commercialization of MIP-based drug delivery platforms is a promising avenue for pharmaceutical companies seeking innovative solutions for patient care (Bodoki et al. 2021). Moreover, MIPs are increasingly finding applications in the food and beverage industry, particularly in the detection and removal of contaminants or undesirable compounds (Basak et al. 2022). They are employed in the production of high-quality and safe food products by selectively capturing toxins, allergens, or unwanted flavors. The commercial use of MIPs in food safety ensures compliance with regulatory standards and meets consumer demands for quality assurance (Lehotay and Chen 2018). In forensic science, MIPs play a crucial role in the identification and detection of specific substances, such as drugs or explosives. These polymers can be tailored to recognize the molecular signatures of illicit substances, aiding law enforcement agencies in forensic investigations. The commercialization of MIPs in forensic applications enhances the precision and efficiency of substance analysis (Karadurmus et al. 2022). MIPs are also integrated into various industrial processes for purification, catalysis, and separation. Their use in chromatography and solid-phase extraction contributes to the efficiency of manufacturing processes across industries such as petrochemicals, cosmetics, and materials science. The commercial adoption of MIPs in industrial applications promotes cost-effectiveness and sustainability (Saylan et al. 2024). As research continues to advance, MIPs are likely to play an increasingly pivotal role in addressing complex problems and meeting the evolving needs of various industries (Ostovan et al. 2022). As the commercialization of MIPs continues to progress, collaborations between research institutions and industry players are crucial for developing scalable and cost-effective production methods (Ayankojo et al. 2022). The versatility of MIPs across different sectors highlights their potential to revolutionize various aspects of technology, healthcare, and manufacturing (Leibl et al. 2021).

1.4.3 Applications of MIPs

In the realm of analytical procedures, it is typically required to possess the capability of selectively and sensitively detecting the target analyte. These requirements have been addressed by employing natural antibodies, receptors, or enzymes (Justino et al. 2015). Regrettably, the utilization of biomolecules is constrained to moderate circumstances due to the susceptibility of protein denaturation at severe temperatures, pH levels, and exposure to organic solvents. Furthermore, the process of immobilization is characterized by its demanding nature and the subsequent obstacles it poses in terms of coating surfaces, such as transducers. Moreover, it is commonly observed that they tend to possess a significant cost or require a substantial investment of time for their production. Artificial receptors have been designed in order to overcome these constraints (Wackerlig and Schirhagl 2016). MIPs have garnered significant interest and have emerged as a compelling choice in various domains, including separation, chemosensing, biosensing, drug delivery, catalysis and degradation. This interest stems from the distinctive features exhibited by MIPs, such as specificity in recognizing target structures and universality of application. These polymers are highly physically stable, can be easily prepared, exhibit remarkable robustness, and are cost-effective when compared to other recognition systems (Chen et al. 2016).The aim of this part is to offer a concise overview of MIP-based composite systems and the specific investigations conducted in this domain, focusing on research papers published after 2020.

Sensing Applications

Sensor systems combine a physical transducer (i.e., electrochemical, optical or piezoelectric) with a sensing component. This integration allows for the translation of interactions between the target and recognition molecules into a signal that can be measured. The utilization of sensor systems facilitates the expeditious and precise detection of labeled or label-free substances, hence diminishing the duration of assays and minimizing the requirement for preliminary sample pre-processing procedures. Therefore, these systems serve as robust alternatives to conventional analytical methodologies (Saylan et al. 2020a). The integration of sensors with MIPs has been successfully achieved for the purpose of screening in several fields, such as medical diagnostics (Ayankojo et al. 2020; Kidakova et al. 2020; Li et al. 2022; Raziq et al. 2021), food safety (Caldara et al. 2022; Işık et al. 2021; Wang et al. 2023b; Xu et al. 2020b) and environmental applications (Arabi and Chen 2022; Singh et al. 2020).

A recent study focused on the synthesis of MIPs that exhibit a high selectivity towards lactate and the key interaction sites between the MIPs and the lactate-monomer complex was also investigated. For the purpose of rapid detection of variations from established "normal" lactate levels (1 mM), a linear range of 0.1–1.7 mM was selected in this study. Furthermore, in order to assess the selectivity capabilities of synthesized MIPs, malic acid and sodium 2-hydroxybutyrate were employed as competitive interferents owing to their chemical similarity to the target compound. The thermal characterizations provided confirmation of interactions occurring between the template-monomer complex. Additionally, these characterizations indicated that lactate could be integrated into the polymer matrix without causing substantial changes to the thermal characteristics of the material. The analysis of surface area provided evidence for the effective incorporation of lactate cavities, as indicated by a 1.3-fold increase in comparison to the MIP relative to the NIP. Additionally, the analysis revealed the presence of interconnected networks and potential merging of cavities, which may impose constraints on the binding capacity of MIPs. The results indicate that the MIP possesses a notable capacity to recapture lactate molecules, with an efficiency reaching up to 63.5%. Furthermore, the selective properties were examined by studying structurally analogous compounds such as malic acid and sodium 2-hydroxybutyrate, which acted as interfering substances in lactate-containing cavities (Mustafa and Leese 2023). The development of a non-invasive, material-based electrochemical sensor that provides continuous monitoring of fenamiphos (FMS) levels in vegetable samples is of great importance for the regulation and management of agricultural products. The present study involved the fabrication of an electrochemical sensor utilizing gold stabilization on sensitive MIP/metal–organic framework/graphite carbon nitride (MIP-Au@MOF-235@g-C_3N_4). This sensor was designed for the purpose of continuous examination of the FMS level in real samples, exhibiting a notable degree of sensitivity. The MIP-Au@MOF-235@g-C_3N_4 composite had a significantly enhanced specific surface area and catalytic performance. Moreover, the sensor demonstrated a remarkable sensitivity of 1.07 μA/μM and a wide linear range spanning from 0.01 to 16.4 μM. Furthermore, the technique that has been put out demonstrates recoveries that are deemed satisfactory, ranging from 94.7% to 107.9% (Mehmandoust et al. 2023). *Pseudomonas aeruginosa (P. aeruginosa)* is a widely distributed bacteria that exhibits resistance to multiple drugs, and has the potential to induce severe infections. Another study was carried out with the objective of developing a thermal sensor for the indirect detection of *P. aeruginosa* contamination utilizing MIPs. The developed sensor was based on the utilization of MIPs for pyocyanin recognition, which acts as the predominant toxin produced by *P. aeruginosa.* Phenazine was employed as a mimic template to assess various polymeric compositions in order to develop a specific MIPs for the recognition of pyocyanin. Based on the findings derived from the UV–vis study conducted to evaluate the sensitivity of MIPs, the synthesized MIPs had an imprinting factor of 1.59 and a rebinding capacity of 30 μmol/g. The immobilization of MIPs onto planar aluminum chip was carried out using an adhesive layer in order to conduct thermal resistance tests at pyocyanin concentrations that are clinically relevant (ranging from 1.4 to 9.8 μM). This immobilization process allowed for the achievement of a limit of detection (LOD) of 0.347 ± 0.027 μM. In order to assess the practicality of utilizing the MIP-based sensor in clinical diagnostics, saliva samples spiked with pyocyanin were used, resulting in the successful detection of the target analyte at a detection limit of 0.569 ± 0.063 μM. The sensor that has been developed is capable of detecting the presence of toxins, even at low concentrations (Frigoli et al. 2023). Insulin is a peptide hormone responsible for the regulation of blood glucose levels and serves as a marker for diabetes. A novel electrochemical insulin sensor was fabricated by employing a gold electrode that was surface-modified with carboxylated multiwalled carbon nanotubes (f-MWCNTs) and MIP cryogel. The MIP exhibited distinct binding sites for insulin, whilst the macroporous structure of the cryogel facilitated the efficient transport of insulin molecules towards these binding sites (Fig. 1.1). The utilization of f-MWCNTs resulted in an increase of the sensor's effective surface area and conductivity, while concurrently weakening the potential necessary for insulin oxidation. The oxidation of insulin was quantitatively assessed in a continuous flow system with the application of square wave voltammetry. The cryogel/f-MWCNTs sensor demonstrated a linear range spanning from 0.050 to 1.40 pM, exhibiting an exceptionally low LOD of 33 fM. The sensor demonstrated a notable level of selectivity and maintained its stability for an extended duration of 10 weeks during dry condition at ambient temperature. The findings of insulin quantification in human serum utilizing the sensor shown a strong concordance with the outcomes obtained from the Elecsys insulin assay. The created sensor presents a potentially advantageous option for the diagnosis and management of diabetes (Wardani et al. 2023).

Sharma et al. outlines a method to detect cortisol in biological samples and aqueous solutions. It involves the use of MIPs-coated graphite electrodes that have been modified by silver nanoparticles through the technique of differential pulse voltammetry (DPV). The imprinted graphite electrode features an expanded surface area due to the incorporation of silver nanoparticles, resulting in enhanced electroactivity. The responses of DPV, with its exceptional electrical conductivity, was elucidated. The study applied different cortisol concentrations (between 0.395 and 3.96 nM) to the MIPs-coated graphite electrodes, and the DPV method yielded signal responses with

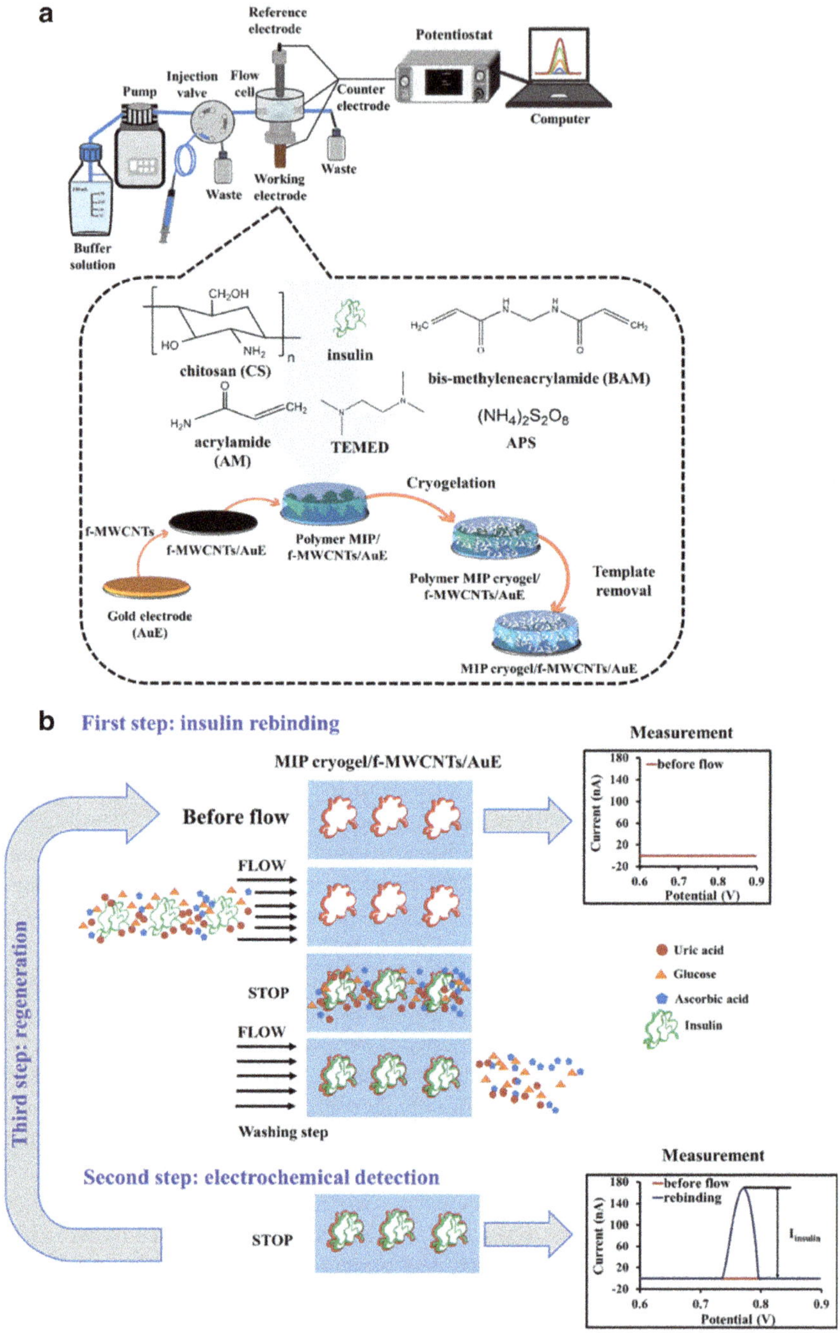

Fig. 1.1 Scheme of the MIP cryogel/f-MWCNTs modified gold electrode preparation and its use as a working electrode in a flow system (**a**) and steps of the insulin detection (**b**) (Wardani et al. 2023)

a high regression R^2 value of 0.9951. The tailored electrode exhibited strong electrocatalytic activity with a low LOD recorded at 0.214 nM. These results underscore the significant potential of this sensor for the determination of cortisol in biological samples (Shama et al. 2023). In this study, Saylan introduces the development of a MIPs-based optical sensor designed to detect *Enterococcus faecalis (E. faecalis)*, a fecal contaminant found in water. The optical sensor employs an imprinted nanofilm produced with combining gold nanoparticles. Thorough characterization results were conducted to assess the sensor's performance, encompassing aspects such as kinetics, reusability, selectivity, and real sample analysis for *E. faecalis* detection. The results indicate that the optical sensor shows exceptional determination capabilities, boasting a low LOD of 6.6×10^3 cfu/mL and a high correlation coefficient ($R^2 = 0.9886$) over the concentration range of 5×10^4–1×10^8 cfu/mL. Crucially, the sensor's effectiveness is not confined to buffer solution but extends to water solutions, underscoring its suitability for environmental applications. According to the findings, it is reasonable to conclude that the developed optical sensor holds substantial promise for other bacteria detection, giving a cost-effective solution with heightened sensitivity (Saylan 2023). Saylan et al. also synthesized magnetic nanoparticles which are combined with optical surfaces for detecting a clinically significant biomarker, interferon-α, in both buffer and artificial plasma sample solutions. Initially, they synthesized the magnetic nanoparticles and conducted a thorough characterization of their properties and then scrutinized potential external factors affecting these magnetic nanoparticles to establish a benchmark for detection performance (Fig. 1.2). Upon identifying the optimal conditions, they modified the optical sensor surface with these magnetic nanoparticles for detection of interferon-α. The kinetic performance of the sensor, enhanced by magnetic nanoparticles, exhibited a high correlation coefficient ($R^2 = 0.99$) and a low LOD (41 IU/mL) across a wide interferon-α concentrations range. As a result, this sensor demonstrates significant potential for the highly sensitive and interference-free detection of other biomolecules. It could serve as an alternative method for biomarker detection across various applications (Saylan et al. 2022).

As an updated study, Erdem et al. designed a micro-reactor for real-time and continuous synthesis of an exceptionally large quantity of MIP nanoparticles that carry the unique characteristics of bovine serum albumin. This synthesis process is accomplished rapidly, typically within a short time frame of 5–30 min. They first employed COMSOL simulations to assess the mixing efficiency by adjusting flow rates and then conducted practical experiments to validate the micro-reactor's performance in generating nanoparticles with 52–106 nm sizes. They also delved into the molecular interactions between monomers and protein by employing molecular docking and dynamics simulations. Following these steps, they assessed the micro-reactor's parameters by evaluating the uniformity and concentration of the molecularly imprinted polymers, utilizing principal component analysis. The sensory capabilities of these MIPs were then put to the test on a metamaterial sensor, yielding an impressive precision rate of 81%, along with a notably high level of selectivity (4.5 times greater than alternatives), all while maintaining usability over three consecutive cycles. The micro-reactor distinguishes itself through its remarkable productivity, offering a significant reduction in assay time (48–288 times faster) and a substantial

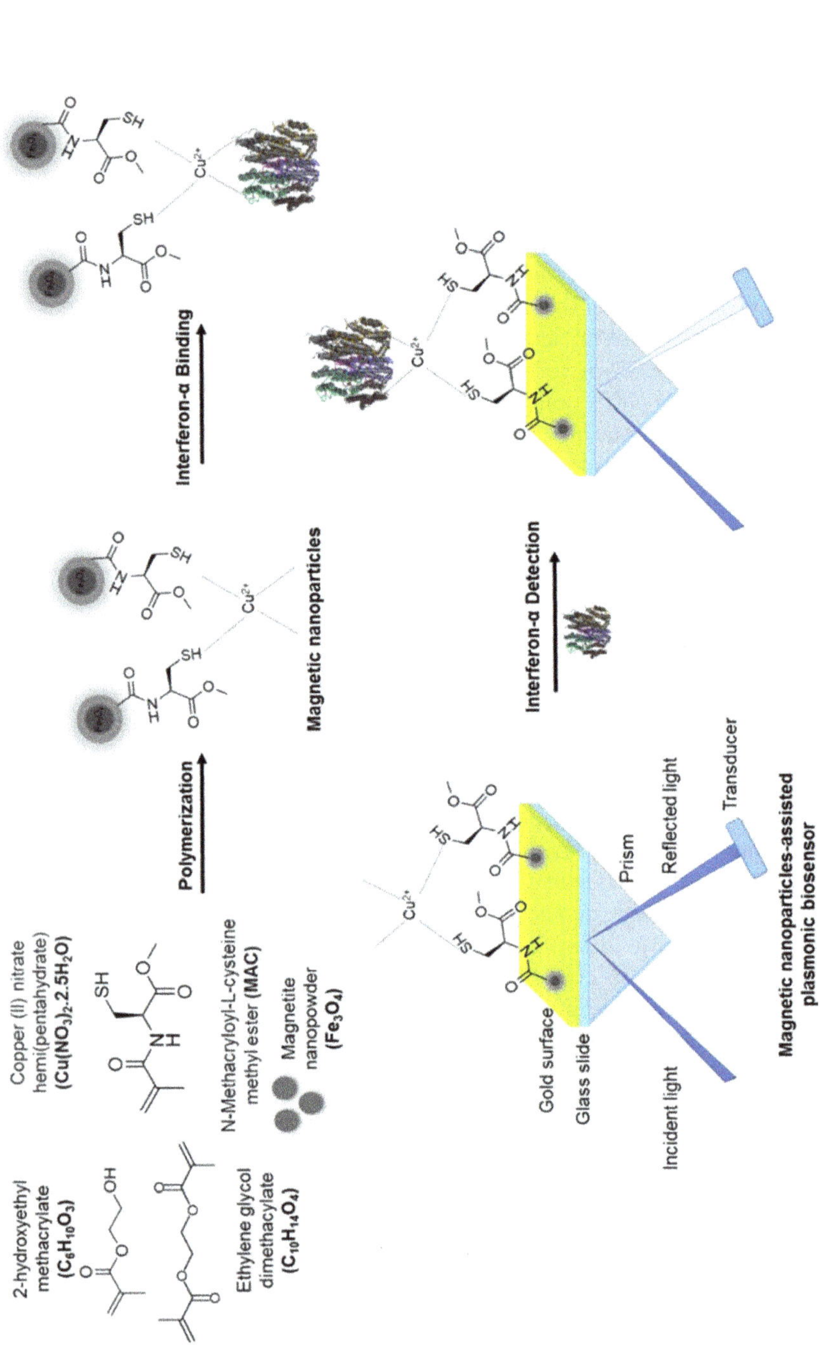

Fig. 1.2 Scheme of magnetic nanoparticles preparation and interferon-α detection with magnetic nanoparticles-assisted optical sensor (Saylan et al. 2022)

decrease in reagent volume (2 times less) when compared to conventional methods. This enables the production of 1.4–1.5 times more MIPs in a single step, ensuring a continuous and efficient production process (Erdem et al. 2023).

Adsorption and Separation Applications

The selective cavities of MIPs possess the ability to separate between target analytes and analogs because of their size, shape, and functional groups. These materials also exhibit significant potential in the isolation of biomacromolecules, small organic molecules, metal ions, and other substances (Bagheri et al. 2021). Their versatile nature allows for extensive utilization across various domains, including pollutant separation and treatment (Decompte et al. 2020; Diab et al. 2021; Sun et al. 2019; Vargas-Berrones et al. 2023; Villar-Navarro et al. 2017).

Quercetin, a highly physiologically active natural flavonoid, is abundantly present in a diverse range of plants. In a recent study, quercetin-MIPs were created using surface molecular imprinting method and the sol–gel polymerization on SiO_2. Synthesized MIPs were analyzed for characterization using scanning electron microscopy (SEM), Fourier transform and infrared spectroscopy (FT-IR) in order to confirm its functionality, surface morphology and structure. SEM scans showed that MIPs exhibited rough and spherical structure, measuring 260 nm diameter size and the FTIR spectra of MIP show bands N–H and C–H groups (Fig. 1.3). A static adsorption experiment was assessed quercetin-adsorbing MIP. The results showed that adsorption equilibrium could be attained in 90 min and that 35.70 mg/g was the maximum capacity. Following the pseudo-second-order kinetics and Freundlich isotherm model, chemical adsorption and heterogeneous surface with multilayer adsorption governed the MIP adsorption kinetics and isotherm for quercetin. Unlike NIPs, MIPs showed strong selectivity and precise detection for quercetin, with 1.61 selectivity coefficient over biochanin A. MIP's adsorption capability remains above 90% after five adsorption–desorption cycles, demonstrating excellent reusability and promise for selective quercetin adsorption (Zhi et al. 2023).

There is a strong demand for cost-effective and efficient alternatives to high-priced ethanol fractionation and affinity chromatography techniques for the extraction of IgG from human plasma. MIPs designed for IgG molecules exhibit promise as prospective candidates. However, these polymers encounter significant challenges, including the laborious removal of templates and inadequate printing efficiency. In particular research, a novel approach was implemented in order to address the aforementioned challenges. In the process of producing MIPs, poly (L-glutamic acid) (PLGA) peptide cross-linkers were employed as an alternative to conventional cross-linkers like *N,N*-methylenebisacrylamide (BIS). The expansion of imprinting cavities in polymers can be achieved using the pH-induced helix–coil transition and accurate folding of peptide segments. By altering the pH from 5.0 to 7.0, the cavities can be extended, but their original size and shape can be recovered by readjusting the pH. Hence, it is possible to obtain complete elution of the IgG template using mild conditions, leading to a notable enhancement in imprinting efficiency. In comparison

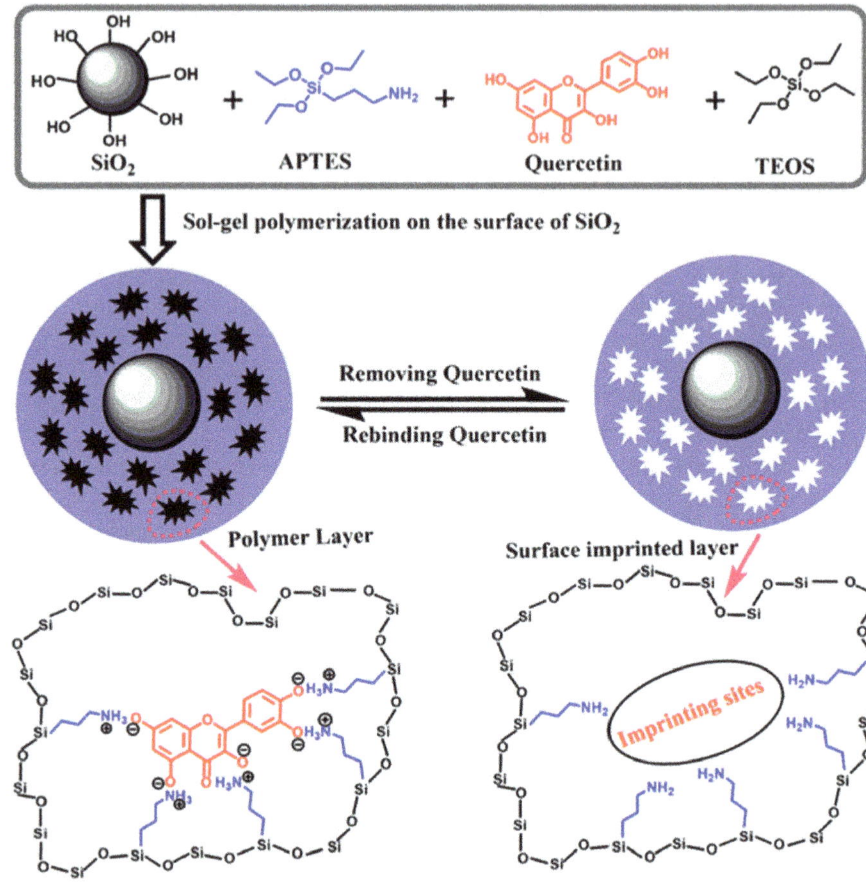

Fig. 1.3 Scheme of MIP preparation (Zhi et al. 2023)

to BIS cross-linked MIPs, it was observed that molecular imprinting of PLGA cross-linked MIPs resulted in the creation of 8.6 times more binding sites. According to the results, the synthesized MIPs exhibited notable characteristics, including a substantial adsorption capacity of 612.5 mg/g, a high imprinting factor of 4.92, and a remarkable level of selectivity. The study suggests that the MIPs developed in this research possess a high adsorption capacity and selectivity, which may enable the effective separation of IgG from human serum, hence facilitating the production of high purity products (Dong et al. 2023). Apurinic/apyrimidinic endonuclease 1 (APE1) is a multifunctional DNA repair protein that is found in various subcellular locations. The precise mechanisms underlying the tightly controlled subcellular localization and interactomes of this protein remain incompletely explained, however they are intricately related to post-translational modifications occurring in diverse biological scenarios. In recent research, a bio-nanocomposite possessing antibody-like characteristics was successfully created. This bio-nanocomposite was able to effectively capture APE1 from cellular matrices, thereby facilitating a thorough investigation of this particular protein. At first, the APE1 template was immobilized onto the surface of silica-coated magnetic nanoparticles that had been modified using avidin. In the present experimental setting, the initial step involved the addition of 3-aminophenylboronic acid, serving as the functional monomer, to facilitate a reaction with the glycosyl residues present in avidin. Subsequently, 2-acrylamido-2-methylpropane sulfonic acid was introduced. In order to improve the affinity and selectivity of the binding sites, a secondary imprinting process was conducted utilizing dopamine as a functional monomer. The bio-nanocomposite based on MIPs exhibited notable characteristics such as strong affinity, specificity, and capacity towards template APE1. This enabled the efficient APE1 extractions from cell lysates, resulting in high recovery (78–92%) and purity levels. The highest binding capability of MIP for APE1 was observed to be 90 $\pm$ 4 nmol/g. Moreover, the bio-nanocomposite has the ability to efficiently release protein that is tied to specific cavities (Xie et al. 2023). The potential problem of water contamination resulting from the discharge of toxic heavy metals by diverse industrial sectors has emerged as a prominent concern for environmentalists. Mercury in the form of Hg (II) is widely recognized as a highly hazardous heavy metal owing to its capacity to induce carcinogenicity and several adverse health effects. In light of the adverse impacts of mercury on both the environment and human health, a study was undertaken to develop a MIP/nanoporous carbon (NC) nanocomposite that is environmentally sustainable. This nanocomposite was synthesized with the specific purpose of effectively adsorbing Hg (II) from aqueous solutions. Various characterizations, such as morphological analysis, thermal degradation analysis, functional analysis, and surface area analysis, were conducted on the synthesized nanocomposite. The efficacy of a MIP/NC nanocomposite, possessing a substantial specific surface area of 884.9 m^2/g, was assessed in terms of its ability to adsorb and remove Hg (II) from aqueous solution. Analyzes showed that a maximum adsorption efficiency of 116 mg/g Hg(II) was achieved at pH 4. Furthermore, the equilibrium sorption result complied with to the Freundlich model, indicating its suitability for pseudo-second order kinetics. The adsorption effectiveness of the MIP/NC was assessed through a comparative analysis with a real condensate sample derived from the oil and gas sector. The results revealed that Hg(II) was successfully recovered at a rate of 87.4%. The synthesized MIP/NC has demonstrated potential as an adsorbent for Hg(II) in contaminated settings. This implies that a range of composite absorbents including diverse precursors should be considered for the assessment of heavy metal and pharmaceutical removals (Abubakar et al. 2023). Beyond the essential task of comprehending and elucidating protein functions, there is a growing demand in protein purification for methods that are both cost-effective and efficient. This need is particularly pronounced in the context of enzyme production. In response to this demand, an innovative method has been demonstrated by Özbek et al. for the thrombin purification, a protein critical in the hemostatic process. This method leverages thrombin-imprinted microcryogels, which offer the advantages of reusability and high selectivity. The microcryogels were subjected to comprehensive characterization studies, utilizing techniques. Through an exploration of various parameters influencing thrombin adsorption, the maximum adsorption capacity (Q_{max}) was determined to be 55.86 mg/g. Additionally, the microcryogels' selectivity was assessed through competitive agent testing, and their reusability was evaluated. The purity of the thrombin was confirmed using the fast performance liquid chromatography method. The experimental results revealed that the adsorption of thrombin by the microcryogels adhered to the Langmuir isotherm (Q_{max}: 55.86 mg/g, R^2: 0.9505) and followed the pseudo-second-order kinetics for various thrombin concentrations (Özbek et al. 2023). Öztürk et al. developed MIP-based magnetic silica microspheres that are incorporated into cryogels to create a novel composite for the tyrosine adsorption (Fig. 1.4). Tyrosine is a vital chemical involved in numerous cellular processes, including brain regulation, stress reduction, mood control, melanin pigment production, and the regulation of thyroid functions. Through the characterizations, they optimized the conditions for tyrosine adsorption on these composite cryogels, achieving a maximum adsorption capacity of 62.27 mg/g in a pH 8.0 buffer solution. The reusability of the composite cryogel carrier was assessed through multiple adsorption and desorption cycles using 0.1 M NaCl, and it demonstrated the ability to be used more than 8 times without a decline in performance. Additionally, competitive tests with various amino acids revealed the high selectivity of the cryogel. The Langmuir isotherm provided an excellent fit to the adsorption data, highlighting the efficiency of the adsorption process (Öztürk et al. 2021).

Lastly, Saylan et al. engineered magnetic bacterial cellulose nanofibers serving as a new adsorbent with the capability for thymidine adsorption. The process involved silanizing magnetic bacterial cellulose nanofibers with 3-(trimethoxysilyl) propyl methacrylate, followed by polymerization with a monomer to template thymidine through metal chelate coordination. These modified nanofibers were subjected to various characterization methods, and their effectiveness in thymidine recognition was assessed to optimize adsorption conditions. During this evaluation, several factors were considered, and the maximum adsorption capacity was 431.3 mg/g, observed under pH 9.0 and at a temperature of 25 °C. Furthermore, the selectivity of these magnetic bacterial cellulose nanofibers was evaluated by conducting competitive experiments with other nucleosides. Impressively, these nanofibers exhibited high selectivity, with adsorption levels for thymidine being 4.13 and 3.80 times higher compared to other nucleosides (cytidine and uridine) at equal concentrations. The nanofibers also displayed a strong reusability capacity after multiple adsorption–desorption cycles. This innovative approach, combining molecular imprinting with magnetic bacterial cellulose nanofibers, demonstrates excellent potential for nucleoside adsorption and holds promise for various applications (Saylan et al. 2020b).

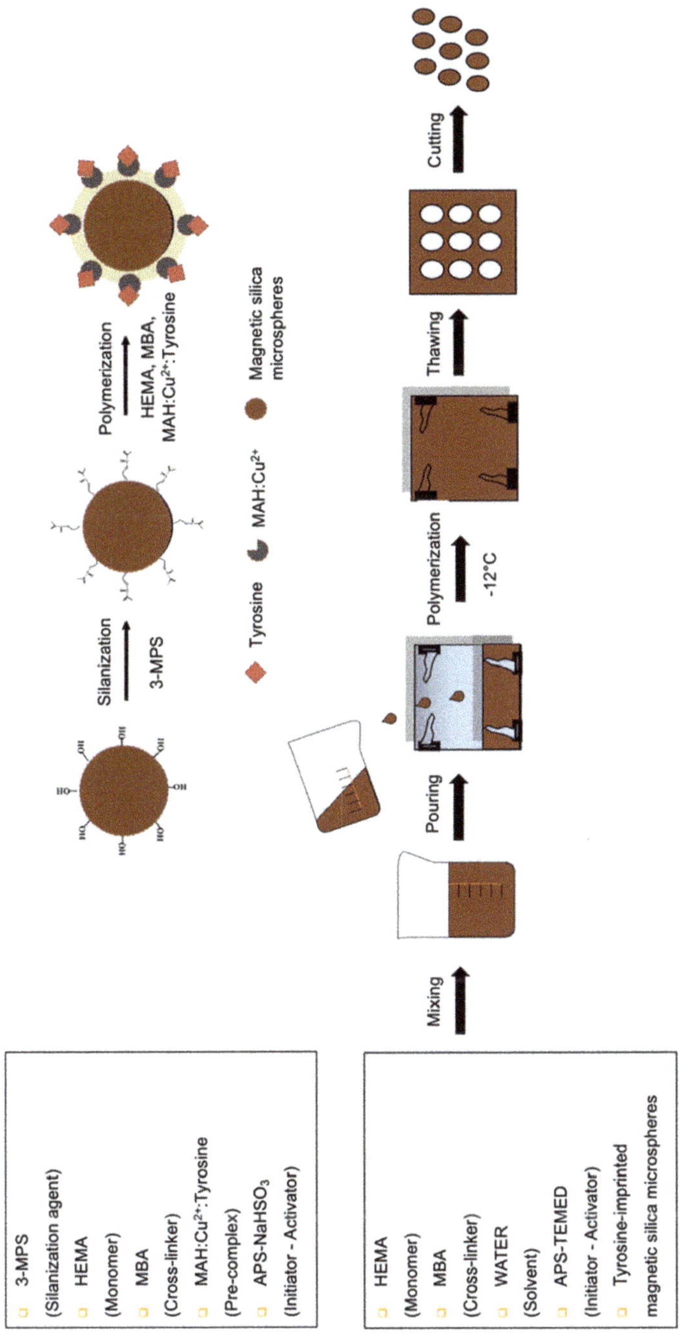

Fig. 1.4 Scheme of composite cryogels preparation (Öztürk et al. 2021)

1.5 Conclusion and Future Perspectives

In the realm of molecular recognition and molecular imprinting, significant progress has been made, leading to the development of highly specialized and selective materials for molecular detection and separation. This technology has allowed for the precise targeting and binding of molecules of interest, offering a broad range of applications in several fields, including chemistry, biology, and medicine. The studies and research presented here demonstrate the effectiveness of molecular imprinting in creating tailored recognition sites within polymers, nanoparticles, and other materials. These molecularly imprinted polymers have shown remarkable potential in selective adsorption, catalysis, sensing, and separation processes. The ability to design materials with specific binding sites for target molecules has opened up avenues for improved analytical techniques, drug delivery systems, and environmental remediation. However, challenges still exist in optimizing molecular imprinting techniques, such as enhancing the selectivity and stability of these materials and broadening their applicability across a wider range of molecules and conditions. Advances in nanotechnology, computational modeling, and novel imprinting strategies are likely to address these challenges and further propel the field. The future of molecular recognition and molecular imprinting holds exciting prospects. As technology continues to advance, several key developments can be anticipated. For instance, the integration of molecular imprinting with cutting-edge nanomaterials will lead to the creation of highly efficient and versatile recognition systems. These materials will offer enhanced selectivity, sensitivity, and reusability in various applications. The development of responsive and adaptive molecularly imprinted polymers, capable of adjusting their recognition properties in response to environmental changes or stimuli, will enable dynamic molecular recognition with broader applicability. In addition, the use of machine learning and computational modeling will accelerate the design and optimization of molecularly imprinted materials, reducing the trial-and-error process and increasing their specificity. Molecular imprinting will continue to play a crucial role in drug delivery, diagnostics, and targeted therapies, offering more precise and personalized medical treatments. The use of molecular imprinting for selective removal of pollutants and contaminants from the environment will become more prevalent, contributing to cleaner water, air, and soil. Molecularly imprinted polymers will be integrated into sensor technologies, enabling rapid and accurate detection of target molecules, with applications in fields like food safety, environmental monitoring, and security. Also, collaborations between chemists, biologists, materials scientists, and engineers will drive innovation in molecular imprinting, leading to the development of multifunctional materials with a wide range of applications. In conclusion, molecular recognition and molecular imprinting are poised to make significant contributions to diverse scientific and technological domains. With ongoing research, technological advancements, and cross-disciplinary collaborations, this field promises to revolutionize our ability to selectively manipulate and harness molecular interactions, addressing complex challenges in the modern world.

Acknowledgements Dr. Yeşeren Saylan and Özge Altıntaş thank to Scientific and Technological Research Council of Turkey (TÜBİTAK) Directorate of Science Fellowships and Grant (BİDEB) 2247-D (Project No: 121C226) Programme. Dr. Özgecan Erdem gratefully acknowledges the support of the TÜBİTAK 2218-National Postdoctoral Research Fellowship Program (Project No:121C431). Dr. Fatih Inci gratefully acknowledges the support of TÜBİTAK 2232–International Fellowship for Outstanding Researchers (project no: 118C254), GEBIP Award from Turkish Academy of Sciences (TÜBA), and BAGEP Award from Science Academy. This publication has been produced benefiting from the 2232 International Fellowship for Outstanding Researchers Program of TÜBİTAK (Project No: 118C254). However, the entire responsibility for the article belongs to the owner of the article. The financial support received from TÜBİTAK does not mean that the content of the publication is approved in a scientific sense by TÜBİTAK.

References

Abubakar L, Yusof NA, Abdullah AH, Mohammad F, Wahid MH, Ismail S, Abdul RZ, Al-Lohedan HA, Soleiman AA (2023) Molecularly imprinted polymer-based nanoporous carbon nanocomposite for effective adsorption of Hg (II) ions from aqueous suspensions. Separations 10:454

Aghoutane Y, Diouf A, Österlund L, Bouchikhi B, El Bari N (2020) Development of a molecularly imprinted polymer electrochemical sensor and its application for sensitive detection and determination of malathion in olive fruits and oils. Bioelectrochemistry 132:107404

Akgönüllü S, Kılıç S, Esen C, Denizli A (2023) Molecularly imprinted polymer-based sensors for protein detection. Polymers 15:629

Ali GK, Omer KM (2022) Molecular imprinted polymer combined with aptamer (MIP-aptamer) as a hybrid dual recognition element for bio (chemical) sensing applications. Review. Talanta 236:122878

Arabi M, Chen L (2022) Technical challenges of molecular-imprinting-based optical sensors for environmental pollutants. Langmuir 38:5963–5967

Arabi M, Ostovan A, Li J, Wang X, Zhang Z, Choo J, Chen L (2021) Molecular imprinting: green perspectives and strategies. Adv Mater 33:2100543

Arshady R, Mosbach K (1981) Synthesis of substrate-selective polymers by host-guest polymerization. Macromol Chem 182:687–692

Ayankojo AG, Reut J, Ciocan V, Öpik A, Syritski V (2020) Molecularly imprinted polymer-based sensor for electrochemical detection of erythromycin. Talanta 209:120502

Ayankojo AG, Reut J, Nguyen VBC, Boroznjak R, Syritski V (2022) Advances in detection of antibiotic pollutants in aqueous media using molecular imprinting technique—a review. Biosensors 12:441

Ayivi RD, Obare SO, Wei J (2023) Molecularly imprinted polymers as chemosensors for organophosphate pesticide detection and environmental applications. TrAC, Trends Anal Chem 117231

Babine RE, Bender SL (1997) Molecular recognition of protein-ligand complexes: applications to drug design. Chem Rev 97:1359–1472

Bagheri AR, Aramesh N, Khan AA, Gul I, Ghotekar S, Bilal M (2021) Molecularly imprinted polymers-based adsorption and photocatalytic approaches for mitigation of environmentally-hazardous pollutants-a review. J Environ Chem Eng 9:104879

Basak S, Venkatram R, Singhal RS (2022) Recent advances in the application of molecularly imprinted polymers (MIPs) in food analysis. Food Control 139:109074

Behncken SN, Waters MJ (1999) Molecular recognition events involved in the activation of the growth hormone receptor by growth hormone. J Mol Recognit 12:355–362

BelBruno JJ (2018) Molecularly imprinted polymers. Chem Rev 119:94–119

Bodoki AE, Iacob B-C, Dinte E, Vostinaru O, Samoila O, Bodoki E (2021) Perspectives of molecularly imprinted polymer-based drug delivery systems in ocular therapy. Polymers 13:3649
Bonatti AF, De Maria C, Vozzi G (2021) Molecular imprinting strategies for tissue engineering applications: a review. Polymers 13:548
Bozal-Palabiyik B, Erkmen C, Uslu B (2020) Molecularly imprinted electrochemical sensors: analytical and pharmaceutical applications based on ortho-phenylenediamine polymerization. Curr Pharm Anal 16:350–366
Bräuer B, Unger C, Werner M, Lieberzeit PA (2021) Biomimetic sensors to detect bioanalytes in real-life samples using molecularly imprinted polymers: a review. Sensors 21:5550
Caldara M, Lowdon JW, Royakkers J, Peeters M, Cleij TJ, Diliën H, Eersels K, van Grinsven B (2022) A molecularly imprinted polymer-based thermal sensor for the selective detection of melamine in milk samples. Foods 11:2906
Cao Y, Feng T, Xu J, Xue C (2019) Recent advances of molecularly imprinted polymer-based sensors in the detection of food safety hazard factors. Biosens Bioelectron 141:111447
Chatterji D (2016) Basics of molecular recognition. CRC Press, pp 1–15
Chen L, Wang X, Lu W, Wu X, Li J (2016) Molecular imprinting: perspectives and applications. Chem Soc Rev 45:2137–2211
Chen W, Tian X, He W, Li J, Feng Y, Pan G (2020) Emerging functional materials based on chemically designed molecular recognition. BMC Mater 2:1–22
Ciferri A (2021) Critical issues in molecular recognition: the enzyme-substrate association. Soft Matter 17:8585–8589
Cormack PA, Elorza AZ (2004) Molecularly imprinted polymers: synthesis and characterisation. J Chromatogr B 804:173–182
Cowen T, Stefanucci E, Piletska E, Marrazza G, Canfarotta F, Piletsky SA (2020) Synthetic mechanism of molecular imprinting at the solid phase. Macromol 53:1435–1442
Decompte E, Lobaz V, Monperrus M, Deniau E, Save M (2020) Molecularly imprinted polymer colloids synthesized by miniemulsion polymerization for recognition and separation of nonylphenol. ACS Appl Polym Mater 2:3543–3556
Diab KE, Salama E, Hassan HS, Abd E-M, Elkady MF (2021) Biocompatible MIP-202 Zr-MOF tunable sorbent for cost-effective decontamination of anionic and cationic pollutants from waste solutions. Sci Rep 11:6619
Ding S, Lyu Z, Niu X, Zhou Y, Liu D, Falahati M, Du D, Lin Y (2020) Integrating ionic liquids with molecular imprinting technology for biorecognition and biosensing: a review. Biosens Bioelectron 149:111830
Divya M, Rajput Y, Sharma R, Divya K (2023) Molecularly imprinted polymers for the extraction and sensing of vitamins. In: Molecularly imprinted polymers for environmental monitoring: fundamentals and applications. IOP Publishing Bristol, UK, pp 8–26
Dong C, Shi H, Han Y, Yang Y, Wang R, Men J (2021) Molecularly imprinted polymers by the surface imprinting technique. Eur Polym J 145:110231
Dong Q, Yang M, Wang Y, Guan Y, Zhang W, Zhang Y (2023) Peptide-crosslinked molecularly imprinted polymers for efficient separation of immunoglobulin G from human serum. Biomater Sci 11:1398–1407
Du W, Lei C, Zhang S, Bai G, Zhou H, Sun M, Fu Q, Chang C (2014) Determination of clenbuterol from pork samples using surface molecularly imprinted polymers as the selective sorbents for microextraction in packed syringe. J Pharm Biomed Anal 91:160–168
Elaine AA, Krisyanto SI, Hasanah AN (2022) Dual-functional monomer MIPs and their comparison to mono-functional monomer MIPs for SPE and as sensors. Polymers 14:3498
Ensafi AA, Nasr-Esfahani P (2021) Fundamental aspects of molecular imprinting. In: Molecularly imprinted polymer composites. Elsevier, pp 5–20
Ensafi AA, Kazemifard N, Dehkordi ZS (2021) Parameters that affect molecular imprinting polymers. In: Molecularly imprinted polymer composites. Elsevier, pp 21–48

Erdem Ö, Eş I, Saylan Y, Atabay M, Gungen MA, Ölmez K, Denizli A, Inci F (2023) In situ synthesis and dynamic simulation of molecularly imprinted polymeric nanoparticles on a micro-reactor system. Nat Commun 14:4840

Esfandyari-Manesh M, Javanbakht M, Atyabi F, Badiei A, Dinarvand R (2011) Effect of porogenic solvent on the morphology, recognition and release properties of carbamazepine-molecularly imprinted polymer nanospheres. J Appl Polym Sci 121:1118–1126

Frigoli M, Lowdon JW, Caldara M, Arreguin-Campos R, Sewall J, Cleij TJ, Diliën H, Eersels K, Van Grinsven B (2023) Thermal pyocyanin sensor based on molecularly imprinted polymers for the Indirect detection of Pseudomonas aeruginosa. ACS Sens 8:353–362

Fuchs Y, Soppera O, Haupt K (2012) Photopolymerization and photostructuring of molecularly imprinted polymers for sensor applications—a review. Anal Chim Acta 717:7–20

Gellman SH (1997) Introduction: molecular recognition. Chem Rev 97:1231–1232

Gkika DA, Tolkou AK, Lambropoulou DA, Bikiaris DN, Kokkinos P, Kalavrouziotis IK, Kyzas GZ (2024) Application of molecularly imprinted polymers (MIPs) as environmental separation tools. RSC Appl Polym 2:127–148

Hasanah AN, Safitri N, Zulfa A, Neli N, Rahayu D (2021) Factors affecting preparation of molecularly imprinted polymer and methods on finding template-monomer interaction as the key of selective properties of the materials. Molecules 26:5612

Haupt K (2010) Plastic antibodies. Nat Mater 9:612–614

Haupt K, Linares AV, Bompart M, Bui BTS (2011) Molecularly imprinted polymers. In: Molecular imprinting. Springer, Berlin, Heidelberg, pp 1–28

He Q, Liang J-J, Chen L-X, Chen S-L, Zheng H-L, Liu H-X, Zhang H-J (2020) Removal of the environmental pollutant carbamazepine using molecular imprinted adsorbents: molecular simulation, adsorption properties, and mechanisms. Water Res 168:115164

Heath JR, Ratner MA (2003) Molecular electronics. Phys Today 56:43–49

Huang C, Wang H, Ma S, Bo C, Ou J, Gong B (2021) Recent application of molecular imprinting technique in food safety. J Chromatogr A 1657:462579

Idil N, Bakhshpour M, Perçin I, Mattiasson B, Denizli A (2021) Molecular imprinting-based sensing platforms for recognition of microorganisms. In: Molecular imprinting for nanosensors and other sensing applications. Elsevier, pp 255–281

Işık D, Şahin S, Caglayan MO, Üstündağ Z (2021) Electrochemical impedimetric detection of kanamycin using molecular imprinting for food safety. Microchem J 160:105713

Iturralde I, Paulis M, Leiza J (2014) The effect of the crosslinking agent on the performance of propranolol imprinted polymers. Eur Polym J 53:282–291

Janiak DS, Kofinas P (2007) Molecular imprinting of peptides and proteins in aqueous media. Anal Bioanal Chem 389:399–404

Justino CI, Freitas AC, Pereira R, Duarte AC, Santos TAR (2015) Recent developments in recognition elements for chemical sensors and biosensors. TrAC, Trends Anal Chem 68:2–17

Kadhem AJ, Gentile GJ, Fidalgo de Cortalezzi MM (2021) Molecularly imprinted polymers (MIPs) in sensors for environmental and biomedical applications: a review. Molecules 26:6233

Kandimalla VB, Ju H (2004) Molecular imprinting: a dynamic technique for diverse applications in analytical chemistry. Anal Bioanal Chem 380:587–605

Karadurmus L, Bilge S, Sınağ A, Ozkan SA (2022) Molecularly imprinted polymer (MIP)-based sensing for detection of explosives: current perspectives and future applications. TrAC, Trends Anal Chem 155:116694

Kidakova A, Boroznjak R, Reut J, Öpik A, Saarma M, Syritski V (2020) Molecularly imprinted polymer-based SAW sensor for label-free detection of cerebral dopamine neurotrophic factor protein. Sens Actuators B Chem 308:127708

Krengel U, Bousquet PA (2014) Molecular recognition of gangliosides and their potential for cancer immunotherapies. Front Immunol 5:325

Lanza F, Sellergren B (2004) Molecularly imprinted polymers via high-throughput and combinatorial techniques. Macromol Rapid Commun 25:59–68

Lee JH, Prud'Homme RK, Aksay IA (2001) Cure depth in photopolymerization: experiments and theory. J Mater Res 16:3536–3544
Lehotay SJ, Chen Y (2018) Hits and misses in research trends to monitor contaminants in foods. Anal Bioanal Chem 410:5331–5351
Leibl N, Haupt K, Gonzato C, Duma L (2021) Molecularly imprinted polymers for chemical sensing: a tutorial review. Chemosensors 9:123
Li R, Feng Y, Pan G, Liu L (2019) Advances in molecularly imprinting technology for bioanalytical applications. Sensors 19:177
Li Y, Luo L, Nie M, Davenport A, Li Y, Li B, Choy K-L (2022) A graphene nanoplatelet-polydopamine molecularly imprinted biosensor for ultratrace creatinine detection. Biosens Bioelectron 216:114638
Li Y, Luo L, Kong Y, Li Y, Wang Q, Wang M, Li Y, Davenport A, Li B (2024) Recent advances in molecularly imprinted polymer-based electrochemical sensors. Biosens Bioelectron 116018
Lusina A, Cegłowski M (2022) Molecularly imprinted polymers as state-of-the-art drug carriers in hydrogel transdermal drug delivery applications. Polymers 14:640
Ma W, Row KH (2018) Solid-phase extraction of chlorophenols in seawater using a magnetic ionic liquid molecularly imprinted polymer with incorporated silicon dioxide as a sorbent. J Chromatogr A 1559:78–85
Madikizela LM, Ncube S, Nomngongo PN, Pakade VE (2022) Molecular imprinting with deep eutectic solvents: synthesis, applications, their significance, and benefits. J Mol Liq 362:119696
Mahony J, Nolan K, Smyth M, Mizaikoff B (2005) Molecularly imprinted polymers—potential and challenges in analytical chemistry. Anal Chim Acta 534:31–39
Malik MI, Shaikh H, Mustafa G, Bhanger MI (2019) Recent applications of molecularly imprinted polymers in analytical chemistry. Sep Purif Rev 48:179–219
Mayes AG, Mosbach K (1997) Molecularly imprinted polymers: useful materials for analytical chemistry? TrAC, Trends Anal Chem 16:321–332
McCluskey A, Holdsworth CI, Bowyer MC (2007) Molecularly imprinted polymers (MIPs): sensing, an explosive new opportunity? Org Biomol Chem 5:3233–3244
Mehmandoust M, Erk N, Naser M, Soylak M (2023) Molecularly imprinted polymer film loaded on the metal–organic framework with improved performance using stabilized gold-doped graphite carbon nitride nanosheets for the single-step detection of Fenamiphos. Food Chem 404:134627
Morsi SM, Abd El-Aziz ME, Mohamed HA (2023) Smart polymers as molecular imprinted polymers for recognition of target molecules. Int J Polym Mater Polym Biomater 72:612–635
Moyer BA, Custelcean R, Hay BP, Sessler JL, Bowman-James K, Day VW, Kang S-O (2013) A case for molecular recognition in nuclear separations: sulfate separation from nuclear wastes. Inorg Chem 52:3473–3490
Muhammad T, Nur Z, Piletska EV, Yimit O, Piletsky SA (2012) Rational design of molecularly imprinted polymer: the choice of cross-linker. Analyst 137:2623–2628
Mujahid A, Afzal A, Dickert FL (2023) Transitioning from supramolecular chemistry to molecularly imprinted polymers in chemical sensing. Sensors 23:7457
Muratsugu S, Shirai S, Tada M (2020) Recent progress in molecularly imprinted approach for catalysis. Tetrahedron Lett 61:151603
Murdaya N, Triadenda AL, Rahayu D, Hasanah AN (2022) A review: using multiple templates for molecular imprinted polymer: is it good? Polymers 14:4441
Mustafa YL, Leese HS (2023) Fabrication of a lactate-specific molecularly imprinted polymer toward disease detection. ACS Omega 8:8732–8742
Ndunda EN (2020) Molecularly imprinted polymers—a closer look at the control polymer used in determining the imprinting effect: a mini review. J Mol Recognit 33:e2855
Nishchaya K, Rai VK, Bansode H (2023) Methacrylic acid as a potential monomer for molecular imprinting: a review of recent advances. Results Mater 18:100379
Ostovan A, Arabi M, Wang Y, Li J, Li B, Wang X, Chen L (2022) Greenificated molecularly imprinted materials for advanced applications. Adv Mater 34:2203154

Özbek MA, Özgür E, Bereli N, Denizli A (2023) Molecular imprinted based microcryogels for thrombin purification. J Chromatogr B 1228:123848
Öztürk G, Saylan Y, Denizli A (2021) Designing composite cryogel carriers for tyrosine adsorption. Sep Purif Technol 254:117622
Paleos CM, Sideratou Z, Tsiourvas D (2001) Molecular recognition of complementary liposomes in modeling cell-cell recognition. ChemBioChem 2:305–310
Pandey A, Singh S, Singh SP (2023) Molecularly imprinted polymers: applications and challenges in biological and environmental sample analysis. In: Singh M (ed) Molecularly imprinted polymers (MIPs). pp 321–344
Paruli EI, Soppera O, Haupt K, Gonzato C (2021) Photopolymerization and photostructuring of molecularly imprinted polymers. ACS Appl Polym Mater 3:4769–4790
Peng Y, Su H (2015) Recent innovations of molecularly imprinted electrochemical sensors based on electropolymerization technique. Curr Anal Chem 11:307–317
Persch E, Dumele O, Diederich F (2015) Molecular recognition in chemical and biological systems. Angew Chem Int Ed 54:3290–3327
Piletsky S (2006) Molecular imprinting of polymers. CRC Press Boca Raton, FL, pp 64–70
Piletsky SA, Alcock S, Turner AP (2001) Molecular imprinting: at the edge of the third millennium. Trends Biotechnol 19:9–12
Pirondini L, Dalcanale E (2007) Molecular recognition at the gas–solid interface: a powerful tool for chemical sensing. Chem Soc Rev 36:695–706
Pratama KF, Manik MER, Rahayu D, Hasanah AN (2020) Effect of the molecularly imprinted polymer component ratio on analytical performance. Chem Pharm Bull 68:1013–1024
Rahman Au, Iqbal M, Rahman Fu, Fu D, Yaseen M, Lv Y, Omer M, Garver M, Yang L, Tan T (2012) Synthesis and characterization of reactive macroporous poly (glycidyl methacrylate-triallyl isocyanurate-ethylene glycol dimethacrylate) microspheres by suspension polymerization: effect of synthesis variables on surface area and porosity. J Appl Polym Sci 124:915–926
Rahman S, Bozal-Palabiyik B, Unal DN, Erkmen C, Siddiq M, Shah A, Uslu B (2022) Molecularly imprinted polymers (MIPs) combined with nanomaterials as electrochemical sensing applications for environmental pollutants. Trends Environ Anal Chem 36:e00176
Rajpal S, Singh S, Mishra P, Bhakta S (2023) Role of monomer compositions for molecularly imprinted polymers (MIPs). In: Molecularly imprinted polymers (MIPs). Elsevier, pp. 81–99
Raziq A, Kidakova A, Boroznjak R, Reut J, Öpik A, Syritski V (2021) Development of a portable MIP-based electrochemical sensor for detection of SARS-CoV-2 antigen. Biosens Bioelectron 178:113029
Refaat D, Aggour MG, Farghali AA, Mahajan R, Wiklander JG, Nicholls IA, Piletsky SA (2019) Strategies for molecular imprinting and the evolution of MIP nanoparticles as plastic antibodies—synthesis and applications. Int J Mol Sci 20:6304
Roushani M, Zalpour N (2021) Selective detection of Asulam with in-situ dopamine electropolymerization based electrochemical MIP sensor. React Funct Polym 169:105069
Saylan Y (2023) Unveiling the pollution of bacteria in water samples through optic sensor. Microchem J 193:109057
Saylan Y, Denizli A (2019) Molecularly imprinted polymer-based microfluidic systems for point-of-care applications. Micromachines 10:766
Saylan Y, Denizli A (2020) Advances in molecularly imprinted systems: materials, characterization methods and analytical applications. Curr Anal Chem 16:196–207
Saylan Y, Erdem Ö, Inci F, Denizli A (2020a) Advances in biomimetic systems for molecular recognition and biosensing. Biomimetics 5:20
Saylan Y, Göktürk l, Pospiskova K, Safarik I, Denizli A (2020b) Magnetic bacterial cellulose nanofibers for nucleoside recognition. Cellulose 27:9479–9492
Saylan Y, Akgönüllü S, Denizli A (2022) Preparation of magnetic nanoparticles—assisted plasmonic biosensors with metal affinity for interferon-α detection. Mater Sci Eng B 280:115687
Saylan Y, Altıntaş Ö, Denizli A (2023) Microfluidic-based molecularly imprinted polymers-integrated optic sensors. Results Opt 13:100541

Saylan Y, Kılıç S, Denizli A (2024) Biosensing applications of molecularly imprinted-polymer-based nanomaterials. Processes 12:177
Schmidt RH, Belmont A-S, Haupt K (2005) Porogen formulations for obtaining molecularly imprinted polymers with optimized binding properties. Anal Chim Acta 542:118–124
Shama NA, Aşır Sl, Göktürk I, Yılmaz F, Türkmen D, Denizli A (2023) Electrochemical detection of cortisol by silver nanoparticle-modified molecularly imprinted polymer-coated pencil graphite electrodes. ACS Omega 8:29202–29212
Shimizu KD, Stephenson CJ (2010) Molecularly imprinted polymer sensor arrays. Curr Opin Chem Biol 14:743–750
Sibrian-Vazquez M, Spivak DA (2004) Molecular imprinting made easy. J Am Chem Soc 126:7827–7833
Singh M, Singh S, Singh SP, Patel SS (2020) Recent advancement of carbon nanomaterials engrained molecular imprinted polymer for environmental matrix. Trends Environ Anal Chem 27:e00092
Skerra A (2007) Alternative non-antibody scaffolds for molecular recognition. Curr Opin Biotechnol 18:295–304
Song Z, Li J, Lu W, Li B, Yang G, Bi Y, Arabi M, Wang X, Ma J, Chen L (2022) Molecularly imprinted polymers based materials and their applications in chromatographic and electrophoretic separations. TrAC, Trends Anal Chem 146:116504
Soufi GJ, Iravani S, Varma RS (2021) Molecularly imprinted polymers for the detection of viruses: challenges and opportunities. Analyst 146:3087–3100
Spivak DA (2005) Optimization, evaluation, and characterization of molecularly imprinted polymers. Adv Drug Deliv Rev 57:1779–1794
Sun L, Li J, Li X, Liu C, Wang H, Huo P, Yan sheng Y (2019) Molecularly imprinted Ag/Ag_3VO_4/g-C_3N_4 Z-scheme photocatalysts for enhanced preferential removal of tetracycline. J Colloid Interface Sci 552:271–286
Suzaei FM, Daryanavard SM, Abdel-Rehim A, Bassyouni F, Abdel-Rehim M (2023) Recent molecularly imprinted polymers applications in bioanalysis. Chem Pap 77:619–655
Tarannum N, Khatoon S, Dzantiev BB (2020) Perspective and application of molecular imprinting approach for antibiotic detection in food and environmental samples: a critical review. Food Control 118:107381
Tashiro S, Shionoya M (2020) Novel porous crystals with macrocycle-based well-defined molecular recognition sites. Acc Chem Res 53:632–643
Tse Sum Bui B, Mier A, Haupt K (2023) Molecularly imprinted polymers as synthetic antibodies for protein recognition: the next generation. Small 19:2206453
Turiel E, Martín-Esteban A (2010) Molecularly imprinted polymers for sample preparation: a review. Anal Chim Acta 668:87–99
Vargas-Berrones K, Ocampo-Perez R, Rodríguez-Torres I, Medellín-Castillo NA, Flores-Ramírez R (2023) Molecularly imprinted polymers (MIPs) as efficient catalytic tools for the oxidative degradation of 4-nonylphenol and its by-products. Environ Sci Pollut Res 30:90741–90756
Villar-Navarro M, Martín-Valero MJ, Fernández-Torres RM, Callejón-Mochón M, Bello-López MÁ (2017) Easy, fast and environmental friendly method for the simultaneous extraction of the 16 EPA PAHs using magnetic molecular imprinted polymers (mag-MIPs). J Chromatogr B Biomed Appl 1044:63–69
Wackerlig J, Schirhagl R (2016) Applications of molecularly imprinted polymer nanoparticles and their advances toward industrial use: a review. Anal Chem 88:250–261
Wang X, Liu H, Sun Z, Zhao S, Zhou Y, Li J, Cai T, Gong B (2020) Monodisperse restricted access material with molecularly imprinted surface for selective solid-phase extraction of 17β-estradiol from milk. J Sep Sci 43:3520–3533
Wang L, Pagett M, Zhang W (2023a) Molecularly imprinted polymer (MIP) based electrochemical sensors and their recent advances in health applications. Sens Actuators Rep 5:100153
Wang M, Qiu J, Zhu C, Hua Y, Yu J, Jia L, Xu J, Li J, Li Q (2023b) A fluorescent molecularly imprinted polymer-coated paper sensor for on-site and rapid detection of glyphosate. Molecules 28:2398

Wang S, Zhang L, Zeng J, Hu X, Wang X, Yu L, Wang D, Cheng L, Ahmed R, Romanovski V (2023c) Multi-templates molecularly imprinted polymers for simultaneous recognition of multiple targets: from academy to application. TrAC, Trends Anal Chem 166:117173

Wardani NI, Kangkamano T, Wannapob R, Kanatharana P, Thavarungkul P, Limbut W (2023) Electrochemical sensor based on molecularly imprinted polymer cryogel and multiwalled carbon nanotubes for direct insulin detection. Talanta 254:124137

Whitcombe MJ, Kirsch N, Nicholls IA (2014) Molecular imprinting science and technology: a survey of the literature for the years 2004–2011. J Mol Recognit 27:297–401

Włoch M, Datta J (2019) Synthesis and polymerisation techniques of molecularly imprinted polymers. In: Comprehensive analytical chemistry. Elsevier, pp 17–40

Wulff G (2013) Fourty years of molecular imprinting in synthetic polymers: origin, features and perspectives. Mikrochim Acta 180:1359–1370

Xie H, Sun Y, Zhang R, Zhang Y, Zhao M (2023) Surface imprinted bio-nanocomposites for affinity separation of a cellular DNA repair protein. Biopolymers 114:e23537

Xu J, Miao H, Wang J, Pan G (2020a) Molecularly imprinted synthetic antibodies: from chemical design to biomedical applications. Small 16:1906644

Xu Y, Hassan MM, Ali S, Li H, Chen Q (2020b) SERS-based rapid detection of 2, 4-dichlorophenoxyacetic acid in food matrices using molecularly imprinted magnetic polymers. Mikrochim Acta 187:1–9

Yan H, Row KH (2006) Characteristic and synthetic approach of molecularly imprinted polymer. Int J Mol Sci 7:155–178

Zaidi SA (2016) Latest trends in molecular imprinted polymer based drug delivery systems. RSC Adv 6:88807–88819

Zaidi SA (2020) Molecular imprinting: a useful approach for drug delivery. Mater Sci Energy Technol 3:72–77

Zembrzuska D, Kalecki J, Cieplak M, Lisowski W, Borowicz P, Noworyta K, Sharma PS (2019) Electrochemically initiated co-polymerization of monomers of different oxidation potentials for molecular imprinting of electroactive analyte. Sens Actuators B Chem 298:126884

Zhang Y, Song D, Lanni LM, Shimizu KD (2010) Importance of functional monomer dimerization in the molecular imprinting process. Macromol 43:6284–6294

Zhang W, Zhang Y, Wang R, Zhang P, Zhang Y, Randell E, Zhang M, Jia Q (2022) A review: development and application of surface molecularly imprinted polymers toward amino acids, peptides, and proteins. Anal Chim Acta 1234:340319

Zhi K, Li Z, Luo H, Ding Y, Chen F, Tan Y, Liu H (2023) Selective adsorption of quercetin by the Sol-Gel surface molecularly imprinted polymer. Polymers 15:905

Zhou Q, Xu Z, Liu Z (2022) Molecularly imprinting–aptamer techniques and their applications in molecular recognition. Biosens 12:576

Chapter 2
Synthesis Methods and Strategies for MIPs

Zahra Mirzaei Karazan and Mahmoud Roushani

2.1 Introduction

Molecular printing is a technology in which a polymer matrix of highly cross-linked is created around the target molecule with selective molecular diagnostic capability. The polymer matrix structure is macroporous with a microsite that complements the target molecule which is the most main parameter for molecular diagnosis (Chen et al. 2011; Haupt et al. 2012). The original components of the synthesis of MIPs are the target, initiator, functional monomer, cross-linker, and porogen. MIPs are supplied via a complex polymerization reaction (Wackerlig and Lieberzeit 2015). To ideal MIPs, it is critical to choose proper target molecules, cross-linkers, functional monomers, and solvents. Two main mechanisms, for the preparation of MIPs, are the sol–gel process (Han et al. 2005a) and free-radical polymerization (Sulitzky et al. 2002a). Free-radical polymerization contains suspension polymerization (Lin et al. 2016), bulk polymerization (Lenain et al. 2012), precipitation polymerization (Gutierrez-Climente et al. 2016), seed polymerization (Zhang and Yao 2014), emulsion polymerization (Han et al. 2020), in-situ polymerization (Mirzaei Karazan and Roushani 2021), and microwave polymerization (Lu et al. 2018). The sol–gel processes include embedding, copolymerization, template pore-forming, and multi-stage imprinting (Xiao et al. 2020). On the other hand, combining this technique with the surface imprinting method can improve the reproducibility, responsibility, and selectivity of MIP particles. The scheme of the MIP process is indicated in Fig. 2.1 (Vasapollo et al. 2011). The first stage is the dissolving of the target, cross-linker, initiator, and functional monomer in a porogen. The target molecule bonds with the selected functional monomer, which results in the creation of a stable target-monomer complex. The functional monomer is placed around the target molecule and is fixed via copolymerization usage cross-linkers. Sequentially, eluting of the

Z. M. Karazan · M. Roushani (✉)
Department of Chemistry, Faculty of Sciences, Ilam University, 69315-516, Ilam, Iran
e-mail: m.roushani@Ilam.ac.ir

S. Patra and M. Sillanpaa (eds.), *Molecularly Imprinted Polymers as Artificial Antibodies for the Environmental Health*, https://doi.org/10.1007/978-3-031-58995-9_2

target for example through washing solvent. The binding cavities are complementary in size to the target, so achieve polymer matrix diagnostics and bind selectively to the target (Yan and Row 2006). The functional monomers construct a complex with an analyte via covalent and non-covalent, and semi-covalent strategies, which are compared in Table 2.1 (Włoch and Datta 2019). The most common strategy is the non-covalent approach because rapidity and simplicity of the technique. The semi-covalent approach consisted durability of the covalent approach with rapidity and simplicity of target binding and elution. This technique comprises covalent imprinting of functional monomers, whereas target rebinding is based on non-covalent bonds (Beltran et al. 2010).

Generally, three common general strategies in the preparation of MIPs are covalent, non-covalent, and semi-covalent approaches (Bagheri et al. 2019; Canfarotta et al. 2018). Figure 2.2 (Lusina and CegLowski 2022) indicates covalent and non-covalent molecular imprinting procedures. In the covalent imprinting approach, the

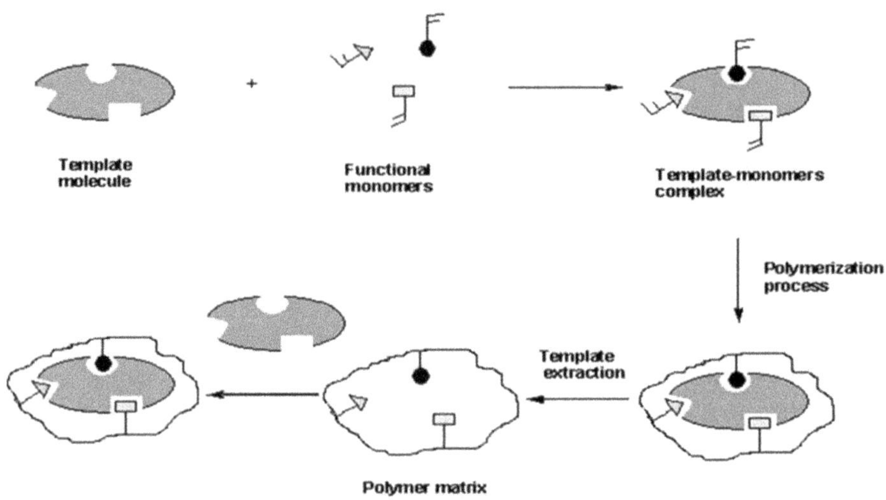

Fig. 2.1 Scheme of molecular imprinting process (Vasapollo et al. 2011)

Table 2.1 Comparison of covalent and non-covalent approach (Włoch and Datta 2019)

Type	Covalent imprinting	Non-covalent imprinting
Interactions	Reversible condensation reactions (boronate esters, Ketals/acetals, Schiff's bases)	Ionic interactions, hydrogen bonding, van der Waals forces, π–π interactions
Characteristics	• More rigid and durable type of interaction • Strong covalent interactions results in slow binding and rebinding	• More flexible type of interactions • Simple and fast binding and removal of template molecule • Non-covalent interactions are sensitive to disruptions (for example by presence of water)

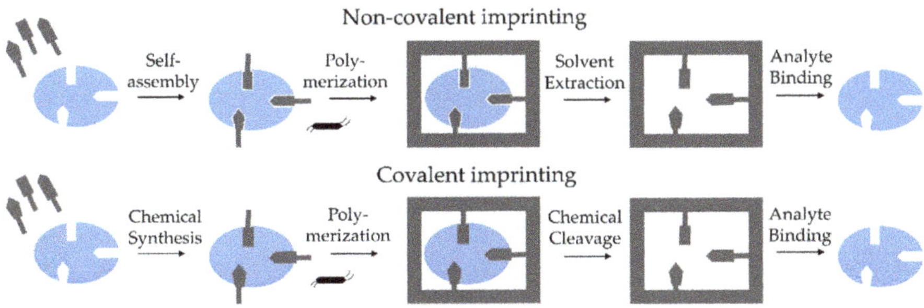

Fig. 2.2 Schematic diagram of covalent and non-covalent approach of MIPs (Lusina and CegLowski 2022)

polymerizable monomer is bound to an analyte through a labile bond of covalent. The monomer and analyte are copolymerized in the presence of a cross-linker in the solvent. The cross-linked film surrounds the target that is fixed in the imprinting polymers through the chemically labile bond. Sequentially, the formed interaction is broken for analyte elution, leaving binding sites that are complementary functional groups in placement. Nonetheless, the number of literature that has employed the covalent approach is limited. Maybe the reason for this is that the covalent approach requires multiple heteroatom functionalities available in the analyte (Chen et al. 2017).

On the other, due to fewer restrictions and extensible application, a non-covalent approach is the most usual imprinting approach. For non-covalent imprinting, the common process of MIP preparation can be summarized as cross-linker polymerization with a monomer around an analyte in a solvent. The analyte interaction and MIP ordinarily are a combination of non-covalent interactions including H-bonding, π–π, or electrostatic bonds. Afterward, the analyte is removed from the imprinting polymer and selective sites complementary to the analyte in size, shape, and functional groups, which can specially rebind the analyte. The main deficiency of this strategy is the non-diagnostic binding cavities achieved arising from the aggregation of complexes with various monomer-target stoichiometry. Its pre-polymerization stage is inaccessible from a complete process defined and it leads to the increase of the monomer compared to the analyte in the system of the polymerization. The free functional monomers are completely distributed in the polymer and this leads to the creation of non-diagnostic binding cavities (Ashley et al. 2017).

The semi-covalent approach couples the advantage of both non-covalent and covalent strategies. In the copolymerization process, the functional monomer bonds with the analyte through a reversible covalent interaction, just like the covalent strategy. After that, the analyte is eluted, revealing functional groups. In the rebinding stage, non-covalent interaction is applied, just as in the non-covalent strategy. Hence, the semi-covalent strategy is a useful technique to achieve MIPs with higher particular selection ability because it is an achievable method to control the tailoring of homogeneous binding cavities in polymerization (Ansari 2017).

This chapter provides an overview of the area of components, synthesis techniques, and strategies of MIPs.

2.2 Main Components of MIPs

2.2.1 Template Molecules

The molecular imprinting aim is for the formation polymer structure with high affinity toward the selected target. The target should consist of functional groups, which are fixed in the polymerization process and can create stable monomer-target complexes. In addition, targets are characterized by excellent thermal and chemical stability in polymerization. Generally, template molecules include four categories, i.e., viruses and cells (e.g. bovine leukemia virus), biomacromolecules (e.g. bovine serum albumin, adenosine), organic molecules (e.g. explosives, pesticides, amino acids and peptides, pharmaceuticals, sugars, endocrine disrupting chemicals), and ions (e.g. Pb^{2+}, Cu^{2+}, Hg^{2+}, Cd^{2+}) (Madikizela et al. 2017; Rodrıguez et al. 2010; Bunte et al. 2007).

2.2.2 Functional Monomers

The proper choice of functional monomers is a main parameter in the molecular imprinting process that can interact with the target, in a non-covalent strategy i.e., hydrogen bonding (choice of H-bond acceptor and donor). Some common monomers are displayed in Fig. 2.3 (Kang et al. 2023). The functional monomers affect the structure and the number of diagnostic binding cavities. Currently, functional monomers are applied in the molar excess to targets that permit the fabrication of functional complexes of the template monomer. Functional monomer leads to achieving highly particular sites complementary to the target and also uniform distribution of binding cavities in MIP (Vasapollo et al. 2011).

2.2.3 Cross-Linkers

The cross-linkers fixed the functional monomers around the target in the polymerization, resulting in the fabricating of a secure polymer network. The popular cross-linkers in the MIPs synthesis are depicted in Fig. 2.4 (Moreno-Bondi et al. 2008). Those factors influence the mechanical properties and structure of the polymer matrix, as well as the binding capability and selectivity of MIPs. The crucial factor

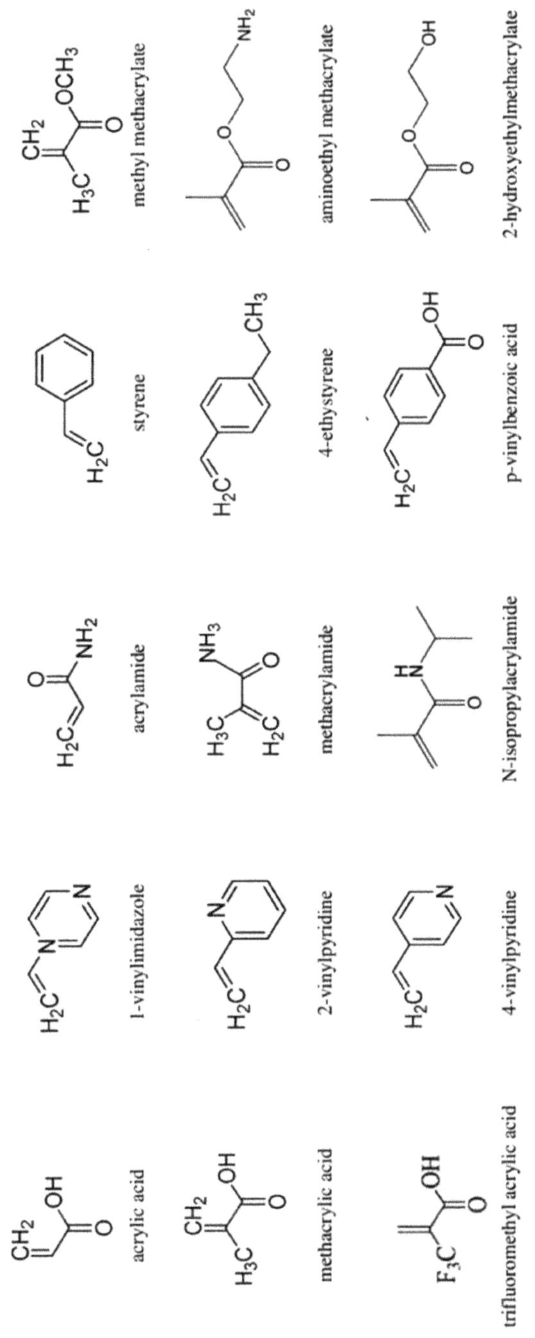

Fig. 2.3 Common functional monomers in MIP synthesis (Kang et al. 2023)

is the quantity and type of cross-linker used during polymerization. Increasing the amount of cross-linker results in a more stable porous material (Li et al. 2014).

2.2.4 Initiators

Azo and Peroxy compounds are usually applied as initiators via free radical polymerization in the MIPs synthesis (Fig. 2.5) (Włoch and Datta 2019). The azobisisobutyronitrile (AIBN) is the most common initiator that can decompose through photolysis and thermolysis. The polymerization is initiated by the generated stabilized and carbon-centred radicals. Free radical polymerization should be carried out in a gas atmosphere to avoid possible side reactions with oxygen (Yan and Row 2006).

2.2.5 Porogen Solvents

In the polymerization process, porogenic agents are employed as reaction media. The main of them are *N*,*N*-dimethylformamide, chloroform, acetonitrile, toluene, dichloroethane, 2-methoxy ethanol, methanol, and tetra-hydrofuran. In the non-covalent strategy, less polar or non-polar solvents are applied. The solvent type affects the interaction between the functional monomers and template, morphology, absorption properties, and imprinting ability of polymers. The enhancing value of the solvent applied results in enhanced pore size. Its suitable amount will improve the creation of particular sites designed toward binding targets. It is worth mentioning, that solvent nature plays a significant role. The more polar solvents disturb interactions in the template-monomer complexes, while less polar solvents improve the formation of functional complexes (Saloni et al. 2013).

2.3 Main Aspects of Molecular Imprinting

2.3.1 Covalent Approach

A typical method for producing MIPs is the covalent approach or preorganized approach (Anderson 2000). It can create regular homogeneous diagnostic sites in the polymer solution. The target is chemically interacted with the monomer via a labile and reversible covalent interaction. Then, the monomer and template complex are polymerized via the cross-linker resulting in a rigid structure. In the next step, the target is removed, and rebinding to the MIP is performed through cleavage and reversible formation of the covalent reactions (Wulff and Sarhan 1972).

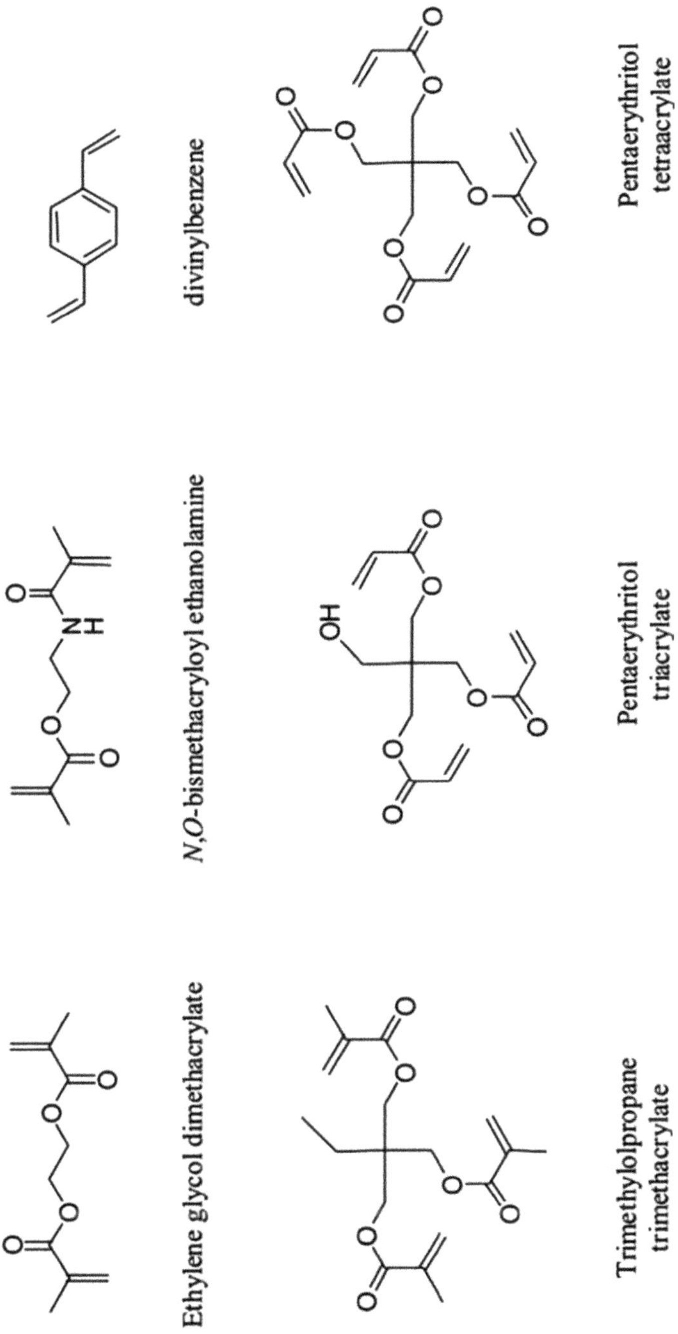

Fig. 2.4 Common cross-linkers in MIP synthesis (Moreno-Bondi et al. 2008)

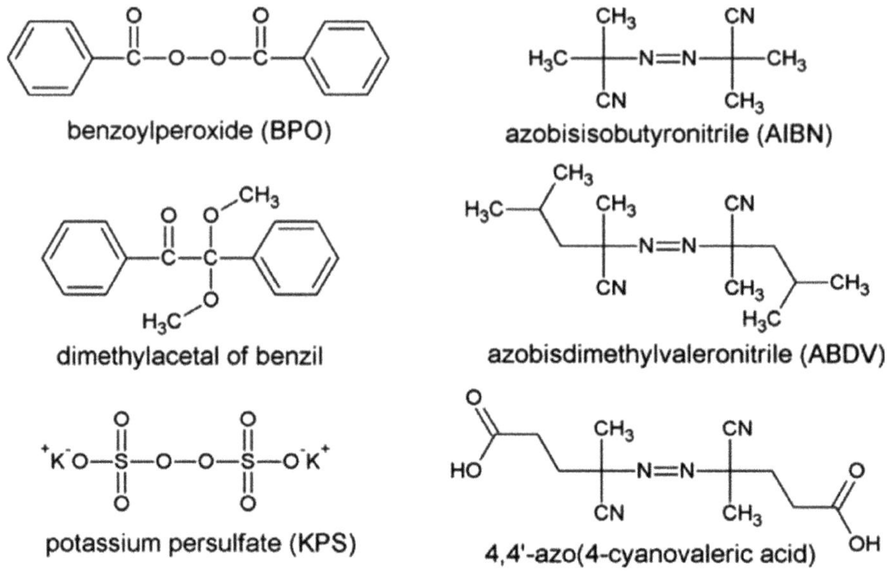

Fig. 2.5 Common initiators in MIP synthesis (Włoch and Datta 2019)

Few compounds are imprinted using this strategy such as aldehydes, alcohols (diols), ketones, carboxylic acids, and amines. For this purpose, reversible condensation reactions including Schiff's base to the synthesis of enantioselective MIPs of amino acids, ketal/acetal to the format of conjugate mono- and diacetone/mono alcohols, and boronate esters toward fructose, galactose, and nucleotides are used. Since the number of reversible condensation reactions is limited, the covalent approach is not a common strategy for preparing the MIPs.

The covalent strategy offers a significant advantage of being able to clearly define the monomer-template complex and control the imprinting process. This control allows for precise stoichiometry, minimizing non-specific interactions and resulting in the creation of highly uniform binding cavities. However, there are some drawbacks to this strategy, including a limited selection of functional monomers, the need to synthesize the monomer-template complex before polymerization, and its limited applicability.

2.3.2 *Non-covalent Approach*

Up until now, the most widely used method for preparing MIPs is through a non-covalent strategy. This method, also known as the self-assembly approach, allows for a wide range of target molecules to be focused on. It's considered the easiest strategy for introducing functional groups into the polymer cavities (Anderson 2000). The weak non-covalent bonds of target-monomers are applied to select the target. In this approach, the non-covalent bonds including ionic interactions, hydrogen bonding, hydrophobic interactions, and π–π interactions are used and one of the most specific is the hydrogen bond because of the particular geometric directionality.

Its main deficiency is the heterogeneous cavities that are achieved from the non-well-defined stage of pre-polymerization. The creation of monomer-target complexes with various ratios leads to various cavities and additional monomers are applied to fabricate the complexes of monomer-target also leads to non-recognized cavities (Caro et al. 2002). Lastly, the low efficiency of high-affinity cavities related to the amount of target molecules exists in the pre-polymerization step. These challenges can solved via tailor-made monomers and a stoichiometric approach of non-covalent based on the creation of stable template-monomer complexes to confirm that most of the monomer is allied with the target in the imprinting process (Sellergren 2001). Its major advantages are an easy procedure and avoiding the requirement of target chemical derivatization before polymerization. This strategy is employed for the development of recognized receptors for optical sensing.

2.3.3 Semi-covalent Approach

In this strategy, a hybrid approach, the target is reversible covalent interaction with a monomer to synthesize of polymer. After template removal usually via hydrolysis, target rebinding achieves place via non-covalent bounds (Fig. 2.6 (Moreno-Bondi et al. 2008)) (Joshi et al. 1998). Since the synthesis process is covalent, the distributed monomers functional groups are limited and every binding cavity is uniform. In addition, the target rebinding step is hardly affected by kinetic restrictions. In this strategy, the monomer and target can be interacted via an amide or ester bond (Cheong et al. 1997).

The advantages of non-covalent and covalent techniques are combined for semi-covalent imprinting. Its advantage is the agreement with the wide window of polymerization conditions which has permitted its employment in the imprinted synthesis of dendrimers, emulsions, and ring-opening metathesis polymers. Its disadvantages are needed for some synthetic chemistry levels to perform on the target before polymerization and chemical analysis to target the release of the polymer. This approach is also limited to a small functional group number. Furthermore, monomer-template hydrolysis is not simple. Steric congestion is the main impediment in the rebinding step due to the steric needs of the target to monomer in hydrogen bonding contact various from the corresponding target-monomer.

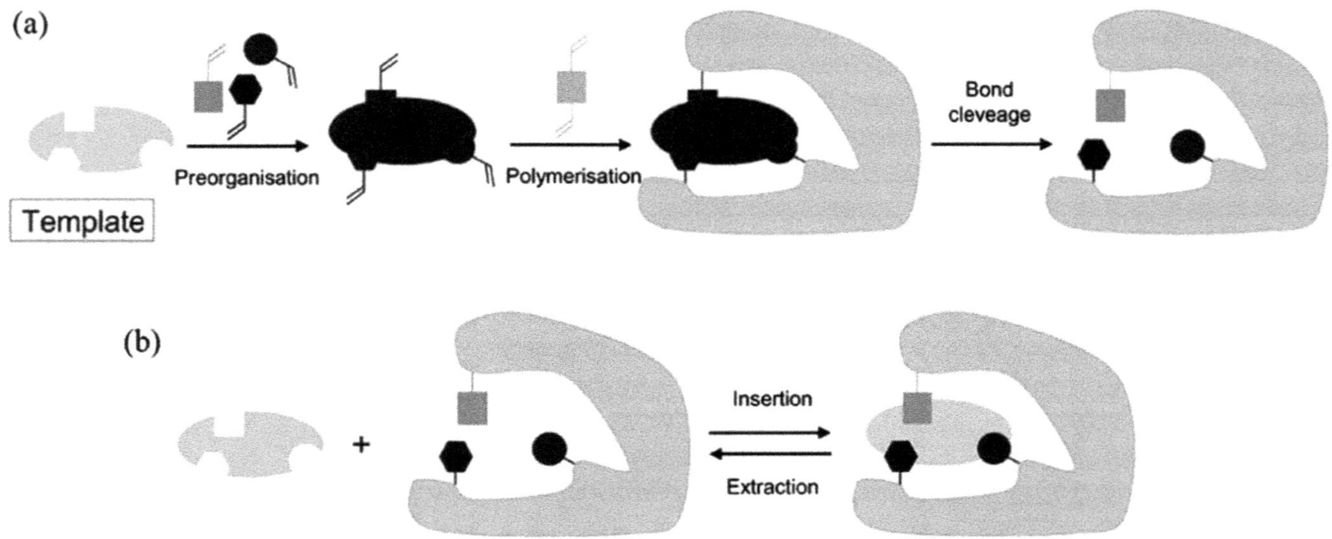

Fig. 2.6 Schematic diagram of the semi-covalent strategy for MIP production. MIP synthesis via the covalent strategy (**a**). Template molecular diagnostic via non-covalent strategies (**b**) (Moreno-Bondi et al. 2008)

2.4 Main Mechanisms for MIP Synthesis

The most important method for producing MIPs is synthesis. The target interacts with functional monomers for the formation of a complex network with non-covalent or covalent bonds. Based on the rebinding step between the functional monomer and the target in the polymer, it is classified into three types of imprinting: covalent, non-covalent, and semi-covalent (Huang et al. 2018). The choice of a suitable preparation method is critical to the production of MIPs with satisfactory properties. The mechanisms of MIP supply consist of sol–gel processes and free radical polymerization (Han et al. 2005b; Sulitzky et al. 2002b).

2.4.1 *Free Radical Polymerization*

The most common synthetic technique to convert monomers into polymer is free radical polymerization which is applied widely in industry to the product. Abundant vinyl monomers can polymerize effectively in acceptable yields via free radical polymerization techniques such as styrene, methyl methacrylate, and ethylene which are of specific industrial significance. Free radical polymerization can carried out in mild conditions of the reaction (e.g. atmospheric pressure and room temperature) in solution. Many vinyl monomers exist commercially at inexpensive, and free radical polymerization is currently the technique of selection to prepare MIPs.

The mechanism of free radical polymerization (Fig. 2.7 (Włoch and Datta 2019)) is defined through initiation, propagation, and termination steps (Cormack et al. 2004). Initiation step: decompose of initiators (under UV/visible irradiation, heating) with create of the free radical ($R^{\bullet}$) that attacks the monomer (M) double bond and $R^{\cdot}$ is created ($R\text{-}M_n^{\cdot}$), propagation step: it connected with the chain growth caused via addition of monomers on the growing macro radical ($R\text{-}M_n^{\cdot}$), and termination stage: recombination of two macro radicals resulting in the construction of C–H and C=C.

Free radical polymerization is begun via monomeric free radicals including unsaturated double bonds. Ontology polymerization is one of the most widespread techniques of slupping MIPs via usual free radical polymerization that has advantages such as easy and fast preparation and does not require expensive and complex tools and purification. Nevertheless, the whole polymer achieved by ontopolymerization should be crushed, sieved, and ground to a suitable size that significantly decreases production yield. Furthermore, grinding operations can result in irregular particle sizes and shapes that can disrupt some of the binding cavities with high affinity. Ontopolymerization polymers have heterogeneous binding cavity distributions that limit the employment of MIPs as chromatography adsorbents. To solve these deficiencies, different polymerization methods including precipitation polymerization, emulsion polymerization, suspension polymerization, and seed polymerization have been introduced.

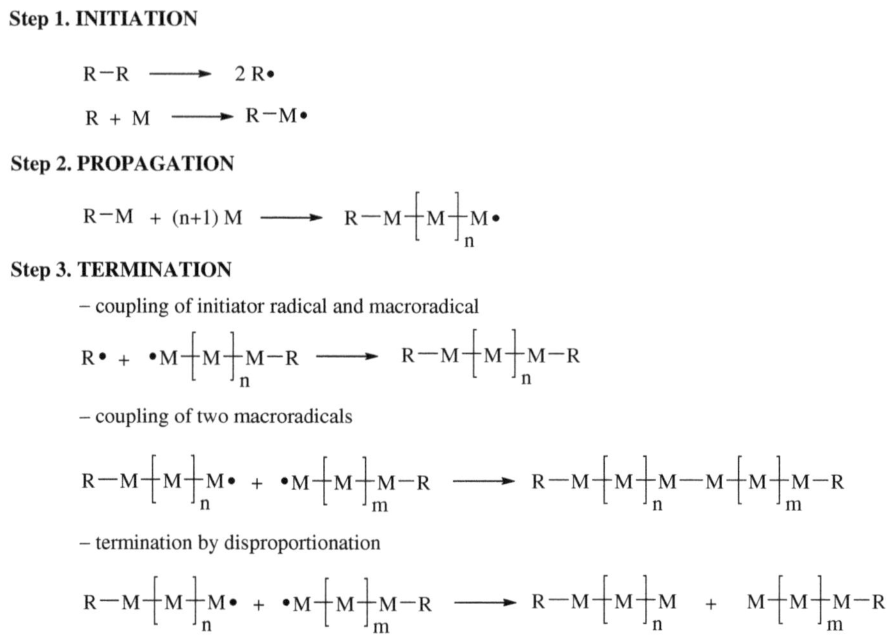

Fig. 2.7 Mechanism of free radical polymerization (Włoch and Datta 2019)

Iniferter Polymerization

An iniferter is a chemical compound that simultaneously acts as initiator, transfer agent, and terminator (hence the name ini-fer-ter) in controlled free radical iniferter polymerizations, the most common is the dithiocarbamate type (Otsu and Yoshida 1982).

Iniferters are initiators that induce radical polymerization that proceeds through initiation, propagation, primary radical termination, and transfer to initiator. Since bimolecular termination and other transfer reactions are negligible, these polymerizations are carried out via introducing monomers into the iniferter bond, resulting in polymers with two iniferter fragments at the end of chain (Patra et al. 2015). If the end groups of the polymers achieved through a suitable iniferter act more as a polymeric iniferter, these polymerizations are carried out proceed via a living radical polymerization mechanism in a homogeneous system. In these cases, the iniferters are considered as dormant species of the initiating and propagating radicals. According to controlled polymer synthesis, they can be divided into several types such as photoiniferters, monomeric/polymeric/gel iniferters, mono or polyfunctional iniferters, monomer or macromonomer iniferters. These create different polymers including telechelic, monofunction, graft, block, cross linked, and star (Otsu 2000).

2.4.2 *Sol–Gel Method*

One of the most common methods to prepare MIPs is the sol–gel technique because one-step preparation step, material homogeneity at the molecular amount, thermal stability, and good solvent (Beqqali et al. 2017; Guoning et al. 2020). The most famous sol–gel technique applies ethyl or methyl orthosilicate (TEOS/TMOS) as a precursor to propose the imprinted target in the inorganic structure that is homogeneously mixed in the solution state and subjected for hydrolyzation and condensation chemistry for the formation of crosslinked polymers with high stable (Barsbay and Guven 2018). The most important properties of this technique are easy preparation methods (Liu and Wulff 2008), improving the selectivity of the materials in the structure of the network via the introduction of particular chemical substances (Ma et al. 2015), and the controllability of surface area and porosity of the polymer film/powder (Wei et al. 2015).

2.5 Synthesis Methods of MIPs

MIPs are strong absorbents to select a template in an identical approach to an antibody-to-antigen mechanism found in natural bio-recognition. The synthesis of MIPs including the monomer, template, and cross-linking in the presence of a porogen solvent (Guc and Schroeder 2017). The choice of the polymerization technique is an important parameter and it depends on the format and size of MIPs along with their thermal and natural stability. Generally, there are many polymerization methods for the synthesis of MIPs, as depicted in Fig. 2.8 (Metwally et al. 2021). A summary of MIP prepared via various synthesis techniques is shown in Table 2.2 (Yan and Row 2006).

2.5.1 *Bulk Polymerization*

Bulk polymerization is the most commonly applied method, which has been depicted to be very simple, rapid, effective, and easy. It is prepared only using a functional monomer, template, initiator, solvent, and cross-linker. The fabricated polymer can be grained and then sieved according to the particle arrangement desired in the template application. Firstly, a pre-polymerization complex between the functional monomer and target is a fabrication for the formation of non-covalent bonds followed by the addition of an initiator and the suitable cross-linker in a porogen-with specific ratio (Roushania et al. 2021). An inadequate cross-linker amount can decrease the stability of the polymer structure which leads to the shedding of functional monomers. Its excess amount can reduce the number of MIP diagnostic cavities and thus reduce the binding efficiency. After the degassing step and employing photoinitiation or

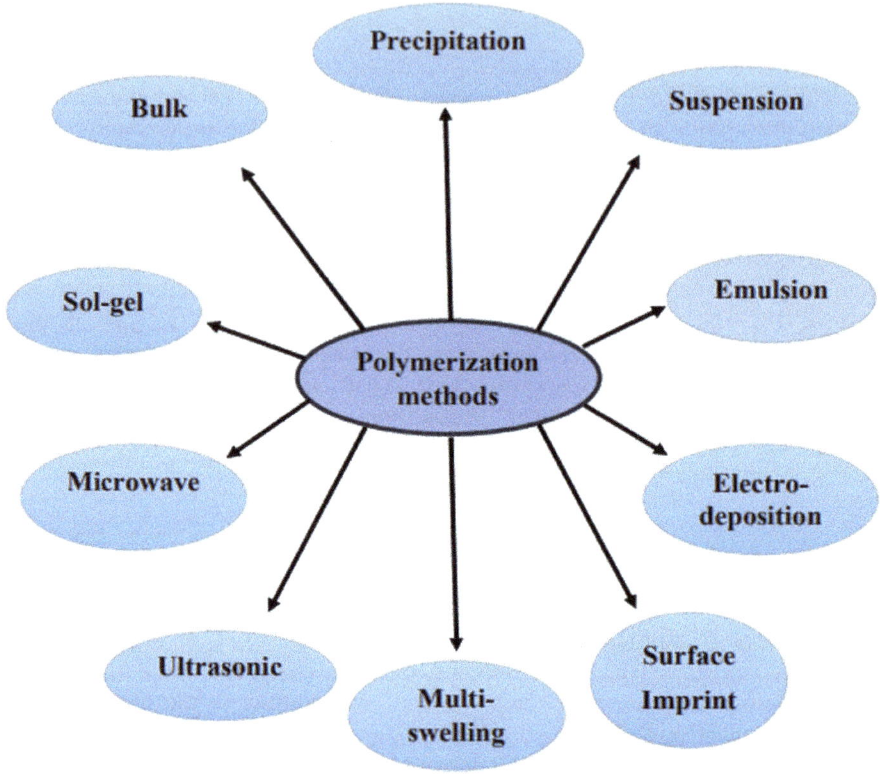

Fig. 2.8 The most common methods of polymerization applied to MIP synthesis (Ma et al. 2015)

thermal, the polymer is created. Then, it can be sieved to achieve the size of desired particles according to the intended application.

This method is simple and quick in its practical implementation, does not require special operator skills or complex tools, and condition optimization is easy, however, it displays many disadvantages. Its major deficiencies are (1) the irregularity in the size and shape of the obtained particles due to the sieving and grinding stage, which can decrease the detection capacity and be unfavorable toward chromatographic applications, however, it is known as an effective modifier for various types of electrochemical sensors (Zhou 2019). (2) The washing step is long and consumes large values of washing solvents (Olcer et al. 2017). (3) It is characterized by poor selectivity and reproducibility (Zhang et al. 2017). (4) Another disadvantage is the relatively high amounts of target needed to synthesize, which is not economically viable for expensive targets (Rachkov et al. 1998).

Table 2.2 Brief of MIP prepared through various techniques (Yan and Row 2006)

MIP format	Benefits	Limitations
Bulk polymerization	Polymerization simplicity and universality, no require particular skills or sophisticated instrumentation	Tedious procedures of grinding, sieving, and column packing, irregular particle in size and shape, low performance
Suspension polymerization	Spherical particles, highly reproducible results, large scale possible	Phase partitioning of complicates system, Water is incompatible with most imprinted procedures, Specialist surfactant polymers required
Seed polymerization	Monodisperse beads of controlled diameter, excellent particle for HPLC	Complicated procedures and reaction conditions, need for aqueous emulsions
Precipitation polymerization	Imprinted microspheres, uniform size and high yields	Large amount of template high dilution factor
Surface polymerization	Monodisperse product, thin imprinted layers	Complicated system, time consuming
In-situ polymerization	One-step, in-situ preparation, cost-efficient, good porosity	Extensive optimization required for each new template system

2.5.2 *Precipitation Polymerization*

Precipitation polymerization is an almost easy one-step technique in the form of sub-micrometer polymeric beads under suitable conditions and belongs to the high-efficiency polymerization technique (Chen et al. 2011). In this technique, the polymerization process takes place in an excess of porogen. The growing polymer continues to take up monomers from the solution until they reach a critical size that results in precipitation (Jing et al. 2009). The polymer beads are filtered via washing and centrifugation (Vasapollo et al. 2011). The major difference between bulk polymerization and this method is the large amount of use of porogen in the case of the precipitation technique. Notably, this polymerization has been chiefly employed to prepare high-quality MIP beads without the need for surfactants or stabilizers (surfactant-free polymerization) (Farooq et al. 2018). This polymerization allows to achievement of regular, spherical, and uniform particles. Many parameters affect the size of achieved particles such as the solvent polarity, stirring speed, and reaction temperature, so the conditions must be optimized (Mohajeri et al. 2011). The major drawbacks include the large value of solvent in the polymerization and the large amount of target needed to synthesize. However, precipitation polymerization is one of the most commonly employed methods to overcome the irregularity in the size and shape of obtained particles of bulk polymerization.

2.5.3 Suspension Polymerization

Suspension polymerization is also an easy method due to one-step polymerization, which permits the achievement of uniform and spherical particles but is characterized via almost big size with a range between micrometers and millimeters. This method is carried out in an aqueous phase, but mineral oils and perfluorocarbon liquids can also be employed as the continuous phase. This type of polymerization is one of the effective techniques to produce uniform beads due to the employ of water-insoluble or partially water-soluble monomers. Therefore, this method is carried out via monomer droplets suspended in a water medium (Pichon and Chapuis-Hugon 2008).

The advantages of this method include a homogeneous transmission of heat and usability on a large scale e.g., to industrial scale-up (Geng et al. 2012). Its disadvantages are (1) the small number of binding cavities because of the employ of water and aqueous phase, which may delay the kinetics of the reaction between the monomer and target and (2) the MIPs achieved through this method are characterized via poor diagnostic because of influence of dispersing phase (Sarpong et al. 2019).

2.5.4 Emulsion Polymerization

This type of polymerization is a biphasic system, which includes emulsifying the functional monomer, template, and cross-linkers in a water medium (Speltini et al. 2017). It is a polymerization technique with high efficiency, which allows for achieving mono-dispersed particles. Particles achieved in the size range between tens and hundreds of nanometers. Stabilizers are added to the non-aqueous phase for the production of homogeneous, stabilized, and small emulsion droplets. Water is commonly employed as the continuous phase, in which surfactant and water can affect the imprinting process yield by affecting the interaction stability between monomer and target. A soluble redox initiator in water is usually employed in this method. The initiator decomposes under heating or illumination conditions to produce free radicals. First, the pre-polymerization of the target and monomer is done, which takes about 12 h and ultrasonication is applied to improve the effect of pre-polymerization. Thereafter, initiators and cross-linkers are added to perform the polymerization reaction. Then, the achieved polymer is washed to remove the target under suitable conditions.

Although the emulsion polymerization creates uniform particles, it has drawbacks including indicating low product performance, needing complex conditions that make it expensive and difficult, and the existence of the water phase in the system that may have a negative effect on the polymer performance, which is supposed to be based on desirable hydrogen bonding to the target rather than the existing of water in the synthesis process (Ashley et al. 2017; Zhang et al. 2012). Then, this method is not very attractive to the synthesis of MIP (Yang et al. 2015).

2.5.5 Electropolymerization

In situ polymerization, electropolymerization, is a particular class of imprinting on the surface of an electrode such as platinum, gold, glassy carbon, etc. rather than a non-conductive support. In situ production of MIP films introduced a very easy and fast one-step technique to achieve selective layers directly at the working electrode. The polymerization solution commonly includes a template and monomer dissolved in a suitable electrolyte (Mirzaei Karazan et al. 2022). This type of polymerization depends on the production of free radicals by oxidation of monomers onto the surface of the sensor due to the employed potential. Then, free radicals are adsorbed on the electrode and create a polymeric structure. By optimizing the current and potential, a thin polymer layer with controlled density and thickness can be coated on the electrode in an easy and fast method (Roushani and Mirzaei Karazan 2022). In addition, this technique is recommended to direct the pattern and deposit of thin films on the transducer surface with high mass transfer and accessibility.

Its advantage is to control the thickness of the polymer film by the optimize of different deposition parameters including the number of deposition cycles, used voltage, and current density, which takes a more uniform film of the imprinted polymer on the surface of the electrode (Kamal Ahmed et al. 2021). The printed polymers can be also placed onto the electrode surface through either electrospraying or spin coating of the template/monomer mixture on the surface. Electropolymerization is particularly useful for very expensive targets, as little target is consumed in the printing step. In addition, this method is very rapid compared to other chemical polymerization processes (Ashley et al. 2017).

2.5.6 Surface Imprinting Polymerization

In this type of polymerization, the monomer and target are mixed for the formation pre-polymerization solution. The polymerization is conducted onto the solid substrate surface such as chitosan, magnetite, silica, etc. in the presence of a cross-linker and initiator for the formation of the printed film. The target knocking out of the polymer film through chemical or physical techniques results in the construction of three-dimensional sites onto the substrate surface (Farooq et al. 2018). The modified surfaces are employed in sensing or separation applications (Yang et al. 2018).

Its advantages are wide usability, inexpensive and simple preparation, wide surface area with very selective binding cavities, and controllable shape and size (Chen et al. 2016). It increases the transfer of the charge, resulting in complete elution of the target and offering excellent accessibility to the template. Surface imprinting is acceptable for imprinting macromolecules consisting of cells, viruses, and proteins. It is usually carried out on support beads that are chiefly silica beads and are employed to deposit imprinted films. This method benefits core–shell nanostructure for the final polymer, so designing a MIP shell on the support material of the core. Other support materials consisting of chitosan and polystyrene can be applied. Its limitations include the possibility of MIP layer detaching after repeated use the difficulty in controlling its thickness, and the time-consuming method.

2.5.7 Seed Polymerization

Seed polymerization, multi-step swelling polymerization, is ascertained via complex time-consuming methods consisting of particular reaction conditions that allow obtaining uniform monodispersed spherical particles in size and shape that can be directly supplied as beads of controlled diameter (Nakamura et al. 2005). It is worth mentioning, that reaction conditions and relatively complex methods are needed, and the aqueous suspensions employed can interfere with the imprinting and thus lead to reduced selectivity. MIPs achieved through this method are suitable for applications of chromatography. The realization of seed polymerization is related to the preparation of aqueous solutions that can affect the selectivity and the printing process of the achieved MIPs (Nakamura et al. 2017).

2.5.8 Smart Imprinting

A smart polymer can be considered as a polymer that is able to respond consistently and reliably to the application of external stimuli. On the other hand, MIPs have attracted much attention due to their strong polymer network, reusability, cost-effectiveness, and inherent simplicity. Therefore, by integrating the imprinted polymers and smart materials technology, "smart" MIPs with self-control capabilities have been successfully introduced and provided (Ge et al. 2013). Smart imprinting is an intelligent MIP that can respond to an external stimulus that changes its affinity for the template through enabling conversion for release or binding. A general approach to prepare smart MIPs is to incorporate a stimuli-responsive material in the molecular imprinting method. Therefore, the MIP takes a switchable releasing or binding capacity through smart material. The common stimuli can be changes in pH, incident light, temperature, ionic strength, and the presence of a specific template (Morsi et al. 2023).

Molecularly imprinted nanoparticles (NanoMIPs) have been used as an alternative to antibodies and enzymes in sensors and assays, which show excellent selectivity and sensitivity in biological medium (Garcia-Cruz et al. 2020). The use of NanoMIPs enables the production of uniform detection sites, leading to greater overall specificity and performance than traditional methods (Haq et al. 2021). In addition, NanoMIPs are easily synthesized and integrated into a variety of transducers, such as electrochemical apparatus (Alanazi et al. 2021). The smart or stimuli-responsive nanomaterials are known as nano-actuators. These nano-actuators were supplied through electroactive polymer nanoparticles, prepared through molecularly imprinting (Haq et al. 2023). It is worth mentioning, these nanoparticles are employed as sensors because they respond to a physical stimulus and transfer an obtaining impulse to electrical signals (Turner 2013). Its mechanism is particle actuation (Algamili et al. 2021). The template detection process causes a change in the polymer conformation and produces an electrical signal. Consequently, these nanoparticles are actuators because they cause molecular motion upon template recognition and the motion of the nanoparticles is converted into an electric signal. Thus, NanoMIP actuation is target-specific as no change was observed in the presence of other templates (Haq et al. 2023). Compared to the bulk polymerization, in addition to the ability to respond to an external stimulus, this method has advantages such as lower consumption of washing solvent and template molecule, easier removal of template molecule, more regular particles and thus higher detection capacity, better selectivity and reproducibility, and higher access to detection sites.

2.5.9 Epitope Imprinting Technology

Epitope imprinting is a promising technique to create particular selective sites. This type of imprinting has many advantages such as the easy access to analytes, the flexibility in choosing monomers, their resistance to various environments, and the excellent specificity toward analytes. The epitope imprinting is used in various fields including pharmacy, clinical therapy, and bioanalysis. Here, the discussion is focused on the epitope imprinting technologies such as epitope surface imprinting and epitope bulk imprinting (Yang et al. 2019).

Epitope Bulk Imprinting

In epitope bulk imprinting, the epitope, crosslinkers, initiators, and functional monomers are mixed with solvents. Thereafter, the analytes are extracted from the bulk, followed through some post-treatment, including crushing, grinding, and sieving, to achieve the demanded particle size and expose the recognition holes. This technique has advantages such as simple and quick preparation, inexpensive instrumentation, and good purity is obtained in the imprinted polymers.

Epitope Surface Imprinting

The purpose of epitope surface imprinting is to provide imprinted polymers through controlling the epitope analyte so that it is placed on the surface of the material. In compared with epitope bulk polymer, this technique results in the complete extract of analytes and prepares good accessibility to analytes, especially for large cells and target molecules. It is worth mentioning, a supporting matrix for epitope immobilization is always produced during the create of surface-imprinted polymers.

2.6 Summary

MIPs are widely used in various fields, including sensing, separation, catalysis, and drug delivery. The most impressive parameter in the providing of MIPs is the selection of proper: components (cross-linkers, initiator, functional monomer, and solvent), strategies of polymerization (covalent, non-covalent, and semi-covalent approach), mechanisms type of polymerization (sol–gel and free radical polymerization), and method of polymerization (suspension, bulk, precipitation, emulsion, seed, surface imprinting, and electropolymerization). The declared parameters are related to MIP applications and affect MIP properties such as porosity, size and shape of particles, surface area, binding capability, mechanical performance of polymer, imprinting yield, and diagnostic efficiency. Recently, much research has been performed and these parameters have been optimized.

References

Alanazi K, Cruz AG, Di Masi S, Voorhaar A, Ahmad OS, Cowen T, Piletska E, Langford N, Coats TJ, Sims MR, Piletsky SA (2021) Sensor Actuat B: Chem 329:129128
Algamili AS, Khir MHM, Dennis JO, Ahmed AY, Alabsi SS, Ba Hashwan SS, Junaid MM (2021) Nanoscale Res Lett 16:1
Anderson LI (2000) J Chromatogr B 745:3
Ansari S (2017) Trends Anal Chem 93:134
Ashley J, Shahbazi M-A, Kant K, Chidambara VA, Wolff A, Bang DD (2017) Biosens Bioelectron 91:606–615
Bagheri AR, Arabi M, Ghaedi M, Ostovan A, Wang X, Li J (2019) Talanta 195:390
Barsbay M, Guven O (2018) Radiat Phys Chem 169:107816
Beltran A, Marce RM, Cormack PA, Borrull F (2010) Anal Chim Acta 1:72
Beqqali AE, Andersson LI, Jeppsson AD, Abdel-Rehim M (2017) J Chromatogr B 1063:130
Bunte G, Hurttlen J, Pontius H, Hartlieb K, Krause H (2007) Anal Chim Acta 1:49
Canfarotta F, Rapini R, Piletsky S (2018) Curr Opin Electrochem 7:146
Caro E, Masque N, Marce RM, Borrull F, Cormack PAG, Sherrington DC (2002) J Chromatogr A 963:169
Chen L, Xu S, Li J (2011) Chem Soc Rev 40:2922
Chen L, Wang X, Lu W, Wu X, Li J (2016) Chem Soc Rev 45:2137
Chen Y, Zhou S, Li L, Zhu J-J (2017) Nano Today 12:98
Cheong SH, McNiven S, Rachkov AE, Levi R, Yano K, Karube I (1997) Macromolecules 30:1317
Cormack PAG, Zurutuza Elorza A (2004) J Chromatogr B 804:173
Farooq S, Nie J, Cheng Y, Yan Z, Li J, Bacha SAS, Mushtaq A, Zhang H (2018) Analyst 143:3971
Garcia-Cruz A, Cowen T, Voorhaar A, Piletska E, Piletsky SA (2020) Analyst 145:4224
Ge Y, Butler B, Mirza F, Habib-Ullah S, Fei D (2013) Macromol Rapid Commun 34:903
Geng L, Kou X, Lei J, Su H, Ma G, Su Z (2012) J Chem Technol Biotechnol 87:635
Guc M, Schroeder G (2017) J Res Rev 5:36
Guoning C, Hua S, Wang L, Qianqian H, Xia C, Hongge Z (2020) J Pharm Biomed Anal 190:113511
Gutierrez-Climente R, Gomez-Caballero A, Goicolea M, Barrio RJ (2016) J Mol Recognit 3:106
Han DM, Fang GZ, Yan XP (2005a) J Chromatogr A 2:131
Han DM, Fang GZ, Yan XP (2005b) J Chromatogr A 1100:131
Han S, Lyu S, Chen Z, Fu F, Wang S (2020) Carbohydr Polym 234:115923

Haq I, Alanazi K, Czulak J, Di Masi S, Piletska E, Mujahid A, Hussain T, Piletsky SA, Garcia-Cruz A (2021) Nanoscale Adv 3:4276
Haq I, Cruz AG, Di Masi S, Cowen T, Allcock NS, Malitesta C, Mujahid A, Hussain T, Piletska E, Piletsky SA (2023) Sensor Actuat B: Chem 376:132928
Haupt K, Linares AV, Bompart M, Tse Sum Bui B (2012) Top Curr Chem 325:1
Huang S, Xu J, Zheng J, Zhu F, Xie L, Ouyang G (2018) Anal Bioanal Chem 410:3991
Jing T, Gao XD, Wang P, Wang Y, Lin YF, Hu XZ, Hao QL, Zhou YK, Mei SR (2009) Anal Bioanal Chem 2009:393
Joshi VP, Karode SK, Kulkarni MG, Mashelkar RA (1998) Chem Eng Sci 53:2271
Kamal Ahmed R, Saad EM, Fahmy HM, El Nashar RM (2021) Microchem J 171:106771
Kang MS, Cho E, Choi HE, Amri C, Lee J, Kim KS (2023) Biomater Res 27:45
Lenain P, Mavungu J, Dubruel P, Robbens J, Saeger S (2012) Anal Chem 23:10411
Li X, Li M, Li J, Lei F, Su X, Liu M, Li P, Tan X (2014) Anal Methods 6:6397
Lin Z-Z, Zhang H, Li L, Huang Z-Y (2016) React Funct Polym 98:24
Liu J-Q, Wulff G (2008) J Am Chem Soc 130:8044
Lu Z, Yu Z, Dong J, Song M, Liu Y, Liu X, Ma Z, Su H, Yan Y, Huo P (2018) Chem Eng J 337:228
Lusina A, CegLowski M (2022) Polymers 14:640
Ma Y, Xu S, Wang S, Wang L (2015) Trends Anal Chem 67:209
Madikizela L, Tavengwa N, Pakade V (2017) IntechOpen Limited 47
Metwally MG, Benhawy AH, Khalifa RM, El Nashar RM, Trojanowicz M (2021) Molecules 26:6515
Mirzaei Karazan Z, Roushani M (2021) Anal Methods 13:4904
Mirzaei Karazan Z, Roushani M (2022) Talanta 246:123491
Mohajeri SA, Karimi G, Aghamohammadian J, Khansari MR (2011) J Appl Polym Sci 12:3590
Moreno-Bondi MC, Navarro-Villoslada F, Benito-Peña E, Urraca JL (2008) Curr Anal Chem 4:316
Morsi SM, Abd El-Aziz ME, Mohamed HA (2023) Int J Polym Mater Po 72:612
Nakamura M, Ono M, Nakajima T, Ito Y, Aketo T, Haginaka J (2005) J Pharm Biomed Anal 37:231
Nakamura Y, Masumoto S, Kubo A, Matsunaga H, Haginaka J (2017) J Chromatogr A 1516:71
Olcer YA, Demirkurt M, Demir MM, Eroglu AE (2017) RSC Adv 7:31441
Otsu T (2000) J Polym Sci Part A: Polym Chem 38:2121
Otsu T, Yoshida M (1982) Makromol Chem, Rapid Commun 3:127
Patra S, Roy E, Madhuri R, Sharma PK (2015) Biosens Bioelectron 66:1
Pichon V, Chapuis-Hugon F (2008) Anal Chim Acta 622:48
Rachkov AE, Cheong SH, El'skaya AV, Yano K, Karube I (1998) Polym Adv Technol 9:511
Rodrıguez JJS, Padron MET, Aufartova J, Ferrera ZS (2010) IntechOpen Limited 305
Roushani M, Mirzaei Karazan Z (2022) IEEE Sens J 22:7
Roushania M, Saedia Z, Mirzaei Karazana Z, Azadbakht A (2021) Anal Bioanal Chem Res 8:493
Saloni J, Walker K, Hill G Jr (2013) J Phys Chem A 117:1531
Sarpong KA, Xu W, Huang W, Yang W (2019) J Anal Chem 10:202
Sellergren B (2001) J Chromatogr A 906:227
Speltini A, Scalabrini A, Maraschi F, Sturini M, Profumo A (2017) Anal Chim Acta 974:1
Sulitzky C, Rückert B, Hall AJ, Lanza F, Unger K (2002a) Macromolecules 1:79
Sulitzky C, Ruckert B, Hall AJ, Lanza F, Unger K, Sellergren B (2002b) Macromolecules 35:79
Turner AP (2013) Chem Soc Rev 42:3184
Vasapollo G, Del Sole R, Mergola L, Rosaria Lazzoi M, Scardino A, Scorrano S, Mele G (2011) Int J Mol Sci **12**:5908
Wackerlig J, Lieberzeit PA (2015) Sensors Actuators B Chem 207:144
Wei S, Hu X, Liu H, Wang Q, He C (2015) J Hazard Mater 294:168
Włoch M, Datta J (2019) Compr Anal Chem 86. ISSN 0166-526X
Wulff G, Sarhan A (1972) Angew Chem 84:364
Xiao D, Su L, Teng Y, Hao J, Bi Y (2020) Microchim Acta 187:399
Yan H, Row KH (2006) Int J Mol Sci 7:155
Yang J, Li Y, Wang J, Sun X, Cao R, Sun H, Huang C, Chen J (2015) Anal Chim Acta 872:35

Yang Y, Meng X, Xiao Z (2018) RSC Adv 8:9802
Yang K, Li S, Liu L, Chen Y, Zhou W, Pei J, Liang Z, Zhang L, Zhang Y (2019) Adv Mater 31:1902048
Zhang Y, Yao X (2014) Polym Sci Ser B 4:538
Zhang H, Dramou P, He H, Tan S, Pham-Huy C, Pan H (2012) J Chromatogr Sci 50:499
Zhang W, She X, Wang L, Fan H, Zhou Q, Huang X, Tang JZ (2017) Materials 10:475
Zhou T (2019) Lund University: Lund, Sweden. ISBN 9789174224566

Chapter 3
Molecularly Imprinted Polymers: Cutting-Edge Characterization Strategies

Raif İlktaç

3.1 Introduction

3.1.1 General Aspects of Molecularly Imprinted Polymers (MIPs)

Molecularly imprinted polymers (MIPs), in the simplest terms, can be defined as synthetic polymers with specific molecular recognition capabilities for the targeted analyte (Haupt et al. 2012). The five main components of an MIP can be classified as the template molecule, functional monomer, cross-linker, initiator, and solvent, which is generally referred to as the porogen in the MIP process.

MIPs can be synthesized using various polymerization methods involving a process known as Molecular Imprinting Technology (MIT). The polymer structure is formed through the polymerization of functional monomer(s), template molecule(s), and cross-linker(s) within a suitable porogen, in the presence of an initiator (Zhou et al. 2019). The recognition mechanism relies on the interactions occurring between the functional monomer and the template molecule (Leibl et al. 2021). The different main components employed in MIT are presented in Table 3.1. MIT is centered on generating specific binding sites by removing the template molecule from the polymeric network. This process creates binding sites that exhibit a shape and chemical functionality complementary to the template, thereby imparting selectivity to MIPs (Dong et al. 2020).

R. İlktaç (✉)
Application and Research Center for Testing and Analysis, University of Ege, Bornova, İzmir, Turkey
e-mail: raif.ilktac@ege.edu.tr

S. Patra and M. Sillanpaa (eds.), *Molecularly Imprinted Polymers as Artificial Antibodies for the Environmental Health*, https://doi.org/10.1007/978-3-031-58995-9_3

Table 3.1 Commonly used chemicals in MIP

Functional monomers	Cross-linker	Initiator	Porogen
Acrylic acid (AA)	Ethylene glycol dimethacrylate (EGDMA)	Azobisisobutyronitrile (AIBN)	Chloroform
Methacrylic acid (MAA)	Trimethylolpropane trimethacrylate (TRIM)	Benzoyl peroxide (BPO)	Acetonitrile
Acrylamide (AM)	Divinylbenzene (DVB)	Ammonium persulfate (APS)	Tetrahydrofuran
2-Vinylpyridine (2-VP)	Pentaerythritol trimethacrylate (PETA)	Photo initiator	Ionic liquids
4-Vinylpyridine (4-VP)		Thermal initiator	

MIPs offer a multitude of advantages that contribute to their widespread usage. Firstly, they can be precisely engineered to exhibit remarkable selectivity and specificity towards specific target molecules, a crucial trait for accurate detection and separation processes. This not only enables accurate and sensitive detection but also facilitates the separation of analytes in complex matrices. Another key factor is their adaptability, as MIPs can be tailored for an extensive array of target molecules, encompassing a broad spectrum from small organic compounds to intricate biomolecules. This adaptability makes them applicable in fields ranging from pharmaceuticals to environmental monitoring (Liu et al. 2018).

Notably, the excellent stability of MIPs makes them attractive adsorbents for the separation or detection of analytes in environmental samples. Additionally, their stability and reusability significantly enhance the cost-effectiveness of the developed method or system, which is crucial for applications requiring repeated analyses (Zare et al. 2022). By fine-tuning polymerization conditions, MIPs can be custom-designed to possess specific binding properties, catering to the needs of various analytes. Their ability to work effectively with various formats, such as sensors, chromatographic columns, and solid-phase extraction cartridges, makes it easy to smoothly incorporate MIPs into existing analytical methods (Yuksel and Tektas 2022; Martín-Esteban 2016; Mpupa et al. 2021).

In summary, the cost-effectiveness, reusability, stability, versatility, and superior affinity of MIPs towards target molecules establish them as the preferred choice across domains such as analytical chemistry, biotechnology, medicine, and environmental monitoring (Kryscio and Peppas 2012; Mahmoudpour et al. 2020).

Above all, the determination of the chemical, physical, and mechanical properties of the produced MIP using different instruments and the clear presentation of its structure are highly effective in enhancing the binding capacity and selectivity of the MIP, optimizing the synthesis conditions, as well as demonstrating that the production process is repeatable and consistent.

The development and optimization of imprinted polymers require a profound understanding of their structural, morphological, and functional characteristics. This chapter provides a comprehensive introduction to the characterization methods employed in imprinted polymer studies. As these polymers continue to find applications in diverse fields such as sensors, separations, and drug delivery, a detailed exploration of advanced characterization techniques becomes essential for harnessing their full potential.

Thus, the characterization of the produced MIP and the clear presentation of its chemical, physical, and mechanical structure are of utmost importance. In this perspective, this chapter focuses on the general aspects and applications of the most widely used techniques for the characterization of MIPs.

3.1.2 Basic Principles of Characterization Techniques

Scanning Electron Microscopy-Energy Dispersive X-ray (SEM–EDX) Spectroscopy

The Scanning Electron Microscope (SEM) serves as a powerful imaging tool across various fields, allowing the observation of surface morphology and micro/nanostructures of materials at high resolution. Its imaging principle relies on the interaction between an electron beam and the sample under examination. The SEM employs an electron gun to generate a focused and highly accelerated beam of electrons, reaching accelerating voltages of up to 30.0 kV (Mohammed and Abdullah 2019).

Upon interaction with the sample's surface, high-energy primary electrons induce the emission of secondary electrons (SEs), backscattered electrons (BSEs), or X-rays. A detector collects the SEs or BSEs, and the signals obtained are converted into electrical signals, ultimately forming an image on a display screen (Zhou et al. 2007).

Backscattered electrons (BSEs) result from the elastic scattering of primary electrons bouncing off the atomic nuclei of the sample's atoms. The intensity of BSEs depends on the atomic number of the sample's elements, with higher atomic numbers generating brighter areas in the image. This imaging technique provides information about the elemental composition of the sample's surface, with brighter areas indicating higher atomic number elements and darker areas representing lower atomic number elements (Subramanian et al. 2018).

On the other hand, SEs are emitted from the sample's surface upon interaction with primary electrons, revealing the sample's surface morphology. The intensity of SEs is used to create the image, with brighter areas corresponding to regions emitting more SEs, indicating higher surface features, and darker areas representing lower surface features. SEs play a crucial role in gathering detailed information about the surface morphology and topography of the sample (Abd Mutalib et al. 2017).

Additionally, as SEs are emitted, vacancies in the atoms on the sample surface are created, leading to instability. Outer-shell electrons then fill these vacancies, emitting X-rays with specific energies. Each element produces characteristic X-rays, which are collected by an X-ray detector. This process identifies the element by measuring the energies of the characteristic X-rays. The intensities of the X-ray peaks provide information about the relative concentrations of elements within the range of boron to uranium (Akhtar et al. 2018).

Transmission Electron Microscopy (TEM)

Transmission Electron Microscopy (TEM) is a powerful imaging technique used to visualize the internal structure and morphology of materials at a very high resolution, down to the nanoscale (Fan and Ellisman 2000).

TEM operates by directing a focused beam of electrons through a thin sample, engaging with the material to generate high-resolution images that unveil intricate aspects of the material's structure and composition. The electrons that traverse the sample are captured by a detector and transformed into an image displayed on a monitor (Tang and Yang 2017).

With accelerating voltages of electrons reaching up to 300.0 kV, TEM can achieve extremely high magnifications and resolve structures down to the atomic level, providing detailed information about crystal lattices, nanoscale particles, and molecular arrangements (Kannan 2018).

Atomic Force Microscopy (AFM)

Atomic Force Microscopy (AFM) operates on the principle of using a sharp tip attached to a flexible cantilever to scan a sample's surface. The tip interacts with the surface at a nanometer scale, and the interaction forces between the tip and the sample are measured. As the cantilever moves over the sample's surface, the interaction forces, primarily including van der Waals and electrostatic forces, cause the cantilever to deflect (Seo and Jhe 2008).

The information about the deflection is used to generate a topographic image of the sample's surface with extremely high resolution. By continuously monitoring the cantilever's deflection, AFM can create detailed maps of surface features, including variations in height, roughness, and even molecular interactions. AFM is used for investigating and analyzing the size, shape, and distribution of nanomaterials (Giessibl 1985).

AFM can be operated in several modes, including contact mode, non-contact mode, and tapping mode. In contact mode, the tip remains in constant contact with the sample's surface, useful for obtaining high-resolution images. In non-contact mode, the tip oscillates slightly above the surface, minimizing any damage to the sample. Tapping mode involves oscillating the cantilever, allowing intermittent contact between the tip and the sample (Shinato et al. 2020; Senthil Kumar et al. 2019).

Fourier Transform Infrared (FTIR) Spectroscopy

When a sample is exposed to infrared (IR) light, its molecules absorb specific wavelengths of the light due to their vibrational modes. Vibrational modes of the molecules can be defined as stretching and bending vibrations, responsible for the absorption of IR (Wellner 2013).

Stretching vibrations involve the displacement of an atom along the bond axis, without altering the bond angle. In symmetrical stretching, both atoms move in the same direction-either away from or towards the molecule's center. In asymmetrical stretching, one atom moves closer to the central atom, while the other moves farther away. In bending vibrations, the movement occurs among atoms bonded to a central atom, resulting in a change in both bond axis and angle. This change occurs without affecting the bond length. Rocking and scissoring vibrations take place within the molecular plane, while wagging and twisting motions occur out of the molecular plane (Ismail et al. 1997; Guerrero-Perez and Patience 2020).

For the absorption of IR light by a molecule, the energy of its vibrational transition must be equal to that of the radiation. Furthermore, a net change in the dipole moment should occur upon the absorption of IR light. If the net change in the dipole moment of the molecule is absent, then no absorption in the IR region is detected (Dutta 2017).

Traditional dispersive IR spectroscopy uses a prism or grating to separate the different wavelengths of IR light. However, FTIR spectroscopy uses an interferometer that splits the IR beam into two beams: the sample beam and the reference beam. The sample beam passes through the sample, whereas the reference beam is used as a baseline for comparison. These beams undergo mirror reflections and then recombine. The resultant interference between the beams generates a pattern of constructive and destructive interference, recorded as the interferogram. A mathematical process, Fourier transform, is applied to the interferogram to transform it into an IR absorption spectrum (Mohamed et al. 2017; Petit and Madejova 2013).

The total measurement range in wavenumbers for FTIR spectroscopy is approximately 4000–400 cm^{-1}. This range encompasses the mid-infrared (MIR) region of the electromagnetic spectrum, widely utilized for identifying molecular functional groups, analyzing chemical bonds, and studying various types of materials.

Nuclear Magnetic Resonance (NMR) Spectroscopy

NMR spectroscopy is a significant analytical technique based on the concept that atomic nuclei with specific magnetic properties can absorb energy when subjected to a strong magnetic field and radiofrequency (RF) radiation.

An RF pulse is applied perpendicular to the magnetic field, with a frequency corresponding to the resonant frequency of the nuclei being studied. This pulse provides energy that matches the energy difference between nuclear spin states, causing some nuclei to transition from their lower-energy to higher-energy states. When the radiation ceases, the nuclei return to their original state and release the absorbed energy as RF signals, detected by the NMR instrument, providing valuable information about the sample's composition and structure (Jackman and Sternhell 1969).

The absorbed energy is proportional to the strength of the magnetic field and the nucleus's inherent characteristics. The resonant frequencies at which these signals manifest offer insights into the chemical environments surrounding the nuclei, facilitating the analysis of molecular properties (Becker 2000).

By carefully controlling the magnetic field and analyzing the frequency and intensity of the emitted signals, NMR spectroscopy provides information about the sample's chemical environment, molecular structure, and interactions. Different nuclei resonate at unique frequencies due to their local chemical surroundings, allowing NMR to elucidate molecular details with remarkable precision (Jin 1962; Yao et al. 2020).

NMR spectroscopy can study elements with nuclei possessing nonzero nuclear spin. The most commonly studied elements are ^{1}H (hydrogen/proton) and ^{13}C (carbon), abundant in organic compounds. NMR provides insights into the conformational flexibility and secondary structure of polymer chains. Other elements with magnetic nuclei, such as ^{15}N (nitrogen), ^{17}O (oxygen), and ^{31}P (phosphorus), can also be studied using NMR (Luther et al. 2003). Changes in chemical shifts and coupling constants can indicate variations in conformation resulting from template binding or environmental conditions (Li and Row 2018; Cheong et al. 2013).

X-Ray Photoelectron Spectroscopy (XPS)

X-ray Photoelectron Spectroscopy (XPS), also known as Electron Spectroscopy for Chemical Analysis (ESCA), is a surface analysis technique that provides insights into the elemental composition, chemical states, and electronic structure of materials (Brigiano et al. 2022; Tougaard 2010).

The principle underlying XPS revolves around the interaction of X-rays with a material's surface and the subsequent measurement of emitted photoelectrons. When a beam of X-rays interacts with the material's surface, it has sufficient energy to penetrate the outer layers of the material. Photoelectrons, initially bound to atoms or molecules within the material, acquire sufficient energy from absorbed photons and are subsequently released from the atom. The energy of a photoelectron is linked to the difference between the energy of the incident photon and the binding energy of the electron within the atom. Analyzing the kinetic energy of the emitted photoelectrons enables the determination of the electron's binding energy in its initial state, providing information about the chemical composition and electronic structure of the material's surface (Chan and Weng 2016; Bubert et al. 2011).

The interaction of X-rays with the sample primarily results in the expulsion of core-level electrons from the innermost energy levels (e.g., 1s, 2s, 2p) of the atoms. When a core-level electron is ejected, it generates a 'core hole' within the atom's energy levels. To compensate for this core hole, an electron from a higher energy level (such as a valence level) may transition to the lower energy level. During this transition, an electron is emitted from the atom, referred to as an Auger electron. Auger electrons are emitted as part of the relaxation process subsequent to the expulsion of core-level electrons. The energies of the emitted Auger electrons correspond to specific electronic transitions within the atom, providing characteristics indicative of the element present and its chemical environment (Venezia 2003; Peredkov et al. 2007).

XPS captures the emitted photoelectrons and Auger electrons, and through the analysis of their energies, it provides comprehensive insights into the surface composition and electronic structure of the material (Ghobeira et al. 2022; Ayiania et al. 2020).

Thermogravimetric Analysis (TGA)

Thermogravimetric Analysis (TGA) is a thermal analysis technique used to study changes in the weight of a sample as it is heated under controlled conditions. The primary principle behind TGA is to measure the weight changes of a sample as it undergoes thermal decomposition, oxidation, reduction, or other chemical reactions over a range of temperatures. TGA experiments can be conducted under various heating conditions, such as isothermal (constant temperature) or dynamic modes, and under controlled atmospheres, such as inert gas (nitrogen or argon) or reactive gases (oxygen) (Nurazzi et al. 2021; Akash and Rehman 2019).

The weight of the sample is continuously measured throughout the heating process as it undergoes the aforementioned reactions. Its weight can change due to the loss or gain of volatile components, gases, or other materials (Mansfield et al. 2014; Bach and Chen 2017).

Differential Thermogravimetry (DTG) is another technique that illustrates the variation in the rate of weight change with temperature. DTG is a derivative analysis technique applied to TGA data to highlight the rate of weight change at various temperatures. It offers insights into different weight loss or gain steps in the thermogram and the corresponding temperatures at which these changes occur (De Blasio 2019; Loganathan et al. 2017).

Both techniques are employed to determine the thermal stability and behavior of materials, including phase transitions, polymorphic changes, glass transition temperatures, melting points, and decomposition temperatures (Ng et al. 2018; Menczel and Prime 2008).

Differential Scanning Calorimetry (DSC)

Differential Scanning Calorimetry (DSC) is another thermal analysis technique that measures the heat flow into or out of a sample as it undergoes temperature changes. The basic principle of DSC involves comparing the heat flow to a reference material as both the sample and reference are subjected to the same temperature program (Byrn et al. 2017; Almoselhy 2020).

In DSC analysis, a controlled temperature program is applied to both the sample and reference cells. By comparing the heat flow between the sample and reference cells, DSC captures the energy absorbed or released during processes like phase transitions, reactions, and thermal events. As temperature changes, the DSC instrument records the heat flow difference between the sample and reference, creating a thermogram (Jin and Wunderlich 1993; Zheng et al. 2019).

DSC is widely used to investigate phase transitions, thermal stability, heat capacities, reaction kinetics, and other thermal properties such as melting, freezing, and crystallization points, glass transition temperature, and heat capacity of materials (Leyva-Porras et al. 2019; Gabbott 2008).

Brunauer Emmett Teller (BET) Analysis

Brunauer–Emmett–Teller (BET) analysis is a widely used method for determining the specific surface area of porous materials, especially solids like powders and porous materials. The principle of BET analysis is based on the adsorption of gas molecules onto the surface of a solid material, proposing that the first layer of gas molecules forms a complete monolayer on the surface (Sinha et al. 2019; Tian and Wu 2018).

The analysis is conducted using a gas adsorption isotherm, a plot of the amount of gas adsorbed on the material's surface against the pressure at a constant temperature. By plotting the amount of adsorbed gas against relative pressure, the BET equation provides a linear relationship from which the specific surface area can be calculated (Irwansyah et al. 2024; Sing 2001).

In BET analysis, typically inert gases are used, with nitrogen being the most commonly chosen gas. Nitrogen is preferred due to its non-reactive nature and availability, allowing accurate measurements of surface area and adsorption properties without chemical reactions occurring between the gas and the material being analyzed. Other inert gases like argon can also be used in certain cases, with the choice depending on factors such as the specific properties of the material and the available equipment. The primary consideration is to use a gas that does not chemically interact with the material and can provide accurate surface area measurements (Naderi 2015).

BET analysis aids in understanding the porosity and adsorption properties of various materials, making it a valuable tool in materials science and surface characterization.

Vibrating Sample Magnetometer (VSM)

The Vibrating Sample Magnetometer (VSM) operates based on the fundamental principle of measuring the magnetic properties of materials, particularly their magnetization as a function of an applied magnetic field. In this technique, a sample is attached to a thin rod or wire and is placed within a uniform magnetic field. The sample is then subjected to a varying magnetic field strength, causing the sample to undergo magnetization changes. As the magnetic field strength changes, the sample's magnetic moments align and adjust accordingly, leading to changes in its magnetic properties (Niazi et al. 2015; Kirupakar et al. 2016).

During the measurement, the sample is made to vibrate or oscillate within the magnetic field, enabling the instrument to record changes in magnetization as a function of the applied magnetic field. The induced voltage or current generated by the sample's changing magnetic moments is detected and translated into magnetization data (Phillips et al. 2022).

The data derived from VSM analysis offers significant insights into the magnetic characteristics of the material, encompassing aspects such as its magnetic moment, magnetic susceptibility, coercivity, and remanence (Jordán et al. 2018; El-Alaily et al. 2015).

3.2 Applications

Characterization techniques can be divided into four sub-categories which can be classified as physical, chemical, thermal and functional characterization. The classification of these characterization techniques is illustrated in Fig. 3.1.

3.2.1 Physical Characterization

Physical characterization delves into the examination of a material's physical properties, encompassing aspects like size, shape, structure, magnetic behavior, and mechanical traits. As a result, it offers valuable insights into the material's morphology, surface roughness, mechanical strength, and various other physical attributes (Fig. 3.2).

Transmission Electron Microscopy (TEM) serves as a crucial tool for visualizing imprinted cavities, surface morphology, and the dispersity of nanoparticles, particularly in the context of nanocomposites and core–shell structures.

A noteworthy example by Bonyadi and Ghanbari (2023) involved the fabrication of an MIP sensor using a carbon paste electrode modified with ZnO nanoparticles. TEM images unveiled a homogeneous dispersion of ZnO nanoparticles within the polymer structure, with spherical polymer particles measuring less than 100 nm in diameter.

Foroughirad et al. (2021) crafted a magnetic halloysite-based molecularly imprinted polymer for the specific detection of sunset yellow. TEM analysis provided insights into a thin polymeric shell, approximately 10 nm in thickness, coating the surface of modified magnetic nanotubes.

Fig. 3.1 Classification of the characterization techniques

Similarly, Boitard et al. (2019) synthesized a core–shell compound comprising MIP-coated magnetic iron oxide nanoparticles. TEM analysis uncovered an amorphous coating around the nanoparticles, with a coating thickness ranging from 1 to 10 nm.

In a study conducted by Li et al. (2009), a core–shell structure of MIP film-silica-coated magnetic nanoparticles was produced, featuring a 10 nm thin layer of polymer on the surface of the core structure.

SEM emerges as a widely employed instrument for determining the particle size and morphology of synthesized MIPs. Numerous researchers have utilized SEM to observe the surface morphology and discern the nano/microstructure of MIPs. Generally, SEM is instrumental in revealing the shape of MIPs and detecting any structural variations arising from surface modifications or template removal processes.

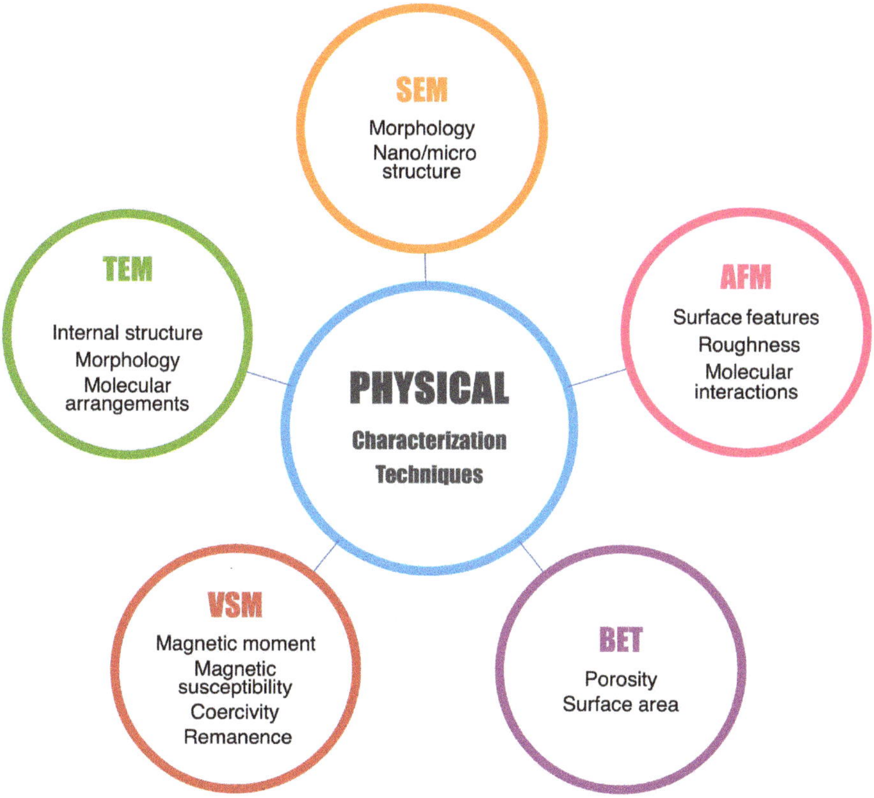

Fig. 3.2 Physical characterization techniques used in MIP studies

For instance, Zhou et al. (2014) advocated the use of MIPs for the extraction and analysis of low-abundance steroid compounds in aqueous samples. SEM was employed to verify the spherical shape of the MIP beads and to analyze their surface morphology post the removal of silica nanoparticles. The synthesized imprinted polymer beads exhibited a small diameter with a narrow size distribution (18.9 ± 2.3 μm).

In another study by Ye et al. (2006), MIPs were fabricated in supercritical carbon dioxide (sCO_2) using propranolol as a model template, methacrylic acid as a functional monomer, and divinylbenzene as a crosslinker. SEM was utilized to investigate the effect of the solvent on the size of the MIPs, demonstrating that producing MIPs in sCO_2 resulted in much smaller particle sizes (100 nm) compared to using acetonitrile (up to 2 mm in diameter).

Lu et al. (2020) explored the impact of preparation methods on the size and shape of obtained MIPs. SEM images in published papers were reviewed, revealing that the diameter and size of MIP particles are influenced by the preparation method and morphology. Mesoporous MIPs prepared by one-pot methods generally have a particle size below 100 nm. The size of core–shell structured MIPs slightly larger, distributed in the range of 100–1000 nm. Hollow MIPs prepared by the seed swelling method have a size of about 3–5 μm, while spherical MIPs obtained by precipitation polymerization fall in the range of 0.5–5 μm. Particle sizes from Pickering polymerization and suspension polymerization are larger, exceeding 10 μm.

Shahar et al. (2016) explored the impact of multiple parameters, encompassing synthesis volume and temperature, cross-linker concentration, the choice of solvent, type of template, and template concentration. Through the analysis of SEM images, they discovered that meticulous control over these parameters enables the synthesis of MIP particles with diameters spanning from tens of nanometers to a few microns.

As AFM can provide atomic-scale resolution, it is primarily used to determine the formation of cavities on the MIP surface and the surface roughness of the material. Werner et al. (2022) employed AFM to compare the effects of two different imprinting techniques: stamp imprinting and polymerization of Pickering emulsions. The investigated techniques utilized *Escherichia coli* bacteria as a model template and poly(styrene-co-divinylbenzene) as a model polymer for both approaches. AFM findings showed that the cavities created by stamp imprinting had smooth surfaces and channels that resembled the bacteria's negative pattern on the stamp and their filaments. On the other hand, when synthesizing imprinted polymers through the polymerization of Pickering emulsions, the result was the accumulation of globular nanostructures within the imprints.

Surya et al. (2020) employed AFM to explore the surface topology and roughness differences between chitosan-gold nanoparticles-based imprinted and non-imprinted polymers. The combination of gold nanoparticles and chitosan in the MIP led to the development of a sensor for detecting ciprofloxacin with high sensitivity and selectivity. After template removal, the imprinted polymer exhibited a more porous surface morphology, an increased surface area, and more cavities compared to the non-imprinted polymer. These features suggested a higher affinity of the imprinted polymer towards the analyte molecule.

Nguyen et al. (2022) investigated the topographic profile of MIP synthesized using 3,4-ethylene dioxythiophene as a functional monomer and confirmed the presence of a 180 nm deep pore on the surface of the MIP designed for diclofenac detection. In another study, Akgönüllü et al. (2020) demonstrated both the homogeneous coating of MIP nanofilm on the surface of surface plasmon resonance chips and the surface roughness values for unmodified, allyl-modified, and aflatoxin B-MIP films as 0.8 nm, 4.8 nm, and 9.4 nm, respectively, using AFM.

BET analysis, on the other hand, is a physical characterization technique used to determine the specific surface area and porosity of materials, especially porous materials like MIPs. Meléndez-Marmolejo et al. (2022) investigated the specific surface area of various synthesized MIPs and NIPs using BET analysis. The obtained specific surface area values (111–410 $m^2\ g^{-1}$) indicated that both MIPs and NIPs are highly porous materials. Moreover, the specific surface area of MIPs was found to be larger than that of NIPs, suggesting the effectiveness of the imprinting process.

Sun et al. (2022) proposed a novel amino-functionalized titanium dioxide/surface molecularly imprinted polymer for methylene blue detection. BET analysis was utilized to determine the surface area, mean pore diameter, and pore volume of the imprinted and non-imprinted polymers. After template removal, the values for these parameters in MIPs were higher than those in NIPs, indicating the presence of mesopores on the adsorbent.

Moradi Falah Langeroodi et al. (2022) also used BET analysis to determine the specific surface area for MIP and NIP. The specific surface area of MIP was higher than that of NIP, and MIP's pore volume and pore size were larger, suggesting the removal of the template from the polymer structure.

Vibrating Sample Magnetometer (VSM) analysis was employed by Khan et al. (2022) to investigate the magnetic properties of magnetic molecularly imprinted polymers (MMIPs) synthesized using the core–shell method. The VSM results indicated a significant decrease in the saturation magnetization value of MMIP compared to magnetite, suggesting the introduction of a polymeric layer.

In a study by Ilktac (2022), VSM analysis was conducted to assess the magnetic properties of magnetite and MMIP particles. Magnetite exhibited a higher saturation magnetization value than MMIP, indicating the development of a diamagnetic polymer layer covering the magnetite particle surface.

Guoning et al. (2020) also used VSM to determine the magnetic properties of magnetite and MMIPs. The saturation magnetization values of magnetite and MMIPs were found to be 78.2 and 45.6 emu g^{-1}, respectively. Comparatively, the magnetic saturation value of the MMIPs was lower than that of magnetite, confirming the presence of a thin polymer layer on the surface of magnetite. Nonetheless, the MMIPs retained significant magnetism and could be rapidly separated from the solution when exposed to an external magnetic field.

3.2.2 Chemical Characterization

Chemical characterization focuses on identifying and analyzing the chemical composition, functional groups, and molecular interactions within the MIP structure. It provides information about the types of atoms present, their arrangement, bonding, and the chemical reactions that the material can undergo (Fig. 3.3).

FTIR is the most common technique used to identify the presence of specific functional groups, chemical bonds, and molecular structures. It is also employed to confirm the successful removal of the template molecule from the MIP after the polymerization process.

In FTIR analysis, the main aim is to detect the characteristic peaks that could be assigned to the main components, including the functional groups present in the functional monomer, template, and cross-linker, that constitute the MIP structure.

Hou et al. (2020) synthesized an MIP using erythromycin as the template molecule, acrylic acid as the functional monomer, and *N*,*N*′-methylenebisacrylamide as the crosslinking agent. To confirm the formation of the polymer structure, the research group used FTIR to detect characteristic peaks present in the polymer. Upon the formation of intermolecular hydrogen bonds, the peak at 3349 cm^{-1} was

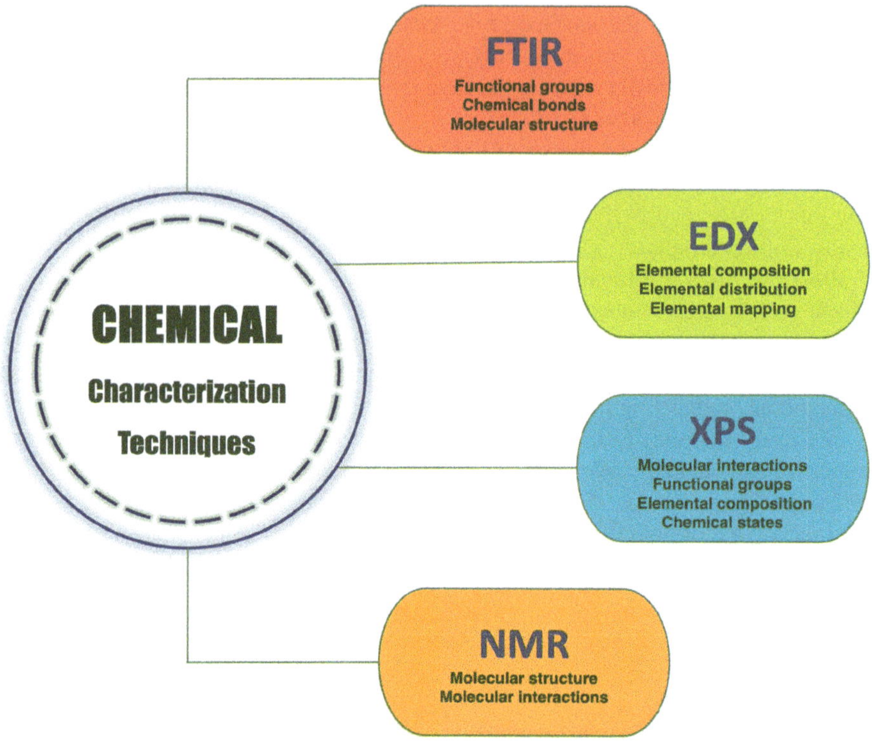

Fig. 3.3 Techniques used for the chemical characterization of MIP

detected due to the stretching of the O–H bond. The peak at 2950 cm^{-1} was attributed to the stretching vibration of the saturated alkane, and the peak at 1587 cm^{-1} was attributed to the stretching vibration of the alkenes. The bands appearing at 1345 and 1172 cm^{-1} can be considered as tensile vibrations of C–N and C–C bonds.

Silva et al. (2021) synthesized MIP using methacrylic acid (MAA) as the monomer and ethylene glycol dimethacrylate (EGDMA) as the cross-linker for the detection of thiamethoxam. The FTIR spectrum of the MIP exhibited characteristic bands, including C–H stretching (2892 cm^{-1}), C–H asymmetric deformation (1450 cm^{-1}), C=O stretching of MAA and EGDMA (1719 cm^{-1}), axial deformation of C–O–C in EGDMA (1151 cm^{-1}), and C–C stretching (946 cm^{-1}). Additionally, the typical absorption band related to the NO_2 group (1500–1600 cm^{-1}) of thiamethoxam was removed upon the removal of template.

Itaconic acid and EGDMA were used as the functional monomer and cross-linker, respectively, by Gupta et al. (2023) for the separation of chlorogenic acid from food waste. The FTIR spectrum shows the characteristic peaks of itaconic acid for C=O stretching and OH bending at 1721 cm^{-1} and 1393 cm^{-1}, respectively. The peak at 1150 cm^{-1} is related to C–O–C stretching corresponding to EGDMA, which provides evidence of the crosslinked polymer structure.

Retnaningtyas et al. (2020) also used FTIR to characterize the polymer structure and verify its components. MIP was prepared using acrylamide as the functional monomer and EGDMA as the cross-linking agent for the selective adsorption of D-arabinitol from spiked urine. Peaks at approximately 3400 and 1725 cm^{-1} were related to the presence of N–H and C=O stretching vibrations of acrylamide. The existence of peaks at 2953 cm^{-1} corresponded to C–H stretching arising from aliphatic compounds in the polymer structure whereas the presence of D-arabinitol was characterized by the peaks at 2956 cm^{-1} and 600 cm^{-1} related to C–H stretching and O–H bending vibrations, respectively.

EDX is another technique widely used for the chemical characterization of MIPs. It is often coupled with SEM analyses, providing valuable information about the composition and distribution of elements on the surface of MIPs. EDX can detect elements in the range from boron to uranium. EDX is generally used to confirm the presence of the components in the polymer structure.

Shafqat et al. (2020) synthesized MIP using MAA as the functional monomer, EGDMA as the cross-linker, and congo red as the template. The adsorbent was analyzed by EDX without the removal of the template from the structure to evaluate the presence of each component. The EDX spectrum reveals the presence of C, O and N. C and O are derived from all the components, whereas N is present due to Congo red.

Gaho et al. (2022) fabricated MIPs for the removal of norfloxacin from aqueous samples using MAA and EGDMA. EDX indicates the presence of carbon, oxygen, and fluorine on the surface of the MIP. Carbon and oxygen are derived from all the components, with atomic percentages of 67.9% and 29.5%, respectively, whereas fluorine is present at 2.6% due to the presence of norfloxacin.

Phenylalanine-imprinted magnetic core–shell particles were synthesized on the surface of the $Fe_3O_4@SiO_2–NH_2$ particles using MAA as the functional monomer and EGDMA as the cross-linker by Gul et al. (2019). EDX analysis reveals the presence of peaks for C, O, Si, and Fe, confirming the successful synthesis of the core–shell particles.

Jalilian et al. (2021) prepared a magnetic molecularly imprinted polymer (MMIP) based on silica-coated magnetite functionalized with 3-aminopropyltriethoxysilane (APTES) using EGDMA as a cross-linker for the extraction of phenolphthalein in urine samples and herbal slimming capsules. The EDX spectra of unleached MMIP, leached MMIP, and MNIP were investigated. All spectra exhibited the presence of C, N, O, Si, and Fe, with changes in the percentage of elements upon the removal of the template molecule. The highest percentage of carbon (74.5%) was found in the EDX spectrum of unleached MMIP, due to the presence of phenolphthalein molecules in the structure. It was observed that the percentage of carbon decreased to 58.3% in leached MMIP, confirming the removal of the template from the MMIP structure. Also, the percentage of carbon in the EDX spectrum of MNIP (27.2%) was less than that in unleached MMIP, due to the absence of phenolphthalein in the polymer structure.

Especially over the last decade, XPS has become one of the most widely used techniques in MIP characterization for determining, identifying, and quantifying the functional groups, elements, and their chemical states on the MIP's surface.

Ramakers et al. (2019) prepared MIPs for the detection of histamine using organocatalyzed metal-free photo atom transfer radical polymerization. To confirm the formation of the MIP, the existence of characteristic peaks in the polymer structure was investigated using XPS. In the synthesis, MAA was used as the monomer, and EGDMA was used as the cross-linking agent. The existence of two characteristic peaks in the C 1s spectrum at 286.8 and 289.1 eV corresponded to C–O and O=C–O, respectively.

Abdollahi et al. (2018) fabricated MIPs for use as a drug delivery carrier using phenytoin as the template, acrylamide as the monomer, and EGDMA as the cross-linker. The effects of two polymerization reactions, reversible addition fragmentation chain transfer (RAFT) and traditional radical polymerization (TRP), were investigated. The surface chemistry of the synthesized polymers was analyzed using XPS analysis. The C 1s peak at 284.9 eV is attributed to C–C, the peak at 286.6 eV is related to C–O, C–N, and C–S groups, and the peak at 289.0 eV is due to C=O groups. Additionally, the O 1s spectrum presented two peaks at 534 eV, attributed to C=O groups, and the peak at 532.5 eV, due to C–O groups. Considering the peak areas and the position of binding energies, it was observed that small changes were present between the XPS spectra of MIP-RAFT and MIP-TRP. It can be concluded that RAFT polymerization leads to an increase in the surface C=O groups content from 42.9% to 52.7%.

Another MIP was proposed by Söylemez and Güven (2018) using MAA and EGDMA as the functional monomer and crosslinker, respectively, for the analysis of penicillin G. Penicillin G imprinted polymers were synthesized by grafting onto PE/PP non-woven fabrics with the aid of gamma irradiation. Two main peaks of C 1s are related to C–C and C–H (284.3 eV), C–O (285.1 eV) due to the poly(methacrylic acid-co-ethylene glycol dimethacrylate) structure for both NIP and MIP grafted fabrics. In the XPS C 1s spectrum of the MIP structure, a new peak at 288.9 eV was observed due to the C–N group of penicillin G, where the intensity of the peak decreased after the removal of the analyte. Additionally, an N 1s peak was detected at 400 eV, which originated from the nitrogen of penicillin G.

NMR is being increasingly used in recent years to investigate the molecular structure, interactions, and dynamics of atoms within MIPs. Like other chemical characterization techniques, NMR is also used to verify the successful imprinting of target molecules in the polymer, assess the quality of the polymer structure, and investigate molecular interactions within the MIP. ^{1}H NMR characterization is widely used in non-covalent imprinting systems to study the interaction between the monomer and template. The difference in chemical shifts between the mixtures and the pure compounds is considered to estimate interactions.

Using n-butyl acrylate as the monomer, 1,6-hexanediol diacrylate as the cross-linker, and 2-phenoxypropionic acid as the template, Beladghame et al. (2023) synthesized MIP films for the detection of the 2,4-D herbicide analogue, 2-phenoxypropionic acid. They used ^{1}H NMR spectroscopy to analyze the interactions between the template and the monomer. The signal at ^{1}H NMR spectrum at 9.31 ppm is assigned to the deshielded proton of the carboxylic acid group. The addition of the monomer caused a slight shielding effect, shifting the signal to 9.06 ppm. This shift is due to the formation of hydrogen bonds between the monomer's ester group and the carboxylic acid group of the template. Thus, the successful complex formation through hydrogen bonding interaction is confirmed by the researchers.

Xie et al. (2019) used ^{1}H NMR to investigate the interaction between deltamethrin and acrylamide. The peak related to the amino hydrogen atom of acrylamide shifted downfield from 3.10 to 2.89 ppm in the case of the complex. They explained this phenomenon by suggesting that there is a hydrogen bond interaction between the carbonyl group of deltamethrin and the amino group of acrylamide.

Huang et al. (2020) also used ^{1}H NMR to determine the type of interaction in the pre-polymerization mixture of MIP structure. They synthesized MIP for the separation of bilobalide using methacrylamide as the monomer and trimethylolpropane triacrylate as the cross-linker. The ^{1}H NMR spectra of bilobalide and the bilobalide-methacrylamide complex were investigated, and two different types of hydroxyl proton peaks (4.88 and 6.30 ppm) appeared in the low field of bilobalide. In the ^{1}H NMR spectrum of the complex, a slight chemical shift occurred, which may be due to hydrogen bonding, causing the chemical shift to move towards the lower field.

3.2.3 *Thermal Characterization*

Thermal characterization is a valuable tool in MIP analysis, providing insights into the polymer's thermal stability, composition, purity, and suitability (Fig. 3.4).

Armutcu et al. (2020) fabricated MIPs using bovine hemoglobin adsorbed on zinc oxide particles, which were then used to synthesize a polyacrylic acid-based molecular imprinting membrane in the presence of EGDMA as a cross-linker. The research group investigated the thermal properties of the synthesized MIPs via TGA. They observed weight loss in the range of 30–150 °C due to the evaporation of physically adsorbed water. The characteristic main weight loss temperature at around 253 °C arose from the thermal decomposition of acrylic acid chains, causing a reduction in the thermal stability of the membranes.

TGA was also examined by Ata et al. (2022) to investigate the thermal stability of the prepared MIP using MAA and EGDMA for the detection of p-nitrophenol. The MIP and NIP exhibited similar thermal degradation behavior in three steps: in the first step, a 2% weight loss was observed up to 200 °C due to the loss of entrapped water and solvent molecules; in the second step, the polymer degraded from 270 to 420 °C with a 95% weight loss, attributed to the breakdown of the main carbon skeleton of the polymer; and in the last step, the polymer was completely broken down at 510 °C with less than 1% residual mass.

A fluorescence probe based on sulfur-doped carbon dots coated with MIPs using acrylamide and 1-vinyl-3-butylimidazolium tetrafluoroborate as functional monomers and EGDMA as a cross-linker was proposed by Zhang et al. (2019)

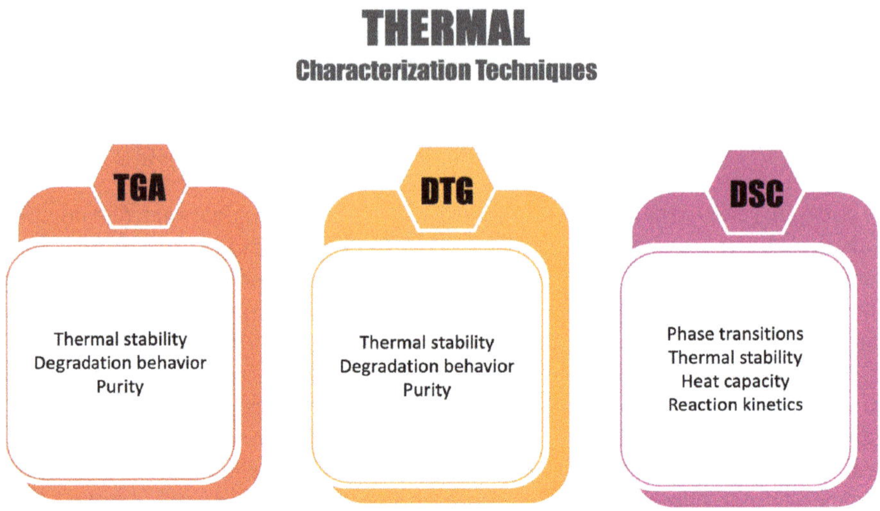

Fig. 3.4 Most widely used thermal techniques in MIP studies

for the detection of lambda-cyhalothrin. The TGA curve indicated that the MIPs possessed high thermal stability and were stable up to ~ 250 °C but decomposed at 349.4 °C. The maximum decomposition rate, determined from the DTG spectrogram, was 6.7% per minute. The DTA curve indicated that the composite completely decomposed at 392.6 °C.

Wei et al. (2020) developed ionic liquid-sensitized molecularly imprinted polymers based on heteroatom co-doped quantum dots functionalized graphene for the sensitive detection of λ-cyhalothrin. They assessed thermal stability using DTG analysis. In the DTG curve, a significant weight loss peak appeared at 76 °C, primarily due to the release of free water and gas from the synthetic materials. In the second stage (200–600 °C), the primary reaction was thermal cracking. During this phase, the macromolecule structure disintegrated, molecular chains broke, and functional groups were cleaved. DTG curves displayed the highest weight loss rate at 387 °C for the adsorbent, mainly caused by the concentrated release of most of the cracking gas. In the third stage (> 550 °C), the condensation stage, both the TGA and DTG curves leveled off, with a weight loss rate of approximately 5%. Pyrolysis was nearly complete, and fewer small molecules were released through molecular condensation. Upon reaching 1000 °C, adsorbent exhibited a total weight loss of 35%. These findings indicate excellent thermal stability in MIPs.

DSC is another thermal analysis technique that measures the heat flow associated with chemical and physical changes occurring within a material as a function of temperature.

Limaee et al. (2020) prepared MIPs using MAA, EGDMA, and 2,4-dichlorophenoxyacetic acid. 1,8-naphthalimide, containing thiourea, and diethylamine-tagged 1,8-naphthalimide were used as receptors to improve the selectivity of the adsorbent. DSC analysis was performed to study the thermal stability of the MIP in the presence and absence of the naphthalimide monomer. In the absence of the naphthalimide monomer, the DSC thermogram of the MIP showed a decrease at about 50 °C, which is related to the endothermic behavior of the sample. Additionally, the glass transition temperature at 338.8 °C is attributed to the destruction and decarboxylation of the polymeric chain. In the presence of the naphthalimide monomer, the increase in the glass transition temperature (390.4 °C) is ascribed to the presence of the naphthalimide monomer in the polymer network, which leads to an increase in the thermal stability of the polymer structure, causing a retardation in decarboxylation and destruction.

Mostafa et al. (2022) also used DSC analysis and observed two endothermic peaks for both MIPs and NIPs synthesized using MAA and EGDMA. The first peak, at 90–100 °C, corresponds to the evaporation of moisture, while the second peak corresponds to the melting of the polymeric layer at around 257 °C.

In the work proposed by Wadie et al. (2022), DSC thermograms were recorded for MIPs and NIPs, showing that they were thermally stable up to 250 °C, after which exothermic transitions occurred. DSC also demonstrated that complete template removal had occurred, as evidenced by the absence of the endothermic peaks observed in the pure drug thermograms of alfuzosin and solifenacin.

3.2.4 Functional Characterization

Functional characterization of MIPs is a critical aspect of their analysis and application, primarily focusing on the selective recognition and binding of the target molecule.

Binding studies investigate the binding characteristics of the target molecule onto synthesized MIPs and NIPs. To determine the binding affinity, adsorption studies are carried out. Adsorption studies are generally referred to as re-binding studies because the template molecule is re-bound to the specific cavities created during the template removal process from the polymer structure.

Re-binding experiments are conducted by using a specific weight of NIPs/MIPs with a defined volume of a target solution at a specific concentration for a certain period of time. After the re-binding (adsorption) process, the remaining concentration of the analyte is detected using an analytical method.

Equilibrium experiments are carried out as batch re-binding studies. The following equations are used to calculate the binding affinity for the MIP/NIPs;

$$Q = \frac{C_0 - C_t}{m} \times V \tag{3.1}$$

$$Sorption\ Efficiency\ (\%) = \frac{C_0 - C_e}{C_0} \times 100 \tag{3.2}$$

$$\textit{Imprinting Factor (IF)} = \frac{Q_{MIP}}{Q_{NIP}} \tag{3.3}$$

$$K = \frac{K_{d(template)}}{K_{d(analogue)}} \tag{3.4}$$

$$K' = \frac{K_{MIP}}{K_{NIP}} \tag{3.5}$$

where Q represents the amount of adsorption, C_0 is the initial concentration, C_e is the remaining concentration after reaching equilibrium, m is the weight of the polymer used in the re-binding experiment, V is the volume of the solution, K_d is the distribution coefficient, K is the selectivity coefficient, and K′ is the relative selectivity coefficient.

A higher Q value indicates stronger binding affinity. Importantly, the Q value for MIPs should consistently surpass that of NIPs. This difference is attributed to the selectivity of MIPs for the target, which outperforms the non-specific binding observed in NIPs. In a similar manner, the sorption efficiency of MIPs is expected to be higher than that of NIPs. Additionally, the Imprinting Factor (IF) value is a key metric used to describe how effectively a polymer has been imprinted with the desired template molecule. A higher IF value indicates the presence and effectiveness of specific cavities or interactions between the template and the functional monomer. The selectivity coefficient, K, quantifies the relative affinity of an MIP/NIP for one analyte compared to another molecule with a similar chemical structure. On the other hand, the relative selectivity coefficient, K′, is a specialized form of the selectivity coefficient that compares the selectivity of the MIP and NIP (Mostafa et al. 2021; Abubakar et al. 2023; Azizi and Bottaro 2020; Zhi et al. 2018).

Xue et al. (2021) examined the Imprinting Factor (IF) of a polymer designed to detect 2,6-dichloro-1,4-benzoquinone. Their findings indicated that the MIP displayed a significantly higher extraction capacity (4.7 times) compared to the NIP, establishing an IF of 4.7. This demonstrates that the MIP, with its selective molecular recognition capability, outperforms the NIP in detecting 2,6-dichloro-1,4-benzoquinone, as the NIP tends to nonspecifically adsorb molecules.

Selectivity coefficients of an MIP synthesized for purifying of tylosin was investigated by Zeng et al. (2020) concerning two main impurities, desmycosin and relomycin. The selectivity coefficients of tylosin to desmycosin and relomycin were 3.3 and 1.5, respectively. Additionally, the highest IF in the study was determined as 3.6. In another study, Ma et al. (2021) revealed the imprinting factor of the MIP synthesized for the selective extraction of poly-γ-glutamic acid as 4.76.

Zouaoui et al. (2020) designed an electrochemical sensor for detecting glyphosate by utilizing MIPs composed of chitosan. Their study revealed an impressive imprinting factor of 14.5. The MIPs also demonstrated selectivity coefficients towards glyfosinate, chlorpyrifos, and phosmet, with values of 7.9, 43.5, and 14.5, respectively.

Yuan et al. (2021) developed an MIP for analyzing congo red. In comparison to direct competitive analogs, the selectivity coefficients of the MIP for direct yellow GR and methyl orange were determined to be 2.2 and 4.9, respectively.

Ceylan Cömert et al. (2021) calculated the relative selectivity coefficient of the MIP designed for the detection of melamine in milk samples as 2.4. This value indicates the effective molecular imprinting in terms of selectivity when comparing the binding of melamine to its competitor molecule, cyromazine.

Functional characterization also involves the study of binding isotherms, which describe the equilibrium binding of analyte molecules to MIPs at various concentrations. In MIP studies, a binding isotherm refers to a graphical or mathematical representation that illustrates the relationship between the concentration of a target analyte and the extent of binding or adsorption by the MIP/NIP. These isotherms are crucial tools for characterizing the binding behavior of MIPs and evaluating their affinity and selectivity for specific molecules.

Langmuir isotherm, Freundlich isotherm, and Scatchard plots are the most widely used binding isotherms in MIP studies. The type of plot may vary depending on the nature of the binding interactions and the mechanism involved. The information obtained from binding isotherms provides insights into the binding kinetics, capacity, and specificity of MIPs. Researchers use binding isotherms to optimize MIP design and assess their performance.

Langmuir and Freundlich isotherms are mathematical models used to describe the adsorption of molecules onto the surface of a solid material. These isotherms help to understand and quantify the relationship between the concentration of molecules in a gas or liquid phase and their adsorption onto a solid surface under specific conditions.

In basic terms, the Langmuir model describes the adsorbent surface as homogeneous, and it assumes monolayer adsorption. In other words, the Langmuir isotherm is based on the assumption that adsorption occurs at specific, localized sites on the surface. Once an adsorbate occupies a site, further adsorption cannot take place at that site, and adsorbate molecules do not interact with each other on the surface (Musah et al. 2022; Sun et al. 2019, 2020).

On the other hand, the Freundlich isotherm model is employed to describe adsorption on heterogeneous surfaces, implying that the adsorption sites on the adsorbent are not presumed to be identical. This model allows for multilayer adsorption, where adsorbate molecules can form multiple layers on the adsorbent's surface. Thus, the Freundlich isotherm is an empirical model that describes adsorption on surfaces where adsorption is not limited to specific sites and can occur with multiple layers of adsorbate on the adsorbent's surface. It is often used when the surface is heterogeneous or when multilayer adsorption is involved (Parisi et al. 2020; Mohamed Idris et al. 2020).

In addition to these two isotherm models, the Scatchard analysis method is particularly widely used in MIP studies to determine the binding capacities and the maximum number of binding sites. The sorption isotherm models and related equations can be found in Table 3.2.

In Freundlich isotherm model, q_e is the amount of analyte adsorbed by the adsorbent (mg g^{-1}), C_e is the equilibrium concentration of analyte (mg L^{-1}), K_F (mg g^{-1})

Table 3.2 Sorption isotherm models and related equations

Isotherm	Equation
Langmuir	$\frac{C_e}{q_e} = \frac{1}{K_L Q_m} + \frac{C_e}{Q_m}$
Freundlich	$\ln q_e = \ln K_F + \frac{1}{n} \ln C_e$
Scatchard	$\frac{Q}{C} = \frac{Q_{\max} - Q}{K_d}$

and n (dimensionless) are Freundlich constants related to the adsorption capacity and intensity of adsorption, respectively. Freundlich isotherm was obtained by plotting ln q_e versus ln C_e.

In Langmuir isotherm model, C_e is the equilibrium concentration of analyte (mg L^{-1}), q_e is the adsorption capacity at equilibrium (mg g^{-1}), Q_m is maximum adsorption capacity (mg g^{-1}) and K_L is the Langmuir adsorption constant (L mg^{-1}). Langmuir isotherm was obtained by plotting C_e/q_e versus C_e.

In Scatchard analysis model, Q is the amount of analyte adsorbed by adsorbent (μmol g^{-1}), C is the equilibrium concentration (μmol L^{-1}), Q_{max} is the maximum number of binding sites (μmol g^{-1}) and K_d is the equilibrium dissociation constant (μmol L^{-1}). Scatchard plot was obtained by plotting Q/C versus Q.

In Scatchard analysis, the plot may exhibit varying numbers of linear regions. If the plot features only a single linear region, it suggests that a homogeneous binding site is responsible for the sorption of the analyte. Conversely, if the plot displays two distinct linear regions, this signifies a heterogeneous distribution of binding sites on the polymer (Cho et al. 2021).

When generating graphs related to the isotherms using the respective equations, the isotherm model that exhibits the highest correlation coefficient (R^2 close to 1) is employed to describe adsorption (Çelekli et al. 2019).

Çorman et al. (2023) utilized various isotherm models to investigate the sorption process of benzo[a]pyrene onto MIPs. Four different adsorption models, namely Langmuir, Freundlich, Langmuir–Freundlich, and Scatchard adsorption models, were evaluated, resulting in R^2 values of 0.9962, 0.9585, 0.9519, and 0.9545, respectively. Subsequently, the Langmuir model was identified as the most suitable for describing the adsorption process.

Surapong and Burakham (2021) assessed the Langmuir and Freundlich isotherm models for the enrichment of glyphosate, glufosinate, and aminomethylphosphonic acid adsorption onto MIP. The results indicated that the Langmuir model was more suitable than the Freundlich model, and the recognition sites were uniformly distributed in a monolayer on the surface of the adsorbent. Researchers also employed Scatchard analysis to investigate the binding characteristics of MIP and NIP toward the analytes. They discovered that the Scatchard plot of MIP for the analytes consisted of two linear parts, revealing the presence of two types of binding sites on the MIP surface. In contrast, NIP showed only one straight line, corresponding to a single homogeneous binding site for adsorbing analytes non-specifically.

Cui et al. (2021) employed Langmuir and Freundlich isotherm adsorption models to investigate the adsorption mechanism of sulfonamides onto imprinted polymers. It was observed that the Langmuir isotherm model provided a better fit (with R^2 values of 0.998 for Langmuir and 0.920 for Freundlich isotherm); therefore, the adsorption process can be regarded as a monolayer adsorption process.

Fu et al. (2020) determined the number of binding sites for the MIP synthesized for the identification of zearalenone in grains using the Scatchard model. After plotting the graphs, it was revealed that MMIPs exhibited two linear segments, indicating the presence of two types of binding sites (heterogeneous binding type), whereas MNIPs displayed only one linear segment (homogeneous binding type) on the surface of MMIPs. By employing the equations, the capacities were calculated, and the results demonstrated that MMIPs possessed a higher adsorption capacity for the analyte.

To determine the imprinting mechanism of the MIP designed for resveratrol, Cao et al. (2020) employed Scatchard analysis. The findings revealed that the Scatchard curve of the MIP comprised two lines with distinct slopes, representing high and low-affinity sites. In contrast, the Scatchard curve of the NIP formed a straight line, indicating the presence of low-affinity sites.

3.3 Conclusion

The techniques commonly used in the characterization of MIPs are categorized into four main groups: physical, chemical, thermal, and functional characterization. This chapter focuses on the fundamental principles of these techniques and their applications in characterizing MIPs. Characterization plays a crucial role not only in the fundamental development of MIPs but also in their successful application. This chapter has comprehensively explored spectroscopic, microscopic, and computational techniques, aiming to illuminate the diverse tools available for investigating the molecular architecture and behavior of imprinted polymers. The data generated through these characterization techniques enable researchers to fine-tune polymerization conditions, optimize MIP design, and ensure repeatability in production processes. As MIPs continue to showcase their versatility, stability, and superior affinity for target molecules, their role in various domains, including analytical chemistry, biotechnology, medicine, and environmental monitoring, is poised to expand.

In the future, computational modeling and simulation will possibly play a vital role in characterizing MIPs. Molecular dynamics simulations and quantum mechanical calculations may provide insights into the thermodynamics and energetics of the imprinting process, aiding in the design of more effective MIPs. With advancements in technological facilities and high-resolution imaging, finer details of imprinting sites and the overall nanoscale morphology of MIPs will be identified. Furthermore, progress in analytical techniques for quantifying binding affinities and kinetics of MIPs will be crucial for optimizing their performance and for a more accurate and detailed analysis of molecular interactions. Additionally, the integration of artificial intelligence may lead to more efficient optimization and characterization strategies in

MIP studies. In this context, the clear determination of the structure of the produced material will always be of paramount importance.

Acknowledgements The author would like to thank Dr. Ece Bayır for her great support and valuable comments on this chapter.

References

Abd Mutalib M, Rahman MA, Othman MHD, Ismail AF, Jaafar J (2017) Scanning electron microscopy (SEM) and energy-dispersive X-ray (EDX) spectroscopy. In: Membrane characterization. Elsevier B.V., 161–179p. https://doi.org/10.1016/B978-0-444-63776-5.00009-7

Abdollahi E, Khalafi-Nezhad A, Mohammadi A, Abdouss M, Salami-Kalajahi M (2018) Synthesis of new molecularly imprinted polymer via reversible addition fragmentation transfer polymerization as a drug delivery system. Polymer 143:245–257

Abubakar L, Yusof NA, Abdullah AH, Mohammad F, Wahid MH, Ismail S et al (2023) Molecularly imprinted polymer-based nanoporous carbon nanocomposite for effective adsorption of Hg(II) ions from aqueous suspensions. Separations 10(8)

Akash MSH, Rehman K (2019) Essentials of pharmaceutical analysis. 1–222p

Akgönüllü S, Yavuz H, Denizli A (2020) SPR nanosensor based on molecularly imprinted polymer film with gold nanoparticles for sensitive detection of aflatoxin B1. Talanta 219(March):121219. https://doi.org/10.1016/j.talanta.2020.121219

Akhtar K, Khan SA, Khan SB, Asiri AM (2018) Scanning electron microscopy: principle and applications in nanomaterials characterization. In: Handbook of materials characterization. 113–145p

Almoselhy RIM (2020) Applications of differential scanning calorimetry (DSC) in oils and fats research. A review. Am Res J Agric 6(1):1–9

Armutcu C, Özgür E, Çorman ME, Uzun L (2021) Interface imprinted polymers with well-oriented recognition sites for selective purification of hemoglobin. Colloids Surfaces B Biointerfaces 2021:197(October 2020)

Ata S, Feroz M, Bibi I, Mohsin I ul, Alwadai N, Iqbal M (2022) Investigation of electrochemical reduction and monitoring of p-nitrophenol on imprinted polymer modified electrode. Synth Met 287(April):117083. https://doi.org/10.1016/j.synthmet.2022.117083

Ayiania M, Smith M, Hensley AJR, Scudiero L, McEwen JS, Garcia-Perez M (2020) Deconvoluting the XPS spectra for nitrogen-doped chars: an analysis from first principles. Carbon NY 162:528–544. https://doi.org/10.1016/j.carbon.2020.02.065

Azizi A, Bottaro CS (2020) A critical review of molecularly imprinted polymers for the analysis of organic pollutants in environmental water samples. J Chromatogr A 1614:460603. https://doi.org/10.1016/j.chroma.2019.460603

Bach QV, Chen WH (2017) Pyrolysis characteristics and kinetics of microalgae via thermogravimetric analysis (TGA): a state-of-the-art review. Bioresour Technol 246:88–100. https://doi.org/10.1016/j.biortech.2017.06.087

Becker ED (2000) High resolution NMR theory and chemical applications, 3rd edn. Academic Press

Beladghame O, Bouchikhi N, Lerari D, Charif IE, Soppera O, Maschke U et al (2023) Elaboration and characterization of molecularly imprinted polymer films based on acrylate for recognition of 2,4-D herbicide analogue. Iran Polym J (English Ed.) 32(4):483–497. https://doi.org/10.1007/s13726-023-01143-2

Boitard C, Curcio A, Rollet AL, Wilhelm C, Ménager C, Griffete N (2019) Biological fate of magnetic protein-specific molecularly imprinted polymers: toxicity and degradation. ACS Appl Mater Interfaces 11(39):35556–35565
Bonyadi S, Ghanbari K (2022) Application of molecularly imprinted polymer and ZnO nanoparticles as a novel electrochemical sensor for tartrazine determination. Microchem J 187(August 2022):108398. https://doi.org/10.1016/j.microc.2023.108398
Brigiano FS, Bazin D, Tielens F (2022) Peculiar opportunities given by XPS spectroscopy for the clinician. Comptes Rendus Chim 25(1):149–163
Bubert H, Rivière JC, Werner WSM (2011) X-ray photoelectron spectroscopy (XPS). In: Surface and thin film analysis: a compendium of principles, instrumentation, and applications, 2nd edn. pp 7–41
Byrn SR, Zografi G, Chen XS (2017) Differential scanning calorimetry and thermogravimetric analysis. In: Solid state properties of pharmaceutical materials. pp 124–141
Cao J, Shen C, Wang X, Zhu Y, Bao S, Wu X et al (2021) A porous cellulose-based molecular imprinted polymer for specific recognition and enrichment of resveratrol. Carbohydr Polym 251(June 2020)
Çelekli A, Al-Nuaimi AI, Bozkurt H (2019) Adsorption kinetic and isotherms of Reactive Red 120 on Moringa oleifera seed as an eco-friendly process. J Mol Struct 1195:168–178. https://doi.org/10.1016/j.molstruc.2019.05.106
Ceylan Cömert Ş, Özgür E, Uzun L, Odabaşı M (2022) The creation of selective imprinted cavities on quartz crystal microbalance electrode for the detection of melamine in milk sample. Food Chem 372(May 2021)
Chan CM, Weng LT (2016) Surface characterization of polymer blends by XPS and ToF-SIMS. Materials (Basel) 9(8)
Cheong WJ, Yang SH, Ali F (2013) Molecular imprinted polymers for separation science: a review of reviews. J Sep Sci 36(3):609–628
Cho MG, Hyeong S, Park KK, Chough SH (2021) Characterization of hydrogel type molecularly imprinted polymer for creatinine prepared by precipitation polymerization. Polymer 237(November)
Çorman ME, Armutcu C, Karasu T, Özgür E, Uzun L (2023) Highly selective benzo[a]pyrene detection even under competitive conditions with molecularly imprinted surface Plasmon resonance sensor. Polycycl Aromat Compd 43(5):3896–3909. https://doi.org/10.1080/10406638.2022.2080725
Cui Y, Jiang L, Li H, Meng D, Chen Y, Ding L et al (2021) Molecularly imprinted electrospun nanofibre membrane assisted stir bar sorptive extraction for trace analysis of sulfonamides from animal feeds. J Ind Eng Chem 96:382–389
De Blasio C (2019) Fundamentals of biofuels engineering and technology. In: Thermogravimetric analysis (TGA). Green Energy Technology, pp 91–102
Dong C, Shi H, Han Y, Yang Y, Wang R, Men J (2021) Molecularly imprinted polymers by the surface imprinting technique. Eur Polym J 145(December 2020)
Dutta A (2017) Fourier transform infrared spectroscopy. In: Spectroscopic methods for nanomaterials characterization, vol.2. Elsevier Inc, 73–93p. https://doi.org/10.1016/B978-0-323-46140-5.00004-2
El-Alaily TM, El-Nimr MK, Saafan SA, Kamel MM, Meaz TM, Assar ST (2015) Construction and calibration of a low cost and fully automated vibrating sample magnetometer. J Magn Magn Mater 386:25–30. https://doi.org/10.1016/j.jmmm.2015.03.051
Fan GY, Ellisman MH (2000) Digital imaging in transmission electron microscopy. J Microsc 200(1):1–13
Foroughirad S, Haddadi-Asl V, Khosravi A, Salami-Kalajahi M (2021) Magnetic halloysite-based molecularly imprinted polymer for specific recognition of sunset yellow in dyes mixture. Polym Adv Technol 32(2):803–814
Fu H, Xu W, Wang H, Liao S, Chen G (2020) Preparation of magnetic molecularly imprinted polymers for the identification of zearalenone in grains. Anal Bioanal Chem 412(19):4725–4737

Gabbott P (2008) A practical introduction to differential scanning calorimetry. In: Principles and applications of thermal analysis. pp 1–263
Gaho MM, Memon GZ, Memon J ur R, Arain JB, Arain AJ, Shah A et al (2022) Synthesis of novel magnetic molecularly imprinted polymers by solid-phase extraction method for removal of norfloxacin. Chinese J Anal Chem 50(6)
Ghobeira R, Esbah Tabaei PS, Morent R, De Geyter N (2022) Chemical characterization of plasma-activated polymeric surfaces via XPS analyses: a review. Surfaces Interfaces. 31(February):102087. https://doi.org/10.1016/j.surfin.2022.102087
Giessibl FJ (1985) Principle of NC-AFM 1–2
Guerrero-Pérez MO, Patience GS (2020) Experimental methods in chemical engineering: Fourier transform infrared spectroscopy—FTIR. Can J Chem Eng 98(1):25–33
Gul S, Shah N, Arain MB, Rahman N, Rehan T, Ul-Islam M et al (2019) Fabrication of magnetic core shell particles coated with phenylalanine imprinted polymer. Polym Test 75(February):262–269. https://doi.org/10.1016/j.polymertesting.2019.02.023
Guoning C, Hua S, Wang L, Qianqian H, Xia C, Hongge Z et al (2020) A surfactant-mediated sol–gel method for the preparation of molecularly imprinted polymers and its application in a biomimetic immunoassay for the detection of protein. J Pharm Biomed Anal 190:113511. https://doi.org/10.1016/j.jpba.2020.113511
Gupta Y, Beckett LE, Sadula S, Vargheese V, Korley LSTJ, Vlachos DG (2023) Bio-based molecular imprinted polymers for separation and purification of chlorogenic acid extracted from food waste. Sep Purif Technol 327(August):124857. https://doi.org/10.1016/j.seppur.2023.124857
Haupt K, Linares AV, Bompart M, Bui BTS (2012) Molecular imprinting. Topics in current chemistry. In: Molecularly imprinted polymers, vol 325
Hou L, Han X, Wang N (2020) High performance of molecularly imprinted polymer for the selective adsorption of erythromycin in water. Colloid Polym Sci 298(8):1023–1033
Huang X, Zhang W, Wu Z, Li H, Yang C, Ma W et al (2020) Computer simulation aided preparation of molecularly imprinted polymers for separation of bilobalide. J Mol Model 26(8)
Ilktaç R (2022) A fluorimetric method based on magnetic molecularly imprinted polymer for selective detection of chlorferon. J Fluoresc 32(1):369–379
Irwansyah FS, Amal AI, Diyanthi EW, Hadisantoso EP (2024) How to read and determine the specific surface area of inorganic materials using the Brunauer-Emmett-Teller (BET) method. ASEAN J Sci Eng 4(1):61–70
Ismail AA, van de Voort FR, Sedman J (1997) Chapter 4 Fourier transform infrared spectroscopy: principles and applications. Tech Instrum Anal Chem 18(C):93–139
Jackman LM, Sternhell S (1969) Application of nuclear magnetic resonance spectroscopy in organic chemistry. In: International series in organic chemistry, 2nd edn. Pergamon
Jalilian N, Ebrahimzadeh H, Asgharinezhad AA, Khodayari P (2021) Magnetic molecularly imprinted polymer for the selective dispersive micro solid phase extraction of phenolphthalein in urine samples and herbal slimming capsules prior to HPLC-PDA analysis. Microchem J 160(PB):105712. https://doi.org/10.1016/j.microc.2020.105712
Jin JM (1962) Electromagnetic analysis and design in magnetic resonance imaging, 1st edn. Routledge
Jin Y, Wunderlich B (1993) Single-run heat capacity measurement by DSC: principle, experimental and data analysis. Thermochim Acta 226(C):155–161
Jordán D, González-Chávez D, Laura D, León Hilario LM, Monteblanco E, Gutarra A et al (2018) Detection of magnetic moment in thin films with a home-made vibrating sample magnetometer. J Magn Magn Mater 456:56–61
Kannan M (2018) Transmission electron microscope-principle, components and applications. In: A textbook on fundamentals and applications of nanotechnology. pp 93–102
Khan S, Wong A, Rychlik M, Sotomayor M del PT (2022) A novel synthesis of a magnetic porous imprinted polymer by polyol method coupled with electrochemical biomimetic sensor for the detection of folate in food samples. Chemosensors 10(11)

Kirupakar BR, Vishwanath BA, Padma SM, Deenadayalan (2016) Vibrating sample magnetometer and its application in characterisation of magnetic property of the anti cancer drug magnetic microspheres. Int J Pharm Drug Anal 4(5):227–233

Kryscio DR, Peppas NA (2012) Critical review and perspective of macromolecularly imprinted polymers. Acta Biomater 8(2):461–473

Leibl N, Haupt K, Gonzato C, Duma L (2021) Molecularly imprinted polymers for chemical sensing: a tutorial review. Chemosensors 9(6):1–19

Leyva-Porras C, Cruz-Alcantar P, Espinosa-Sol V, Saavedra-Leos MZ (2019) Application of differential scanning calorimetry (DSC) and modulated differential scanning. Polymers (Basel) 12(5):1–21

Li G, Row KH (2018) Recent applications of molecularly imprinted polymers (MIPs) on micro-extraction techniques. Sep Purif Rev 47(1):1–18. https://doi.org/10.1080/15422119.2017.1315823

Li N, He X, Chen L, Zhang Y (2009) Preparation of core-shell magnetic molecularly imprinted polymer nanoparticles for recognition of bovine hemoglobin. Chem—An Asian J 4(2):286–293

Limaee NY, Rouhani S, Olya ME, Najafi F (2020) Selective recognition of herbicides in water using a fluorescent molecularly imprinted polymer sensor. J Fluoresc 30(2):375–387

Liu Q, Wan J, Cao X (2018) Synthesis of core-shell molecularly imprinted polymers (MIP) for spiramycin I and their application in MIP chromatography. Process Biochem 70(April):168–178

Loganathan S, Valapa RB, Mishra RK, Pugazhenthi G, Thomas S (2017) Thermogravimetric analysis for characterization of nanomaterials. In: Thermal and rheological measurement techniques for nanomaterials characterization, vol 3. Elsevier Inc, 67–108p. https://doi.org/10.1016/B978-0-323-46139-9.00004-9

Lu H, Tian H, Wang C, Xu S (2020) Designing and controlling the morphology of spherical molecularly imprinted polymers. Mater Adv 1(7):2182–2201

Luther TA, Stewart FF, Budzien JL, LaViolette RA, Bauer WF, Harrup MK et al (2003) On the mechanism of ion transport through polyphosphazene solid polymer electrolytes: NMR, IR, and Raman spectroscopic studies and computational analysis of 15N-labeled polyphosphazenes. J Phys Chem B 107(14):3168–3176

Ma X, Li M, Zhang J, Wang R, Jin S (2021) Recognition and selective extraction of poly-γ-glutamic acid based on molecular imprinting technology. Int J Biol Macromol 172:1–9

Mahmoudpour M, Torbati M, Mousavi MM, de la Guardia M, Ezzati Nazhad Dolatabadi J (2020) Nanomaterial-based molecularly imprinted polymers for pesticides detection: recent trends and future prospects. TrAC—Trends Anal Chem 129

Mansfield E, Tyner KM, Poling CM, Blacklock JL (2014) Determination of nanoparticle surface coatings and nanoparticle purity using microscale thermogravimetric analysis. Anal Chem 86(3):1478–1484

Martín-Esteban A (2016) Recent molecularly imprinted polymer-based sample preparation techniques in environmental analysis. Trends Environ Anal Chem 9:8–14

Meléndez-Marmolejo J, Díaz de León-Martínez L, Galván-Romero V, Villarreal-Lucio S, Ocampo-Pérez R, Medellín-Castillo NA et al (2022) Design and application of molecularly imprinted polymers for adsorption and environmental assessment of anti-inflammatory drugs in wastewater samples. Environ Sci Pollut Res. 29(30):45885–45902. https://doi.org/10.1007/s11356-022-19130-0

Menczel JD, Prime RB (2008) Thermal analysis of polymers: fundamentals and applications. 1–688p

Mohamed MA, Jaafar J, Ismail AF, Othman MHD, Rahman MA (2017) Fourier transform infrared (FTIR) spectroscopy. In: Membrane characterization. Elsevier B.V., 3–29p.https://doi.org/10.1016/B978-0-444-63776-5.00001-2

Mohamed Idris Z, Hameed BH, Ye L, Hajizadeh S, Mattiasson B, Mohd Din AT (2020) Amino-functionalised silica-grafted molecularly imprinted polymers for chloramphenicol adsorption. J Environ Chem Eng 8(5)

Mohammed A, Abdullah A (2019) Scanning electron microscopy (SEM): a review. In: Proceedings of the 2018 international conference on hydraulics pneumatics—HERVEX. pp 77–85

Moradi Falah Langeroodi S, Kazemipour M, Eslaminejad T, Naghipour A, Ansari M (2022) Molecular imprinted polymer with dorzolamide for contact lens applications assisted by computational and experimental design. React Funct Polym 178(March):105355. https://doi.org/10.1016/j.reactfunctpolym.2022.105355

Mostafa AM, Barton SJ, Wren SP, Barker J (2021) Review on molecularly imprinted polymers with a focus on their application to the analysis of protein biomarkers. TrAC—Trends Anal Chem 144:116431. https://doi.org/10.1016/j.trac.2021.116431

Mostafa AM, Barton SJ, Wren SP, Barker J (2022) Development of magnetic molecularly imprinted polymers for the extraction of salivary pepsin prior to analysis by a novel HPLC-SEC method. Polymer 261(August):125417. https://doi.org/10.1016/j.polymer.2022.125417

Mpupa A, Selahle SK, Mizaikoff B, Nomngongo PN (2021) Recent advances in solid-phase extraction (Spe) based on molecularly imprinted polymers (mips) for analysis of hormones. Chemosensors 9(7)

Musah M, Azeh Y, Mathew J, Umar M, Abdulhamid Z, Muhammad A (2022) Adsorption kinetics and isotherm models: a review. Caliphate J Sci Technol 4(1):20–26

Naderi M (2015) Surface area: Brunauer-Emmett-Teller (BET). In: Progress in filtration and separation. pp 585–608

Ng HM, Saidi NM, Omar FS, Ramesh K, Ramesh S, Bashir S (2018) Thermogravimetric analysis of polymers. Encycl Polym Sci Technol 13:1–29

Nguyen DHN, Le QH, Nguyen TL, Dinh VT, Nguyen HN, Pham HN et al (2022) Electrosynthesized nanostructured molecularly imprinted polymer for detecting diclofenac molecule. J Electroanal Chem 921(August):116709. https://doi.org/10.1016/j.jelechem.2022.116709

Niazi A, Islamia JM, Poddar P, Rastogi AK (2000) A precision, low-cost vibrating sample magnetometer (December 2015)

Nurazzi NM, Asyraf MRM, Rayung M, Norrrahim MNF, Shazleen SS, Rani MSA et al (2021) Thermogravimetric analysis properties of cellulosic natural fiber polymer composites: a review on influence of chemical treatments. Polymers (Basel) 13(16)

Parisi OI, Ruffo M, Malivindi R, Vattimo AF, Pezzi V, Puoci F (2020) Molecularly imprinted polymers (MIPs) as theranostic systems for sunitinib controlled release and self-monitoring in cancer therapy. Pharmaceutics 12(1)

Peredkov S, Öhrwall G, Schulz J, Lundwall M, Rander T, Lindblad A et al (2007) Free nanoscale sodium clusters studied by core-level photoelectron spectroscopy. Phys Rev B—Condens Matter Mater Phys 75(23):1–8

Petit S, Madejova J (2013) Fourier transform infrared spectroscopy. In: Developments in clay science, 2nd edn, vol 5. Elsevier Ltd, 213–231p. https://doi.org/10.1016/B978-0-08-098259-5.00009-3

Phillips JP, Yazdani S, Highland W, Cheng R (2022) A high sensitivity custom-built vibrating sample magnetometer. Magnetochemistry 8(8):1–7

Ramakers G, Wackers G, Trouillet V, Welle A, Wagner P, Junkers T (2019) Laser-grafted molecularly imprinted polymers for the detection of histamine from organocatalyzed atom transfer radical polymerization. Macromolecules 52(6):2304–2313

Retnaningtyas Y, Supriyanto G, Nyoman Tri Puspaningsih N, Irawan R, Siswodihardjo S (2020) A novel molecular imprinting polymer for the selective adsorptionof D-arabinitol from spiked urine. Turkish J Chem 44(6)

Senthil Kumar P, Grace Pavithra K, Naushad M (2019) Characterization techniques for nanomaterials. In: Nanomaterials for solar cell applications. Elsevier Inc., 97–124p.https://doi.org/10.1016/B978-0-12-813337-8.00004-7

Seo Y, Jhe W (2008) Atomic force microscopy and spectroscopy. Rep Prog Phys 71(1)

Shafqat SR, Bhawani SA, Bakhtiar S, Ibrahim MNM (2020) Synthesis of molecularly imprinted polymer for removal of Congo red. BMC Chem 14(1):1–15. https://doi.org/10.1186/s13065-020-00680-8

Shahar T, Tal N, Mandler D (2016) Molecularly imprinted polymer particles: formation, characterization and application. Colloids Surfaces A Physicochem Eng Asp 495:11–19. https://doi.org/10.1016/j.colsurfa.2016.01.027
Shinato KW, Huang F, Jin Y (2020) Principle and application of atomic force microscopy (AFM) for nanoscale investigation of metal corrosion. Corros Rev 38(5):423–432
Silva CF, Menezes LF, Pereira AC, Nascimento CS (2021) Molecularly imprinted polymer (MIP) for thiamethoxam: a theoretical and experimental study. J Mol Struct 1231
Sing K (2001) The use of nitrogen adsorption for the characterisation of porous materials. Colloids Surfaces A Physicochem Eng Asp 187–188:3–9
Sinha P, Datar A, Jeong C, Deng X, Chung YG, Lin LC (2019) Surface area determination of porous materials using the Brunauer-Emmett-Teller (BET) method: limitations and improvements. J Phys Chem C 123(33):20195–20209
Söylemez MA, Güven O (2018) Preparation and detailed structural characterization of Penicillin G imprinted polymers by PALS and XPS. Radiat Phys Chem 159(September 2018):174–180. https://doi.org/10.1016/j.radphyschem.2019.02.050
Subramanian KS, Janavi GJ, Marimuthu S, Kannan M, Raja K, Haripriya S et al (2018) Scanning electron microscopy: principle, components and applications. Textb Fundam Appl Nanotechnol 91–90
Sun Y, Zhang Y, Ju Z, Niu L, Gong Z, Xu Z (2019) Molecularly imprinted polymers fabricated by Pickering emulsion polymerization for the selective adsorption and separation of quercetin from Spina Gleditsiae. New J Chem 43(37):14747–14755
Sun Y, Yao C, Xie Z, Zhang Y (2021) Lotus seedpod-like molecularly imprinted polymers fabricated by MOF-808 stabilized Pickering emulsion and their specific recognition of hemoglobin. Colloids Surfaces B Biointerfaces 197(July 2020):111446. https://doi.org/10.1016/j.colsurfb.2020.111446
Sun Y, Bai L, Han C, Lv X, Sun X, Wang T (2022) Hybrid amino-functionalized TiO_2/sodium lignosulfonate surface molecularly imprinted polymer for effective scavenging of methylene blue from wastewater. J Clean Prod 337(January):130457. https://doi.org/10.1016/j.jclepro.2022.130457
Surapong N, Burakham R (2021) Magnetic molecularly imprinted polymer for the selective enrichment of glyphosate, glufosinate, and aminomethylphosphonic acid prior to high-performance liquid chromatography. ACS Omega 6(41):27007–27016
Surya SG, Khatoon S, Lahcen AA, Nguyen ATH, Dzantiev BB, Tarannum N et al (2020) A chitosan gold nanoparticles molecularly imprinted polymer based ciprofloxacin sensor. RSC Adv 10(22):12823–12832
Tang CY, Yang Z (2017) Transmission electron microscopy (TEM). In: Membrane characterization. Elsevier B.V., 145–159p.https://doi.org/10.1016/B978-0-444-63776-5.00008-5
Tian Y, Wu J (2018) A comprehensive analysis of the BET area for nanoporous materials. AIChE J 64(1):286–293
Tougaard S (2010) Energy loss in XPS: fundamental processes and applications for quantification, non-destructive depth profiling and 3D imaging. J Electron Spectros Relat Phenomena 178–179(C):128–153. https://doi.org/10.1016/j.elspec.2009.08.005
Venezia AM (2003) X-ray photoelectron spectroscopy (XPS) for catalysts characterization. Catal Today 77(4):359–370
Wadie M, Marzouk HM, Rezk MR, Abdel-Moety EM, Tantawy MA (2022) A sensing platform of molecular imprinted polymer-based polyaniline/carbon paste electrodes for simultaneous potentiometric determination of alfuzosin and solifenacin in binary co-formulation and spiked plasma. Anal Chim Acta 1200:339599. https://doi.org/10.1016/j.aca.2022.339599
Wei J, Yuan X, Zhang Y, Liu H, Sun B (2020) Ionic liquid-sensitized molecularly imprinted polymers based on heteroatom co-doped quantum dots functionalized graphene for sensitive detection of λ-cyhalothrin. Anal Chim Acta 1136:9–18. https://doi.org/10.1016/j.aca.2020.08.041
Wellner N (2013) Fourier transform infrared (FTIR) and Raman microscopy: principles and applications to food microstructures. In: Food microstructures: microscopy, measurement and

modelling. Woodhead Publishing Limited, 163–191p. https://doi.org/10.1533/9780857098894.1.163

Werner M, Glück MS, Bräuer B, Bismarck A, Lieberzeit PA (2022) Investigations on sub-structures within cavities of surface imprinted polymers using AFM and PF-QNM†. Soft Matter 18(11):2245–2251

Xie L, Xiao N, Li L, Xie X, Li Y (2019) An investigation of the intermolecular interactions and recognition properties of molecular imprinted polymers for deltamethrin through computational strategies. Polymers (Basel) 11(11)

Xue W, Li N, Zhang Z, Li G (2022) Dummy template based molecularly imprinted solid-phase microextraction coating for analysis of trace disinfection by-product of 2,6-dichloro-1,4-benzoquinone using high-performance liquid chromatography. Talanta 239(November 2021)

Yao HYY, Wang JQ, Yin JY, Nie SP, Xie MY (2021) A review of NMR analysis in polysaccharide structure and conformation: progress, challenge and perspective. Food Res Int 143(December 2020)

Ye L, Yoshimatsu K, Kolodziej D, Da Cruz FJ, Dey ES (2006) Preparation of molecularly imprinted polymers in supercritical carbon dioxide. J Appl Polym Sci 102(3):2863–2867

Yuan D, Fu D, Wang C (2021) Selective removal of congo red from wastewater using molecularly imprinted polymer. Sep Sci Technol 56(2):233–241. https://doi.org/10.1080/01496395.2020.1715435

Yuksel N, Tektas S (2022) Molecularly imprinted polymers: preparation, characterisation, and application in drug delivery systems. J Microencapsul 39(2):176–196. https://doi.org/10.1080/02652048.2022.2055185

Zare EN, Fallah Z, Le VT, Doan VD, Mudhoo A, Joo SW et al (2022) Remediation of pharmaceuticals from contaminated water by molecularly imprinted polymers: a review. In: Environmental chemistry letters, vol 20. Springer International Publishing, 2629–2664p. https://doi.org/10.1007/s10311-022-01439-4

Zeng H, Yu X, Wan J, Cao X (2020) Rational design and synthesis of molecularly imprinted polymers (MIP) for purifying tylosin by seeded precipitation polymerization. Process Biochem 94(March):329–339. https://doi.org/10.1016/j.procbio.2020.03.025

Zhang D, Tang J, Liu H (2019) Rapid determination of lambda-cyhalothrin using a fluorescent probe based on ionic-liquid-sensitized carbon dots coated with molecularly imprinted polymers. Anal Bioanal Chem 411(20):5309–5316

Zheng Q, Zhang Y, Montazerian M, Gulbiten O, Mauro JC, Zanotto ED et al (2019) Understanding glass through differential scanning calorimetry. Chem Rev 119(13):7848–7939

Zhi K, Wang L, Zhang Y, Zhang X, Zhang L, Liu L et al (2018) Preparation and evaluation of molecularly imprinted polymer for selective recognition and adsorption of gossypol. J Mol Recognit 31(3):1–9

Zhou W, Apkarian R, Wang ZL, Joy D (2007) Fundamentals of scanning electron microscopy (SEM). In: Scanning microscopy for nanotechnology. Techniques and applications. pp 1–40

Zhou T, Shen X, Chaudhary S, Ye L (2014) Molecularly imprinted polymer beads prepared by pickering emulsion polymerization for steroid recognition. J Appl Polym Sci 131(1):1–7

Zhou T, Ding L, Che G, Jiang W, Sang L (2019) Recent advances and trends of molecularly imprinted polymers for specific recognition in aqueous matrix: preparation and application in sample pretreatment. TrAC—Trends Anal Chem 114:11–28

Zouaoui F, Bourouina-Bacha S, Bourouina M, Abroa-Nemeir I, Ben Halima H, Gallardo-Gonzalez J et al (2020) Electrochemical impedance spectroscopy determination of glyphosate using a molecularly imprinted chitosan. Sensors Actuators, B Chem 309(January):1–7

Chapter 4
MIPs in Analytical Instrumentation and Detection

Andréa R. Chaves, Rafael O. Martins, Almir C. Batista, Ricardo A. Bernardo, and Lanaia I. L. Maciel

4.1 Introduction

Throughout history, the world has witnessed a surge in industrialization and the integration of innovative formulations across various societal domains. A notable manifestation of this transformative trend is evident in the evolution of agricultural practices, wherein the application of novel chemical compounds has significantly enhanced crop yields (Li et al. 2023; Nguyen et al. 2023). Despite the progress across various societal domains, the introduction of new formulations and practices has prompted a heightened concern for their environmental effects. Literature extensively illustrates that the continual advancement of humanity has engendered substantial impacts on the environment, underscoring the imperative need for a rigorous assessment of these effects to safeguard environmental well-being (Cherian et al. 2023; Sharma et al. 2023). The ongoing evaluation of these impacts emerges as a critical task, for ensuring environmental safety in the face of evolving human activities.

While there is a pressing necessity for evaluating environmental matrices to trace the effects of human activities, this task is rendered challenging by the inherent complexity of these matrices (Cardoso et al. 2023) which can interfere with the analytical response of the most used analytical strategies, such as Liquid chromatography (LC) and Mass Spectrometry (MS). Thus, consequently impacting the accurate interpretation of environmental samples (Martins et al. 2023). To overcome such a challenge, the literature highlights that molecularly imprinted polymers (MIPs) have emerged as promising sorbent phases, showcasing a capacity for specific recognition with the desired molecule even in the interferent matrix presence, contributing to liquid chromatography and mass spectrometry methods

A. R. Chaves (✉) · R. O. Martins · A. C. Batista · L. I. L. Maciel
Instituto de Química, Universidade Federal de Goiás, Goiânia, GO 74690-900, Brazil
e-mail: andrea_chaves@ufg.br

R. A. Bernardo
Departamento de Química, Universidade Federal do Paraná, Curitiba, PR 81530-900, Brazil

S. Patra and M. Sillanpaa (eds.), *Molecularly Imprinted Polymers as Artificial Antibodies for the Environmental Health*, https://doi.org/10.1007/978-3-031-58995-9_4

sensitivity improvement (Santos et al. 2020). The versatility of these materials in environmental analysis extends to various target compounds, including pesticides (Ayivi et al. 2023), pharmaceuticals (Zare et al. 2022), and analytes originating from personal care products (PCPs) (Lu et al. 2020). Although this category of polymer has found widespread application in traditional analytical methods, especially applied as solid phase material for sample preparation techniques, there remains a dearth of utilization in integrated analytical approaches. Furthermore, there is an untapped potential for employing MIPs as substrates for modification in Ambient Mass Spectrometry (AMS) methods, thereby enhancing the selectivity of these techniques for the comprehensive assessment of environmental samples (Martins et al. 2023). Hence, this chapter will delve into the utilization of novel MIPs as substrates for emerging analytical technologies, encompassing integrated analytical methods and AMS approaches to enhance the effectiveness of evaluating environmental matrices. To offer a comprehensive overview of this chapter, Table 4.1 succinctly outlines the primary advantages and disadvantages of the techniques to be discussed in the following sections.

4.2 Exploring MIPs in Integrated Chromatography for Comprehensive Environmental Matrix Assessment

Effectively separating the compounds within a matrix is crucial for uncovering the sample's composition, thereby furnishing both qualitative and quantitative insights. Such information can be obtained by separation methods, such as LC approaches (Santos et al. 2020). Conventional sorbent phases for LC methodologies include mostly reverse phases, such as C6, C8, and C18, and so on. While these sorbent phases have been extensively used over the decades to separate various compounds within complex samples, their lack of selectivity remains challenging. Such a drawback can compromise the analytical performance of the method, particularly contingent upon the matrix under evaluation (Murtada and Pawliszyn 2023). Therefore, employing MIPs can facilitate a more selective interaction with the target analyte, due to the specific binding sites inherent in their polymeric structure. Besides, adopting integrated LC methods contributes to a more environmentally friendly analytical procedure, as it involves fewer steps and reduces the reliance on organic solvents, consequently minimizing chemical waste generation (Cardoso et al. 2023). Concerning the potential application of MIPs as sorbent phases in integrated LC methodologies, this section seeks to explore the untapped possibilities of incorporating MIPs in column switching and online in-tube SPME. The emphasis will be on their combination for the assessment of environmental analyses.

Table 4.1 Summary of the main advantages and drawbacks of LC and AMS strategies

Analytical technique	Advantages	Drawbacks
Column switching	Enhancing chromatographic selectivity; performing online enrichment of trace-level concentrations; achieving high-throughput performance and sample processing (clean-up and selective trace enrichment); reducing analysis time	Incurring additional expenses for the acquisition of the analytical system due to the necessity of incorporating switching valves, additional columns, and/or pumping systems
In-tube SPME	Automation and seamless integration with analytical instruments to expedite analysis; decreased analysis time; minimized solvent and sample consumption	Low extraction efficiency; clogging of the capillary; limited sample loading capability; coating stability depending on the sample and the main parameters of the method
PSI	Direct analysis requiring minimal or no sample preparation; the use of low solvent volume (microliter range); fast analysis, therefore, high sample throughput; versatility, sensitivity, and cost-effectiveness	Short data acquisition time, mainly linked to quick consumption of the ionization solvent; tedious analysis depending on the number of samples
PESI	A minimal sample consumption (in the picoliter range) is ensured by employing the probe, effectively eliminating the common issue of clogging seen in traditional electrospray analysis (ESI). Additionally, it exhibits greater salt tolerance compared to ESI protocols	The interruption in ionization resulted from the reciprocating movement of the needle along the mass spectrometer inlet, causing a discontinuous ionization process; the laborious optimization of the method; short data acquisition as in PSI
WT-ESI	Streamlined setup configuration, rapid analysis times, and cost-effectiveness characterize this system. Due to its physicochemical properties, the wooden tip proves easily modifiable	The interruption in ionization resulted from the reciprocating movement of the needle along the mass spectrometer inlet, causing a discontinuous ionization process; the laborious optimization of the method; short data acquisition as in PSI

In-tube SPME In-tube solid phase microextraction; *PSI* Paper spray ionization; *PESI* Probe electrospray ionization; *WT-ESI* Wooden-tip electrospray ionization

4.2.1 Column Switching Strategies Empowered by MIPs for Advanced Environmental Analysis

Despite each column-switching configuration possessing unique specifications, the classical approach typically employs two main columns referred to as the first dimension (1D) and the second dimension (2D) in conjunction with a switching valve using six to ten ports (Cruz et al. 2022). The use of column-switching in analytical instrumentation ensures sample preparation (the cleaning of the matrix and pre-concentration of the target analytes) and chromatographic separation of target analytes within complex matrices. By integrating the analytical steps involved in evaluating an unknown sample, the method gains in terms of efficiency, reducing both the time required and the potential for analytical errors introduced by the analyst (Cruz et al. 2022, 2021). In practical application, the method entails percolating the sample through the first dimension (1D), accomplishing purification and preconcentration. Subsequently, the target analytes are transferred for separation through the chromatographic column (2D), with later evaluation by the detection system (Fernández-Ramos et al. 2014).

As outlined earlier, in the context of column-switching, the 1D primarily handles sample preparation. It significantly contributes to improving the method's analytical performance, influencing quantitative information of the method, such as the limit of detection (LOD), the limit of quantification (LOQ), and recovery. Following the activation of a six or eight-way valve for the 2D column, the separation of the analytes is performed based on their interaction with the stationary phase in the 2D. While the analytes are being separated in the 2D, the valve is switched to conduct the washing of the first dimension (1D) and is finally prepared for the subsequent analysis (Kataoka and Saito 2012). In column-switching, elution is carried out through two primary approaches: the straight-flush mode and the back-flush mode, both approaches are illustrated in Fig. 4.1. In the straight-flush mode, the mobile phase moves in alignment with the sorption step, whereas in the back-flush mode, the mobile phase moves in the opposite direction to the sorption step. Each mode, whether straight-flush or back-flush, possesses distinct advantages that take into account the analysis requirements. When the main objective is to ensure separation and resolution the straight-flush mode is most recommended, whereas the back-flush mode is advantageous for diminishing band broadening effect and improving the detection of the method (Cruz et al. 2022; Kataoka and Saito 2012).

As stated by Cruz et al. (2022), it is widely acknowledged that column-switching systems typically demand more complex instrumentation compared to standard on-column focusing systems. However, column switching offers various advantages, including accelerated analyte processing and reduced back pressure. Furthermore, when examining the instrumentation of column-switching methodologies, significant attention is directed toward the compelling prospect of applying diverse sorbent phases in the 1D to perform the sample treatment of the matrix. Therefore, the utilization of MIPs as the1D in a column-switching system has captured the interest of the scientific community. This has led to proposals for selective MIP phases, with a particular emphasis on assessing different target analytes in environmental matrices.

In the optimization of column-switching procedures, paramount focus is placed on understanding the intricate interplay between specific analytical parameters and the response of the 1D system. This involves the analytical response of the MIP used for preconcentration and selectivity extraction on the column-switching system. Although efforts have been directed towards refining the separation of analytes in the 2D system, our forthcoming discussion will delve into the comprehensive optimization of a column-switching system utilizing MIP as the primary component for 1D in environmental analysis.

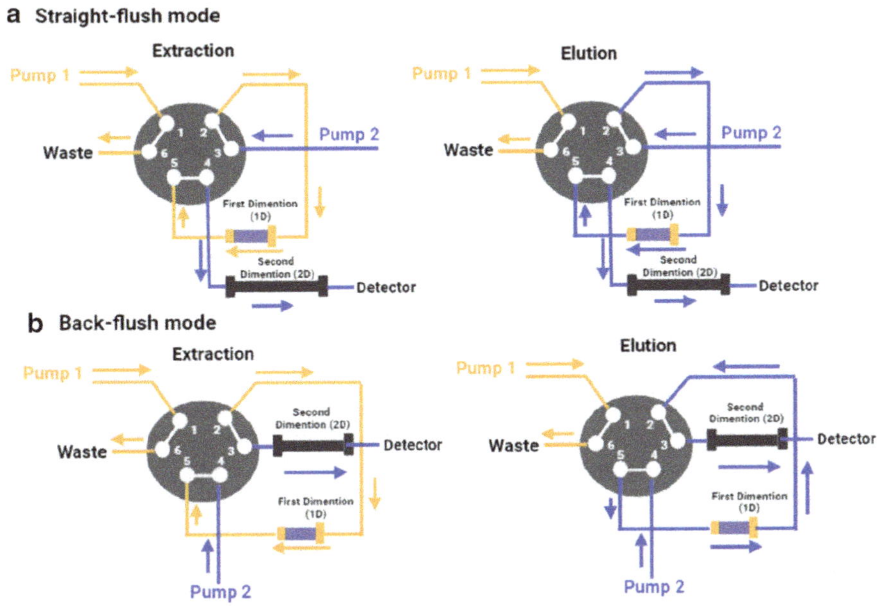

Fig. 4.1 Extraction and elution procedure using **a** Straight-flush mode and **b** Back-flush mode. Adapted from Cruz et al. (2021)

pH Study

In general, it is widely recognized that most MIPs exhibit resistance to variations in pH across a broad range, especially the organic-based ones (Gomes et al. 2022). Nevertheless, the impact of pH on MIPs primarily hinges on the intricacies of polymerization and the binding process with the target analyte. During the polymerization process, meticulous control over pH is desirable as it significantly influences the structure and morphology of the resulting polymer. Hence, it is strongly advisable to maintain precise pH control in the reaction medium. Additionally, regarding template binding affinity, the literature reports studies that inquire into the impact of pH on the development of selective binding sites within the three-dimensional polymeric matrix (Santos et al. 2020). Chen et al. (2001) observed an increase in non-selective sites with the rise in mobile phase pH, thereby influencing the selective binding within the MIP for the extraction of D- and L-phenylalanine anilide. The pH was an important parameter evaluated by Baeza et al. (2022) while studying fluoroquinolones (FQs) in river water using a MIP-silica-based as a pre-concentration stainless-steel column (50 × 4.6 mm i.d.) with a column-switching system. The synthesis of the polymer was conducted considering the use of enoxacin as the template molecule. The authors stated that the main impact of pH resulted from the presence of FQs in an aqueous environment, where they exist in both neutral and anionic forms by the presence of carboxylic and amino groups in their chemical formulations. Furthermore, an observed correlation between acidic conditions with methacrylic acid (MAA) and 2-trifluoromethacrylic acid (TFMAA) revealed protonation under such circumstances, resulting in a weakened interaction with FQs. Conversely, favorable retention results were attained in the pH range of 3–6, enabling an ionic interaction between the protonated amino groups of the FQs and the negatively charged functional monomers. The optimized method demonstrated the ability to achieve a LOD and LOQ within the nanograms per liter (ng L^{-1}) range, along with recovery values exceeding 70%. Notably, the devised column-switching methodology enabled the reuse of the reported MIP in more than 50 assays with adequate analytical performance. Upon meticulous examination of the proposed strategy, this method demonstrates the potential to be an effective approach for assessing FQs and their influence on environmental waters.

Another study provided by Hoshina et al. (2011) conducted a noteworthy pH optimization, introducing the utilization of restricted access material-molecularly imprinted polymer (RAM-MIP) with column-switching system (stainless steel column (100 mm × 4.6 mm i.d)) for assessing non-steroidal and anti-inflammatory drugs (NSAIDs) in river water samples. Flufenamic acid was used as the template molecule for MIP synthesis in this investigation. The findings revealed a reduction in NSAIDs retention when the eluent pH exceeded 7, suggesting a potential dissociation of NSAIDs molecules with pKa values around 4. The authors underscored the significance of pH dependence as the paramount factor in method optimization, emphasizing the need to carefully assess pH control when employing MIPs in column-switching systems.

Regarding the impact of pH on MIP performance, it is crucial to consider not only its interaction with the target molecule but also its influence on the reagents used in synthesis. Commonly, MIP synthesis involves the use of silica particles. The pH plays a significant role in silica activation, responsible for creating reactive sites in silica that enhance molecular imprinting efficacy. Consequently, within specific pH ranges, the silica used in MIP synthesis can establish desired chemical bonds with the monomers. Furthermore, the charge of silica is also subject to pH variations, particularly considering the existence of silanol functions. Altering the pH to induce negative charges at higher pH levels can impact the electrostatic interaction between silica and monomers, thereby influencing the physicochemical properties of the final polymer. Other factors, such as particle shape and surface modification, contribute to the specific recognition between the target analyte and the MIP polymer. These pH-related influences underscore the importance of conducting specific studies to observe the impact of this parameter on MIP-silica-based materials (Kotyrba et al. 2021; Giovaninni et al. 2018).

Sample Loading and Flow Rate

A critical parameter assessed when utilizing a sorbent material is its sample loading capability (Samadifar et al. 2023). This factor holds significant importance in preventing the common issue of overloading the sorbent material, as it can substantially diminish the material's life use (Dai et al. 2009). Moreover, the flow can affect the packing of the material inside the support, thus reflecting the lifetime of the packed material. Particularly for MIPs, overloading the sample can adversely impact the binding sites of the polymer, thereby compromising the overall analytical performance of the method. Careful consideration of sample loading is imperative to ensure the optimal functioning and longevity of the sorbent material, particularly in the context of MIPs (Baggiani et al. 2012). Regarding the importance of such parameters, Han et al. (2005) reported the sample loading and flow on the trace-level evaluation of pentachlorophenol (PCP) in environmental, and wastewater samples by using MIP (1.5 cm × 4 mm i.d.) as sorbent material in column-switching applications. The authors assessed the primary effects of sample loading and loading flow, employing concentrations of 10 $\mu g\ L^{-1}$ and a flow rate of 15 mL/min^{-1}, respectively. The key observations revealed a noteworthy influence on the chromatographic peak area, demonstrating a continuous increase as sample loading time extended, peaking at 12 min. These findings strongly suggest a rapid kinetic adsorption phenomenon between PCP and the MIP polymer. Furthermore, the delicate balance between sample loading and flow holds promise for potentially enhancing the sample's enhancement factor without adversely affecting the extraction efficiency of the column-switching system employing the MIP polymer. The analytical method could ensure a preconcentration capability of 670 and a LOD of 6 $ng\ L^{-1}$ demonstrating the potential of the method in assessing trace-level concentration of PCP in environmental water samples.

Time of Loading and Eluting Sample

In the context of column-switching instrumentation, the assessment often revolves around the loading and elution times of the sample. Both phenomena are intricately linked to the interaction dynamics between the analyte(s) and the sorbent polymer. When MIPs are applied as a sorbent polymer, a critical consideration is determining the equilibrium time of both loading and eluting steps (Suzaei et al. 2022). Both parameters were studied by Guo et al. (2011) in the evaluation of chlorsulfuron (CS) in environmental water samples by using a MIP (stainless steel column of 1.5 cm × 4.0 mm i.d.) as a sorbent material for column-switching back-flush mode. In the loading phase, a concentration of 0.5 $\mu g\ L^{-1}$ of the CS was utilized at a loading rate of 1.0 mL/min^{-1}. The primary findings revealed a consistent rise in the chromatographic peak area, exhibiting an almost linear progression as the sample loading time extended up to 50 min. Additionally, regarding the elution phase, three different times (1.5, 2.5, and 3.5 min) were employed to investigate their real-time impact on the elution process of the CS in the MIP phase. The outcomes demonstrated linearity in the chromatographic peak areas after 1.5 min. Consequently, the authors opted for the shorter elution time, ensuring a more expeditious column-switching evaluation of CS in environmental water matrices. Under the optimum analytical parameters, the column-switching-MIP system could ensure recovery values in the range of 81–110%. Furthermore, the main findings of the study underscore the significant potential of the reported analytical strategy for assessing CS in water matrices through integrated methodologies. This approach proves versatile, not only enabling the detection of this herbicide in the environment but also facilitating the evaluation of other pesticides in similar matrices.

Elution Solvent

In solid–liquid extraction procedures, an elution solvent is frequently employed. The solvent usually promotes the desorption of analytes from the sorbent phase. The selected elution phase must possess an optimal polarity to effectively favors the elution of target analytes from the polymeric matrix of the MIP. Furthermore, the selected elution solvent must demonstrate instrumental compatibility to avoid causing any damage to the analytical instruments. Therefore, the evaluation of the elution solvent is an important step in the execution of column-switching applications. It was evaluated by Kubo et al. (2014) during the application of MIP (stainless steel column (100 mm × 4.6 mm i.d.) as a sorbent material for column-switching evaluation of sulpiride in river water samples. In the synthesis procedure, the authors employed the sulpiride molecule as a template. Notably, the authors emphasized the selection of an elution solvent with a focus on instrumental compatibility. The primary findings, according to the obtained peak areas for sulpiride, revealed improved elution of the target analyte through the utilization of a methanol and acetonitrile mixture (2/8 v/v). Furthermore, meticulous attention was dedicated to controlling the phase pH, maintaining it at a specified level of 3.0. The optimized methodology could evaluate the sulpiride at a 10 ng L^{-1} concentration range, besides, a recovery value of 96% was obtained in such a concentration. Based on the primary findings, the proposed methodology has demonstrated its efficacy in facilitating the assessment of sulpiride in river water samples. This underscores its potential as a valuable analytical instrument for monitoring pharmaceutical compounds, such as sulpiride, in environmental contexts.

4.2.2 Exploring In-tube and Molecularly Imprinted Polymer Strategies for Environmental Monitoring

Solid phase microextraction (SPME) is a miniaturized technique (Pawliszyn et al. 1997) that provides the concurrent separation and preconcentration of both volatile and non-volatile analytes from complex samples. Among its advantages, SPME stands out as a free-solvent technique, being aligned to more sustainable processes. As with many miniaturized techniques, SPME is not an exhaustive extraction method once it relies on the equilibrium between the SPME sorbent phase with the evaluated matrix, continuing until the absorption equilibrium of the analyte is achieved (Jalili et al. 2020; Yamini et al. 2019).

For the automation of SPME, capillary SPME has proven to be effective when coupled with Gas Chromatography (GC) using open tubular trapping, as well as with LC through in-tube SPME (Moliner-Martinez et al. 2015). The concept of in-tube SPME was introduced to address challenges associated with conventional SPME methods. These challenges included issues such as fiber fragility, stability in the presence of LC solvents, low stability, and limited recovery. The introduction of in-tube SPME aimed to overcome these challenges, facilitating the automatization of SPME within LC instrumentations (Eisert and Pawliszyn 1997). According to the literature, in-tube SPME has found extensive use in both traditional and miniaturized LC systems, which demonstrates numerous advantages in terms of sensitivity, efficiency, and great interface with MS detectors (Moliner-Martinez et al. 2015; Eisert and Pawliszyn 1997; Kataoka 2021).

The process of in-tube SPME involves utilizing a capillary column to extract and pre-concentrate analytes before entering the LC system (Souza et al. 2021). The extraction process relies on sample draw/eject cycles or a sample flow, resulting in the adsorption/absorption of analytes onto the inner surface of the capillaries by the stationary phase (Kataoka 2021). This approach hinges on transferring the analyte present in the matrix to the sorbent phase, and the speed of this transfer is contingent upon the sample flow and analyte partition. Following extraction, desorption of the analytes is performed by a desorption solvent or mobile phase, facilitating seamless integration with chromatographic systems. In the realm of in-tube SPME, the extraction duration usually corresponds to the length of the capillary extractive phase, exhibiting an inverse relationship with the sample flow rate and the desorption/mobile phase (Martins et al. 2023; Moliner-Martinez et al. 2015; Kataoka 2021). As mentioned, in-tube SPME can operate in two different approaches, the draw/eject system and the flow-through system (Moliner-Martinez et al. 2015). The original in-tube SPME device employed the draw/eject method, which includes the repetitive draw/eject cycles of the sample into the capillary column. In this setup, the column is situated between an LC injection needle and the loop of a sample injector (Fig. 4.2a) (Moliner-Martinez et al. 2015; Souza et al. 2021). The process comprises four steps: conditioning of the capillary column, analytes extraction by sample draw/eject cycles, injection needle washing, and analytes desorption. The conditioning step typically involves draw/eject cycles of water and methanol into the capillary column, with the introduction of air to create an air band that enhances equilibrium between water and the extractive phase. The introduction of air serves to hinder the blending of the sample solution and the mobile phase. Analytes are then extracted through repetitive draw/eject cycles (usually 25–30 cycles) until equilibrium is achieved between the sample and the sorbent phase. Following the methanol washing of the needle, the six-port valve is toggled to release the extracted analytes from the capillary once equilibrium is established. Using the autosampler injection valve, the analytes are desorbed and carried to an analytical column for further analysis (Moliner-Martinez et al. 2020).

The draw/eject extraction system offers several key advantages. It excels in the seamless handling of drawing and ejecting small sample volumes, enables the processing of a large number of samples without carryover effects, thanks to the integration with an autosampler and needle washing mechanism, and provides precise control over the LC system and data processing. Nevertheless, it is crucial to acknowledge that the draw/eject system has the potential to be time-consuming, particularly when the number of cycles is high (Souza et al. 2021; Moliner-Martinez et al. 2020; Costa Queiroz et al. 2019). Conversely, the in-tube SPME flow-through system relies on a continuous sample flow moving unidirectionally through the capillary column (Moliner-Martinez et al. 2020).

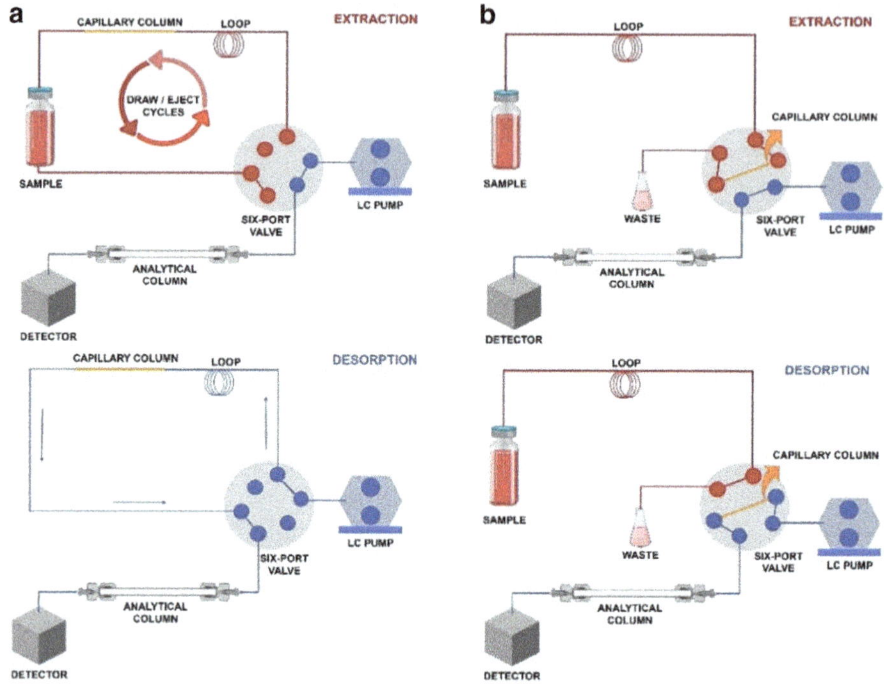

Fig. 4.2 Schematic illustration of the In-tube SPME operation coupled to LC system, in different configurations: **a** Draw/eject cycles system; **b** Flow-through system. Adapted from Kataoka (2021)

This system follows the same four steps of the draw/eject system, but instead of needle washing, in a flow-through system, the capillary column is washed with water to eliminate inherent interferents from the matrix (Moliner-Martinez et al. 2015). In this system, the capillary column serves as the loop of an injector valve, which promotes the analytes extraction during sample loading, followed by its transfer to the analytical column (Kataoka 2021; Costa Queiroz et al. 2019).

This type of approach allows different configurations, such as in-valve in-tube SPME (Fig. 4.2b) (Kataoka 2021; Costa Queiroz et al. 2019). During the extraction procedure, the six-port valve is positioned in the loading position, enabling the sample to enter the capillary column. Typically, sample loading is accomplished using either a syringe pump or an additional pump. Following extraction, the six-port valve is transitioned to the injection state, permitting the analytes held on the capillary column to be directed toward the analytical column by the LC mobile phase. The flow-through system has a drawback as it may be prone to contamination, therefore, the valves could potentially get contaminated with the sample solution. Additionally, achieving equilibrium between the sample and the extractive sorbent is not guaranteed, nevertheless, achieving higher recoveries is possible by using a larger volume of the sample solution, albeit this may result in an extended analysis time.

To address limitations associated with in-tube SPME flow-through systems, an alternative approach involves incorporating two separate pumps. This alternative allows for the utilization of different flow rates during extraction and separation, thereby enhancing control over the sample transfer process (Kataoka 2021; Costa Queiroz et al. 2019).

To enhance the recovery in in-SPME, optimization of certain parameters related to in-tube SPME is crucial. These evaluations encompass the study of the number of draw/eject cycles, sample flow rate, type and volume of the desorption solvent, and the extractive phase (Moliner-Martinez et al. 2020). The optimization of extraction and desorption steps is inherently tied to the specific system being employed. Conventionally, on in-tube SPME, an open tubular fused-silica capillary column is employed, resembling GC columns, through which a small volume of the sample flows. However, contemporary in-tube SPME methodologies can employ various types of capillary columns beyond conventional GC columns, such as fiber-packed, monolith, and sorbent-packed columns (Moliner-Martinez et al. 2015; Costa Queiroz et al. 2019).

Surface-coated capillary columns are extensively utilized in in-tube SPME. These columns have garnered significant popularity due to their well-established presence in the market as stationary phases for GC. One of their primary advantages lies in the accessibility and diverse options available for modifying the extractive phase. Coated columns involve placing an extractive phase on the capillary wall, and synthesis can be achieved through processes, such as polymerization, surface modification, or electrodeposition. Conversely, monolithic capillary columns represent a dynamic and promising aspect of in-tube SPME. They exhibit outstanding features, including exceptional stability, high permeability, fast mass transfer kinetics, and substantial loading capacity. Unlike typical coated capillary columns, monolithic capillary columns provide an increased interaction surface, thereby improving the extraction process. Common approaches for synthesizing monolithic columns involve UV polymerization and thermal polymerization (Moliner-Martinez et al. 2020; Grecco et al. 2021).

Furthermore, contemporary advancements in in-tube SPME center around the utilization of innovative materials such as nanomaterials, ionic liquids, immunosorbents, RAM, and MIPs as extractive phases. This diversified array of extracting phases contributes to the technique's versatility in instrumentation and enables the analysis of a broad spectrum of different compounds (Martins et al. 2023; Moliner-Martinez et al. 2015). MIP materials have become prominent in in-tube SMPE methods for their selectivity and recognition of target molecules, showcasing improved detection limits and higher analytical throughput (Moliner-Martinez et al. 2020). The literature reports MIP capillary columns using different approaches as coated and monolithic capillary columns (Moliner-Martinez et al. 2020; Grecco et al. 2021). Since its inception, MIP-in tube SPME has gained various applications and has been widely applied to biological samples. However, it is possible to find some applications for environmental samples as well, especially regarding the optimization of in-tube SPME procedures for these applications (Grecco et al. 2021).

pH and 'Salting Out' Effects

Within the in-tube SPME system, dependent on the equilibrium between the sample and the extractive phase, adjusting the sample pH and/or ionic strength through the addition of salt (referred to as the 'salting out' effect) can markedly enhance recoveries by modifying the equilibrium. To optimize extraction, it is essential to consider the characteristics of both the sample and the stationary phase, aiming to promote favorable interactions without compromising the operational pH range of the extractive phase, as some extracting phases have specific limitations in this regard (Kataoka 2021; Souza et al. 2021; Costa Queiroz et al. 2019).

The influence of pH and the 'salting out' effect on the MIP-in-tube SPME system was demonstrated in the evaluation of 4-nitrophenol present in water samples (Zare-jousheghani et al. 2013). This molecule demonstrates high toxicity to both the environment and human health, and its occurrence is associated with insecticides, pharmaceuticals, and dyes. The study evaluated the sample pH accounting for the dissociation constant of 4-nitrophenol, which demonstrated that elevated pH values changed the equilibrium of the molecule towards its anionic form, preventing the molecule from fitting into the specific cavities of the polymer created through the molecular form of the same molecule. Additionally, the anionic form demonstrates increased affinity for the aqueous medium, further complicating the molecule's extraction. In this study, the selection of the pH value also accounted for the characteristics of the polymeric material once highly acidic or basic conditions can affect the polymer, resulting in a reduced lifetime and reproducibility. Regarding the 'salting out' effect, the research showed that increasing the ionic strength of the matrix resulted in a reduction in the solubility of 4-nitrophenol. In this sense, the amount of salt must be optimized to avoid its co-extraction, leading to clogging of the capillary extractor tubes. The authors employed a pH of 3 and a concentration of 140 g L^{-1} of NaCl.

An interesting study by Souza et al. (2016) illustrated the impact of pH on the evaluation of parabens using RAM-MIP in-tube SPME. Currently, these molecules are considered emerging pollutants in the environment. The authors considered the parabens pka values (around 8) and the prevailing interactions in each pH value Furthermore, the stability of the synthetized material was considered once under extreme acid pH values, and the RAM-MIP selectivity decreased. Thus, the authors considered that at pH 9, parabens showed good interaction with the material, as pH values above would lead to the hydrolysis of target molecules, and values below would result in hydrogen interactions between the MIP and parabens. Additionally, another study based on MIP-in tube SPME associated with magnetic nanoparticles (Fe_3O_4) to extract 2,4-dinitroaniline from water samples demonstrated the effect of pH and 'salting out effect' on recovery (Song et al. 2023). Evaluating pH from 3 to 10, it was demonstrated that at pH 6, there was a better recovery, as values below generate charged species leading to electrostatic repulsion, and strongly alkaline pH interferes with analyte-extracting phase interactions due to an excess of OH^-. Furthermore, high recovery was observed using 4% of NaCl.

Extraction Time

In a flow-through system, a crucial parameter is the flow rate, directly influencing the amount of sample passing through the capillary column and, consequently, the extraction time. While a higher flow rate can potentially enhance recoveries, it introduces elevated system pressure as a trade-off. The heightened pressure could result in the formation of bubbles within the capillary column, potentially affecting its longevity, and necessitating the optimization of the flow rate value. In a draw/eject system, a higher number of cycles and increased sample volume generally result in higher recovery rates. However, achieving the optimum value for these parameters requires careful consideration of the chromatogram peak. Excessive cycles and larger sample volumes can contribute to peak tailing and enlargement, emphasizing the need to strike a balance. Therefore, the extraction time becomes contingent on factors such as the flow rate, draw/eject cycles, and sample volume (Kataoka 2021; Moliner-Martinez et al. 2020; Costa Queiroz et al. 2019).

To extract 4-nitrophenol from river water samples using MIP-in-tube SPME, Zarejousheghani et al. (2013) conducted a study involving 2 mL of the sample under various flow rates, leading to extraction durations spanning from 30 to 90 min. The investigation revealed that higher extraction times led to increased recoveries, peaking at 60 min. The prolonged equilibration time attributed to a low applied flow rate showed that decreasing the sample flow rate enhanced the effective transfer of the analyte from the matrix to the extractive phase.

Regarding the evaluation of 2,4-dinitroaniline in water samples using the MIP-in-tube SPME method, the study assessed sample volumes ranging from 4 to 4.5 mL and flow rates between 0.06 and 0.14 mL/min^{-1}. Results indicated that sample volumes exceeding 4.5 mL and flow rates surpassing 0.12 mL/min^{-1} did not contribute to optimal analyte recovery efficiency. Taking into account the elevated permeability of the prepared MIP, the authors established the optimal values for sample volume and flow rate as 4.5 mL and 0.12 mL/min^{-1}, respectively (Song et al. 2023).

In a study reporting the extraction of carbamazepine using a nanocomposite of molecularly imprinted polypyrrole on copper oxide (MIP@CuO) in in-tube SPME, the influence of extraction time and flow rate on recovery was investigated. The authors varied the extraction flow rate from 0.5 to 4 mL min^{-1}, observing that an increased extraction flow rate enhanced the extraction of carbamazepine due to an extended contact duration between the sample solution and the surface of the sorbent. However, 3 mL/min^{-1} was chosen to avoid the potential formation of air bubbles at higher flow rates (Kefayati et al. 2021).

As demonstrated in this section, the application of integrated LC methodologies through column-switching and in-tube SPME systems holds significant promise for expeditious and pragmatic analytical assessments. Furthermore, the utilization of MIPs within a 1D column-switching and in-tube SPME framework has proven to be a valuable approach, ensuring selective binding between target molecules in the polymeric network. Such a combination has been extensively documented in the literature for environmental analysis, demonstrating its capability to both concentrate trace-level concentrations and eliminate interfering compounds that contribute to matrix effects. Moreover, the implementation of integrated column-switching and in-tube SPME approach offers several advantages capable of mitigating potential analytical instrumentation errors introduced by the analyst, thereby enhancing the representativeness of matrix evaluations. Despite these mentioned advantages, there remains a need for updated literature on the synergy between MIPs as sorbent phases for both applications described in this section. Furthermore, it is crucial to expand the scope of MIP-column-switching and MIP-in-tube SPME applications to encompass diverse environmental matrices. While our discussion predominantly focuses on water samples evaluation, it is essential to acknowledge that other environmental compartments, including soil and sediment, remain integral aspects of the ecological landscape. Employing the combination of MIP and column-switching/In-tube SPME instrumentations in the assessment of these matrices can unveil potential contaminants, thereby facilitating a comprehensive evaluation of their impact across distinct environmental compartments.

Finally, the fusion of these two strategies has the potential to usher in novel advancements and perspectives in environmental analysis. This is particularly pertinent for monitoring multiresidues that could disrupt the delicate balance of the environment and human society. As such, continued research and exploration in this domain can lead to breakthroughs in addressing challenges related to environmental sustainability and public health.

4.3 From Concept to Application: MIPS and Ambient Mass Spectrometry for Comprehensive Environmental Matrix Exploration

Similar to advancements in chromatographic applications, MS instrumentation has evolved significantly over the decades (Black et al. 2016). Conventional MS systems have undergone modifications for adaptation to ambient applications, resulting in the emergence of ambient mass spectrometry (AMS) methods. Notably, AMS methods seamlessly align with green analytical practices, as many of them eliminate the need for sample preparation steps that traditionally consume significant sample and solvent volumes (Martins et al. 2023). Furthermore, the efficiency and rapid analysis capabilities inherent in AMS methodologies have drawn considerable attention, particularly in the examination of complex matrices (Chaves et al. 2022). Furthermore, the continual evolution of these systems has introduced the incorporation of MIPs as modified substrates for AMS methodologies, in order to enhance the selectivity of these approaches (Santos et al. 2020). Therefore, the literature extensively documents the utilization of MIPs in AMS applications across various matrices, including environmental samples. By exploring the synergies between AMS methods and MIPs, this research contributes to advancing the capabilities of analytical techniques in addressing the unique challenges posed by environmental sample analysis.

Paper spray ionization (PSI)

The paper spray ionization (PSI) approach was reported by Liu et al. (2010) and is recognized as a simple and user-friendly AMS method (Pereira et al. 2018). As PSI is an analog ion source to electrospray (ESI), the ionization mechanism is quite similar. The sample is applied to the surface of a triangular-shaped porous support, typically paper. The system is positioned at a defined distance to the mass spectrometer inlet, $\approx$ 3–5 mm. Therefore, a desorption solvent is applied to the substrate surface, as well as a high voltage (2–5 kV), facilitating the desorption of analytes and their transfer to the gas phase (Chaves et al. 2022). In this AMS technique, the ionization process and analytical performance are influenced by a combination of operational parameters, the analyte's physicochemical properties, and analyte-substrate interaction (Frey et al. 2020).

Complex samples present a formidable challenge in AMS due to the diverse array of constituents, including endogenous compounds and interferent molecules, that can contribute to ion suppression and compromise sensitivity (Han et al. 2018). In the context of environmental chemistry, employing AMS strategies to confront the challenges presented by complex samples emerges as a compelling approach for evaluating these matrices. The complex characteristics of environmental samples introduce a diverse array of components, including both intrinsic compounds inherent to the sample matrix and interfering molecules originating from co-existing species or matrix components (Martins et al. 2021). Additionally, the coexistence of various chemical species in complex samples can lead to ion–molecule reactions, further complicating the ionization dynamics. These factors collectively contribute to decreased ionization efficiency and sensitivity in AMS analyses of complex samples, necessitating innovative approaches and techniques to mitigate these challenges and enhance the reliability of mass spectrometric measurements (Kuo et al. 2020; Feider et al. 2019). By tackling these challenges, researchers can enhance the accuracy of environmental analyses and contribute to a more comprehensive understanding of the complex chemical composition of environmental samples.

In this sense, numerous environmentally friendly sample preparation techniques have been integrated with AMS ion sources to promote sustainability and reduce environmental impact. In this sense, MIPs have been demonstrated as a particularly outstanding and green analytical technique aimed at enhancing sensitivity and selectivity while mitigating ion suppression in PSI methods. MIPs, functioning as sorbents, exhibit the ability to extract species through specific recognition sites embedded within a polymeric network. A notable advancement in this regard is the synthesis of MIPs on a membrane surface, which can serve as a substrate to help the promotion of ionization during the analysis, termed molecularly imprinted polymer-paper spray ionization (MIP-PSI) (Lerma-García et al. 2013). This innovative green technique offers a sustainable solution to overcome classical PSI limitations, as reported in previous studies (Santos et al. 2020). By adopting such green approaches, researchers contribute to the adoption of environmentally friendly analytical strategies that follow green analytical practices, promoting efficiency, reduced waste generation, and overall sustainability in the field of environmental analysis.

Ramalho et al. (2022) employed MIP-PSI-MS to analyze fumunisin B1 (FB1) in maize samples. The concern about fumonisin B1 content in maize samples arises from its association with potential health risks, as this mycotoxin (produced by *Fusarium fungi*) can contaminate maize crops and, upon ingestion, has been linked to negative health impacts in both humans and animals, making accurate and sensitive detection crucial for food safety assessments. The authors synthesized a molecularly imprinted polymer on a cellulose membrane substrate, using the 1,2,3-propanetricarboxylic acid as a template molecule. The reported MIP-PSI-MS method presented linearity ranging from 0.5 to 150 μg mL^{-1} with $R^2 > 0.99$, precision, and accuracy values below 5.5%. The alteration of the membrane surface allowed for the measurement of FB1 in commercial corn samples, revealing concentrations of 0.439, 0.060, and 0.486, 0.061 μg mL^{-1} in two samples.

While the most commonly employed method involves modifying the paper substrate with MIP, Freitas et al. (2022) reported the synthesis of a MIP as a sorbent for the solid extraction phase (MIP-SPE) applied in the sample clean-up procedure before PSI-MS analysis. In this study, the pesticides aminocarb pirimicarb, dimethoate, omethoate, pyridafenthione, and fenitrothion were determined in maize samples. The analytical performance for aminocarb, pirimicarb, and omethoate using MIP-SPE was enhanced when compared to NIP (Non-Imprinted Polymer) before PSI-MS analysis. The method demonstrated a range from 100 to 500 μg mL^{-1}, with $R^2 >$ 0.98 for all pesticides, and LOD and LOQ values about 50 and 100 μg kg^{-1}, respectively. The precision and accuracy values were within the range of less than 10%, and the absolute recovery values spanned from 80 to 120%. MIP-SPE-PSI demonstrated precision and accuracy in identifying carbamates and organophosphorus pesticides, eliminating the necessity for pre-concentration steps.

Probe Electrospray Ionization (PESI)

Probe Electrospray Ionization (PESI) has surfaced as an alternative method in the domain of AMS, revolutionizing the way samples are analyzed in their native environments. PESI enables the direct analysis of intricate samples without requiring complex sample preparation steps (Thirukumaran et al. 2021). This AMS approach uses a sharp-tipped probe that moves vertically. In the lower position, the probe remains in contact with the sample. In the upper position, a sample droplet forms at the tip of the probe, and a high voltage is applied, generating an electrospray plume that transitions ions from the liquid to the gas phase. Subsequently, the generated ions are guided into the mass spectrometer for analysis (Hiraoka et al. 2007).

The advent of PESI represents a promising leap forward in the mass spectrometry analytical field specifically designed to overcome challenges inherent in traditional methods and align with the principles of green chemistry. One notable advantage lies in its ability to address clogging issues, a persistent problem in conventional ESI ion sources that often lead to inefficiencies. The versatility of PESI extends its applications across various domains, including environmental monitoring (Chen et al. 2017), pharmaceuticals (Yang and Deng 2016), and clinical diagnostics (Chung et al. 2020). This widespread applicability underscores the green potential of PESI in transforming and advancing ambient mass spectrometry techniques, reducing the environmental impact associated with sample analysis, especially by reducing solvent consumption. By providing a cleaner and more efficient alternative, PESI contributes to the principles of green chemistry by enhancing analytical capabilities while minimizing resource use and waste production. As research progresses, the integration of PESI holds the promise of further expanding the scope and capabilities of ambient mass spectrometry, offering greener solutions to diverse analytical challenges.

Moreover, PESI serves as a sampling unit through the adsorption process of compounds onto the surface of the probe/needle (Hu et al. 2018). Hence, utilizing MIP technology to modify support materials allows for the effective analysis of both solid and liquid analytes. Integrating MIP polymers as substrates in PESI approaches not only improves analytical capabilities but also promotes the advancement of environmentally sustainable methodologies (Hiraoka et al. 2007).

In this regard, Silva et al. (Bernardo et al. 2019; Silva et al. 2019) applied a molecularly imprinted polymer-Probe Electrospray Ionization-Mass spectrometry (MIP-PESI-MS) method for the analysis of phorbol esters in *Jatropha curcas* leaves for speciation of toxic and non-toxic plants. The MIP-PESI-MS method allowed the detection of phorbol esters with minimal sample preparation, good linearity ($R^2 > 0.99$), and LOD and LOQ about 0.28 and 0.92 $\mu g\ mL^{-1}$, respectively. The precision and accuracy values ranged from 4.06 to 13.49% and from 1.60 to 15.26%, respectively. The phorbol 12,13-diacetate was determined in concentrations ranging from 222.19 ± 23.55 to $528.23 \pm 19.72\ \mu g\ g^{-1}$, phorbol esters, and deoxyphorbol metabolites were identified as toxic genotypic.

Tian et al. (2020) reported a MIP-coated stainless-steel probe for the evaluation of trace fluoroquinolone antibiotics from milk. The authors stated that detecting fluoroquinolone antibiotics in milk samples is of paramount importance due to the potential risks associated with their presence. Continuous exposure to fluoroquinolones through consumption of contaminated milk may contribute to the development of antibiotic-resistant strains of bacteria in humans, posing a significant public health concern. Furthermore, the occurrence of these compounds in the environment due to their presence in food matrices can be regarded as an environmental concern. In this sense, the authors employed methylacrylic acid–base polymer and enrofloxacin as templates for the synthesis of MIP over the stainless-stell surface. The method demonstrated linae range from 1 to 1000 ng mL^{-1}, $R^2 > 0.99$, and LOD and LOQ values ranging from 0.1 to 5 ng mL^{-1}.

The introduction of the wooden-tip electrospray ionization (WT-ESI) represents a groundbreaking advancement in the realm of probe electrospray ionization (PESI). This innovative technique utilizes a wooden substrate as an alternative to traditional metallic probes, introducing a sustainable and cost-effective approach to AMS (Hu and Yao 2022). The porous nature of wood not only facilitates efficient ionization but also offers a unique matrix for the adsorption of analytes, enhancing the selectivity and sensitivity of PESI. As an eco-friendly alternative, the wooden-tip electrospray addresses concerns about environmental impact while maintaining the analytical performance required for comprehensive sample analysis (Hu et al. 2011). This innovative substrate presents considerable potential to broaden the applications of

PESI across diverse fields, providing a more sustainable and resourceful aspect to the advancing realm of ambient ionization techniques in mass spectrometry.

The WT-ESI was initially documented in 2011 by Hu et al. (2011). In this first demonstration, the authors presented the versatility of the technique across a range of analytes, including organic compounds, peptides, and proteins. Additional chemical modifications were designed to enhance the overall analytical performance of the WT. The porous surface of WT also allows changes that increase extraction capacity, expanding its scope of application (Deng et al. 2014).

Liu et al. (2019) reported the MIP-coated-WT method for the analysis of trace macrolide antibiotics in drinking water, honey, and milk samples. This method surface provided a chemically selective, repeatability, and reproducibility even at low concentration levels. It is noteworthy to mention the reusability of the WT substrates, as each substrate consistently exhibited analytical performance for up to five uses.

The use of MIP as a selective substrate for AMS is not only restricted to PSI and PESI-MS. Different materials and geometries may be used for better sensitivity or sample support. Ma et al. (2021) modified the surface of an iron sheet with MIP for selective analysis of sulfonamides. The authors used a triangular-shaped iron sheet, and the ion source setup was very similar to the PSI-MS. The method was capable of accurate and precise detection of five sulfonamide antibiotics in pure water and honey samples within a concentration range from 0.2 to 1000 ng mL^{-1}, $R^2 > 0.99$. The LOQ values ranged from 0.3 to 0.5 ng mL^{-1} in water samples and from 1 to 5 ng mL^{-1} in honey samples.

Bogdanowicz et al. (2023) described the use of molecularly imprinted polymers coupled to ambient plasma mass spectrometry for rapid quantification of 2,4-dichlorophenol. To this aim, the authors reported the successful synthesis of poly(methyl vinyl ether-*alt*-maleic acid)-based molecularly imprinted polymers via amidation, enhancing their affinity for 2,4-dichlorophenol. Thermodynamic studies reveal a spontaneous and entropy-driven adsorption process. When coupled with ambient plasma mass spectrometry, these MIPs enabled highly selective solid-phase extraction of 2,4-dichlorophenol, significantly improving detection limits and showcasing their applicability in environmental sample analysis. This research underscores the potential of tailored adsorbents for developing efficient and rapid analytical techniques for quantifying diverse organic compounds via ambient mass spectrometry.

The integration of versatile materials such as MIPs and AMS methods has ushered in a new era of highly efficient miniaturized sample preparation and direct analysis in mass spectrometry. Beyond its technical merits, the standout attribute lies in its harmonization with the principles green and sustainable green analytical practices. This innovative approach not only minimizes the need for extensive sample preparation but also aligns with sustainable practices by significantly reducing analysis time, sample quantities, and solvent consumption. The versatility of these methods extends their applicability across a diverse range of samples, while their exceptional sensitivity and resolution contribute to unparalleled accuracy in analysis. As we navigate the evolving landscape of analytical methodologies, the synergy between MIP and AMS emerges as a powerful and environmentally conscious solution, paving the way for more streamlined and eco-friendly approaches to sample preparation and analysis.

4.4 Conclusion Remarks

Analytical instrumentations, such as LC and MS approaches have been extensively documented in the literature for their ability to offer qualitative and quantitative assessments of environmental matrices. Nevertheless, to elevate analytical parameters, particularly selectivity and recovery, the use of MIPs has opened new horizons for these instruments, particularly in their application to environmental evaluation. The incorporation of integrated LC strategies, including column-switching and in-tube SPME, has enhanced traditional LC methodologies by capitalizing on their automation potential. Moreover, employing MIPs in these instrumentations can streamline the sample preparation process, encompassing both clean-up and preconcentration of target analytes, along with the separation of compounds within an integrated method. Such a combination not only reduces analysis time but also minimizes the likelihood of potential analytical errors caused by human intervention. Furthermore, the adoption of AMS methodologies has notably highlighted their substantial green analytical potential. These approaches are recognized for their minimal solvent consumption and reduced sample requirements, thereby mitigating chemical waste generation. Therefore, the incorporation of MIPs has been identified as a strategic approach to bolster method selectivity, consequently enhancing the overall analytical performance of AMS methodologies.

While the integration of MIPs into LC and MS instrumentations has advanced the analytical assessment of complex matrices, especially in environmental contexts, there remain challenges pertaining to both MIPs and LC/MS instrumentation aspects. Addressing challenges in MIPs, the majority of synthesized MIPs still rely on traditional and non-eco-friendly synthesis methods. These conventional approaches involve substantial amounts of solvents and reagents, resulting in a significant generation of chemical waste. From the authors' critical standpoint, there is a pressing need to introduce novel synthesis strategies that incorporate biodegradable materials, thus aligning with the green analytical principles of an eco-friendly approach. Furthermore, concerning LC and MS instrumentations, while the incorporation of MIPs has heightened selectivity compared to some traditional sorbent phases in LC approaches and modified substrates in MS analysis, these instruments primarily emphasize Ex-situ analysis. The reliance on Ex-situ analysis intensifies solvent and energy consumption for sample transport and storage. In light of this, the authors contend that well-established concepts from LC and MS strategies in Ex-situ analysis can be extrapolated to propose novel In-situ strategies. Moreover, by deploying these instrumentations directly in real-world environments, such methodologies can achieve point-of-care analysis, thereby enhancing their practical applicability in assessing various matrices, particularly environmental samples. By proffering solutions to the aforementioned challenges, researchers can elevate analytical instrumentations and enhance the application of MIPs in environmental matrices. This involves addressing their complexity and devising intelligent and effective strategies for comprehensive evaluation.

Nevertheless, there remains a need for concerted efforts to stimulate further research studies exploring the application of MIPs in both analytical strategies. There is a pressing need for an update of the literature material concerning environmental analysis applications. Additionally, the application of the developed integrated LC and AMS methods should encompass diverse matrices, extending beyond the focus on common samples, such as environmental waters. The assessment of various environmental matrices through the application of MIPs in both analytical strategies holds the potential to pave new horizons in environmental monitoring, therefore, ensuring a future with broad applicability for the strategies delineated in this chapter.

References

Ayivi RD, Obare SO, Wei J (2023) Molecularly imprinted polymers as chemosensors for organophosphate pesticide detection and environmental applications. TrAC, Trends Anal Chem 167:117231. https://doi.org/10.1016/J.TRAC.2023.117231

Baeza AN, Urriza-Arsuaga I, Navarro-Villoslada F, Urraca JL (2022) The ultratrace determination of fluoroquinolones in river water samples by an online solid-phase extraction method using a molecularly imprinted polymer as a selective sorbent. Molecules 27:8120. https://doi.org/10.3390/MOLECULES27238120/S1

Baggiani C, Giovannoli C, Anfossi L et al (2012) A Connection between the binding properties of imprinted and nonimprinted polymers: a change of perspective in molecular imprinting. J Am Chem Soc 134:1513–1518. https://doi.org/10.1021/JA205632T/SUPPL_FILE/JA205632T_SI_001.PDF

Bernardo RA, da Silva LC, Queiroz MEC et al (2019) Lab-made solid phase microextraction phases for off line extraction and direct mass spectrometry analysis: evaluating the extraction parameters. J Chromatogr A 1603:23–32. https://doi.org/10.1016/J.CHROMA.2019.06.038

Black C, Chevallier OP, Elliott CT (2016) The current and potential applications of ambient mass spectrometry in detecting food fraud. TrAC, Trends Anal Chem 82:268–278. https://doi.org/10.1016/J.TRAC.2016.06.005

Bogdanowicz N, Lusina A, Nazim T, Cegłowski M (2023) Rapid quantification of 2,4-dichlorophenol in river water samples using molecularly imprinted polymers coupled to ambient plasma mass spectrometry. J Hazard Mater 450:131068. https://doi.org/10.1016/J.JHAZMAT.2023.131068

Cardoso AT, Martins RO, Lanças FM, Chaves AR (2023) Molecularly imprinted polymers in online extraction liquid chromatography methods: current advances and recent applications. Anal Chim Acta 1284:341952. https://doi.org/10.1016/J.ACA.2023.341952

Chaves AR, Oliveira Martins R, Ítala L et al (2022) Ambient ionization mass spectrometry: applications and new trends for environmental matrices analysis. Braz J Anal Chem 2022:52–77

Chen Y, Kele M, Quiñones I et al (2001) Influence of the pH on the behavior of an imprinted polymeric stationary phase—supporting evidence for a binding site model. J Chromatogr A 927:1–17. https://doi.org/10.1016/S0021-9673(01)01019-6

Chen R, Deng J, Fang L et al (2017) Recent applications of ambient ionization mass spectrometry in environmental analysis. Trends Environ Anal Chem 15:1–11. https://doi.org/10.1016/J.TEAC.2017.07.001

Cherian T, Ragavendran C, Vijayan S et al (2023) A review on the fate, human health and environmental impacts, as well as regulation of antibiotics used in aquaculture. Environ Adv 13:100411. https://doi.org/10.1016/J.ENVADV.2023.100411

Chung WY, Correa E, Yoshimura K et al (2020) Using probe electrospray ionization mass spectrometry and machine learning for detecting pancreatic cancer with high performance. Am J Transl Res 12:171

Costa Queiroz ME, Donizeti de Souza I, Marchioni C (2019) Current advances and applications of in-tube solid-phase microextraction. TrAC, Trends Anal Chem 111:261–278. https://doi.org/10.1016/j.trac.2018.12.018

Cruz JC, de Souza ID, Grecco CF et al (2021) Recent advances in column switching high-performance liquid chromatography for bioanalysis. Sustain Chem Pharm 21:100431. https://doi.org/10.1016/J.SCP.2021.100431

Cruz JC, de Souza ID, Lanças FM, Queiroz MEC (2022) Current advances and applications of online sample preparation techniques for miniaturized liquid chromatography systems. J Chromatogr A 1668:462925. https://doi.org/10.1016/J.CHROMA.2022.462925

da Silva LC, de Carvalho TC, Pereira I et al (2019) Molecularly imprinted polymer-coated probe electrospray ionization mass spectrometry determines phorbol esters and deoxyphorbol metabolites in *Jatropha curcas* leaves. J Am Soc Mass Spectrom 30:2051–2059. https://doi.org/10.1007/S13361-019-02269-5/SUPPL_FILE/JS8B06219_SI_001.DOCX

Dai J, Carr PW, McCalley DV (2009) A new approach to the determination of column overload characteristics in reversed-phase liquid chromatography. J Chromatogr A 1216:2474–2482. https://doi.org/10.1016/J.CHROMA.2009.01.036

Deng J, Yang Y, Fang L et al (2014) Coupling solid-phase microextraction with ambient mass spectrometry using surface coated wooden-tip probe for rapid analysis of ultra trace perfluorinated compounds in complex samples. Anal Chem 86:11159–11166. https://doi.org/10.1021/AC5034177/SUPPL_FILE/AC5034177_SI_001.PDF

Eisert R, Pawliszyn J (1997) Automated in-tube solid-phase microextraction coupled to high-performance liquid chromatography. Anal Chem 69:3140–3147. https://doi.org/10.1021/ac970319a

Feider CL, Krieger A, Dehoog RJ, Eberlin LS (2019) Ambient ionization mass spectrometry: recent developments and applications. Anal Chem 91:4266–4290. https://doi.org/10.1021/ACS.ANALCHEM.9B00807/ASSET/ACS.ANALCHEM.9B00807.FP.PNG_V03

Fernández-Ramos C, Šatínský D, Šmídová B, Solich P (2014) Analysis of trace organic compounds in environmental, food and biological matrices using large-volume sample injection in column-switching liquid chromatography. TrAC, Trends Anal Chem 62:69–85. https://doi.org/10.1016/J.TRAC.2014.07.008

Freitas C, Machado LS, Pereira I et al (2022) Assessing organophosphorus and carbamate pesticides in maize samples using MIP extraction and PSI-MS analyzes. J Food Sci Technol 59:2510–2515. https://doi.org/10.1007/S13197-022-05464-7/TABLES/1

Frey BS, Damon DE, Badu-Tawiah AK (2020) Emerging trends in paper spray mass spectrometry: microsampling, storage, direct analysis, and applications. Mass Spectrom Rev 39:336–370. https://doi.org/10.1002/MAS.21601

Giovaninni G, Moore CJ, Hall AJ et al (2018) PH-Dependent silica nanoparticle dissolution and cargo release. Colloids Surf B Biointerfaces 169:242–248. https://doi.org/10.1016/J.COLSURFB.2018.04.064

Gomes IC, Martins RO, Machado LS et al (2022) Molecularly imprinted polymer as sorbent phase for disposable pipette extraction: a potential approach for creatinine analysis in human urine samples. J Pharm Biomed Anal 211:114625. https://doi.org/10.1016/J.JPBA.2022.114625

Grecco CF, de Souza ID, Queiroz MEC (2021) Novel materials as capillary coatings for in-tube solid-phase microextraction for bioanalysis. J Sep Sci 44:1662–1693. https://doi.org/10.1002/jssc.202001070

Guo L, Deng Q, Fang G et al (2011) Preparation and evaluation of molecularly imprinted ionic liquids polymer as sorbent for on-line solid-phase extraction of chlorsulfuron in environmental water samples. J Chromatogr A 1218:6271–6277. https://doi.org/10.1016/J.CHROMA.2011.07.016

Han DM, Fang GZ, Yan XP (2005) Preparation and evaluation of a molecularly imprinted sol–gel material for on-line solid-phase extraction coupled with high performance liquid chromatography for the determination of trace pentachlorophenol in water samples. J Chromatogr A 1100:131–136. https://doi.org/10.1016/J.CHROMA.2005.09.035

Han W, Zheng Y, Muyizere T, Zhang Z (2018) Development of paper substrate for paper spray MS in high-sensitivity analysis of biological samples. 10:2047–2062

Hiraoka K, Nishidate K, Mori K et al (2007) Development of probe electrospray using a solid needle. Rapid Commun Mass Spectrom 21:3139–3144. https://doi.org/10.1002/RCM.3201

Hoshina K, Horiyama S, Matsunaga H, Haginaka J (2011) Simultaneous determination of non-steroidal anti-inflammatory drugs in river water samples by liquid chromatography–tandem mass spectrometry using molecularly imprinted polymers as a pretreatment column. J Pharm Biomed Anal 55:916–922. https://doi.org/10.1016/J.JPBA.2011.03.014

Hu B, Yao ZP (2022) Electrospray ionization mass spectrometry with wooden tips: a review. Anal Chim Acta 1209:339136. https://doi.org/10.1016/J.ACA.2021.339136

Hu B, So PK, Chen H, Yao ZP (2011) Electrospray ionization using wooden tips. Anal Chem 83:8201–8207. https://doi.org/10.1021/AC2017713/SUPPL_FILE/AC2017713_SI_001.PDF

Hu B, So PK, Yang Y et al (2018) Surface-modified wooden-tip electrospray ionization mass spectrometry for enhanced detection of analytes in complex samples. Anal Chem 90:1759–1766. https://doi.org/10.1021/ACS.ANALCHEM.7B03675/SUPPL_FILE/AC7B03675_SI_001.PDF

Jalili V, Barkhordari A, Ghiasvand A (2020) A comprehensive look at solid-phase microextraction technique: a review of reviews. Microchem J 152:104319. https://doi.org/10.1016/j.microc.2019.104319

Kataoka H (2021) In-tube solid-phase microextraction: current trends and future perspectives. J Chromatogr A 1636:461787. https://doi.org/10.1016/j.chroma.2020.461787

Kataoka H, Saito K (2012) Recent advances in column switching sample preparation in bioanalysis 4:809–832. https://doi.org/10.4155/bio1228

Kefayati H, Yamini Y, Shamsayei M, Abdi S (2021) Molecularly imprinted polypyrrole@CuO nanocomposite as an in-tube solid-phase microextraction coating for selective extraction of carbamazepine from biological samples. J Pharm Biomed Anal 204:114256. https://doi.org/10.1016/j.jpba.2021.114256

Kotyrba A, Dinc M, Mizaikoff B (2021) Development of silica nanoparticle supported imprinted polymers for selective lysozyme recognition. Nanomaterials 11.https://doi.org/10.3390/NANO11123287

Kubo T, Kuroda K, Tominaga Y et al (2014) Effective determination of a pharmaceutical, sulpiride, in river water by online SPE-LC-MS using a molecularly imprinted polymer as a preconcentration medium. J Pharm Biomed Anal 89:111–117. https://doi.org/10.1016/J.JPBA.2013.10.040

Kuo TH, Dutkiewicz EP, Pei J, Hsu CC (2020) Ambient ionization mass spectrometry today and tomorrow: embracing challenges and opportunities. Anal Chem 92:2353–2363. https://doi.org/10.1021/ACS.ANALCHEM.9B05454/ASSET/IMAGES/MEDIUM/AC9B05454_0008.GIF

Lerma-García MJ, Zougagh M, Ríos A (2013) Magnetic molecular imprint-based extraction of sulfonylurea herbicides and their determination by capillary liquid chromatography. Microchim Acta 180:363–370. https://doi.org/10.1007/S00604-013-0942-6/TABLES/4

Li X, Wang Q, Jiang N et al (2023) Occurrence, source, ecological risk, and mitigation of phthalates (PAEs) in agricultural soils and the environment: a review. Environ Res 220:115196. https://doi.org/10.1016/J.ENVRES.2022.115196

Liu J, Wang H, Manicke NE et al (2010) Development, characterization, and application of paper spray ionization. Anal Chem 82:2463–2471. https://doi.org/10.1021/AC902854G

Liu Y, Yang Q, Chen X et al (2019) Sensitive analysis of trace macrolide antibiotics in complex food samples by ambient mass spectrometry with molecularly imprinted polymer-coated wooden tips. Talanta 204:238–247. https://doi.org/10.1016/J.TALANTA.2019.05.102

Lu YC, Mao JH, Zhang W et al (2020) A novel strategy for selective removal and rapid collection of triclosan from aquatic environment using magnetic molecularly imprinted nano−polymers. Chemosphere 238:124640. https://doi.org/10.1016/J.CHEMOSPHERE.2019.124640

Ma H, Yang M, Wang X et al (2021) Sulfonamide-selective ambient mass spectrometry ion source obtained by modification of an iron sheet with a hydrophilic molecularly imprinted polymer. J Agric Food Chem 69:15425–15433. https://doi.org/10.1021/ACS.JAFC.1C06623/SUPPL_FILE/JF1C06623_SI_001.PDF

Martins RO, de Araújo GL, de Freitas CS et al (2021) Miniaturized sample preparation techniques and ambient mass spectrometry as approaches for food residue analysis. J Chromatogr A 1640:461949. https://doi.org/10.1016/J.CHROMA.2021.461949

Martins RO, Bernardo RA, Machado LS et al (2023) Greener molecularly imprinted polymers: strategies and applications in separation and mass spectrometry methods. TrAC, Trends Anal Chem 168:117285. https://doi.org/10.1016/J.TRAC.2023.117285

Moliner-Martinez Y, Herráez-Hernández R, Verdú-Andrés J et al (2015) Recent advances of in-tube solid-phase microextraction. TrAC, Trends Anal Chem 71:205–213. https://doi.org/10.1016/j.trac.2015.02.020

Moliner-Martinez Y, Ballester-Caudet A, Verdú-Andrés J et al (2020) In-tube solid-phase microextraction. In: Solid-Phase Extraction. Elsevier, pp 387–427

Murtada K, Pawliszyn J (2023) Evolution in understandings of the design, optimization, and application of sorbent-based extraction devices. TrAC, Trends Anal Chem 169:117396. https://doi.org/10.1016/J.TRAC.2023.117396

Nguyen MK, Lin C, Nguyen HL et al (2023) Occurrence, fate, and potential risk of pharmaceutical pollutants in agriculture: challenges and environmentally friendly solutions. Sci Total Environ 899:165323. https://doi.org/10.1016/J.SCITOTENV.2023.165323

Pawliszyn J, Pawliszyn B, Pawliszyn M (1997) Solid phase microextraction (SPME). Chem Educ 2:1–7.https://doi.org/10.1007/s00897970137a

Pereira I, Rodrigues MF, Chaves AR, Vaz BG (2018) Molecularly imprinted polymer (MIP) membrane assisted direct spray ionization mass spectrometry for agrochemicals screening in foodstuffs. Talanta 178:507–514. https://doi.org/10.1016/J.TALANTA.2017.09.080

Ramalho RRF, Pereira I, da S. Lima G et al (2022) Fumonisin B1 analysis in maize by molecularly imprinted polymer paper spray ionization mass spectrometry (MIP-PSI-MS). J Food Compos Anal 107:104362. https://doi.org/10.1016/J.JFCA.2021.104362

Samadifar M, Yamini Y, Khataei MM, Shirani M (2023) Automated and semi-automated packed sorbent solid phase (micro) extraction methods for extraction of organic and inorganic pollutants. J Chromatogr A 1706:464227. https://doi.org/10.1016/J.CHROMA.2023.464227

Santos H, Martins RO, Soares DA, Chaves AR (2020) Molecularly imprinted polymers for miniaturized sample preparation techniques: strategies for chromatographic and mass spectrometry methods. Anal Methods 12:894–911. https://doi.org/10.1039/C9AY02227A

Sharma AK, Sharma M, Sharma AK, Sharma M (2023) Mapping the impact of environmental pollutants on human health and environment: a systematic review and meta-analysis. J Geochem Explor 255:107325. https://doi.org/10.1016/J.GEXPLO.2023.107325

Song X, Li X, Wang J, Huang X (2023) Adoption of new strategy for molecularly imprinted polymer based in-tube solid phase microextraction to improve specific recognition performance and extraction efficiency. Microchem J 194:109224. https://doi.org/10.1016/j.microc.2023.109224

Souza ID, Melo LP, Jardim ICSF et al (2016) Selective molecularly imprinted polymer combined with restricted access material for in-tube SPME/UHPLC-MS/MS of parabens in breast milk samples. Anal Chim Acta 932:49–59. https://doi.org/10.1016/j.aca.2016.05.027

Souza ID, Oliveira IGC, Queiroz MEC (2021) Innovative extraction materials for fiber-in-tube solid phase microextraction: a review. Anal Chim Acta 1165:238110. https://doi.org/10.1016/j.aca.2020.11.042

Suzaei FM, Daryanavard SM, Abdel-Rehim A et al (2022) Recent molecularly imprinted polymers applications in bioanalysis. Chem Papers 77:619–655

Thirukumaran M, Singh V, Arao Y et al (2021) Solid-phase microextraction- probe electrospray ionization devices for screening and quantitating drugs of abuse in small amounts of biofluids. Talanta 231:122317. https://doi.org/10.1016/J.TALANTA.2021.122317

Tian H, Liu T, Mu G et al (2020) Rapid and sensitive determination of trace fluoroquinolone antibiotics in milk by molecularly imprinted polymer-coated stainless steel sheet electrospray ionization mass spectrometry. Talanta 219:121282. https://doi.org/10.1016/J.TALANTA.2020.121282

Yamini Y, Rezazadeh M, Seidi S (2019) Liquid-phase microextraction—the different principles and configurations. TrAC, Trends Anal Chem 112:264–272. https://doi.org/10.1016/j.trac.2018.06.010

Yang Y, Deng J (2016) Analysis of pharmaceutical products and herbal medicines using ambient mass spectrometry. TrAC, Trends Anal Chem 82:68–88. https://doi.org/10.1016/J.TRAC.2016.04.011

Zare EN, Fallah Z, Le VT et al (2022) Remediation of pharmaceuticals from contaminated water by molecularly imprinted polymers: a review. Environ Chem Lett 20:4 20:2629–2664

Zarejousheghani M, Möder M, Borsdorf H (2013) A new strategy for synthesis of an in-tube molecularly imprinted polymer-solid phase microextraction device: selective off-line extraction of 4-nitrophenol as an example of priority pollutants from environmental water samples. Anal Chim Acta 798:48–55. https://doi.org/10.1016/j.aca.2013.08.038

Chapter 5
Molecularly Imprinting Technology and Narrow-Bandgap Semiconductors Application on Photoelectrochemical Sensors Construction

Felipe Augusto Gorla and César Ricardo Teixeira Tarley

5.1 Photoelectrochemical Sensors and Bismuth Vanadate

In recent years, semiconductor materials, especially photoactive ones, have received attention due to their attractive physical and chemical properties and their multiple application possibilities, such as in water treatment, substance degradation studies, organic syntheses, solar panels, electrodes for batteries, and photoelectrochemical (PEC) sensors (Malathi et al. 2018; Tan et al. 2017; Sun et al. 2020; Tang et al. 2020; Isai and Shrivastava 2019; Li et al. 2019a; Zhu and Zhou 2019; Mano et al. 2015; Ferdaous et al. 2019; Ye et al. 2020).

By definition, a semiconductor is a material that exhibits a measurable electrical conductivity that is directly proportional to the temperature increase. These materials have an energetic difference between their valence (VB) and conduction (CB) bands, referred to as the forbidden band. The energy corresponding to the width of this forbidden band is also known as the bandgap energy, and in semiconductors, its value can be considered low, being less than 4 eV (Baccaro and Gutz 2018). After excitation by some external source, which can be thermal, electrical, or luminous for

F. A. Gorla
Federal Institute of Parana, Campus Assis Chateaubriand, Assis Chateaubriand, Parana 85935-000, Brazil

F. A. Gorla · C. R. T. Tarley (✉)
Departament of Chemistry, State University of Londrina (UEL), Londrina, Parana 86057-970, Brazil
e-mail: tarley@uel.br

C. R. T. Tarley
Departament of Chemistry, Institute of Chemistry, National Institute of Science and Technology in Bioanalysis (INCTBio), State University of Campinas (UNICAMP), Campinas, Sao Paulo 13083-970, Brazil

S. Patra and M. Sillanpaa (eds.), *Molecularly Imprinted Polymers as Artificial Antibodies for the Environmental Health*, https://doi.org/10.1007/978-3-031-58995-9_5

photoactive semiconductors, electrons from the top of the valence band are promoted to the conduction band, forming an electron (e^-) in the conduction band and a hole (h^+) in the valence band, known as the electron–hole pair (e^-/h^+) (Tan et al. 2017). This electron–hole pair is responsible for the unique properties of semiconductors and for their widespread application in different areas (Baccaro and Gutz 2018).

Regarding photoelectrochemical sensors, they have proved to be a promising alternative for the development of electroanalytical methods in a wide range of areas (Wang et al. 2014a; Liu et al. 2018a; Alves et al. 2016). For the photoelectrochemical process to take place, it requires an electrode modified with or made up of a photoactive semiconductor, a light excitation source, and a chemical species capable of interacting with the system by receiving or supplying electrons and generating an electrical signal (photocurrent). Thus, the mechanism of a photoelectrochemical sensor (shown schematically in Fig. 5.1) is based on the oxidizing capacity of photogenerated holes (h^+) or the reducing capacity of photogenerated electrons (e^-) and the charge transfer between the analyte, the photoelectrochemically active material, and an electrode (Hun et al. 2017).

This interaction of the analyte with the photoactive material, either by its oxidation, supplying electrons to the gaps, or by its reduction, receiving the photogenerated electrons from the photoactive material, contributes to reducing the recombination rate of the e^-/h^+ pair and increasing the flow of charge in the system (Wang et al. 2009, 2014b).

As mentioned, the performance of the PEC sensor depends mainly on the bandgap energy of the semiconductor used in its preparation.

Several types of semiconductors are mentioned in the literature for the development of PEC sensors, such as tin dioxide (SnO_2), cadmium sulfide (CdS), copper indium selenide ($CuInSe_2$), and cadmium selenide (CdSe), among others (Wang et al. 2009; Grätzel 2001), but the most commonly reported are TiO_2 (Qin et al. 2018; Pang et al. 2017; Liu et al. 2018b; Wang et al. 2018a) and ZnO (Chen et al. 2017; Yang

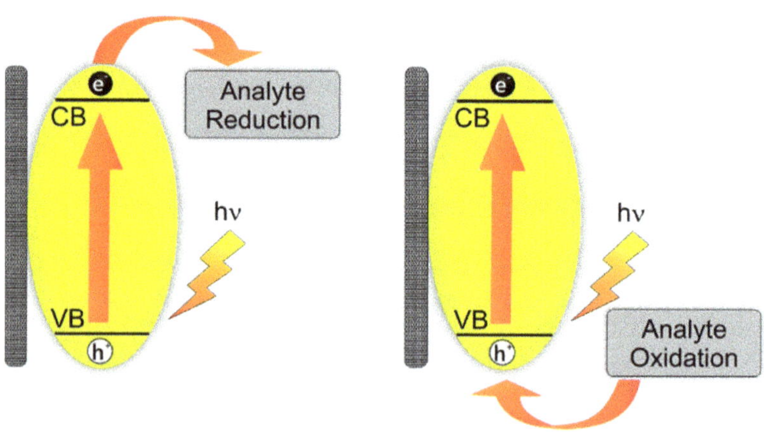

Fig. 5.1 Schematic representation of a PEC sensor working mechanism

et al. 2017a; Kong et al. 2018). These materials offer advantages such as low toxicity, relatively low cost, chemical inertness, and photostability (Wang et al. 2018b). While these materials have wide use, their main drawback is their dependence on high-energy photons for excitation, which are found in the ultraviolet region of the electromagnetic spectrum. This is because the bandgap value for both TiO_2 and ZnO is approximately 3.2 eV ($\approx$ 390 nm) (Yu et al. 2017).

Alternatively, bismuth vanadate ($BiVO_4$) is a highly promising n-type semiconductor for developing PEC sensors (Malathi et al. 2018; Tan et al. 2017; Araújo et al. 2021; Prado et al. 2018). It is advantageous due to its low cost, simple synthesis, stability in aqueous media, and photochemical stability. Moreover, $BiVO_4$ has high dispersion rates in various solvents and is non-toxic. Its photoactivity was first demonstrated by Kudo et al. (1998), wherein O_2 gas evolved from an $AgNO_3$ solution in the presence of $BiVO_4$ irradiated by visible light.

$BiVO_4$ naturally occurs in pucherite minerals, which possess an orthorhombic crystal structure. In laboratory, it can be synthesized in tetragonal and/or monoclinic phases. The photoactivity of $BiVO_4$ is directly related to its crystalline phase. $BiVO_4$ exhibits a bandgap of around 2.9 eV ($\approx$ 428 nm) in the tetragonal phase. Conversely, in the monoclinic phase, it shows photoactivity when excited with photons in the visible region, since its bandgap energy is approximately 2.4 eV ($\approx$ 516 nm) (Tan et al. 2017; Yu et al. 2019). Due to its lower bandgap energy and the ability to be excited with photons in the visible region, $BiVO_4$ in the monoclinic phase is considered the most photoactive and is the most reported in PEC sensors construction.

Monoclinic $BiVO_4$ can be synthesized through several strategies, including combustion synthesis (Afonso et al. 2014), solid-liquid synthesis (Kudo et al. 1999), microwave-assisted synthesis (Yan et al. 2015), precipitation synthesis (Suk Jang et al. 2012), aqueous solution synthesis (Xu et al. 2008), sol–gel synthesis (Liu et al. 2005), and hydrothermal synthesis (Lei et al. 2014). Among these strategies, hydrothermal synthesis has stood out because it makes it possible to work at relatively low temperatures with short reaction times (Lei et al. 2014). Furthermore, the manipulation of pH in hydrothermal synthesis enables the regulation of the morphology and crystalline structure of the acquired materials (Lei et al. 2014; Zhang et al. 2009).

Several recent studies have utilized $BiVO_4$ as a photoactive material in PEC sensors. In 2015, Ribeiro et al. developed a method using a $BiVO_4$-based PEC sensor for the determination of nitrite (Ribeiro et al. 2015). In this paper, $BiVO_4$ was synthesized in the monoclinic phase by depositing a mixture of $Bi(NO_3)_3.5H_2O$ and NH_4VO_3, previously dissolved in polyethylene glycol (PEG 300), on a fluorine-doped tin oxide (FTO)-modified glass substrate, followed by calcination at 500 °C for 1 h in a muffle furnace. The authors found that the NO_3^- anion interacts strongly with the sensor when exposed to visible light, acting as a collector of photogenerated holes even at low potentials and contributing to a significant reduction in the recombination rate of the electron–hole pair. The limit of detection obtained was 1.5 μmol L^{-1} and the method was applied to water samples.

In their work, Yu et al. synthesized $BiVO_4$ microrods in a hydrothermal environment (Yu et al. 2017). The semiconductor, obtained in the monoclinic phase, was suspended in water, deposited onto the surface of an FTO electrode, and used for the

H_2O_2 determination. The photogenerated holes in this material were highly effective in trapping electrons from H_2O_2, allowing this compound to be analyzed at a potential of 0 V relative to the Ag/AgCl (KCl 3.5 mol L^{-1}) reference electrode. They obtained a limit of detection of 8.5 μmol L^{-1}, and the PEC sensor was used to analyze H_2O_2 in milk samples. An important observation about this work is that the absence of interference in the H_2O_2 photocurrent signal by alanine, lysine, glucose, ascorbic acid, lactic acid, lactose, casein, iron (II) and (III), perborate, and percarbonate should be attributed to the 0 V potential used in the photoamperometric analyses and not to the possible specificity of $BiVO_4$ for H_2O_2.

Wang et al. developed a non-enzymatic PEC sensor to detect glucose in 2019 (Wang et al. 2019). The preparation of the nanoporous $BiVO_4$-based sensor involved electrochemical deposition of $Bi(NO_3)_3.5H_2O$ in the presence of KI onto an FTO electrode surface, creating a BiOI film. Thereafter, the reagent vanadyl acetylacetonate was added to the surface of the FTO electrode containing the electrodeposited BiOI film, which was heated in a muffle furnace at 450 °C for 2 h, thermally forming $BiVO_4$, as shown in the scheme in Fig. 5.2. The $BiVO_4$ was obtained in the monoclinic phase and had a bandgap energy of approximately 2.4 eV, consistent with the literature. The sensor exhibited catalytic activity, reproducibility, stability, and the absence of interference from other molecules. These properties allowed the determination of glucose in human serum samples with a limit of detection of 0.13 μmol L^{-1}.

Likewise, Li et al. (2019b) proposed a non-enzymatic photoelectrochemical sensor for ascorbic acid determination. The sensor was prepared by electrodepositing BiOI, followed by the formation of $BiVO_4$ through calcination in the presence of vanadyl acetylacetonate. The resulting $BiVO_4$ was in the clinobisvanite mineral form. The authors argue that ascorbic acid acted as a scavenger of photogenerated holes, reducing electron–hole pair recombination, increasing the photocurrent signal, and consequently enabling detection of ascorbic acid at concentrations as low as 1 μmol L^{-1}. After developing and validating the method in the laboratory, the authors suggested miniaturizing it by utilizing a 20 W LED lamp as the excitation

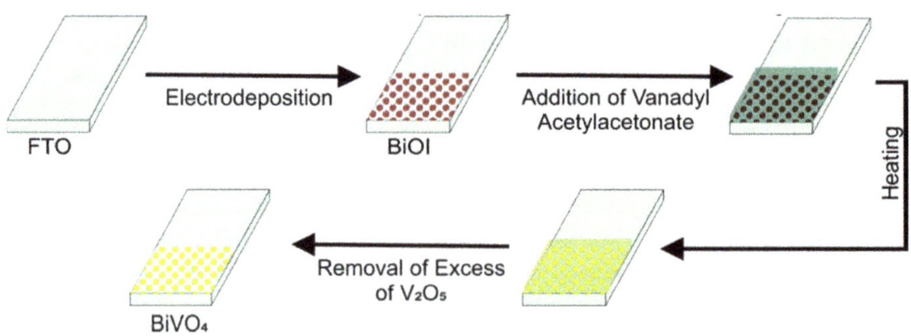

Fig. 5.2 Schematic representation of $BiVO_4$ synthesis preceded by BiOI electrodeposition. 2019 Adapted from Wang et al. ()

source and recording the signal with an ammeter in a portable device. This proposed miniaturization is an interesting aspect of the research.

In contrast to the other studies discussed, which utilized FTO-based electrodes, Araujo et al. (2021) developed a sensor constructed with a $BiVO_4$ film prepared in the monoclinic phase by the combustion synthesis strategy and deposited via drop-casting onto a glassy carbon electrode (GCE). The authors employed a commercial LED lamp as the visible light excitation source, and the analyte was paracetamol. In this study, it was observed that the holes (h^+) generated by the promotion of the electron (e^-) from the valence band of $BiVO_4$ to the conduction band interacted with paracetamol, promoting its oxidation and resulting in a significant increase in photocurrent and a decrease in the recombination rate of the e^-/h^+ pair. The reduction in recombination rate is attributable to the electrode's polarization through the applied potential, which leads the photogenerated electrons towards the electrochemical system and to the paracetamol oxidation promoted by the photogenerated holes in the $BiVO_4$. The limits of detection and quantification recorded were 27 and 91 nmol L^{-1}, respectively, and it was possible to determine paracetamol in pharmaceutical samples without matrix effect.

In addition to the utilization of pure $BiVO_4$ in PEC sensor construction, the literature reports on studies involving the incorporation of this semiconductor with other semiconductors, carbonaceous materials, or even its doping with other transition metals.

In this sense, Prado et al. (2019) conducted a study on the detection of dopamine using an FTO electrode coated with monoclinic $BiVO_4$ and graphene quantum dots (QDG). This study highlights the potential benefits of the association of QDG with $BiVO_4$ in the detection of DA since the results demonstrate that this association contributes to a signal increase of approximately 25% compared to electrodes coated with just $BiVO_4$. The authors attribute the increase in analytical signal to three factors: the increase in the electroactive area of the sensor, the QDG's electroconductive and electrocatalytic properties, and its photosensitizing effect on the $BiVO_4$. Moreover, utilizing QDG fostered greater charge transfer efficiency at the electrode, and the oxidation of DA provided electrons to the photogenerated holes, which decreased the recombination rate of the e^-/h^+ pair. In this study, limits of detection and quantification of 8.2 nmol L^{-1} and 27.3 nmol L^{-1} were obtained, respectively. The linear working range was from 36 to 250,000 nmol L^{-1}, and the method was applied to the determination of DA in urine samples.

In their paper, Okoth et al. (2017) reported a PEC biosensor for streptomycin. The sensor was constructed by modifying an FTO electrode through drop-casting with a suspension consisting of graphene and molybdenum-doped $BiVO_4$ (Mo-$BiVO_4$), which was synthesized in a hydrothermal environment. Afterward, the sensor containing Mo-$BiVO_4$ and graphene was further modified with an aptamer for streptomycin. The authors report that the association with graphene significantly increased the photocurrent signal by increasing the rate of charge transfer in the sensor. Additionally, graphene played a crucial role in immobilizing the streptomycin aptamer through intermolecular interactions via π–π stacking. The method developed with this sensor enabled the determination of streptomycin in samples of veterinary drugs, with a limit of detection of 0.0481 nmol L^{-1}.

The use of other carbonaceous materials was also reported by Prado et al. (2018). The authors proposed a PEC sensor to determine diclofenac in urine, in which $BiVO_4$ was associated with graphene oxide. This association with graphene oxide resulted in a fourfold increase in the photocurrent value compared to the sensor prepared with only $BiVO_4$, which was attributed to the conductive properties of the carbonaceous material and consequently to the increase in charge transfer in the sensor. The authors contend that the capture of electrons from the oxidation of diclofenac by photogenerated holes in $BiVO_4$ suppressed the recombination rate of the e^-/h^+ pair. The method was applied to determine the drug in urine samples without matrix interference.

Feng et al. (2020) investigated the heterojunction between molybdenum-doped porous $BiVO_4$, bismuth sulfide (Bi_2S_3), and an aptamer for the construction of a PEC biosensor for 17β-estradiol. The study determined that the heterojunction enhanced charge separation efficiency, promoted ultrafast electronic transfer, and inhibited recombination of e^-/h^+ pairs.

Considering the aforementioned, it can be affirmed that the use of $BiVO_4$ in the construction of photoelectrochemical sensors is still a field that warrants further exploration. It is evident that monoclinic $BiVO_4$ is the most widely reported for PEC sensor construction. Furthermore, it can be synthesized using various techniques and employed for PEC determination of several analytes. It should also be mentioned that one advantage of hydrothermal synthesis is that it produces $BiVO_4$ in the desired monoclinic phase, eliminating the need for material heat treatment. The association of $BiVO_4$ with other materials is deemed advantageous as it can augment the rate of electron transfer, decrease the e^-/h^+ pair recombination rate, and consequently increase the photocurrent. To provide selectivity and greater sensitivity to the analysis, chemical imprinting technology can be an interesting tool. This technology has been thoroughly described in the literature and is extensively adopted for sensor fabrication (Leibl et al. 2021; Ahmad et al. 2019; Yáñez-Sedeño et al. 2017; Tiwari and Prasad 2015; Blanco-López et al. 2004).

5.2 Chemical Imprinting Technology and Molecularly Imprinting Polymers (MIP)

Chemical imprinting technology plays a crucial role in the development of novel materials for the improvement of sensitivity and selectivity and is generally defined as the process of constructing selective recognition sites in polymeric materials. This is achieved through the use of a template to direct the structure formation of the polymer, and upon removal of the template, a selective cavity is created (Chen et al. 2016). This technology is based on the synthetic replication of lock-and-key biomolecular interactions, such as enzyme–substrate and/or antigen–antibody mechanisms (Tarley et al. 2005) thus producing mimetic materials with high specificity (Belbruno 2019).

Due to their unique properties, chemically imprinted materials have been applied in various fields of chemistry, including adsorption, artificial enzymes, catalysis, drug delivery, drug screening, immunoassays, organic synthesis, separation and sensing (Belbruno 2019; Roy et al. 2016; Pan et al. 2018).

It was Polyakov who introduced this technology in 1931 by reporting the synthesis of silica in the presence of additives, which were later removed by washing with hot water. In subsequent adsorption studies, the synthesized silica exhibited a high capacity for adsorbing the same additives, a characteristic initially described as a memory effect. Further investigations with this material revealed that the high adsorption capacity was due to structural changes in the silica from the presence of additives during synthesis (Alexander et al. 2006). A few years later (1940), Linus Pauling described the preparation of artificial antibodies using antigen molecules as a template (Pauling 1940). In 1949, Dickey (1949), inspired by Pauling and associating lock-and key and molecular imprinting mechanisms, synthesized silica in the presence of methyl, ethyl, n-propyl, and n-butyl orange. He studied the selectivity of this material as an adsorbent, which is considered the first publication describing chemically imprinted silica.

Since then, a variety of chemically imprinted materials have been synthesized using different strategies and applied in a wide range of areas, of which molecularly imprinted polymers (MIP), introduced by Wulff and Sarhan (1972), have been the most widely studied. The advantages of MIPs include simple synthesis, relative low cost, robustness, chemical stability, high reusability, and resistance to a broad pH range (Belbruno 2019; Roy et al. 2016; Cormack and Elorza 2004).

MIPs are commonly formed by the copolymerization, induced by a radical initiator, of a functional monomer, a cross-linking agent, and a target analyte used as a template, in the presence of a porogenic solvent (Tarley et al. 2005; Xie et al. 2016). The analyte interacts with the monomer, typically through covalent or non-covalent bonding (Pan et al. 2018). After the analyte is removed, a three-dimensional cavity is formed within the polymer matrix, maintaining its dimensions and exhibiting a high affinity for the analyte (Chen et al. 2011). During a rebinding process, the target analyte interacts with a selective cavity that is formed through non-covalent interactions, such as van der Waals forces, electrostatic interactions, π–π interactions, or hydrogen bonding (Chen et al. 2016). A schematic representation of the MIP synthesis stages is displayed in Fig. 5.3.

To verify whether chemical imprinting in MIP occurred satisfactorily, a non-imprinted polymer (NIP) is synthesized through the same route, but without the template molecule, and the experimental results obtained with MIP and NIP are compared (Belbruno 2019).

One of the key factors in MIP synthesis is choosing the appropriate functional monomer. The monomer should have a vinyl group in its structure for participation in the polymerization reaction, as well as a functional group capable of interacting with the template (Chen et al. 2016; Pan et al. 2018; Thakur et al. 2023). Thereby, the choice of functional monomer is determined by the functional groups present in the template. Figure 5.4 shows the structures of some functional monomers used in the synthesis of MIPs (Cormack and Elorza 2004).

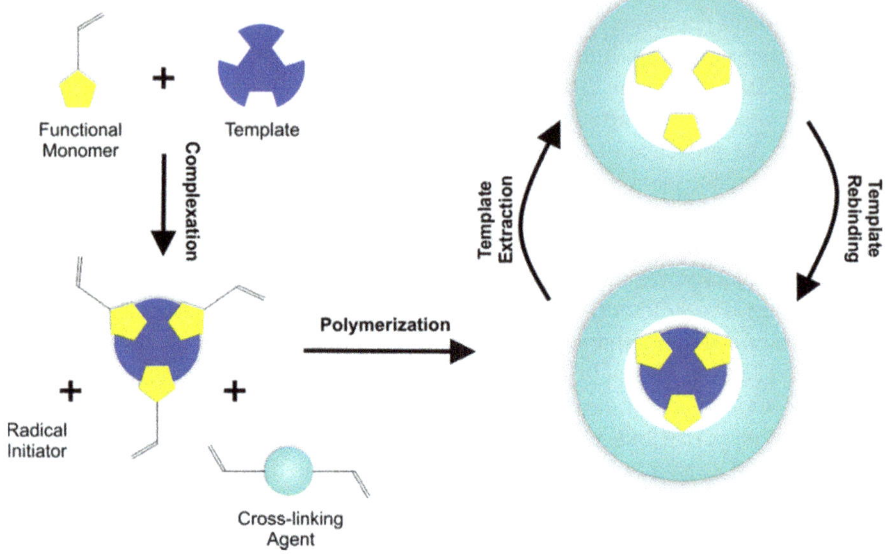

Fig. 5.3 Schematic representation of the formation of a MIP. 2005 Adapted from Tarley et al. ()

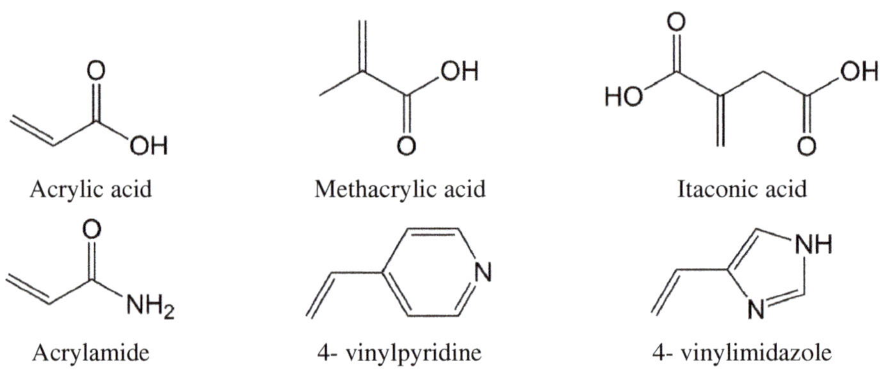

Fig. 5.4 Structure of functional monomers used in the synthesis of MIPs

The cross-linking agent is also essential for synthesizing MIPs. It provides mechanical stability to the material and chemical stability to the binding sites (Elugoke et al. 2021). Multifunctional molecules are commonly used as cross-linking agents in the synthesis of MIPs, with ethylene glycol dimethacrylate (EGDMA) being the most frequently reported (Thakur et al. 2023; Elugoke et al. 2021).

The porogenic solvent utilized in the synthesis of MIP must meet specific criteria. It should dissolve all the synthesis reagents and enable the formation of the monomer-template complex without interfering in the process. Therefore, non-polar solvents with a low dielectric constant are more appropriate when this interaction occurs through electrostatic forces or hydrogen bonds. Solvents with a higher dielectric constant may be used, but they can result in MIPs with reduced interaction between the binding site and the template during rebinding. Because the solvent can affect the formation and size of pores in the material, it is called porogenic (Thakur et al. 2023).

Of all the strategies for synthesizing a MIP, bulk synthesis, also known as polymerization in a homogeneous medium, is the simplest and most widely used. In this approach, all synthesis reagents are mixed in a reaction flask, which is sealed under an inert atmosphere, and polymerization is stimulated through heating, microwave incidence, or UV radiation. After synthesis, the polymeric material is obtained as a monolith and is usually macerated and sieved before use. However, a disadvantage of this approach is the formation of active sites trapped within the material, which are inaccessible to rebinding (Roy et al. 2016).

The bulk synthesis of MIP includes three stages: initiation, propagation, and termination, as schematically shown in Fig. 5.5.

During the initial phase, the radical initiator undergoes homolytic fission, giving rise to two radicals that are typically symmetrical and exceedingly reactive. These radicals then proceed to attack the double bond of the monomer, transferring the active center to it (Canevarolo 2006; Mano and Mendes 2004). In the succeeding propagation stage, which occurs rapidly and with low activation energy, the active center fashioned by the initiation process attacks molecules of the cross-linking agent or functional monomer, initiating a chain reaction that produces a polymer chain. The polymerization reaction concludes once a reactive end of the polymer chain encounters another active center (Canevarolo 2006; Mano and Mendes 2004).

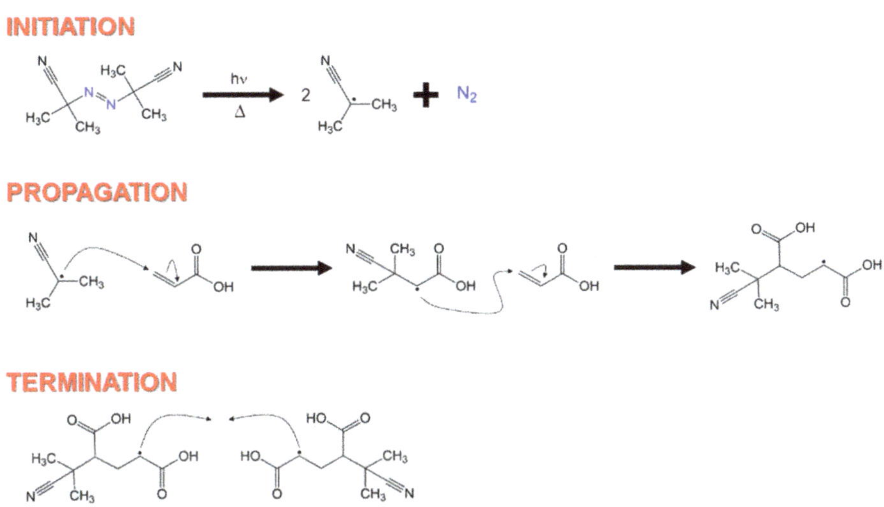

Fig. 5.5 Bulk polymerization steps with the monomer acrylic acid and the radical initiator 2,2′-azo-bis-isobutyronitrile (AIBN). 2006 Adapted from Canevarolo Jr ()

MIPs obtained through the bulk approach typically exhibit insulating properties (Prasad et al. 2010a; Sharma et al. 2019). To circumvent this drawback, MIPs can be synthesized via electropolymerization using conductive monomers like pyrrole or aniline. However, this strategy has the disadvantage of forming a polymer film that covers the electrode surface and, in a PEC sensor, also the photoactive semiconductor material, which can interfere with its photoactivity. MIPs can also be mixed in suspension with carbonaceous materials, including multiwalled carbon nanotubes (MWCNT), which possess significant conductive properties. In this instance, the ratio between the components must be strictly controlled so that excess carbonaceous material does not impair the interaction between the analyte and the MIP, or excess MIP does not interfere with the conductive properties of the carbonaceous material, and a suitable solvent must be used to disperse both the MIP and the carbonaceous material to form a stable suspension.

Another possibility to overcome the lack of conductivity of MIPs is based on the association of these materials with MWCNTs by means of strategies for the synthesis of the material chemically imprinted onto the surface of the carbonaceous material (surface imprinting), forming nanocomposite materials. This method provides the benefit of chemically attaching the MIP to the carbonaceous support.

It should be noted that the widespread use of MWCNTs as carbonaceous supports for the synthesis of MIPs by surface imprinting strategies is mainly based on their high surface area and electrical conductivity, as well as their mechanical and thermal resistance and electrocatalytic properties (Ferreira and Rangel 2009; Herbst et al. 2004; Wang and Arash 2014).

5.3 Surface Imprinting Strategies: Immobilization of Molecularly Imprinted Polymers onto MWCNTs

Two possible methods for immobilizing a MIP onto the surface of MWCNTs will be highlighted herein: grafting to approach and grafting from approach utilizing an INIFERTER reagent (Dai et al. 2015). In the grafting to approach, which is the most widely reported, vinyl or acrylate groups are immobilized onto the surface of the carbonaceous support by covalent bonding or physisorption. Subsequently, these groups are polymerized by the action of the radical initiator, together with the functional monomer and the cross-linking agent, which are dissolved within the solution (Dai et al. 2015). Since the groups react through free radical polymerization (FRP), radicals are also formed within the reaction mixture, similar to bulk polymerization. Therefore, the process of synthesis follows the same mechanism as bulk synthesis, including initiation, propagation, and termination (Canevarolo 2006), as shown in the representative diagram in Fig. 5.6. It should be noted that in the grafting to approach, the radical initiator attacks both the monomers and the immobilized vinyl or acrylate groups on the surface of the carbonaceous support simultaneously and randomly during propagation.

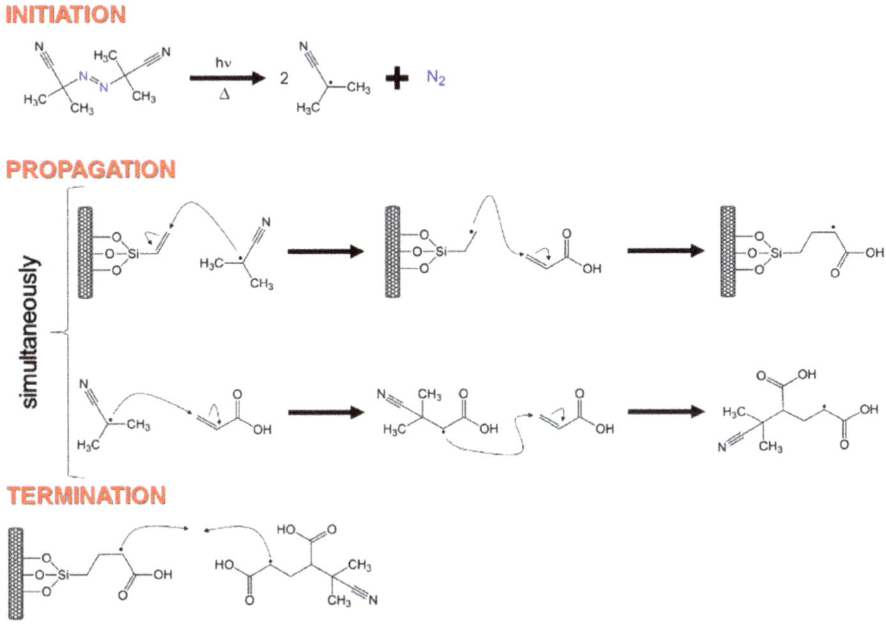

Fig. 5.6 Schematic representation of a free radical polymerization (FRP) reaction involving MWCNT modified with vinyltrimethoxysilane, the acrylic acid monomer, and the radical initiator AIBN. 2006 Adapted from Canevarolo Jr ()

In this situation, the main difference with bulk synthesis is the presence of MWCNTs, which participate in the polymerization through the vinyl or acrylate groups immobilized on their surface, and which can be covered or incorporated into the polymer matrix (Canevarolo 2006). The primary drawback of this approach is the inability to control the polymerization reaction, as parallel reactions can occur within the reaction mixture and not only on the surface of the MWCNTs (Augustine and Mathew 2014; Moretti et al. 2016; Lee et al. 2008; Lee and Kim 2009). For this reason, MIP nanocomposites synthesized using this strategy may have a low amount of carbonaceous material in relation to the polymer matrix.

In contrast, the grafting from approach involves covalent anchoring of the radical initiator to the surface of the carbonaceous support and in situ formation of the MIP by surface-controlled radical polymerization (SCRP). One of the possibilities for SCRP synthesis is the use of an INIFERTER reagent as a radical initiator (SCRP-INIFERTER). The INIFERTER term represents the initiation, transfer, and termination steps carried out concurrently by the initiator during polymerization. Dithiocarbamates constitute the major INIFERTER initiators that undergo decomposition under UV radiation, leading to the generation of two radicals, an active and a dormant one. Figure 5.7 provides a schematic representation of surface-controlled radical polymerization utilizing an INIFERTER reagent.

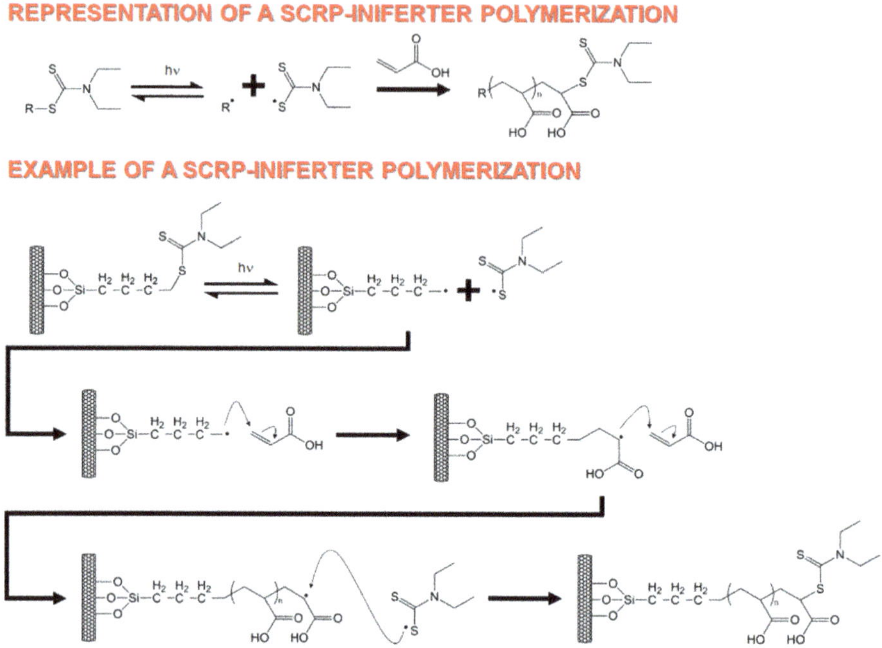

Fig. 5.7 Schematic representation of a PRC-INIFERTER polymerization reaction, where R represents functional groups and R$^{\bullet}$ the active radical. An illustration of SCRP polymerization, which entails MWCNTs modified with an INIFERTER reagent and the acrylic acid monomer. 2017 Adapted from Zoppe et al. ()

During polymerization by SCRP-INIFERTER, the active radical anchored to the carbonaceous support serves as the initiator of the reaction, while the dormant radical, remaining within the reaction mixture, controls the transfer and termination of the reaction (Mishra and Kumar 2012; Rückert et al. 2002). This polymerization mechanism leads to the formation of SCRP-INIFERTER nanocomposites with a polymer matrix onto the surface of the support but with minimal to no polymerization occurring within the reaction mixture. In this strategy, it is noteworthy that the power of the UV light source affects the speed of the polymerization reaction, while the reaction time controls the degree of polymerization (Hattori et al. 2004).

It should be noted that the INIFERTER synthesis strategy of MIPs onto MWCNTs preserves the morphological characteristics of MWCNTs and provides greater surface areas for this type of nanocomposite due to the thin polymer layer that forms. Furthermore, these materials have a higher ratio of imprinted sites per area compared to those synthesized via FRP (Prete et al. 2019).

Several studies have been reported in the literature investigating the use of MIP nanocomposites synthesized on the surface of MWCNT, both by FRP and SCRP-INIFERTER, as adsorbents (Lee and Kim 2009; Prete et al. 2019; Chi et al. 2012; Sajini et al. 2019) and in the construction of electrochemical sensors (Moretti et al. 2016; Capelari et al. 2021; Moretti et al. 2016; Patra et al. 2015; Marques et al.

2021; Prasad et al. 2010b; Zhang et al. 2012; Anirudhan and Alexander 2015; You et al. 2016). However, the use of these strategies in the synthesis of MIPs for the construction of PEC sensors is still underexplored. Regarding SCRP-INIFERTER, no studies that report the use of this strategy for the synthesis of materials for the construction of PEC sensors were found. Thus, this is an area with great potential to be explored.

5.4 Photoelectrochemical Sensors Prepared Out of Chemically Imprinted Materials

To overcome the known limitation in selectivity, various photoelectrochemical devices have been proposed using chemical imprinting technology, either in the synthesis of the imprinted photoactive semiconductor or in association with other chemically imprinted materials, such as molecularly imprinted polymers (Luo et al. 2022).

With regard to molecularly imprinted semiconductors, Luo et al. (2013) used chemical imprinting technology to synthesize a TiO_2/WO_3 nanocomposite with molecular imprinting for 2-nitrophenol and 4-nitrophenol using sol–gel technology. The chemical imprinting occurred within the TiO_2 fraction by means of the hydrolysis of the titanium(IV) butoxide monomer, followed by interaction with the template through hydrogen bonds and/or electrostatic interactions. After synthesis, the template was removed by calcination, resulting in a selective cavity in the nanocomposite, as shown in Fig. 5.8. The researchers observed that TiO_2/WO_3 showed selectivity for the studied template molecules and high adsorption capacity compared to the non-imprinted material, a feature attributed to the chemical interactions and size compatibility between the analytes and the selective cavities. Moreover, the photocatalytic activity of the molecularly imprinted TiO_2/WO_3 was twice that of the non-imprinted material.

Chen et al. (2019) and Yu et al. (2021) also investigated the interactions between the analyte and titanium(IV) butoxide functional monomer. Chen et al. (2019) developed an inorganic matrix using molecularly imprinted CdS/TiO_2 for Di-(2-ethylhexyl)-phthalate, while Yu et al. (2021) synthesized a nanocomposite of carbon quantum dots and molecularly imprinted TiO_2 for methylene blue.

Xu et al. (2014) developed a molecularly imprinted TiO_2 material anchored to the surface of silica-coated magnetic nanoparticles. Liquid-phase deposition was employed as the synthesis strategy, exploiting the semi-covalent approach. The semi-covalent approach discussed herein involves forming a covalent bond between the monomer and the template, which then results in the formation of selective sites after chemical or thermal cleavage capable of interacting with the template through non-covalent interactions in this second step. To synthesize the material depicted in Fig. 5.9, magnetite was initially prepared and covered with silica modified with

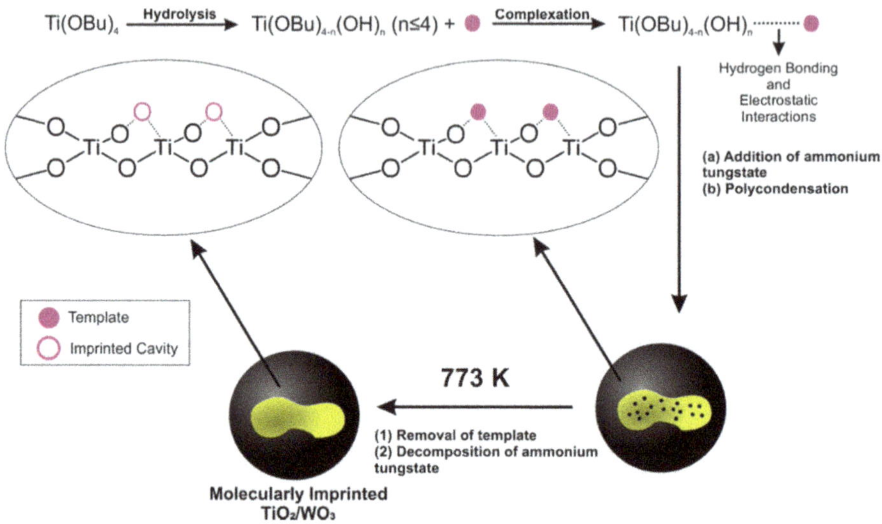

Fig. 5.8 Schematic representation of synthesis of molecularly imprinted TiO_2/WO_3 by Luo et al. 2013 Adapted from Luo et al. ()

3-isocyanatopropyl-trimethoxysilane to form the material named by the authors as $Fe_3O_4@SiO_2$.

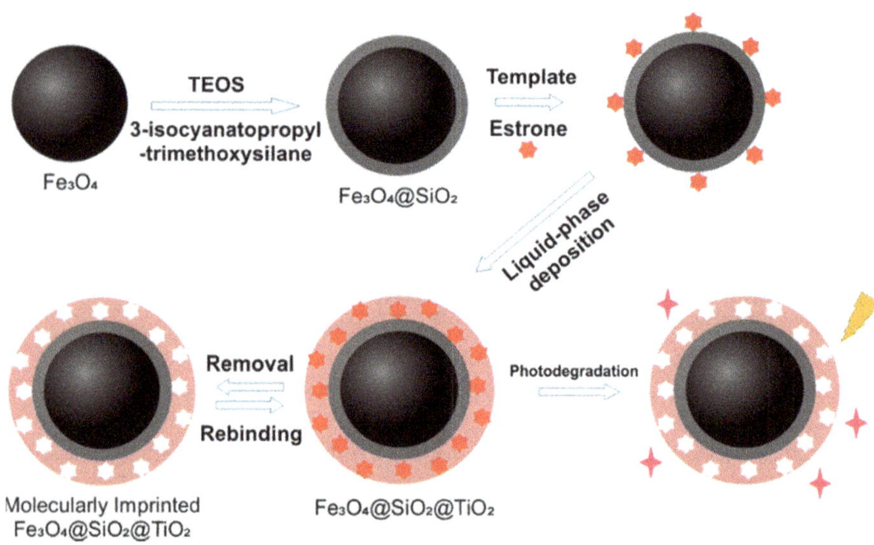

Fig. 5.9 Schematic representation of synthesis of molecularly imprinted $Fe_3O_4@SiO_2@TiO_2$ by Xu et al. 2014 Adapted from Xu et al. ()

Subsequently, estrone, used as the template, was immobilized on the surface of $Fe_3O_4@SiO_2$ through the interaction between the isocyanate groups and the phenolic fraction of estrone, forming a urethane-type covalent bond. Finally, a solution containing $Fe_3O_4@SiO_2$ with immobilized estrone was combined with ammonium hexafluorotitanate, boric acid, and TiO_2 nanoparticles (P25 in anatase form, used as a crystallization director) to undergo liquid-phase deposition of TiO_2, forming the imprinted $Fe_3O_4@SiO_2@TiO_2$. The material underwent calcination to cleave the urethane bond and remove the template. It should be emphasized that estrone was spatially imprinted on the TiO_2 matrix and that chemical interaction occurs between the analyte and $Fe_3O_4@SiO_2@TiO_2$ through hydrogen bonding between the phenolic –OH group of estrone and the nitrogen group that formed after thermal cleavage of the urethane bond. In addition, the authors report selective photodegradation of estrone, even in competitive interaction tests with high concentrations of phenol and bisphenol A.

It is important to highlight that the papers cited so far in this section do not use a carbonaceous phase in association with the photoactive material, which could improve the conductivity and photoactive activity of these semiconductors. Furthermore, chemical imprinting performed directly on the photoactive material reduces the chance of interactions between the monomers and template molecules, and the ratio of imprinted sites per area of the material is lower compared to, for example, a MIP nanocomposite synthesized using surface imprinting strategies onto a carbonaceous support.

Based on the promising results of using molecularly imprinted semiconductors, particularly TiO_2, for selective photodegradation assays, several studies have reported the use of this strategy in the construction of selective PEC sensors (Luo et al. 2022).

In this sense, Yang et al. (2020) proposed a molecularly imprinted TiO_2-based PEC sensor. To assemble the sensor, the researchers used graphene nanofibers decorated with TiO_2 nanoparticles as the support. Then, they employed the sol–gel method to coat the support with a layer of molecularly imprinted TiO_2 for urea, exploiting the interaction between this template molecule and the hydroxyl groups generated during the hydrolysis of the precursor titanium(IV) butoxide. This approach yielded a sensor with a high surface area and improved molecular recognition, which contributed to a high urea adsorption capacity, allowing the proposed method to achieve a limit of detection of 1 nmol L^{-1}.

Zhang et al. (2015a) synthesized molecularly imprinted TiO_2 for uric acid on the surface of Fe_3O_4 nanoparticles previously coated with carbon formed from glucose in a hydrothermal environment. This is illustrated in the scheme of Fig. 5.10.

The precursor utilized in this synthesis was also titanium(IV) butoxide, which underwent hydrolysis and interacted with uric acid molecules through hydrogen bonding during the sol–gel process. As the material had magnetic properties, the researchers were able to build a photoelectrochemical sensor using a magnetic glassy carbon electrode. With this PEC sensor, they developed a method that allowed the determination of uric acid in urine samples with a limit of detection of 20 nmol L^{-1}. Moreover, the sensor was selective for uric acid in competitive analysis with high concentrations of ascorbic acid, glutamic acid, cytosine, glucose, purine, and urea.

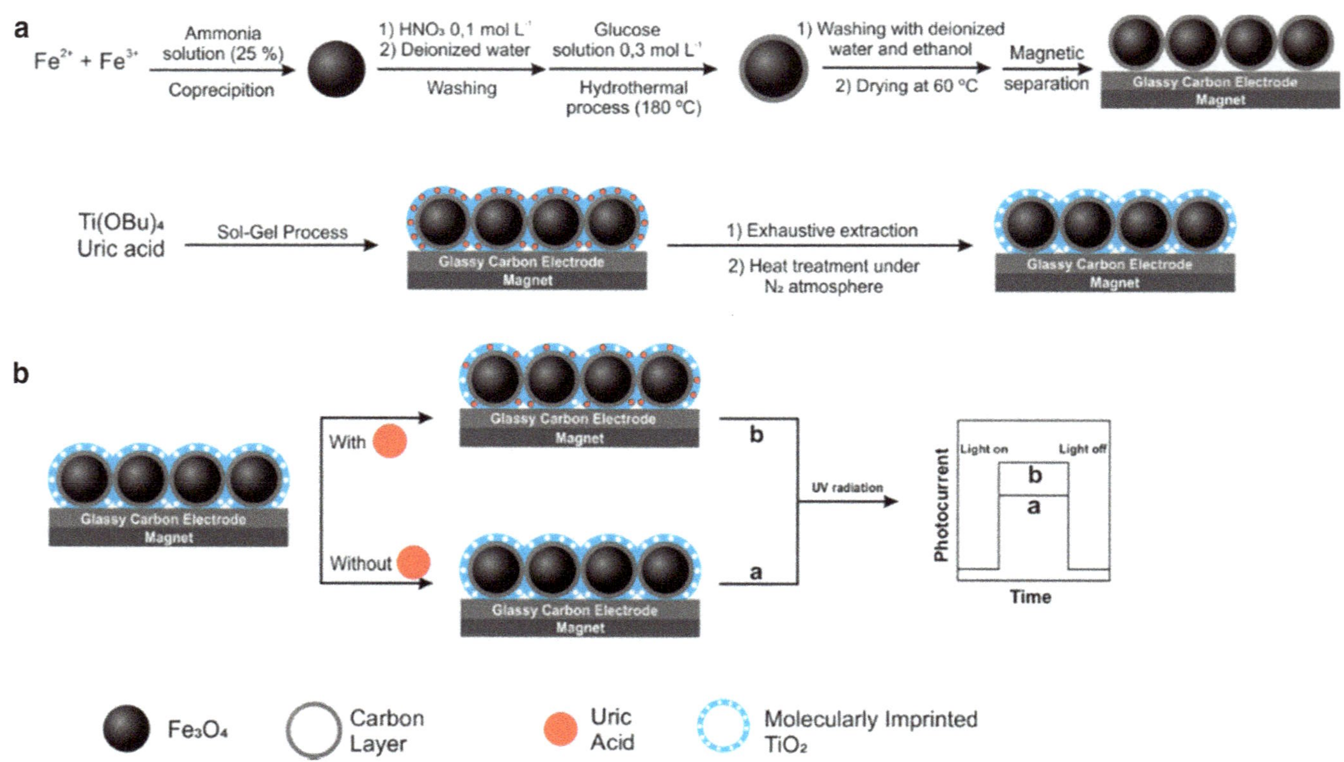

Fig. 5.10 Schematic representation of synthesis of molecularly imprinted TiO_2 modified with a carbon layer and a PEC sensor construction by Zhang et al. 2015a Adapted from Zhang et al. ()

Although the proposal by Zhang et al. is attractive regarding the simplicity of the sensor fabrication, since the magnetic particles coated with the molecularly imprinted TiO_2 easily adhere to the surface of a magnetic glassy carbon electrode, the use of chemical imprinting technology directly on the semiconductor material tends to produce materials with a low ratio of imprinted sites per area.

Other researchers have also reported the use of chemical imprinting technology for synthesizing molecularly imprinted TiO_2 for the development of PEC sensors for atrazine (Zhang et al. 2015b), microcystin-LR (Liu et al. 2017), and PCB 101 (2,2′,4,5,5′-pentachlorobiphenyl) (Shi et al. 2016, 2018).

The use of chemical imprinting technology in a different semiconductor was reported by Xu et al. (2022), who synthesized for the first time molecularly imprinted hydrated tungsten oxide ($WO_3.H_2O$) nanoplates for the herbicide 2,4-dichlorophenoxyacetic acid (2,4-D). The researchers found that synthesis in a hydrothermal environment at pH 3 resulted in a high interaction energy between 2,4-D, which is negatively charged in this environment, and W acting as a Lewis acid, making feasible the chemical imprinting of the $WO_3.H_2O$ matrix. Using this material, the researchers developed a photoelectrochemical sensor that detected 2,4-D within the range of 1 fmol L^{-1} to 10 nmol L^{-1} with a limit of detection of 1 fmol L^{-1}. In relation to this research, the electrostatic interaction between 2,4-D and W is a characteristic that enabled chemical imprinting, but it is also its main limitation. This imprinting strategy requires the template and the material to be imprinted to hold opposite charges at a determined pH value. Additionally, the template must have thermal stability, as chemical imprinting takes place in a hydrothermal environment, a fact that can limit the options of molecules used as templates.

Several studies in the literature have investigated chemical imprinting on semiconductors such as WO_3 and, in most research, TiO_2. The chemical imprinting of these materials is enabled by the interactions between the functional monomers and the template molecules, accomplished either through covalent bond formation or electrostatic interactions, hydrogen bonding or van der Walls interactions. Such interactions between synthesis precursors and template molecules are not observed for $BiVO_4$ (Martinez Suarez et al. 2015; Rodrigues et al. 2020). For this reason, there are no studies in the literature using this molecularly imprinted semiconductor for the construction of PEC sensors.

The association of photoactive materials with molecularly imprinted polymers is a promising and widely explored alternative to overcome this drawback. Furthermore, the association of MIPs with photoactive materials allows the use of different types of functional monomers with acidic, basic, or neutral properties, which considerably expands the possibilities of interaction between the monomer and possible template molecules.

In the meantime, Zhang et al. (2019) developed a PEC sensor based on the physical association between the photoactive material Bi_2S_3 and a MIP for the sensitive and selective determination of sulfadiazine in water samples, as shown in Fig. 5.11.

Bi_2S_3 was synthesized in a hydrothermal environment and exhibited an urchin-shaped morphology due to the stacking of cuboid nanorods, while MIP was synthesized in bulk, using methacrylic acid as the functional monomer, which can interact

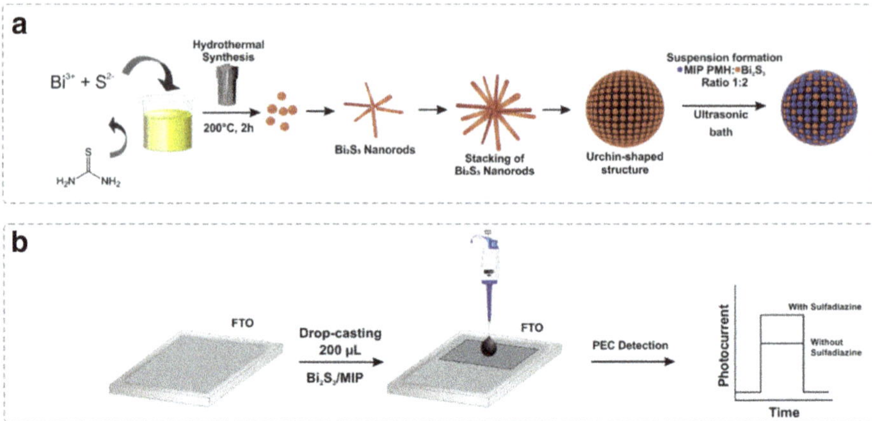

Fig. 5.11 Schematic representation of synthesis and construction of a PEC sensor by Zhang et al. 2019 Adapted from Zhang et al. ()

with the analyte via hydrogen bonding. To build the PEC sensor, a suspension containing 5 mg of MIP and 10 mg of Bi_2S_3 in 1 mL of ethylene glycol was drop-casted onto an FTO electrode. The employment of a MIP in association with Bi_2S_3 resulted in an augmentation of sulfadiazine molecule interactions with photogenerated holes in the photoactive material. This, in turn, contributed to an increase in the photocurrent signal and a decrease in the recombination rate of the e^-/h^+ pair. The physical mixture between the semiconductor and the chemically imprinted material is considered an advantage since it preserves the intrinsic properties of both the semiconductor and the MIP and allows control and optimization of the proportions of each component in the suspension.

Li et al. (2018) synthesized a MIP for perfluorooctane sulfonyl fluoride (PFOSF) using methacrylic acid as the functional monomer and the bulk approach. To construct the PEC sensor, a screen-printed carbon electrode was initially modified with a bismuth oxide iodide (BiOI) film by electrodeposition. Subsequently, the synthesized MIP was suspended in water and deposited by drop-casting onto the BiOI-modified screen-printed electrode surface. By using this association strategy, the researchers had to optimize the thickness of the BiOI film and the amount of MIP on the surface of the PEC sensor. During the photoelectrochemical measurements, ascorbic acid was used to supply electrons to the photogenerated holes in the BiOI, and the analyte was determined indirectly by reducing the photocurrent of the ascorbic acid when the PFOSF interacted with the active sites of the MIP. Using this sensor, the researchers were able to detect PFOSF in water samples without interference, achieving a limit of detection of 0.01 $\mu g\ L^{-1}$. However, it should be noted that the potential utilized in the photoamperometric assays by the researchers has not been disclosed, nor have they presented information regarding the stability and reusability of the PEC sensor.

A comparable approach to exploiting the association of a photoactive material with a MIP was reported by Zhang et al. (2020). The researchers synthesized bismuth oxybromide (BiOBr) microspheres using the hydrothermal method and a MIP for chloramphenicol via the bulk method by employing the monomer methacrylic acid, the cross-linking agent ethylene glycol dimethacrylate, and the radical initiator 2,2′-azo-bis-isobutyronitrile (AIBN). Two suspensions were prepared: one containing BiOBr microspheres and the other containing MIP. To build the PEC sensor, an indium tin oxide (ITO)-modified glass electrode was modified with a BiOBr film, followed by modification with a MIP film, both via drop-casting. The resulting PEC sensor was utilized for the selective determination of chloramphenicol in water samples.

It is worth noting that Li et al. (2018) and Zhang et al. (2020) reported an association strategy where the electrode previously modified with semiconductor material is coated with a film of an insulating MIP, which can increase the resistance to charge transfer of the sensor and decrease light incidence on the semiconductor material, negatively affecting its photoactive properties, depending on the thickness of the film. Furthermore, both studies employed MIP synthesized in bulk, a strategy that leads to the formation of polymers with fewer active sites per area in comparison to surface imprinting strategies on carbonaceous supports.

In their research, Zhang et al. (2021a) coated TiO_2 particles with a MIP for ofloxacin. The researchers prepared the MIL-125(Ti) precursor by hydrothermal method from 1,4-benzenedicarboxylic acid and titanium(IV) butoxide. Subsequently, MIL-125(Ti) was calcined to obtain TiO_2-C, which is TiO_2 functionalized with carboxyl groups. The TiO_2-C, methacrylic acid, ethylene glycol dimethacrylate (EGDMA), and radical initiator AIBN were added to a reaction flask, where polymerization took place under an inert atmosphere using the FRP approach, as schematically demonstrated in Fig. 5.12. This led to the formation of an imprinted polymer chain covering the TiO_2 surface (MIPs@TiO_2-C). The ofloxacin was able to interact with both the TiO_2-C carboxylic groups and the methacrylic acid groups during this process. The photoelectrochemical sensor prepared by drop-casting a suspension of the MIPs@TiO_2-C material onto an ITO electrode was used for the selective determination of ofloxacin with a limit of detection of 2.91 pg mL^{-1}.

Zhang et al. (2021b) also reported on the use of MIPs to coat a photoactive material. In this research, the authors synthesized BiOI microspheres using the hydrothermal method and then modified their surface with vinyl groups by anchoring the reagent 3-(trimethoxysilyl)propyl methacrylate. Afterwards, the BiOI modified with vinyl groups was introduced into a reaction flask along with methacrylic acid, EGDMA, AIBN, and norfloxacin as the template. The polymerization reaction was carried out using the FRP strategy. The PEC sensor was prepared by drop casting and used for the sensitive and selective determination of norfloxacin in water samples.

It is noted in these last two papers that FRP was employed to synthesize the polymer on the semiconductor surfaces. However, neither of the authors utilized a carbon phase to enhance the surface area or conductivity of MIP. Nevertheless, considering the MIP synthesis strategy was FRP, it suggests that there was no control over the polymerization reaction. Semiconductors coated with MIPs prepared using this strategy may exhibit reduced photoactivity as a consequence of reduced light

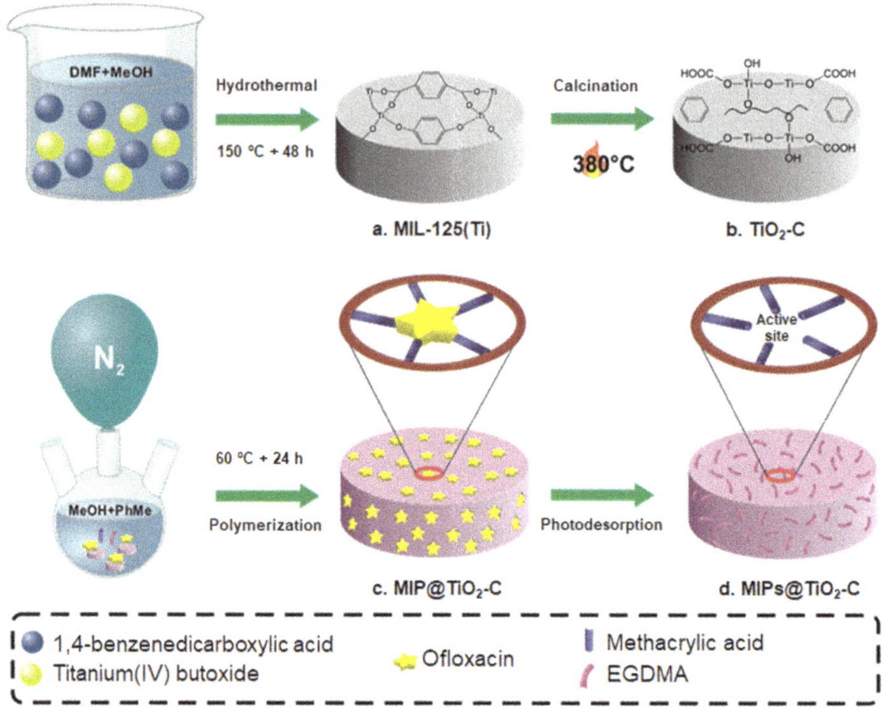

Fig. 5.12 Schematic representation of synthesis of MIPs@TiO_2-C by Zhang et al. 2021a Adapted from Zhang et al. ()

absorption owing to the absence of control over the thickness of the polymer layer on their surface.

On the other hand, Mao et al. used in situ polymerization to prepare MIPs for the construction of PEC sensors. In a first study, the researchers modified an ITO electrode with cadmium sulfide (CdS) quantum dots, graphene oxide, and chitosan (Mao et al. 2019). The solution containing the functional monomer, cross-linking agent, radical initiator, and fumonisin B1 (template) was dropped onto the modified ITO and then exposed to UV radiation, resulting in the direct formation of the MIP on the surface of the sensor. In another study, scientists synthesized an MIP for aflatoxin B1 and prepared a PEC sensor using the same in situ polymerization approach. The sensor was constructed on the surface of an ITO electrode previously coated with a nanocomposite, which was formed by the heterojunction of copper oxide and carbon nitride (Mao et al. 2021).

Gorla et al. recently reported for the first time the association of bismuth vanadate with a molecularly imprinted polymer synthesized onto the surface of MWCNTs (Gorla et al. 2024a). The synthesis strategy used by the authors was SCRP-INIFERTER. $BiVO_4$ was synthesized using the hydrothermal method at pH 8.53 to form the monoclinic phase. To synthesize MIP, the authors first oxidized

MWCNTs and then modified them with the INIFERTER reagent sodium diethyl dithiocarbamate. Acrylic acid was used as the functional monomer and trimethylopropane trimethacrylate (TRIM) as the cross-linking agent. The authors noted that the SCRP-INIFERTER synthesis resulted in the formation of a thin polymeric layer on the surface of the MWCNTs, which gave the MIP a high surface area, good electrical conductivity, and a high ratio of active sites per area. The photoelectrochemical sensor proposed in this study was prepared by physically mixing $BiVO_4$ with MIP-AA (GCE/$BiVO_4$/MIP-AA) was utilized to analyze dopamine in real and simulated urine samples. The dopamine determination with this PEC sensor occurred with a limit of detection of 2.9 nmol L^{-1} and was not affected by epinephrine, norepinephrine, ascorbic acid, citric acid, uric acid, creatine, urea, or other common urine components.

In another work, Gorla et al. proposed a photoelectrochemical sensor for dopamine using $BiVO_4$ combined with poly(acrylic acid-co-TRIM) supported on the surface of MWCNTs (Gorla et al. 2024b). To achieve this, the MWCNTs were oxidized and modified with vinyl groups, followed by polymerization using FRP with the acrylic acid monomer and TRIM cross-linking agent. In contrast to the SCRP-INIFERTER strategy, the authors observed that the MIP obtained by FRP had the polymer matrix incorporated into the MWCNTs, indicating that the polymerization reaction occurred both on the surface of the MWCNTs and within the reaction solution. The sensor prepared in this study (GCE/$BiVO_4$/MWCNT-MIP) achieved a limit of detection of 31.1 nmol L^{-1}, and successfully determined dopamine in simulated urine samples without interference.

Electropolymerization was the strategy used by Lu et al. (2013) to associate MIP and TiO_2 nanotubes in the construction of a PEC sensor. To achieve this, the researchers synthesized TiO_2 nanotubes on the surface of a titanium substrate using an anodic oxidation process. Subsequently, a molecularly imprinted polypyrrole film for bisphenol A was electropolymerized onto this modified substrate. Using the PEC sensor, the researchers eliminated passivation issues and proposed a method for the determination of bisphenol A with a limit of detection of 2.0 nmol L^{-1}. Furthermore, they were able to determine the target analyte without interference, even in the presence of other molecules at concentrations 100-fold higher.

Several PEC sensors have been reported in the literature that associate MIPs, electropolymerized utilizing the pyrrole monomer, along with photoactive semiconductors such as BiOI (Peng et al. 2014), CdS (Wang et al. 2014a), Cu_2O (Chen et al. 2018), SnO_2 (Zhang et al. 2015c), Sn_3O_4 (Wang et al. 2018c), TiO_2 (Shi et al. 2011; Kang et al. 2020; Zhang et al. 2016; Bai et al. 2022; Chen et al. 2012), ZnO (Viter et al. 2020; Zhao et al. 2021), as well as heterostructures such as $AgBiS_2/In_2S_3/Cd^{2+}$ (Xu et al. 2021), MoS_2/rGO/Au (Zhang et al. 2024), and Fe_3O_4@C@TiO_2 (Cai et al. 2020). There are also proposals for PEC sensors built by associating photoactive semiconductors with MIPs electropolymerized with 4-aminobenzenethiol (Yang et al. 2022), o-aminothiophenol (Zeng et al. 2023), p-aminothiophenol (Sun et al. 2017) and the most recently reported, o-phenylenediamine (Wang et al. 2013a, 2013b; Feng et al. 2018; Dashtian et al. 2021; Bai et al. 2020; Yang et al. 2017b, 2021, 2024;

Li et al. 2019c; Zhong et al. 2019; Song et al. 2021; Zhang et al. 2023; Yan et al. 2023; Pan et al. 2023).

It is noteworthy that both studies by Mao et al. detailing the methods for synthesizing MIPs in situ on electrode surfaces and the strategies for electropolymerizing MIPs discussed in the preceding paragraphs share the limitation of forming a polymer layer covering the semiconductor materials. In these cases, the MIP layer may hinder the transmission of the light utilized as an excitation source, resulting in reduced absorption by the semiconductor material and, consequently, diminished photoinduction of electron–hole pairs.

Considering the aforementioned, which papers are also summarized in Table 5.1 at the end of this section, the association of MIPs with different photoactive semiconductor materials constitutes a highly promising approach to developing PEC sensors, as these MIPs offer improved selectivity and sensitivity to the sensors. Of the presented strategies, physically mixing the semiconductor material and MIP in suspension is particularly advantageous. This strategy allows the control and optimization of the proportions between the components and preserves the properties of both the photoactive material and the MIP, which can also be synthesized on the surface of carbonaceous materials using the FRP and SCRP-INIFERTER strategies. Furthermore, after a comprehensive literature review, only two studies were found that reported the association between $BiVO_4$ and MIPs synthesized in bulk or through FRP or SCRP-INIFERTER techniques. This indicates that the development of PEC sensors that associate photoactive semiconductors with chemically imprinted materials is still a field with great potential to be researched.

5.5 Conclusion and Perspectives

Materials synthesized using molecular imprinting technology, particularly MIPs, biomimic the interaction between antigens and antibodies, resulting in improved sensitivity and selectivity. These versatile materials find widespread use in analytical chemistry. Among the many possible applications of molecularly imprinted materials, such as catalysis, controlled drug release, sensor construction, and separation, this review focused on a small and specific fraction: the construction of photoelectrochemical sensors using molecularly imprinted polymers and/or semiconductor materials synthesized by chemical imprinting technology.

Regarding the construction of photoelectrochemical sensors, it is quite clear from the literature that the most commonly used materials are TiO_2 and ZnO. The lack of selectivity of these materials led scientists to look for ways to overcome this drawback by developing chemically imprinted semiconductors or combining them with other molecularly imprinted materials. TiO_2 is particularly attractive and widely reported for synthesizing chemically imprinted semiconductors, due to its synthesis strategy. Other semiconductors with great potential for application in the construction of photoelectrochemical sensors, such as $BiVO_4$, WO_3, Bi_2S_3, BiOI, BiOBr, CdS, Cu_2O, SnO_2, Sn_3O_4, as well as heterostructures such as $AgBiS_2/CdS/In_2S_3$, MoS_2/

Table 5.1 Summary of photoelectrochemical sensors prepared using chemically imprinted materials

Photoelectrochemical sensor base	Chemically imprinted material	Method of synthesis of chemically imprinted material	Target molecule	LOD	References
Graphene nanofibers decorated with TiO_2	TiO_2	Sol–gel	Urea	1 nmol L^{-1}	Yang et al. (2020)
Fe_3O_4@C@MIT	TiO_2	Sol–gel	Uric acid	20 nmol L^{-1}	Zhang et al. (2015a)
MIT/CA	TiO_2	Sol–gel	Atrazine	8 nmol L^{-1}	Zhang et al. (2015b)
MI-TiO_2@CNTs	TiO_2	Sol–gel	Microcystin-LR	0.4 pmol L^{-1}	Liu et al. (2017)
Pd QDs@(SC,MI)-TiO_2 nanorods	TiO_2	Hydrothermal	PCB 101	0.05 pmol L^{-1}	Shi et al. (2018)
(SC, MI)-TiO_2 nanorods	TiO_2	Hydrothermal	PCB 101	0.01 pmol L^{-1}	Shi et al. 2016)
$WO_3.H_2O$ nanoplates	WO_3	Hydrothermal	2,4-D	1 fmol L^{-1}	Xu et al. (2022)
Bi_2S_3/MIP	Poly(methacrylic acid-co-EGDMA)	Bulk	Sulfadiazine	0.9 ng L^{-1}	Zhang et al. (2019)
BiOI/MIP	Poly(methacrylic acid-co-EGDMA)	Bulk	PFOSF	0.01 μg L^{-1}	Li et al. (2018)
BiOBr/MIP	Poly(methacrylic acid-co-EGDMA)	Bulk	Chloramphenicol	3.02 pg mL^{-1}	Zhang et al. (2020)
MIPs@TiO_2-C	Poly(methacrylic acid-co-EGDMA)	Free radical polymerization	Ofloxacin	2.91 pg mL^{-1}	Zhang et al. (2021a)
MIPs@BiOI	Poly(methacrylic acid-co-EGDMA)	Free radical polymerization	Norfloxacin	0.04 nmol L^{-1}	Zhang et al. (2021b)
MIP/ GO-CdS	Poly(methacrylic acid-co-EGDMA)	UV-initiated polymerization	Fumonisin B1	4.7 pg mL^{-1}	Mao et al. (2019)

(continued)

Table 5.1 (continued)

Photoelectrochemical sensor base	Chemically imprinted material	Method of synthesis of chemically imprinted material	Target molecule	LOD	References
CuO-g-C_3N_4/MIP	Poly(methacrylic acid-co-EGDMA)	UV-initiated polymerization	Aflatoxin B1	6.8 pg mL^{-1}	Mao et al. (2021)
$BiVO_4$/MIP-AA	Poly(acrylic acid)	SCRP-INIFERTER	Dopamine	2.9 nmol L^{-1}	Gorla et al. (2024a)
$BiVO_4$/MWCNT-MIP	Poly(acrylic acid-co-TRIM)	Free radical polymerization	Dopamine	31.1 nmol L^{-1}	Gorla et al. (2024b)
MIP PPy@TiO_2 Nanotubes	Polypyrrole	Electropolymerization	Bisphenol A	2.0 nmol L^{-1}	Lu et al. (2013)
MIP@BiOINFs	Polypyrrole	Electropolymerization	2,4-D	0.04 ng mL^{-1}	Peng et al. (2014)
PPy/Cu_2O	Polypyrrole	Electropolymerization	Microcystin-LR	0.23 ng L^{-1}	Chen et al. (2018)
SnO_2/MIP	Polypyrrole	Electropolymerization	Bisphenol A	1.2 nmol L^{-1}	Zhang et al. (2015c)
MIPPy/Sn_3O_4@CFP	Polypyrrole	Electropolymerization	2,4-D	10.8 pmol L^{-1}	Wang et al. (2018c)
PPy@TiO_2NTs	Polypirrole	Electropolymerization	2,4-D	10 nmol L^{-1}	Shi et al. (2011)
MIPs@GQDs/TiO_2 nanotubes	Polypirrole	Electropolymerization	Dopamine	0.018 μmol L^{-1}	Kang et al. (2020)
MIPPy/PTT	Polypirrole	Electropolymerization	Bilirubin	0.001 μmol L^{-1}	Zhang et al. (2016)
PPY/TiO_2 nanoarray	Polypirrole	Electropolymerization	Diclofenac	0.0034 μmol L^{-1}	Bai et al. (2022)

(continued)

Table 5.1 (continued)

Photoelectrochemical sensor base	Chemically imprinted material	Method of synthesis of chemically imprinted material	Target molecule	LOD	References
MIP@TiO_2 nanotubes	Polypirrole	Electropolymerization	Microcystin-LR	0.1 μg L^{-1}	Chen et al. (2012)
ZnO/MIP-Ppy	Polypirrole	Electropolymerization	Bisphenol S	0.7 μmol L^{-1}	Viter et al. (2020)
ZnO/rMIP	Polypirrole	Electropolymerization	Acrylamide	2.147 nmol L^{-1}	Zhao et al. (2021)
MIP-modified $AgBiS_2$//In_2S_3/Cd^{2+}	Polypirrole	Electropolymerization	Bisphenol A	0.18 nmol L^{-1}	Xu et al. (2021)
MoS_2/rGO/Au/rMIP	Polypirrole	Electropolymerization	Acrylamide	1.7 pmol L^{-1}	Zhang et al. (2024)
MI-PPy layer on Fe_3O_4@C@TiO_2	Polypirrole	Electropolymerization	Bisphenol A	0.03 μmol L^{-1}	Cai et al. (2020)
MIP/NiCo-LDHs/Cu_2O	Poly(4-aminobenzenethiol)	Electropolymerization	BSA	0.019 ng L^{-1}	Yang et al. (2022)
MIP/Au/TiO_2	Poly(o-aminothiophenol)	Electropolymerization	Dibutyl phthalate	0.698 nmol L^{-1}	Zeng et al. (2023)
MIP-B-TiO_2 nanorods	Poly(p-aminothiophenol)	Electropolymerization	Chlorpyrifos	7.4 pg mL^{-1}	Sun et al. (2017)
PoPD@TiO_2 nanotubes	Poly(o-phenylenediamine)	Electropolymerization	Lindane	0.03 μmol L^{-1}	Wang et al. (2013a)
PoPD–AuNPs/TiO_2 nanotubes	Poly(o-phenylenediamine)	Electropolymerization	Chlorpyrifos	0.96 nmol L^{-1}	Wang et al. (2013b)
MIP/g-C_3N_4-AuNPs	Poly(o-phenylenediamine)	Electropolymerization	Triclosan	0.601 pmol L^{-1}	Feng et al. (2018)

(continued)

Table 5.1 (continued)

Photoelectrochemical sensor base	Chemically imprinted material	Method of synthesis of chemically imprinted material	Target molecule	LOD	References
Ti/Cu_2O-CIS-DA-IPoPDA	Poly(o-phenylenediamine)	Electropolymerization	Dopamine	2.1 nmol L^{-1}	Dashtian et al. (2021)
ZnS–CdS/rMIP	Poly(o-phenylenediamine)	Electropolymerization	Oxytetracycline	0.10 nmol L^{-1}	Bai et al. (2020)
MIP/NH_2-MWCNTs-Pt–Pd NPs/ MoS_2-P-r-GO	Poly(o-phenylenediamine)	Electropolymerization	Thiamphenicol	0.5 nmol L_{-1}	Yang et al. (2017b)
MIP/Bi_2S_3	Poly(o-phenylenediamine)	Electropolymerization	Dioctyl phthalate	0.1 pmol L^{-1}	Li et al. (2019c)
MIP@AgINPs–TiO_2	Poly(o-phenylenediamine)	Electropolymerization	Benzoyl peroxide	0.253 pmol L^{-1}	Zhong et al. (2019)
MIP/NH_2-MIL-125(Ti)–TiO_2	Poly(o-phenylenediamine)	Electropolymerization	Oxtetracycline	60 pmol L^{-1}	Yang et al. (2021)
MIP/RGO/Ti-Fe-O nanotubes	Poly(o-phenylenediamine)	Electropolymerization	Microcystin-LR	10 pmol L^{-1}	Song et al. (2021)
ZnO-MWCNTs-NH_2/TCPP/ PoPD	Poly(o-phenylenediamine)	Electropolymerization	Chlortetracycline	0.17 nmol L^{-1}	Zhang et al. (2023)
TiO_2 nanorods/NiOOH/rMIP	Poly(o-phenylenediamine)	Electropolymerization	Ciprofloxacin	0.032 μmol L^{-1}	Yan et al. (2023)
rMIP/ZnO/C/	Poly(o-phenylenediamine)	Electropolymerization	Ciprofloxacin	8.41 pmol L^{-1}	Pan et al. (2023)
g-C_3N_4/Zr-MOFs(Fe)/MIPs	Poly(o-phenylenediamine)	Electropolymerization	Norfloxacin	5.7 nmol L^{-1}	Yang et al. (2024)

LOD Limit of detection; *LOQ* Limit of quantification; *NR* Not reported rGO/Au, and Fe_3O_4@C@TiO_2, are very attractive because they have numerous advantages. The main one is the lower bandgap energy when compared to TiO_2 or ZnO, which allows these semiconductors to be excited by photons in lower energy regions of the electromagnetic spectrum.

The chemical imprinting technology applied to the synthesis of molecularly imprinted semiconductors and semiconductor-associated molecularly imprinted polymers has enabled the development of several photoelectrochemical sensors for the determination of several analytes with high selectivity and sensitivity. The very low (or even absence of) interference of unwanted molecules and the low detection and quantification limit values observed in the papers are highlights of the use of this technology.

Particularly for $BiVO_4$, sunlight can be used to excite the e^-/h^+ pair since its band-gap energy is approximately 2.4 eV. The infeasibility of $BiVO_4$ molecular imprinting, as evidenced by the lack of published synthesis reports, is not considered a disadvantage but rather an analytical challenge for scientists.

Regarding the association of $BiVO_4$ with molecularly imprinted materials synthesized onto the surface of carbonaceous materials, this is one of the possible ways to circumvent this property. Although electropolymerization is the most recently reported strategy for constructing more sensitive and selective photoelectrochemical sensors, the association of semiconductors, particularly $BiVO_4$, with MIPs synthesized onto the surface of carbonaceous materials (using FRP or SCRP-INIFERTER strategies) is still a field with great potential to be explored.

Acknowledgements The authors acknowledge the financial support and fellowships of Coordenação de Aperfeiçoamento de Nível Superior (CAPES) (Project Pró-Forenses 3353/2014 Grant no. 23038.007082/2014-03), Finance Code 001, Conselho Nacional de Desenvolvimento Científico e Tecnológico (CNPq) (Grant No 307432/2017-3, 307505/2021-9, 420097/2021-0, 402387/2020-1, 401256/2020-0) and Instituto Nacional de Ciência e Tecnologia de Bioanalítica (INCT) (FAPESP Grant No 2014/50867-3 and CNPq Grant No 465389/2014-7).

References

Afonso R, Serafim JA, Lucilha AC et al (2014) Photoelectroactivity of bismuth vanadate prepared by combustion synthesis: effect of different fuels and surfactants. J Braz Chem Soc 25:726–733. https://doi.org/10.5935/0103-5053.20140026

Ahmad OS, Bedwell TS, Esen C et al (2019) Molecularly imprinted polymers in electrochemical and optical sensors. Trends Biotechnol 37:294–309. https://doi.org/10.1016/j.tibtech.2018.08.009

Alexander C, Andersson HS, Andersson LI et al (2006) Molecular imprinting science and technology: a survey of the literature for the years up to and including 2003. J Mol Recognit 19:106–180. https://doi.org/10.1002/jmr.760

Alves SA, Soares LL, Goulart LA, Mascaro LH (2016) Solvent effects on the photoelectrochemical properties of WO_3 and its application as dopamine sensor. J Solid State Electrochem 20:2461–2470. https://doi.org/10.1007/s10008-016-3144-4

Anirudhan TS, Alexander S (2015) Design and fabrication of molecularly imprinted polymer-based potentiometric sensor from the surface modified multiwalled carbon nanotube for the determination of lindane (γ-hexachlorocyclohexane), an organochlorine pesticide. Biosens Bioelectron 64:586–593. https://doi.org/10.1016/j.bios.2014.09.074

Araújo MS, Barretto TR, Galvão JCR et al (2021) Visible light photoelectrochemical sensor for acetaminophen determination using a glassy carbon electrode modified with $BiVO_4$ nanoparticles. Electroanalysis 33:663–671.https://doi.org/10.1002/elan.202060031

Augustine A, Mathew B (2014) Molecular imprinted specific sorbents based on multiwalled carbon nanotube for the detection of progesterone through chromatographic urine analysis Modification of MWCNTs. Int J Interdiscip Multidiscip Stud 1:140–148

Baccaro ALB, Gutz IGR (2018) Photoelectrocatalysis on semiconductors: from the fundamentals to its conformation at the nanoscale level. Quim Nova 41:326–339. https://doi.org/10.21577/0100-4042.20170174

Bai X, Zhang Y, Gao W et al (2020) Hollow ZnS–CdS nanocage based photoelectrochemical sensor combined with molecularly imprinting technology for sensitive detection of oxytetracycline. Biosens Bioelectron 168:112522. https://doi.org/10.1016/j.bios.2020.112522

Bai X, Gao W, Zhou C et al (2022) Photoelectrochemical determination of diclofenac using oriented single-crystalline TiO_2 nanoarray modified with molecularly imprinted polypyrrole. Microchim Acta 189:90. https://doi.org/10.1007/s00604-022-05206-8

Belbruno JJ (2019) Molecularly imprinted polymers. Chem Rev 119:94–119. https://doi.org/10.1021/acs.chemrev.8b00171

Blanco-López MC, Lobo-Castañón MJ, Miranda-Ordieres AJ, Tuñón-Blanco P (2004) Electrochemical sensors based on molecularly imprinted polymers. Trends Analyt Chem 23:36–48. https://doi.org/10.1016/S0165-9936(04)00102-5

Cai G, Yu Z, Tang D (2020) Actuating photoelectrochemical sensing sensitivity coupling core-core-shell Fe_3O_4@C@TiO_2 with molecularly imprinted polypyrrole. Talanta 219:121341. https://doi.org/10.1016/j.talanta.2020.121341

Canevarolo Jr SV (2006) Síntese de Polímeros. In: Ciência dos Polímeros: Um texto básico para tecnólogos e engenheiros, 2ª. Artliber, pp 110–111

Capelari TB, de Cássia MJ, da Rocha LR et al (2021) Synthesis of novel poly(methacrylic acid)/β-cyclodextrin dual grafted MWCNT-based nanocomposite and its use as electrochemical sensing platform for highly selective determination of cocaine. J Electroanal Chem 880:114791. https://doi.org/10.1016/j.jelechem.2020.114791

Chen L, Xu S, Li J (2011) Recent advances in molecular imprinting technology: current status, challenges and highlighted applications. Chem Soc Rev 40:2922–2942. https://doi.org/10.1039/c0cs00084a

Chen K, Liu M, Zhao G et al (2012) Fabrication of a novel and simple microcystin-LR photoelectrochemical sensor with high sensitivity and selectivity. Environ Sci Technol 46:11955–11961. https://doi.org/10.1021/es302327w

Chen L, Wang X, Lu W et al (2016) Molecular imprinting: perspectives and applications. Chem Soc Rev 45:2137–2211. https://doi.org/10.1039/c6cs00061d

Chen L, Cui J, Sheng X et al (2017) High-performance photoelectronic sensor using mesostructured ZnO nanowires. ACS Sens 2:55. https://doi.org/10.1021/acssensors.7b00477

Chen J, Gao P, Wang H et al (2018) A PPy/Cu_2O molecularly imprinted composite film-based visible light-responsive photoelectrochemical sensor for microcystin-LR. J Mater Chem C Mater 6:3937–3944. https://doi.org/10.1039/c7tc05743a

Chen F, Liu Y, Zhang X et al (2019) Inorganic-framework molecularly imprinted CdS/TiO_2 for selectively photocatalytic degradation of Di(2-ethylhexyl)phthalate. J Chem Soc Pak 41:308–318. https://doi.org/10.52568/000737/jcsp/41.02.2019

Chi W, Shi H, Shi W et al (2012) 4-Nitrophenol surface molecularly imprinted polymers based on multiwalled carbon nanotubes for the elimination of paraoxon pollution. J Hazard Mater 227–228:243–249. https://doi.org/10.1016/j.jhazmat.2012.05.041

Cormack PAG, Elorza AZ (2004) Molecularly imprinted polymers: synthesis and characterisation. J Chromatogr B Analyt Technol Biomed Life Sci 804:173–182. https://doi.org/10.1016/j.jchromb.2004.02.013

Dai H, Xiao D, He H et al (2015) Synthesis and analytical applications of molecularly imprinted polymers on the surface of carbon nanotubes: a review. Microchim Acta 182:893–908. https://doi.org/10.1007/s00604-014-1376-5

Dashtian K, Hajati S, Ghaedi M (2021) Ti-based solid-state imprinted-$Cu_2O/CuInSe_2$ heterojunction photoelectrochemical platform for highly selective dopamine monitoring. Sens Actuators B Chem 326:128824. https://doi.org/10.1016/j.snb.2020.128824

Dickey FH (1949) The preparation of specific adsorbents. Proc Natl Acad Sci U S A 35:227–229. https://doi.org/10.1073/pnas.35.5.227

Elugoke SE, Adekunle AS, Fayemi OE et al (2021) Molecularly imprinted polymers (MIPs) based electrochemical sensors for the determination of catecholamine neurotransmitters—review. Electrochem Sci Adv 1:1–43. https://doi.org/10.1002/elsa.202000026

Feng S, Wei X, Zhong L, Li J (2018) A novel molecularly imprinted photoelectrochemical sensor based on g-C_3N_4-AuNPs for the highly sensitive and selective detection of triclosan. Electroanalysis 30:320–327. https://doi.org/10.1002/elan.201700514

Feng J, Li F, Qian Y et al (2020) Mo-doped porous $BiVO_4/Bi_2S_3$ nanoarray to enhance photoelectrochemical efficiency for quantitative detection of 17β-estradiol. Sens Actuators B Chem 305:127443. https://doi.org/10.1016/j.snb.2019.127443

Ferdaous MT, Shahahmadi SA, Chelvanathan P et al (2019) Elucidating the role of interfacial MoS_2 layer in Cu_2ZnSnS_4 thin film solar cells by numerical analysis. Sol Energy 178:162–172. https://doi.org/10.1016/j.solener.2018.11.055

Ferreira HS, Rangel MDC (2009) Nanotecnologia: Aspectos gerais e potencial de aplicação em catálise. Quim Nova 32:1860–1870. https://doi.org/10.1590/S0100-40422009000700033

Gorla FA, Santos CS, Matos R et al (2024a) Development of novel nanocomposite-modified photoelectrochemical sensor based on the association of bismuth vanadate and MWCNT-grafted-molecularly imprinted poly(acrylic acid) for dopamine determination at nanomolar level. Talanta 266:125044. https://doi.org/10.1016/j.talanta.2023.125044

Gorla FA, Ferreira MP, Santos CS et al (2024b) Highly sensitive and selective dopamine determination using a photoelectrochemical sensor based on the $BiVO_4$ and surface molecularly imprinted poly(acrylic acid-co-TRIM) on vinyl functionalized carbon nanotubes. J Electroanal Chem 953:117988. https://doi.org/10.1016/j.jelechem.2023.117988

Grätzel M (2001) Photoelectrochemical cells. Nature 414:338–344

Hattori K, Hiwatari M, Iiyama C et al (2004) Gate effect of theophylline-imprinted polymers grafted to the cellulose by living radical polymerization. J Memb Sci 233:169–173. https://doi.org/10.1016/j.memsci.2003.12.013

Herbst MH, Macêdo MIF, Rocco AM (2004) Tecnologia dos nanotubos de carbono: Tendências e perspectivas de uma área multidisciplinar. Quim Nova 27:986–992. https://doi.org/10.1590/S0100-40422004000600025

Hun X, Wang S, Wang S et al (2017) A photoelectrochemical sensor for ultrasensitive dopamine detection based on single-layer NanoMoS_2 modified gold electrode. Sens Actuators B Chem 249:83–89. https://doi.org/10.1016/j.snb.2017.04.065

Isai KA, Shrivastava VS (2019) Photocatalytic degradation of methylene blue using ZnO and 2%Fe–ZnO semiconductor nanomaterials synthesized by sol–gel method: a comparative study. SN Appl Sci 1:1–11. https://doi.org/10.1007/s42452-019-1279-5

Kang Q, Zhang Q, Zang L et al (2020) Enhancement anti-interference ability of photoelectrochemical sensor via differential molecularly imprinting technique demonstrated by dopamine determination. Anal Chim Acta 1125:201–209. https://doi.org/10.1016/j.aca.2020.05.063

Kong Q, Wang Y, Zhang L et al (2018) Highly sensitive microfluidic paper-based photoelectrochemical sensing platform based on reversible photo-oxidation products and morphology-preferable multi-plate ZnO nanoflowers. Biosens Bioelectron 110:58–64. https://doi.org/10.1016/J.BIOS.2018.03.050

Kudo A, Ueda K, Kato H, Mikami I (1998) Photocatalytic O_2 evolution under visible light irradiation on $BiVO_4$ in aqueous $AgNO_3$ solution. Catal Letters 53:229–230. https://doi.org/10.1023/a:1019034728816

Kudo A, Omori K, Kato H (1999) A novel aqueous process for preparation of crystal form-controlled and highly crystalline $BiVO_4$ powder from layered vanadates at room temperature and its photocatalytic and photophysical properties. J Am Chem Soc 121:11459–11467. https://doi.org/10.1021/ja992541y

Lee E, Park DW, Lee JO et al (2008) Molecularly imprinted polymers immobilized on carbon nanotube. Colloids Surf A Physicochem Eng Asp 313–314:202–206. https://doi.org/10.1016/j.colsurfa.2007.04.093

Lee HY, Kim B (2009) Grafting of molecularly imprinted polymers on iniferter-modified carbon nanotube. Biosens Bioelectron 25:587–591.https://doi.org/10.1016/j.bios.2009.03.040

Lei BX, Zeng LL, Zhang P et al (2014) Hydrothermal synthesis and photocatalytic properties of visible-light induced $BiVO_4$ with different morphologies. Adv Powder Technol 25:946–951. https://doi.org/10.1016/j.apt.2014.01.014

Leibl N, Haupt K, Gonzato C, Duma L (2021) Molecularly imprinted polymers for chemical sensing: a tutorial review. Chemosensors 9:123. https://doi.org/10.3390/chemosensors9060123

Li X, Wang X, Fang T et al (2018) Disposable photoelectrochemical sensing strip for highly sensitive determination of perfluorooctane sulfonyl fluoride on functionalized screen-printed carbon electrode. Talanta 181:147–153. https://doi.org/10.1016/j.talanta.2018.01.005

Li Q, Zhao T, Li M et al (2019a) One-step construction of Pickering emulsion via commercial TiO_2 nanoparticles for photocatalytic dye degradation. Appl Catal B 249:1–8. https://doi.org/10.1016/j.apcatb.2019.02.057

Li L, Li M, Liu H et al (2019b) A portable non-enzyme photoelectrochemical ascorbic acid sensor based on $BiVO_4$ electrode under 20 W LED light. J Electroanal Chem 855:113573. https://doi.org/10.1016/j.jelechem.2019.113573

Li X, Zhong L, Liu R et al (2019c) A molecularly imprinted photoelectrochemical sensor based on the use of Bi_2S_3 for sensitive determination of dioctyl phthalate. Microchim Acta 186:688. https://doi.org/10.1007/s00604-019-3812-z

Liu H, Nakamura R, Nakato Y (2005) Promoted photo-oxidation reactivity of particulate $BiVO_4$ photocatalyst prepared by a photoassisted sol-gel method. J Electrochem Soc 152:G856. https://doi.org/10.1149/1.2051868

Liu M, Ding X, Yang Q et al (2017) A pM leveled photoelectrochemical sensor for microcystin-LR based on surface molecularly imprinted TiO_2@CNTs nanostructure. J Hazard Mater 331:309–320. https://doi.org/10.1016/j.jhazmat.2017.02.031

Liu Q, Yin Y, Hao N et al (2018a) Nitrogen functionlized graphene quantum dots/3D bismuth oxyiodine hybrid hollow microspheres as remarkable photoelectrode for photoelectrochemical sensing of chlopyrifos. Sens Actuators B Chem 260:1034–1042. https://doi.org/10.1016/j.snb.2018.01.119

Liu XP, Chen JS, Mao jie C et al (2018b) Enhanced photoelectrochemical DNA sensor based on TiO_2/Au hybrid structure. Biosens Bioelectron 116:23–29. https://doi.org/10.1016/j.bios.2018.05.036

Lu B, Liu M, Shi H et al (2013) A Novel photoelectrochemical sensor for bisphenol a with high sensitivity and selectivity based on surface molecularly imprinted polypyrrole modified TiO_2 nanotubes. Electroanalysis 25:771–779. https://doi.org/10.1002/elan.201200585

Luo X, Deng F, Min L et al (2013) Facile one-step synthesis of inorganic-framework molecularly imprinted TiO_2/WO_3 nanocomposite and its molecular recognitive photocatalytic degradation of target contaminant. Environ Sci Technol 47:7404–7412. https://doi.org/10.1021/es4013596

Luo Y, Feng X, Chen Z, Shen X (2022) Molecularly imprinted photocatalysts: fabrication, application and challenges. Mater Adv 8830–8847.https://doi.org/10.1039/d2ma00848c

Malathi A, Madhavan J, Ashokkumar M, Arunachalam P (2018) A review on $BiVO_4$ photocatalyst: activity enhancement methods for solar photocatalytic applications. Appl Catal A Gen 555:47–74. https://doi.org/10.1016/j.apcata.2018.02.010

Mano EB, Mendes LC (2004) Introdução a polímeros, 2nd edn. rev. Edgard Blücher, São Paulo
Mano T, Nishimoto S, Kameshima Y, Miyake M (2015) Water treatment efficacy of various metal oxide semiconductors for photocatalytic ozonation under UV and visible light irradiation. Chem Eng J 264:221–229. https://doi.org/10.1016/j.cej.2014.11.088
Mao L, Ji K, Yao L et al (2019) Molecularly imprinted photoelectrochemical sensor for fumonisin B1 based on GO-CdS heterojunction. Biosens Bioelectron 127:57–63. https://doi.org/10.1016/j.bios.2018.11.040
Mao L, Xue X, Xu X et al (2021) Heterostructured CuO-g-C_3N_4 nanocomposites as a highly efficient photocathode for photoelectrochemical aflatoxin B1 sensing. Sens Actuators B Chem 329:129146. https://doi.org/10.1016/j.snb.2020.129146
Marques GL, Rocha LR, Prete MC et al (2021) Development of electrochemical platform based on molecularly imprinted poly(methacrylic acid) grafted on iniferter-modified carbon nanotubes for 17β-estradiol determination in water samples. Electroanalysis 33:568–578. https://doi.org/10.1002/elan.202060270
Martinez Suarez C, Hernández S, Russo N (2015) $BiVO_4$ as photocatalyst for solar fuels production through water splitting: a short review. Appl Catal A Gen 504:158–170. https://doi.org/10.1016/j.apcata.2014.11.044
Mishra V, Kumar R (2012) Living radical polymerization: a review. J Sci Res 56:141–176
Moretti EDS, De Fátima GJ, Kuceki M et al (2016) A nanocomposite based on multi-walled carbon nanotubes grafted by molecularly imprinted poly(methacrylic acid-hemin) as a peroxidase-like catalyst for biomimetic sensing of acetaminophen. RSC Adv 6:28751–28760. https://doi.org/10.1039/c6ra02150f
Moretti E dos S, Oliveira FM de, Scheel GL et al (2016) Synthesis of surface molecularly imprinted poly(methacrylic acid-hemin) on carbon nanotubes for the voltammetric simultaneous determination of antioxidants from lipid matrices and biodiesel. Electrochim Acta 212:322–332.https://doi.org/10.1016/j.electacta.2016.06.174
Okoth OK, Yan K, Zhang J (2017) Mo-doped $BiVO_4$ and graphene nanocomposites with enhanced photoelectrochemical performance for aptasensing of streptomycin. Carbon 120:194–202. https://doi.org/10.1016/j.carbon.2017.04.079
Pan J, Chen W, Ma Y, Pan G (2018) Molecularly imprinted polymers as receptor mimics for selective cell recognition. Chem Soc Rev 47:5574–5587. https://doi.org/10.1039/c7cs00854f
Pan Q, Peng Y, Yang A, Yang D, Peng (2023) Development of a molecularly imprinted photoelectrochemical sensor for enhanced detection of ciprofloxacin in milk. Int J Electrochem Sci 18:100393. https://doi.org/10.1016/J.IJOES.2023.100393
Pang Y, Xu G, Feng Q et al (2017) Synthesis of α-$Bi_2Mo_3O_{12}$/TiO_2 nanotube arrays for photoelectrochemical COD detection application. Langmuir 33:8933–8942. https://doi.org/10.1021/acs.langmuir.7b01826
Patra S, Roy E, Madhuri R, Sharma PK (2015) Nano-iniferter based imprinted sensor for ultra trace level detection of prostate-specific antigen in both men and women. Biosens Bioelectron 66:1–10. https://doi.org/10.1016/j.bios.2014.10.076
Pauling L (1940) A theory of the structure and process of formation of antibodies. J Am Chem Soc 62:2643–2657. https://doi.org/10.1021/ja01867a018
Peng D, Li X, Zhang L, Gong J (2014) Novel visible-light-responsive photoelectrochemical sensor of 2,4-dichlorophenoxyacetic acid using molecularly imprinted polymer/BiOI nanoflake arrays. Electrochem Commun 47:9–12. https://doi.org/10.1016/j.elecom.2014.07.010
Prado TMd, Cincotto FH, Fatibello-Filho O, Cruz de Moraes F (2018) Bismuth vanadate/reduced graphene oxide nanocomposite electrode for photoelectrochemical determination of diclofenac in urine. Electroanalysis 30:2704–2711.https://doi.org/10.1002/elan.201800524
Prado TM, Carrico A, Cincotto FH et al (2019) Bismuth vanadate/graphene quantum dot: A new nanocomposite for photoelectrochemical determination of dopamine. Sens Actuators B Chem 285:248–253. https://doi.org/10.1016/j.snb.2019.01.059

Prasad BB, Madhuri R, Tiwari MP, Sharma PS (2010a) Imprinted polymer-carbon consolidated composite fiber sensor for substrate-selective electrochemical sensing of folic acid. Biosens Bioelectron 25:2140–2148. https://doi.org/10.1016/j.bios.2010.02.016

Prasad BB, Madhuri R, Tiwari MP, Sharma PS (2010b) Imprinting molecular recognition sites on multiwalled carbon nanotubes surface for electrochemical detection of insulin in real samples. Electrochim Acta 55:9146–9156. https://doi.org/10.1016/j.electacta.2010.09.008

Prete MC, dos Santos DM, Effting L, Tarley CRT (2019) Preparation of molecularly imprinted poly(methacrylic acid) grafted on iniferter-modified multiwalled carbon nanotubes by living-radical polymerization for 17β-estradiol extraction. J Chem Eng Data 64:1978–1990. https://doi.org/10.1021/acs.jced.8b01010

Qin C, Bai X, Zhang Y, Gao K (2018) Photoelectrochemical CdSe/TiO_2 nanotube array microsensor for high-resolution in-situ detection of dopamine. Microchim Acta 185:278. https://doi.org/10.1007/s00604-018-2788-4

Ribeiro FWP, Moraes FC, Pereira EC et al (2015) New application for the $BiVO_4$ photoanode: a photoelectroanalytical sensor for nitrite. Electrochem Commun 61:1–4. https://doi.org/10.1016/j.elecom.2015.09.022

Rodrigues BS, Branco CM, Corio P, Souza JS (2020) Controlling bismuth vanadate morphology and crystalline structure through optimization of microwave-assisted synthesis conditions. Cryst Growth Des 20:3673–3685. https://doi.org/10.1021/acs.cgd.9b01517

Roy E, Patra S, Madhuri R, Sharma PK (2016) Imprinted carbonaceous nanomaterials: a tiny looking big thing in the field of selective and specific analysis. In: Tiwari A, Uzun L (eds) Advanced molecularly imprinting materials. Wiley, Hoboken, NJ, USA, pp 165–216

Rückert B, Hall AJ, Sellergren B (2002) Molecularly imprinted composite materials via iniferter-modified supports. J Mater Chem 12:2275–2280. https://doi.org/10.1039/b203115a

Sajini T, Gigimol MG, Mathew B (2019) Kinetic and thermodynamic studies of molecularly imprinted polymers for the selective adsorption and specific enantiomeric recognition of D-mandelic acid. J Polym Res 26:1–15. https://doi.org/10.1007/s10965-019-1746-0

Sharma PS, Garcia-Cruz A, Cieplak M et al (2019) 'Gate effect' in molecularly imprinted polymers: the current state of understanding. Curr Opin Electrochem 16:50–56. https://doi.org/10.1016/j.coelec.2019.04.020

Shi H, Zhao G, Liu M, Zhu Z (2011) A novel photoelectrochemical sensor based on molecularly imprinted polymer modified TiO_2 nanotubes and its highly selective detection of 2,4-dichlorophenoxyacetic acid. Electrochem Commun 13:1404–1407. https://doi.org/10.1016/j.elecom.2011.08.022

Shi H, Zhao J, Wang Y, Zhao G (2016) A highly selective and picomolar level photoelectrochemical sensor for PCB 101 detection in environmental water samples. Biosens Bioelectron 81:503–509. https://doi.org/10.1016/j.bios.2016.03.023

Shi H, Wang Y, Zhao J et al (2018) Cathodic photoelectrochemical detection of PCB101 in environmental samples with high sensitivity and selectivity. J Hazard Mater 342:131–138. https://doi.org/10.1016/j.jhazmat.2017.07.025

Song M, Sun H, Yu J et al (2021) Enzyme-free molecularly imprinted and graphene-functionalized photoelectrochemical sensor platform for pollutants. ACS Appl Mater Interfaces 13:37212–37222. https://doi.org/10.1021/acsami.1c10242

Suk Jang J, Gyu Kim H, Lee SH (2012) Efficient photocatalytic degradation of acid orange 7 on metal oxide p-n junction composites under visible light. J Phys Chem Solids 73:1372–1377. https://doi.org/10.1016/j.jpcs.2012.07.009

Sun X, Gao C, Zhang L et al (2017) Photoelectrochemical sensor based on molecularly imprinted film modified hierarchical branched titanium dioxide nanorods for chlorpyrifos detection. Sens Actuators B Chem 251:1–8. https://doi.org/10.1016/j.snb.2017.04.130

Sun X, Zhang X, Xie Y (2020) Surface defects in two-dimensional photocatalysts for efficient organic synthesis. Matter 2:842–861. https://doi.org/10.1016/j.matt.2020.02.006

Tan HL, Amal R, Ng YH (2017) Alternative strategies in improving the photocatalytic and photoelectrochemical activities of visible light-driven $BiVO_4$: a review. J Mater Chem A 5:16498–16521. https://doi.org/10.1039/c7ta04441k

Tang S, Wang Z, Yuan D, et al (2020) Enhanced photocatalytic performance of $BiVO_4$ for degradation of methylene blue under LED visible light irradiation assisted by peroxymonosulfate. Int J Electrochem Sci 15:2470–2480. https://doi.org/10.20964/2020.03.09

Tarley CRT, Sotomayor MDPT, Kubota LT (2005) Polímeros biomiméticos em química analítica. Parte 2: Aplicações de MIP ("Molecularly Imprinted Polymers") no desenvolvimento de sensores químicos. Quim Nova 28:1087–1101. https://doi.org/10.1590/s0100-40422005000600025

Thakur N, Das TR, Gupta S et al (2023) Material selection for the fabrication of MIP-based devices: an overview. In: Molecularly imprinted polymers (MIPs). Elsevier, pp 101–114

Tiwari MP, Prasad A (2015) Molecularly imprinted polymer based enantioselective sensing devices: a review. Anal Chim Acta 853:1–18. https://doi.org/10.1016/j.aca.2014.06.011

Viter R, Kunene K, Genys P et al (2020) Photoelectrochemical bisphenol S sensor based on ZnO-nanoroads modified by molecularly imprinted polypyrrole. Macromol Chem Phys 221:1–6. https://doi.org/10.1002/macp.201900232

Wang Q, Arash B (2014) A review on applications of carbon nanotubes and graphenes as nano-resonator sensors. Comput Mater Sci 82:350–360. https://doi.org/10.1016/j.commatsci.2013.10.010

Wang G, Xu J, Chen H (2009) Progress in the studies of photoelectrochemical sensors. Sci China B Chem 52:1789–1800. https://doi.org/10.1007/s11426-009-0271-0

Wang P, Ge L, Li M et al (2013a) Photoelectrochemical sensor based on molecularly imprinted polymer-coated TiO_2 nanotubes for lindane specific recognition and detection. J Inorg Organomet Polym Mater 23:703–711. https://doi.org/10.1007/s10904-013-9836-7

Wang P, Dai W, Ge L et al (2013b) Visible light photoelectrochemical sensor based on Au nanoparticles and molecularly imprinted poly(o-phenylenediamine)-modified TiO_2 nanotubes for specific and sensitive detection chlorpyrifos. Analyst 138:939–945. https://doi.org/10.1039/c2an36266j

Wang R, Yan K, Wang F, Zhang J (2014a) A highly sensitive photoelectrochemical sensor for 4-aminophenol based on CdS-graphene nanocomposites and molecularly imprinted polypyrrole. Electrochim Acta 121:102–108. https://doi.org/10.1016/j.electacta.2013.12.139

Wang GL, Jiao HJ, Liu KL et al (2014b) A novel strategy for the construction of photoelectrochemical sensors based on quantum dots and electron acceptor: the case of dopamine detection. Electrochem Commun 41:47–50. https://doi.org/10.1016/j.elecom.2014.01.014

Wang R, Zu M, Yang S et al (2018a) Visible-light-driven photoelectrochemical determination of Cu^{2+} based on CdS sensitized hydrogenated TiO_2 nanorod arrays. Sens Actuators B Chem 270:270–276. https://doi.org/10.1016/j.snb.2018.05.056

Wang CC, Chou PH, Yu YH, Kei CC (2018b) Deposition of Ni nanoparticles on black TiO_2 nanowire arrays for photoelectrochemical water splitting by atomic layer deposition. Electrochim Acta 284:211–219. https://doi.org/10.1016/j.electacta.2018.07.164

Wang J, Xu Q, Xia WW et al (2018c) High sensitive visible light photoelectrochemical sensor based on in-situ prepared flexible Sn_3O_4 nanosheets and molecularly imprinted polymers. Sens Actuators B Chem 271:215–224. https://doi.org/10.1016/j.snb.2018.05.098

Wang S, Li S, Wang W et al (2019) A non-enzymatic photoelectrochemical glucose sensor based on $BiVO_4$ electrode under visible light. Sens Actuators B Chem 291:34–41. https://doi.org/10.1016/j.snb.2019.04.057

Wulff G, Sarhan A (1972) The use of polymers with enzyme-analogous structures for the resolution of racemates. Angew Chem Int Ed Engl 11:341

Xie X, Bu Y, Wang S (2016) Molecularly imprinting: a tool of modern chemistry for analysis and monitoring of phenolic environmental estrogens. Rev Anal Chem 35:87–97. https://doi.org/10.1515/revac-2016-0005

Xu H, Li H, Wu C et al (2008) Preparation, characterization and photocatalytic properties of Cu-loaded $BiVO_4$. J Hazard Mater 153:877–884. https://doi.org/10.1016/j.jhazmat.2007.09.039

Xu S, Lu H, Chen L, Wang X (2014) Molecularly imprinted TiO_2 hybridized magnetic Fe_3O_4 nanoparticles for selective photocatalytic degradation and removal of estrone. RSC Adv 4:45266–45274. https://doi.org/10.1039/C4RA06632D

Xu R, Du Y, Liu L et al (2021) Molecular imprinted photoelectrochemical sensor for bisphenol A supported by flower-like $AgBiS_2/In_2S_3$ matrix. Sens Actuators B Chem 330:1–7. https://doi.org/10.1016/j.snb.2020.129387

Xu Y, Qin N, Liu Z et al (2022) Functionally imprinted orthorhombic $WO_3.H_2O$ nanoplates for ultrasensitive photoelectrochemical sensing with excellent selectivity. ACS Mater Lett 4:2394–2400. https://doi.org/10.1021/acsmaterialslett.2c00830

Yan M, Yan Y, Wu Y et al (2015) Microwave-assisted synthesis of monoclinic-tetragonal $BiVO_4$ heterojunctions with enhanced visible-light-driven photocatalytic degradation of tetracycline. RSC Adv 5:90255–90264. https://doi.org/10.1039/c5ra13684a

Yan G, Han Z, Hou X et al (2023) A highly sensitive TiO_2-based molecularly imprinted photoelectrochemical sensor with regulation of imprinted sites by photo-deposition. J Colloid Interface Sci 650:1319–1326. https://doi.org/10.1016/j.jcis.2023.07.105

Yáñez-Sedeño P, Campuzano S, Pingarrón JM (2017) Electrochemical sensors based on magnetic molecularly imprinted polymers: a review. Anal Chim Acta 960:1–17. https://doi.org/10.1016/j.aca.2017.01.003

Yang R, Yan X, Li Y et al (2017a) Nitrogen-doped porous carbon-ZnO nanopolyhedra derived from ZIF-8: new materials for photoelectrochemical biosensors. ACS Appl Mater Interfaces 9:42482–42491. https://doi.org/10.1021/acsami.7b10856

Yang G, Chen X, Pan Q et al (2017) A novel photoelectrochemical sensor for thiamphenicol based on porous three-dimensional imprinted film. Int J Electrochem Sci 12:7272–7286. https://doi.org/10.20964/2017.08.34

Yang Z, Qin T, Niu Y et al (2020) Flexible visible-light-driven photoelectrochemical biosensor based on molecularly imprinted nanoparticle intercalation-modulated graphene fiber for ultrasensitive urea detection. Carbon 157:457–465. https://doi.org/10.1016/j.carbon.2019.10.061

Yang Y, Yan W, Wang X et al (2021) Development of a molecularly imprinted photoelectrochemical sensing platform based on NH_2-MIL-125(Ti)–TiO_2 composite for the sensitive and selective determination of oxtetracycline. Biosens Bioelectron 177:113000. https://doi.org/10.1016/j.bios.2021.113000

Yang S, Deng K, Shao S et al (2022) A visible light responsive molecularly imprinted photoelectrochemical sensor for the sensitive detection of BSA. J Solid State Electrochem 26:821–830. https://doi.org/10.1007/s10008-021-05110-w

Yang S, Chen J, Yan L et al (2024) Molecularly imprinted photoelectrochemical sensing platform based on g-C_3N_4/MOFs photoresponsive oxidase mimic for norfloxacin detection. Microchem J 196:109607. https://doi.org/10.1016/j.microc.2023.109607

Ye W, Wu F, Shi N et al (2020) Metal-semiconductor phase twinned hierarchical MoS_2 nanowires with expanded interlayers for sodium-ion batteries with ultralong cycle life. Small 16:1906607. https://doi.org/10.1002/smll.201906607

You M, Yang S, Jiao F et al (2016) Label-free electrochemical multi-sites recognition of G-rich DNA using multi-walled carbon nanotubes-supported molecularly imprinted polymer with guanine sites of DNA. Electrochim Acta 199:133–141. https://doi.org/10.1016/j.electacta.2016.03.151

Yu Z, Lv S, Ren R et al (2017) Photoelectrochemical sensing of hydrogen peroxide at zero working potential using a fluorine-doped tin oxide electrode modified with $BiVO_4$ microrods. Microchim Acta 184:799–806. https://doi.org/10.1007/s00604-016-2071-5

Yu SY, Zhang L, Zhu LB et al (2019) Bismuth-containing semiconductors for photoelectrochemical sensing and biosensing. Coord Chem Rev 393:9–20. https://doi.org/10.1016/j.ccr.2019.05.008

Yu J, Zhang C, Yang Y et al (2021) Facile synthesis of molecularly imprinted black TiO_{2-x}/carbon dots nanocomposite and its recognizable photocatalytic performance under visible-light. Appl Surf Sci 551:149476. https://doi.org/10.1016/j.apsusc.2021.149476

Zeng Y, Zhang M, Peng K et al (2023) Au/TiO_2-based molecularly imprinted photoelectrochemical sensor for dibutyl phthalate detection. Microchim Acta 190:1–9. https://doi.org/10.1007/s00604-023-05824-w

Zhang A, Zhang J, Cui N et al (2009) Effects of pH on hydrothermal synthesis and characterization of visible-light-driven $BiVO_4$ photocatalyst. J Mol Catal A Chem 304:28–32. https://doi.org/10.1016/j.molcata.2009.01.019

Zhang D, Yu D, Zhao W et al (2012) A molecularly imprinted polymer based on functionalized multiwalled carbon nanotubes for the electrochemical detection of parathion-methyl. Analyst 137:2629–2636. https://doi.org/10.1039/c2an35338e

Zhang C, Si S, Yang Z (2015a) A highly selective photoelectrochemical biosensor for uric acid based on core-shell Fe_3O_4@C nanoparticle and molecularly imprinted TiO_2. Biosens Bioelectron 65:115–120. https://doi.org/10.1016/j.bios.2014.10.013

Zhang C, Si S, Yang Z (2015b) Design of molecularly imprinted TiO_2/carbon aerogel electrode for the photoelectrochemical determination of atrazine. Sens Actuators B Chem 211:206–212. https://doi.org/10.1016/j.snb.2015.01.079

Zhang B, Lu L, Huang F, Lin Z (2015c) $[Ru(bpy)_3]^{2+}$-mediated photoelectrochemical detection of bisphenol A on a molecularly imprinted polypyrrole modified SnO_2 electrode. Anal Chim Acta 887:59–66. https://doi.org/10.1016/j.aca.2015.05.051

Zhang C, Bai W, Yang Z (2016) A novel photoelectrochemical sensor for bilirubin based on porous transparent TiO_2 and molecularly imprinted polypyrrole. Electrochim Acta 187:451–456. https://doi.org/10.1016/j.electacta.2015.11.098

Zhang Y, Gao W, Ji S et al (2019) An ingenious molecularly imprinted photo-induced sensor for ultrasensitive determination of sulfadiazine based on the urchin-shaped Bi_2S_3. Electrochim Acta 320:134619. https://doi.org/10.1016/j.electacta.2019.134619

Zhang Z, Zhou H, Jiang C, Wang Y (2020) Molecularly imprinted polymer functionalized flower-like BiOBr microspheres for photoelectrochemical sensing of chloramphenicol. Electrochim Acta 344:136161. https://doi.org/10.1016/j.electacta.2020.136161

Zhang Z, Huang LL, Sheng S et al (2021a) MIL-125(Ti)-derived COOH functionalized TiO_2 grafted molecularly imprinted polymers for photoelectrochemical sensing of ofloxacin. Sens Actuators B Chem 343:130119. https://doi.org/10.1016/j.snb.2021.130119

Zhang Z, Wu T, Zhou H et al (2021b) 3D flower-shaped BiOI encapsulated in molecularly imprinted polymer for hypersensitivity to norfloxacin. Microchem J 164:106017. https://doi.org/10.1016/j.microc.2021.106017

Zhang X, Li T, Gao X et al (2023) Porphyrin and molecularly imprinted polymer double sensitized cathode photoelectrochemical sensor for chlortetracycline. Microchem J 193:109094. https://doi.org/10.1016/j.microc.2023.109094

Zhang C, Ou W, Zeng Z et al (2024) Polypyrrole participates in the construction of a polarity-switchable photoelectrochemical molecularly imprinted sensor for the detection of acrylamide in fried foods. Sens Actuators B Chem 402:135037. https://doi.org/10.1016/j.snb.2023.135037

Zhao D, Zhang Y, Ji S et al (2021) Molecularly imprinted photoelectrochemical sensing based on ZnO/polypyrrole nanocomposites for acrylamide detection. Biosens Bioelectron 173:112816. https://doi.org/10.1016/j.bios.2020.112816

Zhong L, Li X, Liu R et al (2019) A visible-light-driven photoelectrochemical molecularly imprinted sensor based on titanium dioxide nanotube arrays loaded with silver iodide nanoparticles for the sensitive detection of benzoyl peroxide. Analyst 144:3405–3413. https://doi.org/10.1039/c9an00234k

Zhu D, Zhou Q (2019) Action and mechanism of semiconductor photocatalysis on degradation of organic pollutants in water treatment: a review. Environ Nanotechnol Monit Manag 12:100255. https://doi.org/10.1016/j.enmm.2019.100255

Zoppe JO, Ataman NC, Mocny P et al (2017) Surface-initiated controlled radical polymerization: state-of-the-art, opportunities, and challenges in surface and interface engineering with polymer brushes. Chem Rev 117:1105–1318. https://doi.org/10.1021/acs.chemrev.6b00314

Chapter 6
Molecularly Imprinted Polymer-Based Sensors for the Monitoring of Antibiotic Traces and Microorganisms in Water Samples to Combat Antimicrobial Resistance

Oliver Jamieson, Xinlu Liu, Cameron Read, Jack Reeder, Toby Hallam, Kasper Eersels, Bart van Grinsven, Elizabeth Heidrich, Divya Ottoor, and Marloes Peeters

6.1 Introduction

6.1.1 Antimicrobial Resistance

The World Health Organisation (WHO) predicts that by 2050, there will be ten million annual global deaths due to antibiotic-resistant infections (de Kraker 2016). This makes antimicrobial resistance (AMR) a global threat on par with climate change, with a similar number of predicted deaths as cancer by 2050 (see Fig. 6.1). The main

O. Jamieson · J. Reeder · E. Heidrich · D. Ottoor
School of Engineering, Newcastle University, Merz Court, Claremont Road, Newcastle Upon Tyne NE1 7RU, UK

O. Jamieson · X. Liu · M. Peeters (✉)
School of Engineering, Department of Chemical Engineering and Analytical Science, University of Manchester, Booth Street East, Manchester M13 9QS, UK
e-mail: Marloes.Peeters@newcastle.ac.uk

C. Read · T. Hallam
Mathematics, Statistics and Physics, Newcastle University, Herschel Building, Newcastle Upon Tyne NE1 7RU, UK

K. Eersels · B. van Grinsven
Faculty of Science and Engineering, Sensor Engineering, Maastricht University, Duboisdomein 30, 6229 GT Maastricht, the Netherlands

D. Ottoor
Department of Chemistry, Savitribai Phule Pune University, Pune 411007, India

S. Patra and M. Sillanpaa (eds.), *Molecularly Imprinted Polymers as Artificial Antibodies for the Environmental Health*, https://doi.org/10.1007/978-3-031-58995-9_6

drivers for the accelerated development of drug-resistant pathogens include misuse and overuse of antimicrobials, in addition to the lack of clean water and sanitation since inadequate infection control promotes the spread of microbes. Moreover, the COVID-19 pandemic has posed unique challenges, since it has been reported that up to 70% of patients with SARS-CoV-2 receive antimicrobials, thereby favouring the emergence of AMR (Heinze 2005; Langford et al. 2021). Besides this, COVID-19 has affected the availability of standard pharmaceuticals including antimicrobials via affecting the supply chain and global manufacturing (Langford et al. 2021). This has had a disproportionate impact on low and lower-middle-income countries, meaning that AMR becomes an equity and healthcare inequality issue.

Compared to 2017, there was a 15% increase in bloodstream infections due to resistant *Escherichia coli* (*E. coli*), *Salmonella* species, and gonorrhoea strains (Spicer 2015). There are several microorganisms with particularly high resistance levels; for instance, *Klebsiella pneumonia (K. pneumoniae)* and *Acetinobacter* have resistance levels of over 50%. Some of these (8%) are even resistant to last resort antibiotics such as carbapenems, resulting in high mortality for these infections. The WHO has identified carbapenem-resistant gram-negative bacterial infections in its priority pathogen list with the urgent need for new antimicrobials; however, in 2019, only 32 new antibiotics were in clinical development. Moreover, if we do not change our approach to use of antimicrobials and sanitation, the risk remains of accelerated drug resistance that will render any novel treatments ineffective within a matter of years. It will be crucial to monitor selective pressure (e.g. presence of antibiotics) in

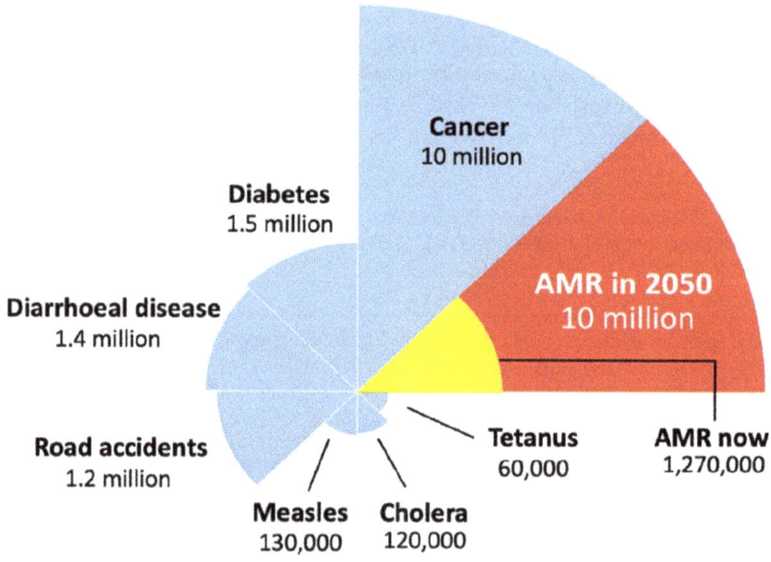

Fig. 6.1 Predicted mortality from AMR compared to common causes of death today (adapted from O'Neill 2016; Murray et al. 2022)

the supply chain and gain insight into how pathogens develop drug resistant properties to tackle this issue from its source as prevention is key to emergence of AMR.

6.1.2 *Antimicrobial Resistance in the Environment in the Environment*

The use of antibiotics in veterinary or clinical practice exerts a selective pressure, which accelerates the emergence of AMR. Thus, the possibility of antibiotics leaching into water systems via the effluents of the pharmaceutical industry, agriculture, and hospitals is a major concern. While wastewater treatment (WWT) systems can remove harmful contaminants from the water systems, it has been reported that on a range of antibiotics screened, the removal rate was typically between 10–70% (Behera et al. 2011). This in stark contrast to caffeine, a common contaminant which used as anthropogenic marker, and has a typical removal efficiency of 99% when the same conditions were used as for treatment of the antibiotics (Paíga et al. 2017). Besides the waterways, other routes of entry include livestock feed, as feed is still supplemented with antibiotics for prophylactic purposes or to promote growth, and excrement. In the farming industry, penicillins and tetracycles cover more than 50% of the total antibiotic consumption worldwide due to their broad spectrum of action (Bottari 2019) (Fig. 6.2).

The European Commission strictly regulates the level of antibiotic residues in food and feed products and in the European Commission Regulation 37/2010, a

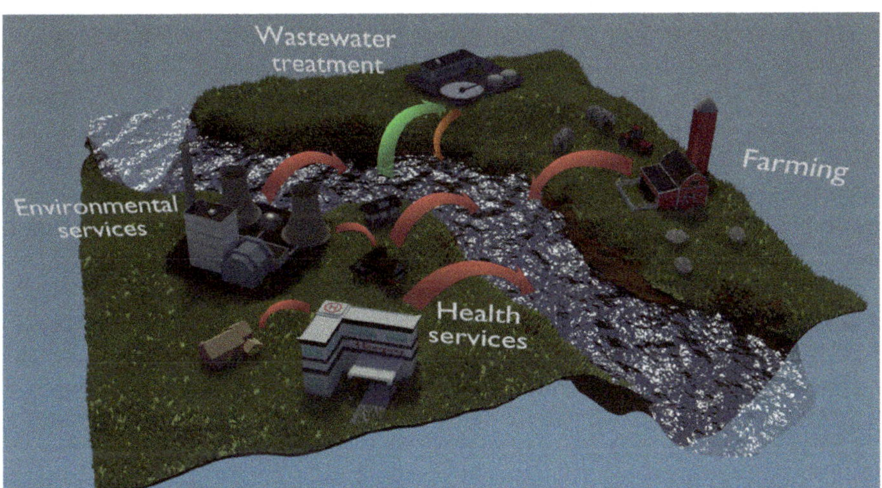

Fig. 6.2 The different ways into which antibiotic traces can enter the waterways. Red arrows indicate how antibiotics enter the system, the green arrow indicates how the antibiotic are removed via wastewater treatment plants. Reprinted from Jamieson et al. (2021), with permission from Wiley

maximum residue limit (MRL) has been set for each antibiotic in different food products. While this is the case for food products, the legislation on antibiotic traces in the environment, particularly in the waterways, is limited. This is surprising given that the European Commission acknowledges that AMR via transmitted via the environment is a well-documented health risk, with ample reports on new resistant MOs found in both soil and water (European Commission 2019).

The European Water Framework Directive, a legislative department set up in 2000, is the main piece of legislation for the protection and sustainable use of European freshwater resources. In 2023, there were 45 priority substances on the watchlist but none of them are pharmaceuticals; however, there are three antibiotics (azithromycin, clarithromycin, and erythromycin) on the watch list. The inclusion of antibiotics or other antimicrobial substances in the EU legislation would therefore be a vital tool to combat accelerated development of AMR (Ågerstrand et al. 2023). An alternative is to establish these standards at national level, with Sweden having quality standards for ciprofloxacin in place to protect against AMR (Sahlin et al. 2018). Current tests for measuring antibiotic traces in food samples include microbial inhibition assays or screening tests based on immunoassays or enzymatic reactions. Microbial inhibition assays are low-cost and easy to use but the interpretation of the outcome is the test is subjective and there is a high incidence of false negative results. Radio receptor assays, where target molecules labelled with radioisotopes to bind to a specific site, are costly, require trained operators, and have complex waste-disposal protocols. Colloidal gold receptor assays are low-cost options and have been commercialised in Europe (Beta Star and Twin Sensor BT) and USA (Charm Rosa MRL); however, they are not able to quantify results and are prone to interference. Enzyme-labelling receptor assays overcome the aforementioned issues and have an operating principle similar to Enzyme-Linked Immunosorbent Assays (ELISA). However, they suffer from poor protein stability, thus limiting large-scale application of this method.

Besides monitoring of pharmaceuticals in the waterways, it is also crucial to monitor the presence of harmful microorganisms as these can also exert additional selective pressure. A key area of concern is the contamination of waterways with faecal matter, which is introduced via various sources such as sewage, agricultural and urban diffuse, and wildlife. Besides implications to AMR, this has also serious public health issues since ingestion of as few as 10 microorganisms can lead to serious illnesses, especially for the vulnerable population (Abbas et al. 2023).

The monitoring of faecal matter is normally achieved via measuring the quantity of faecal indicator microorganisms (FIOs), such as *E. coli* or intestinal *Enterococci*. *E. coli* is the organism of choice as an indicator organism for faecal matter contamination but has also been suggested as an indicator microorganism for development of AMR, biosafety, and disinfection (Dewaele et al. 2011). With new proposals to EU urban wastewater directive seeking to add the systematically monitor of wastewater for several viruses, amongst which CoV-SARS-19, and anti-microbial resistance (European Commission 2022). Current regulations and guidelines are set in place to outline the detection of FIOs, as a measure of faecal contamination. Bathing water regulations ensure and safeguard public health for recreational water-related activities. Under many regulations the recommended frequency for sampling is once a month, with sampling only required during the bathing season. Both the UK and EU regulations define "excellent" bathing waters has having an *E. coli* level less than 500 cfu/100 mL for inland waters and less than 250 cfu/100 mL for coastal waters (European Commission 2006; UK Government 2013). Whilst the United States Environment Protection Agency recommends a limit of 320 cfu/100 mL for fresh water. The monitoring of *E. coli* in water supplies is crucial to maintaining safe drinking water fit for human consumption. Monitoring is undertaken by the responsible water companies with the frequency depending on the volume of water supplied. Due to the significant risk presented by presence of *E. coli* in water supplies, regulations set the limit at 0 bacterium/100 mL for consumer water supply and service reservoirs. A limit of 0 bacterium/250 mL is also set for bottled water (UK Government 2016).

There are currently two international standardised methods by which *E. coli* and coliform bacteria in water are detected, these are membrane filtration method (ISO 9308-1:2014) and most probable number (MPN) method (ISO 9308-2:2014). The method outlined in ISO 9308-1 is for the enumeration of *E. coli* and coliform bacteria based on membrane filtration, subsequent culture on a chromogenic coliform agar medium, and calculation of the number of target organisms in the sample. This method is not suitable for surface waters or shallow well waters due to the low selectivity of the differential agar medium. However, this method is well suited to waters with low bacterial numbers such as drinking water or disinfected pool water.

The method outlined in ISO 9308-2 is designed for the most probable number enumeration of *E. coli* and coliform samples from treated and untreated waters. The method is based on a bacterial enzyme technology that signals the presence of coliform bacteria through the hydrolysis of a fluorogenic substrate diagnostic for the bacterium, allowing simultaneous detection of *E. coli* at limit of detection of 1 cfu/100 mL within 18 h. Although the two methods outlined in the ISO standards are the most frequently utilised other methods do exist but are yet to be standardised or frequently used for routine monitoring of water. New methods seek to improve the detection time and sensitivity of pathogen monitoring. Nucleic acid-based techniques operate on the principle of detecting specific DNA or RNA sequences extracted from the target organism. In combination with amplification techniques such as polymerase chain reaction (PCR), these methods offer rapid and highly sensitive detection able to detect individual pathogens (Girones et al. 2010). Recent advances and new methods such as quantitative PCR, enable identification of different strains as well as microbial source tracking, helping to identify sources of contamination (Gomi et al. 2014). The effectiveness of these methods is hindered by the requirement for an enrichment step as well as the presence of inhibitors affecting nucleic acid extraction. In addition, implementation is limited by the high cost associated with specialist equipment and the requirement for a trained analyst for operation.

The current monitoring regimes for detection of FIOs are not widespread and are commonly only done to fulfil the minimum requirements of regulations. Following current regulations and methods, monitoring is a time-consuming and costly (£45/sample) process, which requires skilled work and equipment. Taken together, there is a critical demand for low-cost and easy to use sensing platforms that can detect both antibiotic traces and bacterial concentration in situ and in continuous fashion. It is expected that biosensors will fill the gap in this market, but issues around interference and stability of the recognition elements will need to be overcome to achieve this goal. Moreover, further challenges include the low concentrations of contaminants in water and that emerging compounds are often overlooked, thus requiring regular updating of the watchlist.

6.1.3 Biosensors

Biosensors are analytical devices that incorporate a biological receptor to provide recognition specificity for a particular analyte, which is coupled to transduction technology to quantify the binding between analyte and detector molecule (Fig. 6.3).

In recent years, there is a move towards replacing these biological receptors, such as antibodies or enzymes, with biomimetic receptors that can be produced at scale and are low-cost and robust. The latter is important to consider given that monitoring of environmental contaminants on site requires the used recognition elements to function in a wide range of conditions, including varying pH and temperature. In this respect, Molecularly Imprinted Polymers (MIPs) are an appealing cost-effective alternative towards traditional receptors as they exhibit superior stability due to their cross-linked nature (Haupt and Mosbach 2000). These synthetic materials can rival the specificity and selectivity of commercial biological receptors once the design is optimised. Their high affinity originates from distinct cavities (e.g., binding sites) in the polymer matrix that are generated via the inclusion of template molecules during the polymerization process (Jamieson et al. 2021). This template molecule can be target itself, which is often the case for small molecules, but for proteins and viruses it is more common to use an epitope of the full target as this reduces costs and is better from a biosafety perspective (Canfarotta et al. 2016). In addition, it is also possible to use a "dummy" template that mimics the functionality or spatial dimensions of the original template; for instance, it is possible to use (functionalized) latex beads that can mimic the size and shape of bacteria. After polymerization and

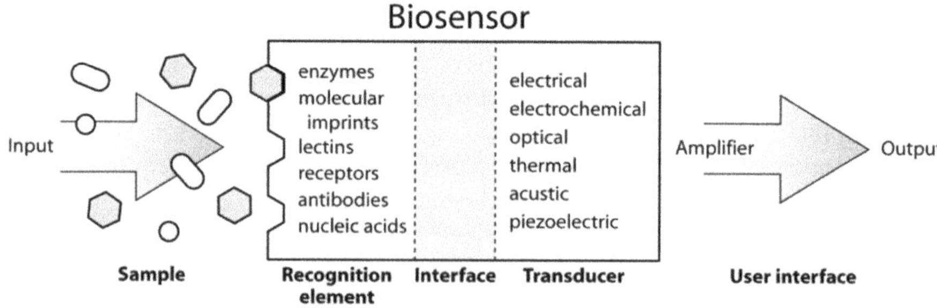

Fig. 6.3 Configuration of a biosensor showing (biological/biomimetic) recognition elements, interface, and transduction elements. Reprinted from Chambers et al. (2008), with permission from MDPI

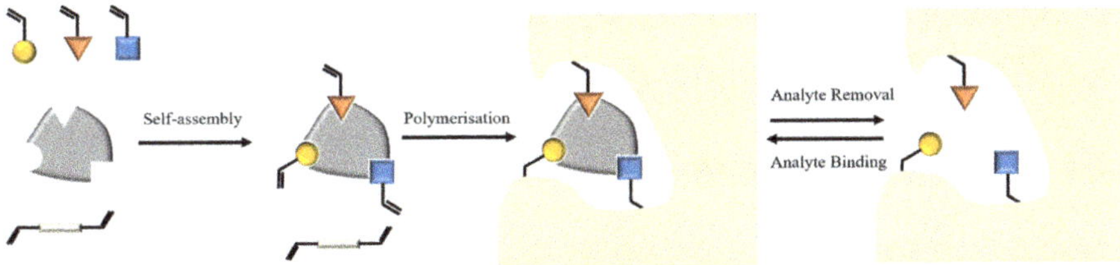

Fig. 6.4 Schematic of imprinting process which involves self-assembly of functional monomers around the target molecule, followed by polymerization and extraction. This process yields binding sites that have the same spatial orientation and chemical functionality as the original template molecule and thus bind it with high affinity

subsequent extraction of the template molecule, the remaining imprints meet the spatial requirements and chemical functionality of the intended target (Fig. 6.4).

A key strength of imprinting technology is its versatility, ensuring it can be used for analytes as small as ions up to larger macromolecules such as viruses and bacteria. However, the use of MIPs in commercial applications are sparse and is mainly limited to materials used for purification and extraction. An example is Sigma Aldrich's SupelMIP® solid phase extraction (SPE) line that is available for selectively extracting a single analyte of interest or a class of structurally related analytes from complex samples, including food and beverages. These MIPs can withstand harsher wash conditions during SPE methodology, resulting in cleaner extracts and thus lower limits of detection. For extraction and purification, binding capacity is the determining factor and it is possible to use (heterogeneous) microparticles that can be easily packed into columns (Lowdon et al. 2020). However, integration of MIPs into commercial sensors, where sensitivity and specificity are critical, remains challenging and there is no standard procedure yet for producing MIP sensors at scale. Despite this, there is high potential for use of MIPs in sensors; for instance, we have reported on a MIP-based sensor platform that can detect COVID-19 concentrations with a 6000× specificity compared to commercial lateral flow tests (McClements et al. 2022).

In this book chapter, we will discuss literature of the last ten years on portable sensors that can monitor compounds associated with antimicrobial resistance in a rapid and low-cost manner. We will focus on sensing of antibiotics and microorganisms, particularly *E. coli*, which is often used as a model organism. Finally, there will be a critical comparison of the described MIP-based sensors vs current state-of-the-art and future outlook of the application of these sensors.

6.2 Sensing of Antibiotic Traces to Prevent Antimicrobial Resistance

6.2.1 Monomer Selection

Antibiotics generally have a wide range of functional groups, such as amide, hydroxyl and amino groups, that can be used for recognition (Song et al. 2020). Therefore, there is a wide variety of MIPs available in literature that can be used for sensing of these compounds. Via tailoring the monomer composition, it is possible to target a family of antibiotics (via focusing on one specific pharma core such as a β-lactam ring) or specifically recognise one antibiotic (via a selection of monomer that can anchor onto different places of the target molecule). Methacrylic acid (MAA) is often used as base monomer since it can act as both a hydrogen bond donor and acceptor. For instance, Xie et al. showed with computational modelling and NMR spectroscopy that MAA was the most suitable functional monomer for binding to the antibiotic chloramphenicol compared to other monomers such as acrylamide, methylmethacrylate or acrylic acid due to strong hydrogen bonding (Xie et al. 2020). MAA is often supplemented with hydrophilic monomers that improve the hydrophilicity of the exposed surface of the MIPs and thus enhance recognition in aqueous media (Song et al. 2020).

An appropriate read-out technique needs to be selected that can monitor binding of the target to the functional monomers. The designed sensor needs to enable on site sensing, and thus provide rapid and low-cost detection. Optical and electrochemical techniques are most suited for this task. MAA can be modified with a fluorescein moiety to enable optical detection of beta lactam antibiotics in water samples (Hudson et al. 2021). Alternative methods include incorporation of quantum dots or custom-designed fluorophores to achieve highly specific and selective detection of antibiotics (Wan et al. 2013; Zhang and Chen 2016). However, all these approaches require bespoke synthesis and are not readily achieved with commercial monomers. Furthermore, many fluorophores are sensitive to changes in the environment, such as pH and presence of contaminants, which poses a significant challenge for in situ measurements. Electrochemical methods are an appealing alternative because they enable direct integration of the MIP onto the transducer surface via electropolymerization of the monomer(s) under potentiostatic or potentiodynamic conditions (Heinze 2005; Naveen et al. 2017). While electropolymerization is gaining in popularity due to its straightforward protocol and ability to precisely control the surface architecture, there are several areas of advancement required. Crapnell et al. (2019) have comprehensively described the use of electropolymerized MIPs (eMIPs) for sensing of bioanalytes such as antibiotics and mentioned areas requiring improvement, such as the solubility and minimization of interactions between the solvent, target, and functional monomer(s). However, the limitation of this method is that most "traditional" MIP monomers, such as MAA, are not conductive, and it can be complicated to electrodeposit a mixture of monomers which is key to achieving selectivity. Common monomers include pyrrole, (thio)phenols, boronic acid, and aniline and carboxyl-substituted anilines (Li et al. 2002). These monomers are often supplemented with nanomaterials, such as graphene, carbon nanotubes (CNTs) or (gold) decorated nanoparticles, to enhance the surface area and thus increase the electrochemical response. An overview of how carbon nanomaterials decorated with MIPs can be used for effective sensing is provided by Singhal et al. (2022).

6.2.2 Influence of Electrode Material for Electrochemical Sensing

Besides the monomer composition, and use of potential additives, the chosen electrode material and electrochemical detection technique have a pivotal influence on sensing. Gold and glassy carbon electrodes (GCEs) remain the most used electrode materials. These electrodes are high cost but have potential to be reused and have a smooth surface, thus facilitating easy characterisation (Cervini and Cavalheiro 2012; Cui and Zhang 2012).

Screen-printed electrodes (SPEs) are a promising alternative that have gained much interest due to their high reproducibility, low-cost, and inherent suitability for mass-production (Li et al. 2012). However, unlike the aforementioned electrode materials they are not re-usable and integration of polymer structures into SPEs is complicated due to their low resistance to traditional organic solvents. Strategies to incorporate MIPs into SPEs include direct electropolymerization, on the surface (Crapnell et al. 2019), incorporation of MIPs into the screen-printing ink (Casadio et al. 2017) or depositing onto nanomaterials decorated on the surface (Jin et al. 2018). Direct electropolymerization struggles from a limited selection of monomers and poor scalability, which can be overcome via other polymerization methods such as stereolitography (Jamieson et al. 2020). In contrast, incorporation of MIPs into the screen-printing ink is a scalable method but leads to sensor with low sensitivity as only ~ 20% of MIPs can be incorporated into the ink, as a key requirement for screen-printing is that the ink is conductive. Devkota et al. (2018) reported on an electrochemical sensor for the detection of tetracycline which was produced via overoxidizing a polypyrrole layer that was subsequently decorated with gold nanoparticles (AuNP) (Fig. 6.5).

Rohani Moghadam et al. (2019) used a combination of gold nano-urchins and graphene oxide (GO) dropcasted onto the surface to enhance the electrochemical signal of the MIP for binding of the antibiotic oxacillin. To avoid the need for multiple functionalisation steps, Dechtrirat et al. (2018) devised a method for inkjet printing of a nanocomposite layer, including AuNPs and poly(3,4-ethylenedioxythiophene)/poly(styrene sulfonate) (PEDOT:PSS), onto the surface of SPEs to optimise sensing for nitrofurantoin. The advantage of this method is that it leads to the manufacturing of a highly reproducible sensor that remained stable over a prolonged period.

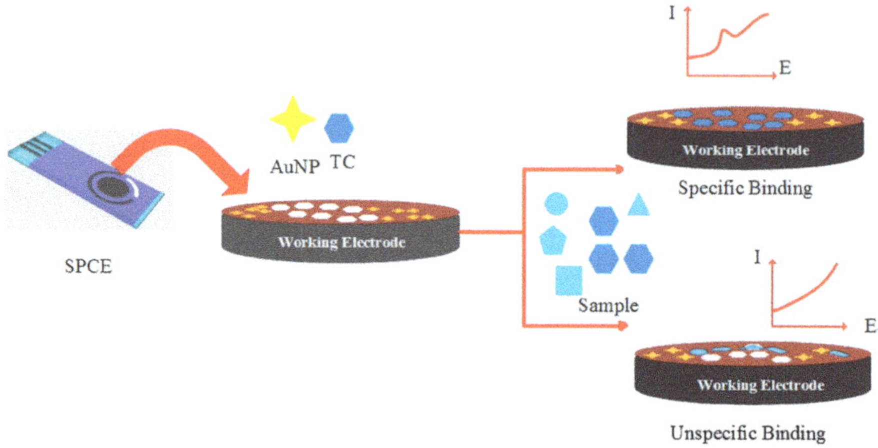

Fig. 6.5 An electrochemical sensor for the detection of tetracycline, involving an overoxidized polypyrrole layer deposited onto a SPCE that was decorated with AuNPs. Figure from Devkota et al. (2018), reproduced with permission from Elsevier

6.2.3 Influence of Electrochemical Sensing Method on Detection of Antibiotics with Portable Devices

Certain antibiotics are electroactive and thus facilitate easy electrochemical sensing. For those that are not electroactive, detection is achieved via indirect direction via a redox probe or via direct binding between the target and an electroactive layer containing MIPs. Electrochemical methods that have been used for MIP-based detection of antibiotics include differential pulse voltammetry (DPV), cyclic voltammetry (CV), square wave voltammetry (SWV), linear sweep voltammetry (LSV), amperometry, and electrochemical impedance spectroscopy (EIS) (Losito et al. 2003; Alizadeh et al. 2010; Zouaoui et al. 2020). CV is a fast and easy to implement method, but generally suffers from lower specificity compared to for instance DPV and SWV. The advantage of EIS is that it is label-free, but it might require more costly equipment compared to the other methods and can be complicated to analyze and time-consuming, which is why it is not often employed for commercial point-of-care electrochemical sensing.

A selection of papers that use conductive MIP materials in combination with electrochemical read-out is summarised for its key performance indicators in Table 6.1. The table demonstrates that the functional monomers, bar acrylamide, used to develop MIP-based sensors for electrochemical detection are electroactive. Moreover, it is common to use nanomaterials on the surface to enhance the electrochemical response, which can complicate scale-up for commercial applications. Finally, a wide range of electrode materials and electrochemical techniques are employed for detection, demonstrating the versatility of the technology.

Table 6.1 Overview of MIP-based sensors for the electrochemical detection of antibiotics

Target	Functional monomers	Electrode material	Detection	LoD	References
Erythromycin	m-phenylenediamine (mPD)	SPE	CV	0.12 nM	Ayankojo et al. (2020)
Ceftizoxime	Polycysteine	MWCNT-modified GCE	DPV	0.1 nM	Ali et al. (2021)
Daptomycin	o-phenylenediamine (oPD)	Au-Pt NPs GCE	DPV, CV	0.16 nM	Ozcelikay et al. (2020)
Cefquinome	4-aminobenzoic acid (4-ABA)	MWCNT graphite-modified SPEs	EIS, CV	50 nM	Moro et al. (2019)
Norfloxacin	4-aminothiophenol (*p*-ATP)	AuNPs SPCEs	EIS	0.15 ng/mL	Vu et al. (2022)
Tetracycline	(Overoxidized) pyrrole	AuNPs SPCEs	DPV	0.65 μmol/dm^3	Devkota et al. (2018)
Ceftazidime	Acrylamide	c-MWCNTs GCE	EIS, CV	0.55 nM	Torkashvand et al. (2016)
Azithromycin	3-thienyl boronic acid (3-TBA)	GCE	EIS, CV	0.85 nM	Stoian et al. (2020)
Levofloxacin	oPD, levodopa	Au-NPs Pencil-graphite electrode	DPV	462 nM	El Azab et al. (2022)
Sulfamethoxazole	oPD	GCE	CV, SWV	50 nM	Zhang et al. (2019)

6.2.4 Optical Sensing of Antibiotics with Portable Devices

The majority of reports on MIPs for antibiotics in literature focus on selective extraction of these targets from water and food samples (Giovanolli et al. 2013; Fernandez-Gonzalez et al. 2006). However, while optical methods have potential for point-of-care sensing in the future, most of these methods are based on fluorescence which requires bulky equipment that needs to be used in a lab environment. In this section, we will describe MIP-based devices that have potential for portable sensing, such as those systems that use smartphone-integrated devices.

A dual fluorescence sensing approach was adopted for the detection of the antibiotic ofloxacin (OFL) using carbon dots (CDs) combined with Eu^{3+} as fluorophore. Energy transfer from OFL to Eu^{3+} though the antenna effect led to enhancement of the fluorescence, which enabled detection of the target with a detection limit of 0.25 nM (Wang et al. 2022). A MIP was prepared by the sol gel method using CDs/EU-MOFS as both fluorophores as the supporting matrix, with 3-aminopropyl triethoxysilinate (APTES) as functional monomer and tetraethyl orthosilicate (TEOS) as crosslinker monomer. Upon binding of the target oxytetracycline (OTC), a dual response in the fluorescence was observed which includes quenching of the blue fluorescence at 445 due to the inner filter effect (IFE) and subsequently the red fluorescence at 620 nm

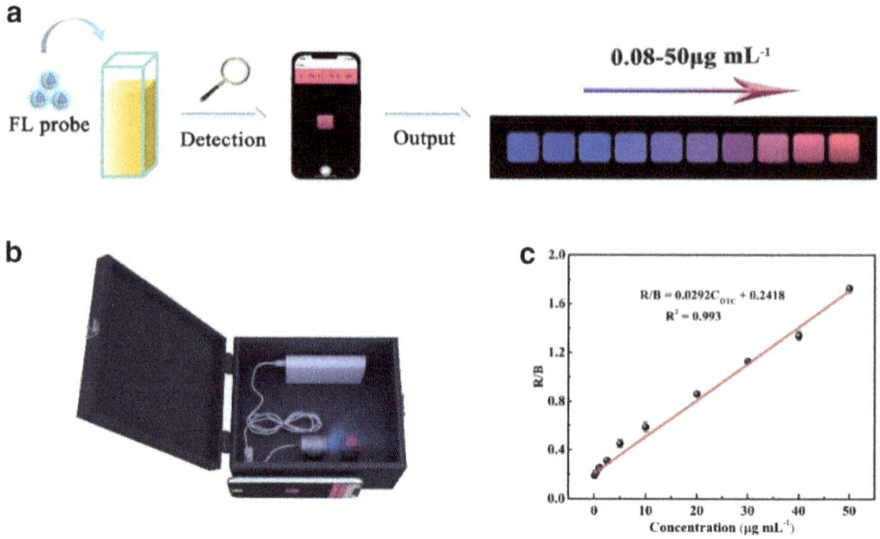

Fig. 6.6 **a** Schematic diagram of using the developed smartphone-integrated sensing platform to detect OTC. **b** The render 3D views of homemade device. **c** Linear relationship between *R*/*B* of the portable sensing platform and the OTC concentrations (OTC concentrations: 0.08, 0.4, 1, 2.5, 5, 10, 20, 30, 40, 50 μg mL^{-1}). Reprinted from Hu et al. (2022), with permission from Elsevier

was enhanced due to the antenna effect. A portable smartphone-integrated optical device was developed, that was able to rapidly detect OTC in real sample including water and milk (Hu et al. 2022) (Fig. 6.6).

MIPs can be combined with other recognition elements such as aptamers (Geng et al. 2018). Geng et al. reported on an acrylic acid-based MIP combined with aptamers and allyl amine modified CdSe QDs as fluorophores for the sensing of the antibiotic kanamycin, achieving a limit of detection of 0.013 μg/mL. The aptamers were introduced to the system via a thiol-ene click reaction. Sustainability is another important aspect that should be considered and this is why CD, in addition to their low-cost, are gaining in popularity. 5-Nitroimidazole antibiotics, such as metronidazole and tinidazole, have been effectively detected in river wate samples using CDs derived from catalpa walnut shells, a food waste product (Sun et al. 2022). Linear discriminant analysis (LDA) was able to discriminate between fluorescent quenching of the different targets in a linear range of 20–5000 nM (Sun et al. 2022).

6.3 Sensing of *Escherichia coli (E. coli)* with Imprinted Polymers

This section solely focuses on *E. coli* as this is often used as indicator molecule for (bacterial) contamination; a high concentration of this MO is associated with a higher chance of the development of accelerated AMR. To our knowledge, there is no MIP-based sensor that can directly discriminate between a MO and its resistant form. However, MIPs would be able to monitor the susceptibility of MOs to antibiotics.

6.3.1 Functional and Crosslinker Monomer Selection for E. coli *Detection*

Imprinting has been widely used for detection of MOs such as *E. coli.* Rather than molecular imprinting, this might involve the use of surface-imprinted polymers (SIPs), where the imprinting is performed with the whole MO (or a substitute such as latex beads) and binding sites are on the surface of the polymer. SIPs have been shown to be selective for detecting its target (Eersels et al. 2013) and beyond sensing, can be used for separation of bacteria (Guo et al. 2012) and cell sorting (Xu et al. 2013). The size and shape of the imprints is important for selective sensing, but Yongabi et al. (2018) have also shown that phospholipid moieties are incorporated in the cavities during the imprinting process and SIP recognition might be mediated via long range hydrophobic forces. Ren and Zare (2012) demonstrated the importance of chemical recognition with cell imprinted polydimethylsiloxane (PDMS) with methylsilane groups resulted in the formation of a cavity with the desired shape of the imprint but lost much of its ability to capture the imprinted bacteria. Moreover, confocal Raman microscopy coupled to Partial Least Squares discriminant analysis to classify the data can be used to distinguish between the polymer surface and imprints, due to differences in surface chemistry (Yongabi et al. 2018).

A portable device, with the dimension of a matchbox, was developed to determine *E. coli* with a concentration as low as 10^2 CFU/mL in relevant food samples (e.g. apple juice) based on changes at the heat-transfer resistance at the solid–liquid interface (Cornelis et al. 2019). Disposable surface-imprinted polyurethane stainless steel sensor chips were produced, which were able to discriminate between *E. coli* and closely related Enterobacteriaceae. While SIPs based on PU can differentiate between different species, there are issues around inclusivity and SIPs imprinted with a certain strain of bacterial species will also bind other strains of the same species. The use of other monomers used in SIP synthesis, such sulphonic acids, carboxylic acids, and heteroatomic bases as discussed by Cui et al. (2020). The recognition affinity for bacteria can be improved via including of other monomers such as MAA, polyvinylpyrrodoline, and polyamines (Pan et al. 2018). Additives such as graphene oxide (GO) can enhance the thermal and electrochemical detection of *E. coli.*

A significant drawback of SIPs is their production time, which can take as long as 18 h due to need for curing of the polymer layer. Electropolymerization is a significantly faster process, which takes minutes rather than hours. Jamieson et al. (2021) formed a PPy-based MIP layer directly on the surface of SPEs and the developed thermal analysis sensor platform detected its target, a model MO (*Saccharomyces cerevisiae*), with high specificity in buffered solutions. Boronic acid-based monomers are also very popular since the hydroxyl groups of boronic covalently react with cis-diols of negatively charged glycoproteins that are abundant on the surface of bacteria (Zhang et al. 2012; Wang et al. 2013). An electrochemical MIP chemosensor for the rapid identification of the *E. coli* K12 strain was developed, which used 2-aminophenyl boronic-acid (2-ABPA) as aniline as functional monomers. An interference study demonstrated that there was a low capacitive change in the presence of strains other than the template bacteria, indicating the unique recognition characteristics of the developed polymer. The thickness of the electrodeposited layer is critical and ultrathin layers (smaller size than the template bacteria) have shown to significantly improve MO binding (Stilman et al. 2022). The crosslinker monomer is an equally important component of the SIPs since it contributes towards the morphology and stability of the imprinted binding sites. For bacteria imprinting, PDMS, polyacrylate, silica and polyurethane are the most common crosslinker monomers (Cui et al. 2020).

6.3.2 *Electrochemical Sensing of* E. coli *with Portable Devices*

Electrochemical methods are compatible with rapid and low-cost detection with portable devices, regardless whether this is for antibiotic or *E. coli* sensing. A layer-by-layer assembly approach, consisting of poly(methacrylic acid) based micro-contact imprinted polymer (PiP) further supplemented with multi-walled carbon nanotubes, nanocellulose films and polyaniline (PANI) doped with phenylboronic acid (PBA) was used to develop a sensor patch with high affinity for *E. coli* in real matrices (Mugo et al. 2020). Binding of the MO to the imprinted cavities resulted in a change in capacitance and impedance, with the sensor recording a limit of detection of *E. coli* of 8 CFU/mL and having a response time of < 5 min. The preparation time of this platform was complex and while it was demonstrated that the sensor could distinguish between Gram positive and Gram negative bacteria, no interference studies with other *E. coli* strains were performed (Fig. 6.7).

Real-life scenarios often involve co-contamination of multiple species of bacterial species and thus the prospect of dual bacteria-imprinted polymers (DBIP) has been studied. There are limited studies to date on this topic, with the main disadvantages of the reported platforms its often-lengthy preparation time (24–48 h) and complex analysis method involving machine learning (Wang et al. 2023). A novel electrochemical platform was developed, which possessed double recognition sites for *E. coli*

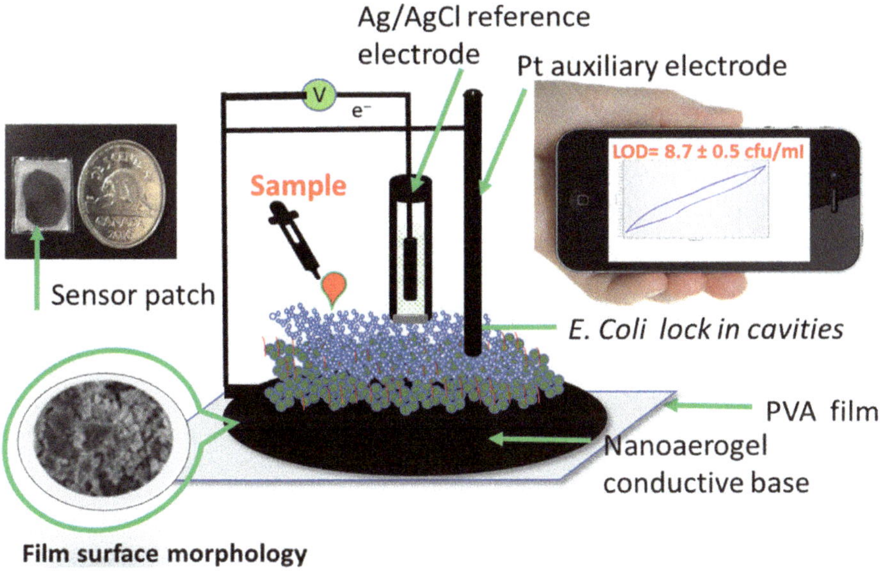

Fig. 6.7 Scheme of the electrochemical sensor for *E. coli*, depicting all its individual components including how it is linked to a smartphone device. Reprinted from Mugo et al. (2020), with permission from Wiley

O157:H7 and *S. aureus*. The in-situ electro-polymerization-based imprinting process yielded a significant reduction in preparation time (~ 20 min), with a measurement time of ~ 15 min (Xu et al. 2023). Detection of *E. coli* O157:H7 and *S. aureus* was possible in the presence of other MOs, but the platform was not able to determine trace levels and was not tested against a wide variety of different species. It has been shown that discrimination between different strains of *E. coli* is complicated with MIPs (Steenredeker et al. 2017), which was not examined in this contribution. Thus, there appears to be a trade-off between sensor complexity and its ability to specifically and selectively detect *E. coli*. For selective detection of MOs, it seems imperative to use rational design, e.g. computational modelling, and a combination of monomers to discriminate between different species and strains of MOs.

6.3.3 Optical Sensing of E. coli *with Portable Devices*

Optical sensing of *E. coli* is less compatible with portable devices compared to integration of electrochemical sensing techniques, thus limiting literature in this field and only a few reports on microfluidic devices were found. For example, a method such as Surface Plasmon Resonance, which is costly and requires bulky equipment, was excluded from this section.

The selective capture of target bacteria involves both shape and chemical recognition; for instance, it was imprinted organosiloxane polymers achieved a 4.5 better selectivity than analogous PDMS imprinted polymers when optical detection was employed (Dulay et al. 2018). This importance of monomer selection was stressed by Mohammadniaei et al. (2020) who compared the performance of MIP-based sensors to aptamers, other emerging receptors, for optical sensing of a range of compounds including *E. coli*. The use of a 96-well plate reader, similar to the format of an ELISA assay, enhances throughput of sensing and will enhance the commercial potential of the technology. For instance, a high-through put system for *E. coli* O157:H7 was developed based on depositing cell-imprinted polypyrrole and nafion film into a gold nanoparticle-coated microplate (Shan et al. 2018). Fluorescent microscopy enabled rapid sensing with a limit of detection of 5.1×10^3 cells, with the selectivity confirmed via testing against a range of different MOs and two other strains of *E. coli*. To enhance the limit of detection of *E. coli* in water, techniques such as magnetic microparticles might be employed. A microfluidic device was developed which created local magnetic to enable the trapping of magnetic cell-imprinted polymers that were coated with fluorescent microparticles. This enabled the sensing of *E. coli* in a dynamic range of 10^2–10^7 CFU/mL, which is a promising approach for rapid detection of MOs in water (Doostmohammadi et al. 2024). It was shown that the sensor's signal towards its original target was twice as high compared to two other strains of bacteria (*E. coli* OP50 and *Sarcina*). However, further work would be required on optimising the cell-imprinted polymers to demonstrate the selectivity of the sensor platform (Fig. 6.8).

An overview of the key characteristics and performance indicators of the MIP-based portable systems for the electrochemical detection of *E. coli* is summarised in Table 6.2.

6.4 Conclusions and Future Outlook

AMR is a global public health threat and key to combatting its development is prevention via monitoring sources of antibiotic and bacterial contamination. This is complicated with traditional sensors that have limited stability, which precludes them for monitoring compounds in situ as the environmental chemistry is highly variable and hostile to biologically derived recognition elements. Thus, due to their low-cost and enhanced robustness, the integration of MIPs into sensors for the monitoring of AMR is an appealing prospect. Moreover, via the adaption of MIP receptors, it is possible to target specific analytes or for instance a family of antibiotics, which is often beneficial for AMR monitoring. Despite this, the application of MIPs in this area remains limited and is traditionally focused on purification and separation. Breakthroughs are needed on large scale production of MIP-based sensors, which are often not compatible with scale-up due to their intricate surface architectures. As antibodies are the "gold standard", it will be a complicated task to move end-users towards MIP-based sensors, which need to provide clear advantages over traditional

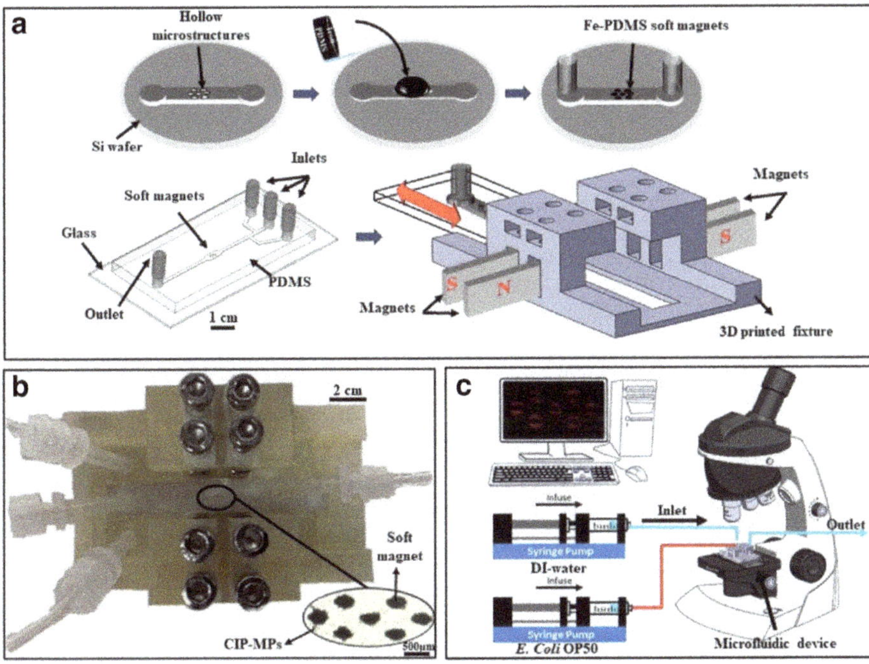

Fig. 6.8 Microfluidic device fabrication procedure and experimental setup. **a** PDMS-glass microfluidic device fabrication and implementation inside the 3d-printed fixture for holding the magnets into position. **b** An actual image of the microfluidic device installed in the 3d-printed fixture. The bottom right inset shows the CIP-MPs accumulating around the oval-shaped soft magnets. **c** Experimental setup consisting of the microfluidic-fixture device mounted on a fluorescence microscope and connected to two syringe pumps for flowing CIP-MPs and bacteria into the channel. Reprinted with permission from Talanta

systems. Therefore, it could be expected that the first commercial application of MIPs for sensors might be for an analyte for which there is currently no antibody available.

We have reviewed portable MIP-based devices for the monitoring of antibiotics and *E. coli*, a common indicator for microbial contamination, in aqueous systems. This was limited to optical and electrochemical sensing methods, which provide the most promise for point-of-care sensing. Optical detection was mainly based on fluorescent techniques, which still require bulky and expensive equipment, and fewer papers were available compared to electrochemical methods. We expect to see more examples of CD-based optical sensors in the future, as these fluorophores are low-cost and can be made from sustainable sources such as commonly disregarded waste products. There are ample examples of electrochemical MIP-based sensors in literature for monitoring of antibiotics but less so for *E. coli* or MO detection. It was demonstrated that electroactive monomers are mostly used for electrochemical sensing, which provides less options to work with and can have a negative impact on selectivity as it can be hard to simultaneously electrodeposit multiple monomers.

Table 6.2 Comparison of MIP-based sensors for electrochemical detection of *E. coli* in terms of functional monomer composition, electrode material, detection technique, limit of detection, and measurement time

Functional monomers	Electrode material	Technique	LoD (CFU/mL)	Measurement time	References
Methacrylic acid	Polyvinyl acetate films modified with doped PANI, CNTs and nanocellulose	CV, EIS	8.7	~ 30 min	Mugo et al. (2020)
3-aminophenylboronic acid	Au disk electrodes	EIS	10	40 min	Wang et al. (2023)
o-phenylenediamine	Glassy carbon electrode (GCE)	CV, EIS	9	10 min	Xu et al. (2023)
N-methacryloyl-L-histidine methyl ester (MAH) and 2-Hydroxyethyl methacrylate (HEMA)	Au electrode	CV	70	30 min	Idil et al. (2017)
Tetraethoxysilane	Ultra-thin inorganic silica coatings on gold electrodes	EIS	1	~ 20 min	Jafari et al. (2019)
Pyrrole	GCE	EIS	21	~ 20 min	Zhu et al. (2023)
3-thiopheneacetic acid	Au electrode	EIS	2	10 min	Wang et al. (2021)

Many examples of electrochemical sensors in literature were diverse in terms of electrode materials used or electrochemical techniques employed. However, most of the sensors employed nanomaterials to enhance the sensor response, which complicates the manufacturing process. Alternative methods for production, such as printing techniques, need to be explored for large scale manufacturing. In that respect, disposable sensors such as SPEs pose an interesting alternative that can be easily modified with MIPs via deposition with robotic handling or via direct integration of the MIPs into the ink. The latter does have requirements for the MIPs in terms of their conductivity.

In summary, MIP-based sensors present a promising alternative for on-site monitoring of AMR, particularly when electrochemical sensors are employed. With the number of publications on electroactive MIPs rapidly increasing, we expect breakthroughs in their employment to battle the ongoing AMR crisis. MIPs have been in commercial use since 1978 for other applications, such as extraction techniques, and there have been efforts to produce commercially viable antibiotic MIP based detection systems. The obstacles which once hindered the commercial use of MIPs for portable, environmental detection have been systematically evaluated and overcame throughout the aforementioned publications so it could be hypothesised that once one commercial application is established for sensors, this is likely to lead to a domino effect and more applications will emerge. In particular, array formatting will significantly enhance the impact of the developed sensor platforms, as there is a dire need to simultaneously measure a range of contaminants.

Acknowledgements & Funding DO would like to thank DST-SERB, Government of India for the SERB international research experience (SIRE) fellowship (SIR/2022/000955). OJ, JR, EH and MP would like to thank the Engineering and Physical Sciences Research Council [grant numbers EP/W031590/1 and EP/S023836/1] for support. TH and CR would like to acknowledge Biotechnology and Biological Sciences Research Council Doctoral Training Partnership for funding for CR's studentship.

References

Abbas K, Ahmed M, Babar ZUD (2023) Trends in prescribing antibiotics between 2012 and 2022: high-income versus low-middle-income countries. In: Encyclopedia of evidence in pharmaceutical public health and health services research in pharmacy, pp 1–12. https://doi.org/10.1007/978-3-030-50247-8_145-1

Ågerstrand M et al (2023) Opportunities to tackle antibiotic resistance development in the aquatic environment through the water framework directive. Ambio 52(5):941–951. https://doi.org/10.1007/s13280-022-01828-7

Ali MR et al (2021) Sensitive MWCNT/P-Cys@MIP sensor for selective electrochemical detection of ceftizoxime. J Mater Sci 56(22):12803–12813. https://doi.org/10.1007/s10853-021-06115-6

Alizadeh T et al (2010) Development of a voltammetric sensor based on a molecularly imprinted polymer (MIP) for caffeine measurement. Electrochim Acta 55(5):1568–1574. https://doi.org/10.1016/j.electacta.2009.09.086

Ayankojo AG et al (2020) Molecularly imprinted polymer-based sensor for electrochemical detection of erythromycin. Talanta 209:120502. https://doi.org/10.1016/j.talanta.2019.120502

Behera SK et al (2011) Occurrence and removal of antibiotics, hormones and several other pharmaceuticals in wastewater treatment plants of the largest industrial city of Korea. Sci Total Environ 409(20):4351–4360. https://doi.org/10.1016/j.scitotenv.2011.07.015

Bottari F (2019) Bio (inspired) strategies for the electro-sensing of β-lactam antibiotics, pp 1–205

Canfarotta F et al (2016) Solid-phase synthesis of molecularly imprinted nanoparticles. Nat Protoc 11:443–455

Casadio S et al (2017) Development of a novel flexible polymer-based biosensor platform for the thermal detection of noradrenaline in aqueous solutions. Chem Eng J 315:459–468. https://doi.org/10.1016/j.cej.2017.01.050

Cervini P, Cavalheiro ÉTG (2012) Strategies for preparation of molecularly imprinted polymers modified electrodes and their application in electroanalysis: a review. Anal Lett 45(4):297–313. https://doi.org/10.1080/00032719.2011.644713

Chambers JP et al (2008) Biosensor recognition elements. Curr Issues Mol Biol 10:1–12

Cornelis P et al (2019) Sensitive and specific detection of *E. coli* using biomimetic receptors in combination with a modified heat-transfer method. Biosens Bioelectron 136(April):97–105. https://doi.org/10.1016/j.bios.2019.04.026

Crapnell RD et al (2019) Recent advances in electrosynthesized molecularly imprinted polymer sensing platforms for bioanalyte detection. Sensors (Switzerland) 19(5). https://doi.org/10.3390/s19051204

Cui F, Zhang X (2012) Electrochemical sensor for epinephrine based on a glassy carbon electrode modified with graphene/gold nanocomposites. J Electroanal Chem 669:35–41. https://doi.org/10.1016/j.jelechem.2012.01.021

Cui F, Zhou Z, Zhou HS (2020) Molecularly imprinted polymers and surface imprinted polymers based electrochemical biosensor for infectious diseases. Sensors (Switzerland) 20(4). https://doi.org/10.3390/s20040996

de Kraker M (2016) Will 10 million people die a year due to antimicrobial resistance by 2050? PLoS Med 13(11):1–6. https://doi.org/10.1371/journal.pmed.1002184

Dechtrirat D et al (2018) A screen-printed carbon electrode modified with gold nanoparticles, poly(3,4-ethylenedioxythiophene), poly(styrene sulfonate) and a molecular imprint for voltammetric determination of nitrofurantoin. Microchim Acta 185(5). https://doi.org/10.1007/s00604-018-2797-3

Devkota L et al (2018) Electrochemical determination of tetracycline using AuNP-coated molecularly imprinted overoxidized polypyrrole sensing interface. Electrochim Acta 270:535–542. https://doi.org/10.1016/j.electacta.2018.03.104

Dewaele I et al (2011) Sampling, prevalence and characterization of methicillin-resistant *Staphylococcus aureus* on two Belgian pig farms. Vet Sci Dev 1(1):1. https://doi.org/10.4081/vsd.2011.e1

Doostmohammadi A et al (2024) Fluorescent bacteria detection in water using cell imprinted polymer (CIP) coated microparticles in a magnetophoretic microfluidic device. Talanta 268(P1):125290. https://doi.org/10.1016/j.talanta.2023.125290

Dulay M et al (2018) Pathogen-imprinted organosiloxane polymers as selective biosensors for the detection of targeted *E.coli*. J Carbon Res 4:29. https://doi.org.https://doi.org/10.3390/c4020029

Eersels K et al (2013) Selective identification of macrophages and cancer cells based on thermal transport through surface-imprinted polymer layers. ACS Appl Mater Interfaces 5(15):7258–7267. https://doi.org/10.1021/am401605d

El Azab NF et al (2022) Point-of-care diagnostics for therapeutic monitoring of levofloxacin in human plasma utilizing electrochemical sensor mussel-inspired molecularly imprinted copolymer. J Electroanal Chem 918:116504

European Commission (2006) Directive 2006/7/EC of the European Parliament and of the Council of 15 February 2006 concerning the management of bathing water quality and repealing Directive 76/160/EEC. In: Official Journal of the European Union, L 064(1882), pp 37–51

European Commission (2019) European Union strategic approach to pharmaceuticals in the environment. In: Communication from the Commission to the European parliament, the council and the European economic and social committee

European Commission (2022) Urban wastewater. Available at: https://environment.ec.europa.eu/topics/water/urban-wastewater_en#review. Accessed 12 Dec 2023

Fernandez-Gonzalez A et al (2006) Mimicking molecular receptors for antibiotics—analytical implications. TrAC, Trends Anal Chem 25(10):949–957

Geng Y et al (2018) A fluorescent molecularly imprinted polymer using aptamer as a functional monomer for sensing of kanamycin. Sens Actuators, B Chem 268:47–54. https://doi.org/10.1016/j.snb.2018.04.065

Giovanolli et al (2013) Solid phase extraction of penicillins from milk using sacrificial silica beads as a support for a molecular imprint. Microchim Acta 180:1371–1377

Girones R et al (2010) Molecular detection of pathogens in water—the pros and cons of molecular techniques. Water Res 44(15):4325–4339. https://doi.org/10.1016/j.watres.2010.06.030

Gomi R et al (2014) Fecal source tracking in water by next-generation sequencing technologies using host-specific *Escherichia coli* genetic markers. Environ Sci Technol 48(16):9616–9623. https://doi.org/10.1021/es501944c

Guo P et al (2012) Microfluidic capture and release of bacteria in a conical nanopore array. Lab Chip 12(3):558–561. https://doi.org/10.1039/c2lc21092d

Haupt, and, Mosbach (2000) Molecularly imprinted polymers and their use in biomimetic sensors. Chem Rev 100(7):2495–2504

Heinze J (2005) Electronically conducting polymers. Electrochemistry IV, pp 1–47
Hu X et al (2022) A signal-amplified ratiometric fluorescence biomimetic sensor based on the synergistic effect of IFE and AE for the visual smart monitoring of oxytetracycline. Chem Eng J 433(P1):134499. https://doi.org/10.1016/j.cej.2022.134499
Hudson AD et al (2021) Dual detection of nafcillin using a molecularly imprinted polymer-based platform coupled to thermal and fluorescence read-out. Mater Adv 2(15):5105–5115. https://doi.org/10.1039/d1ma00192b
Idil N et al (2017) Whole cell based microcontact imprinted capacitive biosensor for the detection of *Escherichia coli*. Biosens Bioelectron 87:807–815. https://doi.org/10.1016/j.bios.2016.08.096
Jafari H et al (2019) Entrapment of uropathogenic *E.coli* cells into ultra-thin sol-gel matrices on gold thin films: a low cost alternative for impedimetric bacteria sensing. Biosens Bioelectron 124–125:161–166. https://doi.org/10.1016/j.bios.2018.10.029
Jamieson O et al (2020) Screen printed electrode based detection systems for the antibiotic amoxicillin in aqueous samples utilising molecularly imprinted polymers as synthetic receptors. Chemosensors 8(1). https://doi.org/10.3390/chemosensors8010005
Jamieson O et al (2021) Electropolymerised molecularly imprinted polymers for the heat-transfer based detection of microorganisms: a proof-of-concept study using yeast. Therm Sci Eng Prog 24(April):100956. https://doi.org/10.1016/j.tsep.2021.100956
Jin H et al (2018) Selective and sensitive electrochemical sensing of gastrodin based on nickel foam modified with reduced graphene oxide/silver nanoparticles complex-encapsulated molecularly imprinted polymers. Sens Actuators, B Chem 277(August):14–21. https://doi.org/10.1016/j.snb.2018.08.156
Langford BJ et al (2021) Antibiotic prescribing in patients with COVID-19: rapid review and meta-analysis. Clin Microbiol Infect 27(4):520–531. https://doi.org/10.1016/j.cmi.2020.12.018
Li XG et al (2002) Novel multifunctional polymers from aromatic diamines by oxidative polymerizations. Chem Rev 102(9):2925–3030. https://doi.org/10.1021/cr010423z
Li M et al (2012) Recent developments and applications of screen-printed electrodes in environmental assays—a review. Anal Chim Acta 734:31–44. https://doi.org/10.1016/j.aca.2012.05.018
Losito I, Palmisano F, Zambonin PG (2003) O-Phenylenediamine electropolymerization by cyclic voltammetry combined with electrospray ionization-ion trap mass spectrometry. Anal Chem 75(19):4988–4995. https://doi.org/10.1021/ac0342424
Lowdon JW et al (2020) MIPs for commercial application in low-cost sensors and assays—an overview of the current status quo. Sens Actuat B Chem 325:128973. https://doi.org/10.1016/j.snb.2020.128973
McClements J et al (2022) Molecularly imprinted polymer nanoparticles enable rapid, reliable, and robust point-of-care thermal detection of SARS-CoV-2. ACS Sensors 7(4):1122–1131. https://doi.org/10.1021/acssensors.2c00100
Mohammadniaei M et al (2020) The use of aptamers and molecularly imprinted polymers in biosensors for environmental monitoring: a tale of two receptors. Chemosensors 8:32. https://doi.og/10/3390/chemosensors8020032
Moro G et al (2019) Conductive imprinted polymers for the direct electrochemical detection of B-lactam antibiotics: the case of cefquinome. Sens Actuators, B Chem 297:126786. https://doi.org/10.1016/j.snb.2019.126786
Mugo SM, Lu W, Dhanjai D (2020) A pathogen imprinted hybrid polymer capacitive sensor for selective Escherichia coli detection. Med Devices Sens 3(2):1–8. https://doi.org/10.1002/mds3.10071
Murray CJ et al (2022) Global burden of bacterial antimicrobial resistance in 2019: a systematic analysis. Lancet 399(10325):629–655. https://doi.org/10.1016/S0140-6736(21)02724-0
Naveen MH, Gurudatt NG, Shim YB (2017) Applications of conducting polymer composites to electrochemical sensors: a review. Appl Mater Today 9:419–433. https://doi.org/10.1016/j.apmt.2017.09.001

O'Neill, (2016) Book review: tackling drug-resistant infections globally. Arch Pharm Pract 7(3):110. https://doi.org/10.4103/2045-080x.186181

Ozcelikay G et al (2020) Au-Pt nanoparticles based molecularly imprinted nanosensor for electochemical detection of the lipopeptide antibiotic drug Daptomycin. Sens Actuators, B Chem 320:128285

Paíga P, Santos LHMLM, Delerue-Matos C (2017) Development of a multi-residue method for the determination of human and veterinary pharmaceuticals and some of their metabolites in aqueous environmental matrices by SPE-UHPLC–MS/MS. J Pharm Biomed Anal 135:75–86. https://doi.org/10.1016/j.jpba.2016.12.013

Pan J et al (2018) Molecularly imprinted polymers as receptor mimics for selective cell recognition. Chem Soc Rev 47(15):5574–5587. https://doi.org/10.1039/c7cs00854f

Ren K, Zare RN (2012) Chemical recognition in cell-imprinted polymers. ACS Nano 6(5):4314–4318. https://doi.org/10.1021/nn300901z

Rohani Moghadam M et al (2019) Voltammetric sensing of oxacillin by using a screen-printed electrode modified with molecularly imprinted polyaniline, gold nanourchins and graphene oxide. Microchim Acta 186(12). https://doi.org/10.1007/s00604-019-3981-9

Sahlin S, Larsson DGJ, Ågerstrand M (2018) Ciprofloxacin EQS data overview. In: ACES REPORT 15, Stockholm University, Stockholm, pp 1–58

Shan X et al (2018) A rapid and specific bacterial detection method based on cell-imprinted microplates. Analyst 143(7):1568–1574. https://doi.org/10.1039/c7an02057k

Singhal A et al (2022) Multifunctional carbon nanomaterials decorated molecularly imprinted hybrid polymers for efficient electrochemical antibiotics sensing. J Environ Chem Eng 10(3):107703. https://doi.org/10.1016/j.jece.2022.107703

Song X et al (2020) Synthesis of molecularly imprinted polymers for the selective extraction of polymyxins from environmental water samples. Polymers 12(1). https://doi.org/10.3390/polym12010131

Spicer N (2015) Global health. JAMA J Am Med Assoc 42(3):89–99

Steen Redeker E et al (2017) Biomimetic bacterial identification platform based on thermal wave transport analysis (TWTA) through surface-imprinted polymers. ACS Infect Dis 3(5):388–397. https://doi.org/10.1021/acsinfecdis.7b00037

Stilman W et al (2022) Low cost, sensitive impedance detection of *E. coli* bacteria in food-matrix samples using surface-imprinted polymers as whole-cell receptors. Phys Status Solidi (A) Appl Mater Sci 219(23):1–13. https://doi.org/10.1002/pssa.202100405

Stoian A et al (2020) Biomimetic electrochemical sensor for the highly selective detection of azithromycin in biological samples. Biosens Bioelectron 155:112098

Sun Q et al (2022) Construction of biomass carbon dots@molecularly imprinted polymer fluorescent sensor array for accurate identification of 5-nitroimidazole antibiotics. Sens Actuators, B Chem 373(September):132716. https://doi.org/10.1016/j.snb.2022.132716

Torkashvand M et al (2016) Construction of a new electrochemical sensor based on molecular imprinting recognition sites on multiwall carbon nanotube surface for analysis of ceftazidime in real samples. Sens Actuators, B Chem 231:759–767

UK Government (2013) The bathing water regulations 2013. Available at: https://www.legislation.gov.uk/uksi/2013/1675/part/5/made. Accessed: 12 Dec 2023

UK Government (2016) The water supply regulations. Available at: https://www.legislation.gov.uk/uksi/2016/614/pdfs/uksi_20160614_en.pdf. Accessed: 12 Dec 2023

Vu OT et al (2022) Highly sensitive molecularly imprinted polymer-based electrochemical sensors enhanced by gold nanoparticles for norfloxacin detection in aquaculture water. ACS Omega. https://doi.org/10.1021/acsomega.2c04414

Wan W et al (2013) Fluorescent sensory microparticles that "light-up" consisting of a silica core and a molecularly imprinted polymer (MIP) shell. Angew Chem Int Edition 52(27):7023–7027. https://doi.org/10.1002/anie.201300322

Wang H et al (2013) Magnetic nanoparticles with dendrimer-assisted boronate avidity for the selective enrichment of trace glycoproteins. Chem Sci 4(11):4298–4303. https://doi.org/10.1039/c3sc51623g

Wang R et al (2021) Rapid, sensitive and label-free detection of pathogenic bacteria using bacteria-imprinted conducting polymer film-based electrochemical sensor. Talanta 226:122135. https://doi.org/10.1016/j.talanta.2021.122135

Wang YQ et al (2022) Sensitive determination of ofloxacin by molecularly imprinted polymers containing ionic liquid functionalized carbon quantum dots and europium ion. ACS Appl Nano Mater 5(6):8467–8474. https://doi.org/10.1021/acsanm.2c01583

Wang C et al (2023) Machine learning-assisted cell-imprinted electrochemical impedance sensor for qualitative and quantitative analysis of three bacteria. Sens Actuators, B Chem 384(March):133672. https://doi.org/10.1016/j.snb.2023.133672

Xie L et al (2020) Theoretical insight into the interaction between chloramphenicol and functional monomer (Methacrylic acid) in molecularly imprinted polymers. Int J Mol Sci 21(11):1–14. https://doi.org/10.3390/ijms21114139

Xu L et al (2013) Specific recognition of tyrosine-phosphorylated peptides by epitope imprinting of phenylphosphonic acid. J Chromatogr A 1293:85–91. https://doi.org/10.1016/j.chroma.2013.04.013

Xu X et al (2023) A novel dual bacteria-imprinted polymer sensor for highly selective and rapid detection of pathogenic bacteria. Biosensors 13:868. https://doi.org/10.3390/bios13090868

Yongabi D et al (2018) Cell detection by surface imprinted polymers SIPs: a study to unravel the recognition mechanisms. Sens Actuators, B Chem 255:907–917. https://doi.org/10.1016/j.snb.2017.08.122

Zhang L, Chen L (2016) Fluorescence probe based on hybrid mesoporous silica/quantum dot/molecularly imprinted polymer for detection of tetracycline. ACS Appl Mater Interfaces 8(25):16248–16256. https://doi.org/10.1021/acsami.6b04381

Zhang X et al (2012) Boronic acid modified magnetic nanoparticles for enrichment of glycoproteins via azide and alkyne click chemistry. J Mater Chem 22(32):16520–16526. https://doi.org/10.1039/c2jm32987e

Zhang H et al (2019) Molecularly imprinted sensor based on o-phenylenediamine for elctrochemical detection of sulfamethoxazole. Int J Electrochem Sci 14(12):10630–10640

Zhu M et al (2023) Bacteria-imprinted impedimetric sensor based on doping-induced nanostructured polypyrrole for determination of *Escherichia coli*. Microchim Acta 190, article number 431. https://doi.org/10.1007/s00604-023-06008-2

Zouaoui F et al (2020) Electrochemical sensors based on molecularly imprinted chitosan: a review. TrAC Trends Anal Chem 130:115982. https://doi.org/10.1016/j.trac.2020.115982

Chapter 7
MIPs-Based Sensors and Biosensors for Environmental Monitoring

Lanqing Yang, Kun Ge, Muhammad Farhan Qadir, Xiaomin Wang, Ying Gu, and Yukun Yang

7.1 Introduction

Pollution is commonly defined as the direct or indirect introduction of substances or energy into the environment by humans, resulting in damage to human health, biological resources, ecological and physical systems of the biosphere (Holdgate 1980; Khanmohammadi et al. 2020). Since the increase in industrialization and manufacturing industries, the problem of environmental pollution has become increasingly serious, for example, the excessive use of pesticides and herbicides, the abuse of antibiotics and so on are the causes of environmental pollution. In addition, heavy metal ions, pathogen contaminants, and other pollutants can enter agricultural products and livestock through river water and soil, and eventually be consumed by humans leading to highest threat and risks to human health. Determining the level of pollution in the environment is therefore a matter of particular importance for ecological balance and human health (Hussain and Kecili 2019; Hussain and Kharisov 2016; Keçili et al. 2021).

It is a great challenge to analyse the wide variety of pollutants in the environment, with their diverse nature and very low concentrations. However, traditional detection methods include instrumental analysis methods, polymerase chain reaction (PCR) (Meece et al. 2010), immunokit method (Wang et al. 2009), culture counting method (Lunau et al. 2005), etc. Instrumental methods based on chromatography (Hoellein

L. Yang · M. F. Qadir · Y. Yang (✉)
School of Life Science, Shanxi University, Taiyuan 030006, China
e-mail: yangyukun@sxu.edu.cn

K. Ge · Y. Gu
Faculty of Food Science and Engineering, Kunming University of Science and Technology, Kunming 650500, China

X. Wang
Institute of Pharmaceutical and Food Engineering, Shanxi University of Chinese Medicine, Yuci 030619, China

S. Patra and M. Sillanpaa (eds.), *Molecularly Imprinted Polymers as Artificial Antibodies for the Environmental Health*, https://doi.org/10.1007/978-3-031-58995-9_7

and Holzgrabe 2014), mass spectrometry (Amulen et al. 2017) and spectroscopy (Cozzolino 2016) can provide high detection sensitivity, but they are expensive, require strict pretreatment of samples and professional operators, and cannot meet the needs of on-site detection. PCR-based methods are costly and require specific gene amplification. The immune kit method has the characteristics of fast detection speed and good specificity, which can lead to rapid detection of a large number of samples, but the experimental results are prone to false positive. Culture counting method is time-consuming and laborious. Therefore, it is of great significance to develop a universal, simple, fast, economical and on-site detection method.

More attention has been paid to the use of low-cost and easy-to-operate analytical equipment, and sensors and biosensors have been developed to detect a wide range of trace hazards. The sensor refers to the identification of target molecules by identifying elements, and the identification reaction is converted into measurable chemical or physical signals by signal converters to achieve quantitative or semi-quantitative detection of the target object. Depending on the difference in the converted quantitative signal, common sensor types are fluorescent sensing (FL) (Guo et al. 2015), colorimetric sensing (CL) (Poornima et al. 2016), surface enhanced Raman scattering (SERS) (Hu et al. 2014), surface plasmon resonance (SPR) (Bergwerff and Knapen 2006), photo-electrochemical sensing (PEC) (Wang et al. 2022a), electrochemical sensing (EC) (Ju and Chen 2015), electrochemiluminescence sensing (ECL) (Guan et al. 2016).

High specific and accurate detection of pollution in complex environmental substrates is a major challenge in sensors and biosensors detection. The purpose of selective detection of target objects was achieved through the combination of biological receptors (enzymes, antibodies, proteins) and biomimetic synthetic materials molecularly imprinted polymers (MIPs) with sensors and biosensors. Compared with biological receptors, MIPs, which has the advantages of simple preparation, low cost, high stability and reusable, has become a famous recognition element for sensor and biosensor development (Ansari and Karimi 2017; Figueiredo et al. 2016; Gui et al. 2018). As a result, the past decade has witnessed an explosion in the development of MIPs-based sensors and biosensors, as well as a focus on their application in the emerging pollutants monitoring. To sum up, MIPs-based sensors and MIPs-based biosensors show surprising advantages in environmental pollution detection.

In this chapter, a comprehensive overview of MIPs-based sensors and biosensors for detecting pollutants in the environment is presented. Firstly, the preparation method of MIPs and the strategy of combining MIPs with different sensors and biosensors are briefly introduced. The recent research progress of MIPs-based sensors and biosensors in the detection of environmental pollutants is comprehensively reviewed. Finally, the challenges and prospects of MIPs-based sensors and biosensors are also discussed.

7.2 Preparation of MIPs and MIPs as Recognition Elements for Sensors and Biosensors

MIPs are specially formulated synthetic materials whose cavities are recognition sites that bind specifically to the target (Chen et al. 2016). There are three main processes for the synthesis of MIPs. First, the target, functional monomer and cross linker are polymerized to form a polymerization. The target is then eluted with an eluent leaving specific cavity. Finally, these cavities can re-selectively trap the target when it is present. In this way, MIPs have a strong ability to selectively recognition target molecules even when the target and structural analogues of the target co-exist (Chen et al. 2016; Madikizela et al. 2018). This process is very similar to the 'lock and key' mechanism of enzymes (Shi et al. 2023; Yang et al. 2021a). Moreover, the high selectivity of MIPs makes MIPs-based sensors and biosensors extremely advantageous in detecting trace pollutants of complex environmental samples.

7.2.1 Preparation of MIPs

The synthesis methods of MIPs include non-covalent chemical interactions (such as π-π interactions, hydrogen bonds, ionic interactions, van der Waals forces, etc.), covalent bonds and semi-covalent bonds between template molecules and functional monomers prior to the polymerization process (Speltini et al. 2017; Wloch and Datta 2019). The first method is the simplest and most economical, however, low selectivity is its fatal disadvantage (Ndunda and Mizaikoff 2016). At the same time, the specificity and sensitivity of the second method are significantly improved, but it is difficult to completely eluate the template. There is also a third method that connects the target to the functional monomer through covalent bonds during the imprinting process, and then re-binds the target using non-covalent bonds (Qi et al. 2010).

The synthesizing technique is a crucial step in determining MIPs qualities. The traditional synthesis methods include bulk polymerization, precipitation polymerization, suspension polymerization, and emulsion polymerization. Lower cost and simpler synthesis methods are the reasons for the widespread use of the native polymerization method. However, a common disadvantage of MIPs prepared by bulk polymerization is the loss of binding sites in crushing the cross-linked material into powder, which can result into low yields (Chen et al. 2011; Farooq et al. 2018). In general, precipitation polymerization, suspension polymerization, and emulsion polymerization for the synthesis of MIPs require large amounts of organic solvents, leading to a variety of undesirable synthetic results (Yadav et al. 2022).

In order to solve the problems of traditional imprinting methods, surface imprinting methods have emerged as an effective way to construct sensing interface of sensors and biosensors. The most common methods include electropolymerization and sol–gel processes. Electropolymerization preparation method is suitable for the direct construction of sensing interface, with the advantages of simple preparation and controllable conditions, which can be used to prepare MIPs by direct electropolymerization on the electrode surface or by electropolymerization after modifying the nanomaterials (Crapnell et al. 2019). Simple and mild sol–gel methods prepare materials with uniform particle size, which can also be modified directly at the functional interface using electrodeposition (Moein et al. 2019; Yang et al. 2013).

In conclusion, nano-MIPs are far superior to bulk-MIPs. The main reasons are that bulk-MIPs tend to lose specific binding sites to target molecules, have lower yields, and have non-specific binding sites, making it more difficult to blot biomolecular. In contrast, the polymerization process of nano-MIPs is similar to that of bulk-MIPs, but with fewer processing steps, and the possibility of the blotting cavity being destroyed is small substantially reduced. In addition, nano-MIPs have good stability under extreme conditions such as high temperature and high pressure, and still have good performance after long time storage. The nano-MIPs have uniform particle size, controllable conditions, high selectivity and good specificity (Roy et al. 2017; Thakur et al. 2023).

In recent years, the concept of smart imprinting has come into public view. Smart imprinting changes its affinity for the target molecule by providing the possibility of conversion to binding or release in response to stimuli such as temperature, incident light, pH, ionic strength, etc. (Morsi et al. 2023).

7.2.2 *MIPs as Recognition Elements for Sensors and Biosensors*

Combining MIPs with different types of sensors enables highly selective, sensitive and low-cost detection of trace contaminants in the environment. The integration of MIPs and the sensing interface is a key scientific problem in constructing MIPs-based sensors (Eersels et al. 2016; Yang et al. 2022a). For the construction MIPs-based electrochemical sensors, electrochemiluminescence sensors, surface plasmon resonance sensor and photoelectrochemical sensors, they all involve modification of the sensing functional interface, and there are several general methods. Firstly, MIPs can be prepared on the electrode surface by the simple electropolymerization and electrodeposition (Roushani and Zalpour 2021). Secondly, the MIPs is mixed with the material of the working electrode to construct the MIPs electrode (Yang et al. 2022b). Thirdly, previously prepared MIPs or MIPs-based composites are sequentially fixed on the electrode surface by a step-by-step modification through dropping or adhesive bonding (Yang et al. 2021b). Nanomaterials with excellent optical, magnetic and electrical properties were extensively applied in the construction of the sensing interface to greatly improve the sensing performance. The emerging nanomaterials are metal nanoparticles (Rao et al. 2017), metal oxide nanoparticles (Farid et al. 2016), carbon nanotubes (Rezaei et al. 2014), graphene (Mao et al. 2011), mesoporous carbon (Yang et al. 2015), etc.

FL, CL, SERS are all optical sensors, the combination of MIPs with different types of sensors to construct MIPs-based sensor are unique. Commonly, the strategies for constructing MIPs-based fluorescent sensors can be divided into two main categories. When the target has fluorescent properties, the target generates fluorescent signal after recombining with MIPs to achieve quantitative detection of the target (Ton et al. 2012). When the target does not have fluorescent properties, fluorescent MIPs are usually prepared using surface molecularly imprinting on the fluorescent materials, and quantitative detection of the target is achieved by the change of fluorescence intensity or the change of fluorescence color after target recognition (Ghani et al. 2020). Fluorescent MIPs can be prepared using fluorescent functional monomers or just embedded fluorescent materials into the MIPs. Familiar fluorescent materials include fluorescent dyes (Xu et al. 2017), metal–organic framework materials (Yang et al. 2024), quantum dots (QDs) (Nsibande and Forbes 2016), metal nanoclusters (Chen et al. 2023a), lanthanide chelates (Donmez et al. 2017), rare earth materials (Qiao et al. 2019). Although fluorescence and colorimetric sensors are both optical sensors, unlike fluorescence sensors based on MIPs, the color development mechanism of MIPs based colorimetric sensor includes surface plasmon resonance color development (Ghasemi et al. 2015), color development with peroxide simulated enzyme activity (Gao et al. 2007) and color development catalyzed by non-enzyme substrate (Xiao et al. 2017). In general, MIPs-SERS sensors are divided into simple MIP-SERS sensors and complex MIPs-SERS sensors. Simple MIPs-SERS sensors include core–shell (SERS substrate as the core and MIPs as the shell) (Bompart et al. 2010), planar (MIPs membrane or microspheres coated on a planar SERS substrate) (Kostrewa et al. 2003; Kamra et al. 2015), and sandwich (MIPs-target molecule-Raman reporter molecule) types (Ye et al. 2014). The so-called complex MIPs-SERS method consists of two main steps (Feng et al. 2013): separation, enrichment and elution of the target with MIP, followed by dispersion on various SERS substrates for SERS detection. MIPs-SPR sensors usually enrich the target after modifying the MIPs film on the metal surface of the SPR sensor, and the accompanying SPR signal will be enhanced.

In theory, MIPs can be designed to identify any target existed. However, there are still many challenges in developing MIPs-based sensors to detect targets of macromolecules due to the difficulty of imprinting and the weak imprinting effect. Especially in the case of high specificity and ultra-sensitivity, only MIPs for target identification limits its application in practical sample detection. Whereas aptamer has the advantages of smaller size, longer storage time and easy modification (Negahdary 2020), antibodies have good antigen specificity and affinity (Zhao et al. 2021). Therefore, the development of biosensors integrating MIPs with aptamer and MIPs with antibody is an inevitable trend in the development of biosensor detection. There are three general strategies for constructing MIPs-Apt sensors. Firstly, the target is first recognized by signal-labeled Apt (for example, methylene blue and ferrocene as electrochemical signal-labeled Apt) to form an Apt-target complex, and then MIPs is wrapped around the outer layer of the Apt-target complex. The target is then eluted from the cavity. Different quantitative signals are used to evaluate the rebinding ability of the target and the cavity (Jolly et al. 2016). Secondly, the target is recognized by

MIPs before elution, and the eluted target binds to the aptamer for quantitative detection (Li et al. 2018). Thirdly, sandwich structure of MIPs-target-Apt is formed a thus reporting the signal. After the target is captured by MIPs, the target is combined with the Apt of the signal marker to produce a quantitative signal (You et al. 2020). MIPs-antibody sensors are usually sandwich structures of antibody-target-MIPs. Similarly, after the antibody is specifically bound to the target, the target can enter the specific cavity of the MIPs (Song et al. 2022a).

It is worth mentioning that improving the sensitivity and selectivity of sensing detection is a challenge for all MIPs-based sensors. Therefore, future research should focus on combining novel nanomaterials, nanoenzymes and some amplification strategies to enhance the signal strength and improve the sensitivity and accuracy of sensing detection. In addition, the combination of cutting-edge technologies in digital communication networks, artificial intelligence (AI) and MIPs-based sensors has made possible for the construction of intelligent analytical tools, which greatly improves the possibility of pollutant detection in environmental samples.

7.3 MIPs-Based Sensors and Biosensors for Contaminants Analysis

7.3.1 Antibiotics

Since the first discovery of penicillin by Alexander Fleming in 1928, various antibiotics have been widely used in lives saving. However, the antibiotic residues induced by massive use of antibiotics is becoming the serious threat to environment and human health (Jin et al. 2022). For example, the antibiotics are widely used in food-producing animal breeding for ensuring the health of animal and improving production of meat. The serious antibiotic residues yield drug-resistant pathogens (super bacteria), as well as accumulation through whole food chain. In recent years, the establishment of various MIPs-based sensors have become more popular trend for antibiotic monitoring due to their rapidity, portability, high accuracy, and specificity.

MIPs-based electrochemistry sensors have been considered as a powerful tool for antibiotic monitoring due to its coupling advantages of high selectivity and sensitivity (Stoian et al. 2020; Afsharara et al. 2023). Li et al. (2022) designed and fabricated a MIPs voltammetric sensor for selective determination of norfloxacin (NOR) in pharmaceutical and environmental samples based on MIPs/Au nanoparticles decorated black phosphorus nanosheet (MIPs/BPNS-AuNPs). As expected, the BPNS-AuNPs exhibited two main effects, firstly, providing large surface area for more template molecules locating and secondly, ensuring synergistic catalytic effect for NOR oxidation. As a consequence, the sensing performance for NOR with limit of detection (LOD) of 0.012 nM and linear range of 0.1 nM–10 μM can be obtained. These results provide a new strategy for design and fabrication of MIPs electrochemical sensors with high performance using effective support substrate of versatile BPNS-AuNPs composites. Another MIPs-quartz crystal microbalance (QCM) sensors were proposed by Ayankojo's team for quantitative analysis of amoxicillin (AMO) in aqueous samples (Ayankojo et al. 2018a). The prepolymerization reaction among the monomer molecules (meta-phenylenediamine, mPD) and the template (AMO) was achieved. Comparing with reference film, developed AMO-MIPs QCM sensor exhibited better adsorption capacity and sensitivity, indicating an important step towards the construction of MIPs-based electrochemical sensor for environmental analysis. Furthermore, the same research team also constructed a portable environmental sensor based on sulfamethizole (SMZ)-imprinted polymer modified screen-printed electrodes (SPE) (Fig. 7.1a) (Ayankojo et al. 2020). The electrochemical technique of differential pulse voltammetry (DPV) was used to confirm the analytical performance, exhibiting rapid response, good sensitivity (LOD = 0.9 nM), high selectivity. This study provides a promising potentials in fabrication of commercial viable platform with peculiarities of low cost, portability, and ease of operation for antibiotic monitoring in environmental water. Wang et al. (2022b) proposed a novel array sensor (ET) for simultaneous analysis of fluoroquinolones (FQs) including ciprofloxacin (CFX), moxifloxacin (MFX) and levofloxacin (LFX) based on electropolymerized MIPs film on Au nanoparticle-decorated multiwall carbon nanotubes (MIP(FQs)@Au-fMWCNTs) and machine learning (ML) technology. This fabricated MIPs(FQs)@Au-fMWCNT sensors exhibited different responses for three FQs with LOD of 1.0 μM and the limitation of MIP's fcross-response can be significantly tackled by ML approach. In addition, the strategy of dual functional monomers was adapted by Nguyen et al. (2023) for fabrication of macrolide MIPs-SPE sensors in quantitative analysis of erythromycin (ERY), azithromycin (AZI), and clarithromycin (CLA) in aqueous media. This study exhibited inspiring and vital fabrication of MIPs-SPE sensor's outcome in providing great possibility for broad recognition of individual molecules of this class. The MIPs-based electrochemistry sensors were recognized as valuable and powerful tool for antibiotic determination in environment with high sensitivity, selectivity, and rapidity.

Antibiotic monitoring in the environment has been done through MIPs fluorescence sensor based technology, so in this regard Tang et al. (2017) proposed a shell modified up-conversion particles (UCPs, $NaYF_4$: Er, Yb) by MIPs fluorescence sensor based technology for quantitative analysis of enrofloxacin (ENR). The UCPs were firstly covered by SiO_2 layer, modified and followed by polymerization of MIPs shell (UCPs@MIP). With the increasing of ENR concentration in range of 63.0 ng/L–60.0 μg/L, the fluorescence intensity decreased linearly with LOD of 8.0 ng/L. Further, a dual functional fluorescence sensor based MIPs coated study was proposed by Wang et al. (2020a) for the visual and fluorescent analysis of NOR in aqueous solutions. The LOD of ENR for visual and fluorescent quantitative can be reached to 0.1 μg/mL and 2.6 ng/mL, indicating the good potentials in on-site semi-quantitative (visualization) and quantitative (fluorescence) analysis of antibiotics. Another thermosensitive magnetic fluorescence MIPs sensor was fabricated for highly analysis of doxycycline (DOX) based on carbon dots (CDs) covered Fe_3O_4 combined with MIP layer (Li et al. 2023a). The rapid and selective enrichment of DOX result in

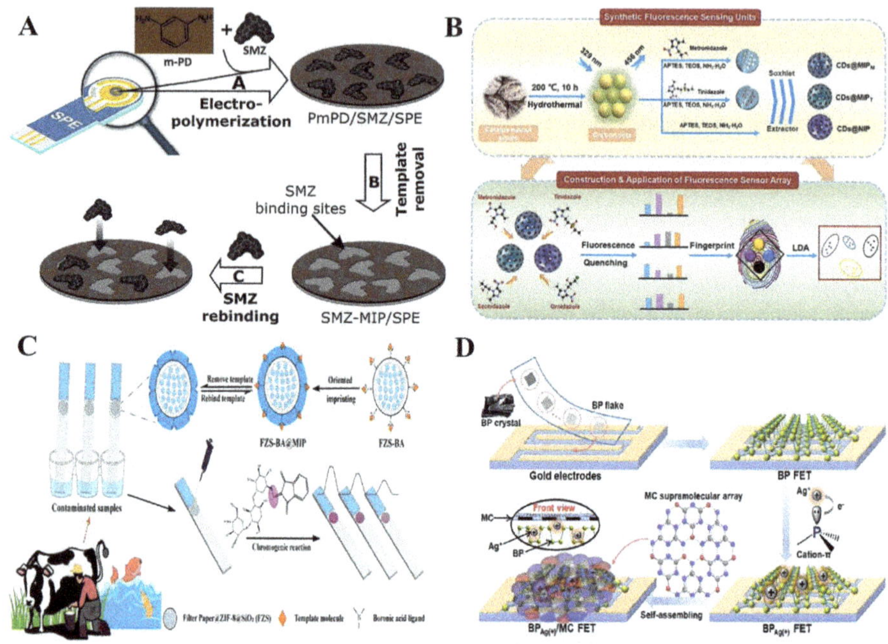

Fig. 7.1 **a** Scheme for SMZ-MIPs/SPE preparation. Reproduced with permission (Ayankojo et al. 2020). Copyright 2020, Elsevier. **b** Construction strategy of fluorescent sensor array and application to recognition of 5-nitroimidazole antibiotics. Reproduced with permission (Sun et al. 2022). Copyright 2022, Elsevier. **c** Schematic diagram of the preparation and operation of the integrated analytical platform for KAN detection. Reproduced with permission (Song et al. 2022b). Copyright 2022, American Chemical Society. **d** Schematic of the BP FET fabrication process: micromechanical exfoliation, Ag^+ coordination, and MC supramolecular passivation (Chen et al. 2023b). Copyright 2023, American Chemical Society

the significant fluorescence quenching of CDs due to internal filtration effect (IFE) and separation and enrichment effect of Fe_3O_4. The good analytical performance with linear range of 0.5–30 μg/Land LOD of 0.2 μg/L can be obtained. In addition a ratiometric fluorescent sensor based on MIPs containing europium ions (Eu^{3+}) and ionic liquid functionalized CDs was constructed for ofloxacin (OFL) quantification (Wang et al. 2022c). The sensing mechanism of OFL can be attributed to the antenna effect induced energy transfer from OFL to Eu^{3+}, exhibiting significant fluorescence enhancement with increasing OFL concentration. The analytical performance with linear range of 0.83–40.0 nM and LOD of 0.25 nM can be achieved, demonstrating the good prospects in rapid, sensitive and selective detection of OFL in environment samples. Simultaneous determination of 5-nitroimidazole antibiotics (metronidazole, ornidazole, tinidazole, secnidazole) was reported by Sun et al. (2022) based on fluorescence array platform of CDs@MIPs (Fig. 7.1b). The fluorescence variation degree was applied to construct the fingerprint information, followed by analysis with linear discriminant analysis (LDA). During prediction analysis, the accuracies of 100% can be achieved for test datasets (buffer solutions and spiked water samples). These results provide a great inspiration for rapid and simultaneous identification of antibiotics by combining fluorescent CDs@MIPs array platform with ML tools. The MIPs-based fluorescent sensors will play an essential role in antibiotic determination in environment.

In addition, there are other analytical technologies including colorimetric, surface-enhanced Raman scattering (SERS), surface plasmon resonance (SPR) and field-effect transistor (FET) were applied to construct MIPs sensors. Song et al. (Song et al. 2022b) proposed a boronate affinity metal–organic framework/MIPs microfluidic chip (FZS-BA@MIPs) for rapid and colorimetric detection of residual kanamycin (KAN) (Fig. 7.1c). The metal–organic framework of ZIF-8 provides significant effect of scaffold and enrichment, and MIPs layer ensure the selective recognition for KAN. Thanking to synergistic effects of MIPs recognition, boronate affinity, and specific chromogenic effect, the good analytical results was obtained with linear range of 0.1–25 mg/L and LOD of 4.7 μg/L. Another colorimetric MIPs sensor was fabricated based on the superior oxidase-mimetic activity of alkali-etched imprinted Mn-based Prussian blue analogues (Mn-PBA_{NaOH}@MIPs) for selective and sensitive sensing of tetracycline (TC) (Liu et al. 2023a). The addition of TC will mask the active sites on the MIPs layer, causing decrease of enzymatic activity and making color lighter. Carrasco et al. (2016) prepared a multibranched gold–silica–MIPs (bAu@mSiO_2@MIPs) sensor for specific SERS recognition of ENR. The multibranched gold nanoparticles provide dense "hot spot" for enhancing Raman signals and MIPs shell ensure the specific enrichment of ENR molecules, achieving high sensitivity (LOD = 1.5 nM) and selectivity. A similar surface plasmon resonance (SPR) based MIPs sensor (AMO-MIP) was fabricated by Ayankojo et al. (2018b) using sol–gel and spin-coating method for label-free sensing of AMO. The as-fabricated AMO-MIPs sensor has good stability and uniformity for SPR detection of AMO with LOD of 73.0 pM, suggesting the promising potentials in rapid, label-free, sensitive and selective analysis of antibiotics in aqueous environment. In addition, Chen et al. (2023b) proposed field-effect transistor (FET) based MIP sensors for rapid, sensitive and selective analysis of TC (Fig. 7.1d). The Ag^+ ions were modified on the surface of 2D BPNS through cation–π interaction, followed by self-assembly of supramolecular platform (melamine and cyanuric acid) onto the above BP surface. This fabrication process not only enhance the stability of BPNS, but also supply supramolecular platform for sequential MIPs decoration. As expected, the as-prepared FET based MIPs sensors show high selectivity for TC, quick response (within 6 s), and low detection limit (7.94 nM), showing good potentials in antibiotics quantification.

The past few years witnessed the flourishing development of MIPs based chemosensros in various fields. In case of antibiotics sensing, the various MIPs based-sensors are among the potential candidates due to their particular properties, such as promising rapidity, portability, high accuracy, and specificity. In the future, the MIPs based-sensors would play an important role in antibiotics quantification and more MIPs sensors will be designed and fabricated according to the specific tasks.

At the same time, the MIPs technology should combine more analytical method for expanding their analytical applications in antibiotics monitoring.

7.3.2 Pesticides

With the increase in population, the pesticides were considered as the considerable tool for increasing the crop yields and ensuring the food safety (Mahmoudpour et al. 2020; Wang et al. 2020b). Generally, on the basis of uses, pesticides can be categorized into fungicides, insecticides, rodenticides, herbicides, nematicides, and molluscicides (Fang et al. 2021; Majdinasab et al. 2021). However, the pollution - problems caused by pesticide residues can be attributed to the wide and massive use in agricultural, requiring efficient method for rapid and sensitive monitoring (Samsidar et al. 2018). Similar with antibiotics, various MIPs-based sensors have been reported for pesticides detection according to same advantages of rapidity, portability, high accuracy, and specificity.

Considering the unique advantages of electrochemical method, the combination of electrochemistry sensors with MIPs tools have been widely used in pesticides detection (Mostafiz et al. 2021; Li et al. 2023b). Sun et al. (2017) proposed a photoelectrochemical (PEC) sensor based on MIPs decorated titanium dioxide nanorods with hierarchical branch structure (MIPs-B-TiO_2 NRs) for efficient determination of chlorpyris (CPF). The B-TiO_2 NRs were in-situ grown on fluorine doped tin oxide platform, followed by formation of MIPs layer through electropolymerization process between CPF and *p*-aminothiophenol (*p*-ATP). The quantitative analysis of CPF can be achieved based on the specific recognition of MIPs layer, and the detection linear range in 0.01–100 ng/mL with low LOD of 7.4 pg/mL can be obtained. A hybrid electrochemical sensor composed by composite of Ag–N co-doped zinc oxide/coconut husk prepared activated carbon (Ag–N@ZnO/CHAC) and MIPs film (dual functional monomers, DM) was fabricated for cypermethrin (CYP) detection (Li et al. 2019a) (Fig. 7.2a). Comparing with one-monomer involving MIPs, the as-fabricated DM-MIPs-Ag–N@ZnO/CHAC sensors show better selectivity and higher response, showing good analytical performance with linear range of 0.2 pM–1.0 nM and LOD of 0.07 pM. Zhang et al. (2019) proposed a screen-printed carbon electrode (SPCE) based MIPs sensor for highly analysis of cyhexatin (CYT). The CYT and *o*-aminothiophenol (*o*-ATP) were electropolymerized on the AuNPs/graphene oxide (electrochemical reduction) modified electrode (MIPs/AuNPs/ERGO-SPCE). Under the optimal condition, the satisfied analytical performance with LOD of 0.20 ng/mL can be obtained, attributing to the synergistic effect of AuNPs and ERGO. Another MIPs-based QCM sensors were designed by Dayal et al. (2019) for quantitative analysis of hydrophilic organophosphate trichlorfon (TCF) using hydrophilic PVDF-based MIPs layer. These results eliminate the requirement in complicated polymerization reaction and reduces the amount of solvents for removing template molecules. In addition, an electrochemical impedance sensor based on Fe_3O_4 and polydopamine MIPs (PDA@Fe_3O_4-MIP) was designed and fabricated for detection of dichlorodiphenyltrichloroethane (DDT) (Miao et al. 2020). The high stability, specific recognition and efficient extraction of DDT ensuring good analytical performance with linear range of 60.0 pM–1.0 mM and LOD of 6.0 pM. Further, Liu et al. (2021) prepared MIPs electrochemiluminescence (ECL) sensor for aminotriazole determination using MIPs shell and multiple $Ru(bpy)_3Cl_2$-encapsulated liposomes immobilized by light-triggered click reaction (Fig. 7.2b). The relationships between ECL response and aminotriazole concentration were in range of 5.0 aM–1.0 pM, established with LOD of 1.2 aM. The great developments of MIPs-based electrochemical sensors in pesticides analysis will significantly control the pesticide residues in environment.

The MIPs-based fluorescent sensors provided a strategy for high detection of pesticide residues in complex samples, which is comparable analytical performance with MIPs-based electrochemical sensors (Tan et al. 2019; Kumar and Kim 2022). Huang et al. (2018a) fabricated a MIPs-based fluorescent sensors based on MIPs layer encapsulated luminescent perovskite quantum dots (MIPs@$CsPbBr_3$ QDs) for omethoate (OMT) detection (Fig. 7.2c). The fluorescence of $CsPbBr_3$ QDs can be significantly quenched due to the charge transfer from QDs to OMT. The imprinting

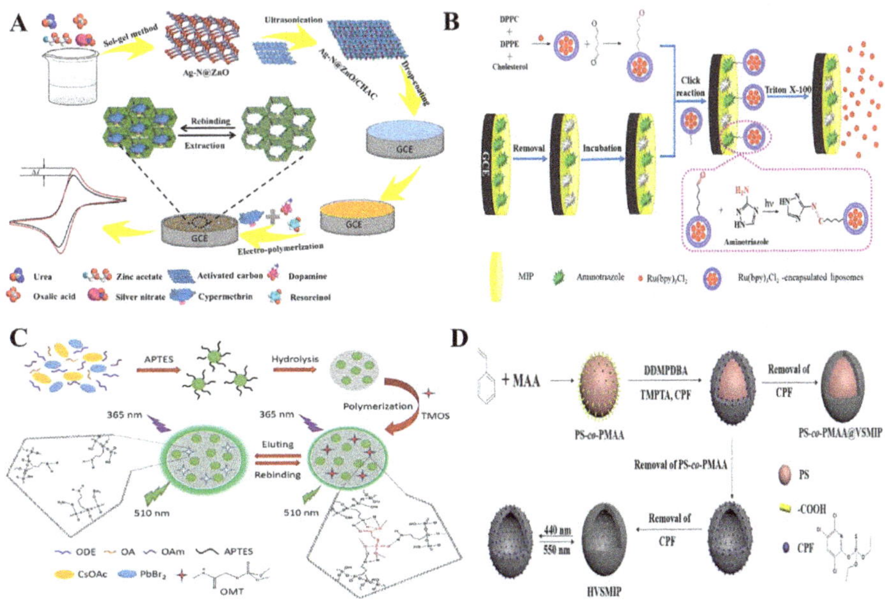

Fig. 7.2 **a** Schematic preparation process of DMMIPs-Ag-N@ZnO/CHAC/GCE. Reproduced with permission (Li et al. 2019a). Copyright 2018, Elsevier. **b** Schematic diagram of the $Ru(bpy)_3Cl_2$-encapsulated liposome signal amplification MIPs-ECL sensor based on the click reaction for aminotriazole determination (Liu et al. 2021). Copyright 2021, American Chemical Society. **c** Schematic representation of the MIPs@$CsPbBr_3$ QD sensor. Reproduced with permission (Huang et al. 2018a). Copyright 2018, American Chemical Society. **d** Schematic representation of the preparation of the HVSMIP. Reproduced with permission (Chen et al. 2021). Copyright 2021, Elsevier

factor (IF) can reach 3.2, indicating the good specificity of MIPs layer for inorganic perovskites, as well as promising linear range (50.0–400.0 ng/mL) and LOD (18.8 ng/mL) for OMT quantitative analysis. Another fluorescent strip-based sensor was prepared by combining nitrogen-doped graphene quantum dots with MIPs technology (N-GQDs@MIPs) for thiacloprid determination (Liu et al. 2018). The N-GQDs were firstly immersed into filter paper, followed by been self-polymerized between dopamine (DA) and thiacloprid on surface of test strip. The as-prepared strip-based sensor for thiacloprid detection showed a linear ranging from 0.1 to 10.0 mg/L with LOD of 0.03 mg/L, exhibiting portable potential in pesticide detection with high sensitivity and selectivity. The similar portable MIPs-based fluorescent platforms were also reported for pesticides monitoring with high analytical performance (Yang et al. 2019; Yan et al. 2021). In addition, a mesoporous MIPs-decorated luminescent MOF-76 (mMIPs@MOF-76) was fabricated for specific recognition of CPF (Eskandari et al. 2020). Comparing with traditional MIPs-based sensors, the mesoporous structure of MIPs layer and high porosity of MOF-76 significantly improve the adsorptivity and sensitivity for target molecules. The sensing mechanism can be attributed to the remarkable quenching effect of CPF for MOF-76, and the linear range of 10.0 ng/mL–1.0 μg/mL and LOD of 3.4 ng/mL can be obtained. In addition, Wei et al. (2020) fabricated ionic liquid-sensitized MIPs based on Mn, Cu doped QDs ($ZnIn_2S_4$) functionalized graphene (QDs-G@ILMIPs) for quantitative analysis of λ-cyhalothrin. The IL played an obvious role in improving selectivity and adsorption capacity, as well as enhancing fluorescence stability and sensitivity. After λ-cyhalothrin addition, the fluorescence of $ZnIn_2S_4$ QDs can be quenched and the quantitative properties can be confirmed in linear range of 1.0–350.0 μg/g (LOD = 0.25 μg/g). The fluorescence MIPs sensors can achieve the goal in quantitative analysis of pesticides with high analytical performance.

Apart from fluorescence and electrochemistry based MIPs sensors, there are some MIPs based sensors have been applied to detect pesticide residues. Ye et al. (2018) proposed a simple MIPs-based colorimetric sensor for visual detection of 3-phenoxybenzaldehyde (3-PBD). The MIPs layer was deposited on monodispersed silica nanoparticles with dual monomers of 3-aminopropyltriethoxysilane (APTES) and phenyltrimethoxysilane (PTES). After absorption of 3-PBD by MIPs-based sensor, the distinctive color change can be found due to the reduction effect of $KMnO_4$. The as-developed method exhibited linear range for 3-PBD from 0.1 μg/mL to 1.0 μg/mL with low LOD of 0.052 μg/mL. Moreover, a portable label-free MIPs-based colorimetric sensor was fabricated based on MIPs shell coated inverse opal photonic hydrogel particles for colorimetric detection of methanephosphonic acid (MPA) (Huang et al. 2018b). The reflection peak shift of as-fabricated sensors can be obviously recognized by naked eyes after the addition of MPA. Another hollow visible-light-responsive surface MIP (HVSMIP) was prepared for determine trace CPF with PS-co-PMAA as template (Fig. 7.2d) (Chen et al. 2021). The as-fabricated HVSMIP sensors exhibited larger surface area, dissociation constant, pore volume, imprinting effect, binding capacity, and photo-isomerization rate compared to PS-co-PMAA@VSMIP. The detection linear range was 0.2 μM–10.0 μM, suggesting the good feasibility of HVSMIP sensor in quantitative analysis of CPF. Furthermore,

Qi et al. (2021) proposed a novel dual-recognized strategy using acetylcholinesterase (AChE) and MIPs as affinity molecules for detection of acephate (AP). The MIPs layer can capture the certain moieties but not phosphorus of AP and AChE will selectively recognize phosphorus of AP captured by MIPs for quantitative analysis. The relationships between absorption intensity and AP concentrations were established in range of 2.5–200.0 μg/kg with LOD of 0.9 μg/kg. The proposed dual-recognized strategy show unique potentials in improving detection accuracy compared to classical AChE inhibition assay.

Pesticides have been proven to show significant risk to environment and human health, as well as plant and animal species (Mostafiz et al. 2021). Chemosensors based on MIPs technology have been widely applied to monitor the environmental pollutions due to its unique features of broad-spectrum applications, cost-effectiveness, rapidity, selectivity, sensitivity, and real-time analysis (Ayivi et al. 2023). MIPs-based sensors have proven to be the powerful tools in conventional pesticides testing, which will gain the great development in the future. However, the limitations of low selectivity and reusability for MIPs-based sensors still need to be overcame.

7.3.3 Heavy Metals

Heavy metals are an important class of pollutants due to their persistent nature and high toxicity for environment and human health (Rao et al. 2007). The rapid and accurate quantification of heavy metals is of utmost requirement. Thus, the effective analytical method for trace detection of heavy metals in environmental samples is still required. MIPs-based sensors show particular features of rapidity, sensitivity and selectivity for quantitative analysis of heavy metals in environment (Rebelo et al. 2021; Keçili et al. 2023). The MIPs-based sensors including electrochemistry and fluorescence have been widely used in heavy metals monitoring on the basis of their promising rapidity, sensitivity, and selectivity.

In terms of MIPs-based electrochemistry sensors, Wang et al. (Zhihua et al. 2011) proposed MIPs modified carbon paste electrode (CPE) for direct Cu^{2+} determination. The calibration curve can be plotted in range of 70.0 nM–1.0 μM and 1.0 μM–0.1 mM with LOD of 23.0 nM. The similar MIPs-based electrochemistry sensors for heavy metals (U^{2+}, Ce^{3+}, Pb^{2+}, Ag^{+} and Pd^{2+}) detection were also reported (Bojdi et al. 2014a, b; Shamsipur et al. 2014; Güney and Güney 2016; Chen et al. 2018). In addition, the carbon-based materials were applied to prepare ion-imprinted polymer (IIP) on CPE for determination of Ce^{3+} and Bi^{3+} based on the improved signals (Alizadeh et al. 2016, 2017). Another magnetic IIP decorated on CPE was adapted for efficient extraction of Ag^{+} to improve the detection sensitivity (Ghanei-Motlagh and Taher 2017). Alizadeh et al. (2014) prepared IIP on the surface of CPE for competitive detection of Gd^{3+} in the presence of Pb^{2+} (Fig. 7.3a). The DPV signal of oxidative stripping of Pb^{2+} was indirectly used to Gd^{3+} quantification. The quantity of electroactive Pb^{2+} can be reduced in the presence of Gd^{3+}, resulting in dropping of DPV signal. The linear detection ranges for Gd^{3+} was 6.0 nM–48 μM with LOD of 4.5 nM. Another IIP impregnated carbon ionic liquid paste electrode (CILE) was prepared using dithizone as suitable ligand for selective and trace analysis of Hg^{2+} (Bahrami et al. 2015). The promising analytical performance in sensitivity (0.1 nM), selectivity and accuracy can be achieved, indicating good potentials in Hg^{2+} analysis. Li et al. (2015) fabricated a novel ECL sensor based MIPs technology for ultra-trace analysis of Be^{2+}. The complex of 4-(2-pyridylazo)-resorcinol-Be^{2+} (Be-PAR) was chosen as the template and the ECL signal can be monitored by luminol-H_2O_2 system. In the range of 70.0 pM to 8.0 nM, the ECL signal exhibited linear relationship with Be^{2+} concentration and LOD was 24.3 pM. Another ECL MIPs sensor was designed and fabricated for trace analysis of Co^{2+} based on dual recognition strategy (Fig. 7.3b) (Li et al. 2019b). The multi-walled carbon nanotube/Cu nanoparticles/CDs (MWCNT/Cu/CDs) was used as ECL device, and coordination of bovine serum albuminmetal-Co^{2+} (BSA-Co^{2+}) acted template for ECL signal quenching. The low LOD of 0.3 nM ensured the practical analysis of Co^{2+} in agricultural products, environment water, and soil samples. The continuous focus on developing highly sensitive and accurate MIPs-based electrochemistry sensors for heavy metals analysis is still under consideration.

The fluorescence MIPs sensors were also fabricated for trace heavy metals determination (Keçili et al. 2023). Shao et al. (2017) proposed a fluorescent IIP sensor

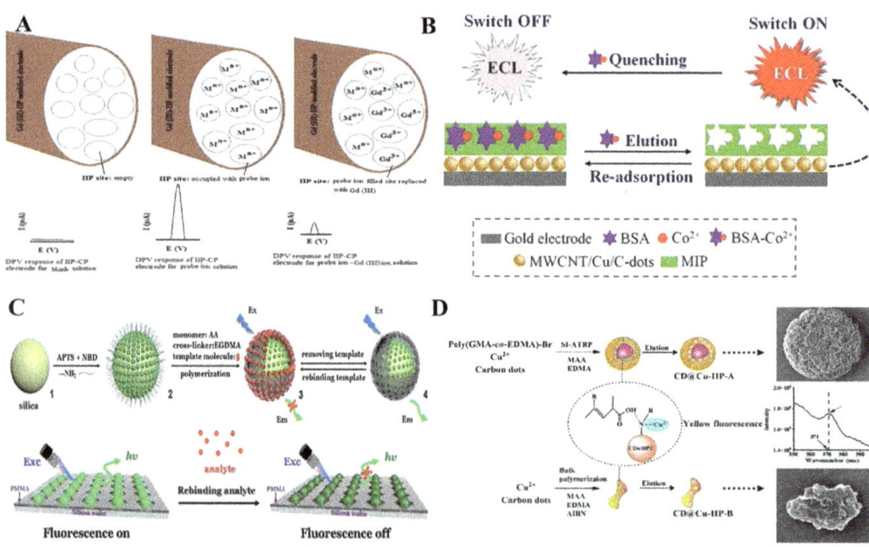

Fig. 7.3 **a** Schematic representation of the determination mechanism of the developed method for Pb^{2+} detection. Reproduced with permission (Alizadeh et al. 2014). Copyright 2015, Nature Springer. **b** Procedure for switch sensor fabrication for the detection of Co^{2+}. Reproduced with permission (Li et al. 2019b). Copyright 2019, Elsevier. **c** Schematic illustration for the synthesis of SiO_2@MIPs based microsensor arrays for heavy metals detection. Reproduced with permission (Xia et al. 2021). Copyright 2021, American Chemical Society. **d** Schematic preparation of spherical CD@Cu-IIP-A via SI-ATRP and amorphous CD@Cu-IIP by bulk polymerization. Reproduced with permission (Bai et al. 2023). Copyright 2023, Elsevier

($CQD@SiO_2@IIP$) using carbon quantum dots (CQDs) as luminophore for detection Li^+ in water samples. The fluorescence of CQDs can be effectively quenched through photoinduced charge transfer (PET) effect from Li^+ to CQDs. The linear range in 0.5–50 μM with LOD as low as 0.2 μM can be obtained. The similar fluorescent IIP sensor (CQDs@Cu-IIP) was fabricated by Wang et al. (2021) with existence of SBA-15 mesoporous molecular sieve for Cu^{2+} sensing. Additionally, a IIP based microsensor arrays were fabricated for trace heavy metal ions (Pb^{2+}, Cu^{2+}, Cd^{2+}, and Hg^{2+}) detection (Fig. 7.3c) (Xia et al. 2021). The SiO_2 nanoparticles were firstly modified by 3-aminopropyltriethoxysilane conjugated with 4-chloro-7-nitro-1,2,3-benzoxadiazole (NBD-Cl), followed by been encapsulated by heavy metal ions imprinted layer. The final microsensor arrays can be formed through self-assembly process of as-fabricated SiO_2@MIPs. The addition of heavy metal ions can reduce the fluorescence intensity of NBD-Cl through the nonemission FRET effect. Using Pb^{2+} as an example, the low LOD of 1.0 nM (4.1 μg/L) can be achieved, providing visual potential in heavy metal ions detection. Bai et al. (2023) proposed a novel strategy to prepare CDs based fluorescence MIPs sensor (CD@Cu-IIP-A) through atom transfer radical polymerization (ATRP) for Cu^{2+} sensing (Fig. 7.3d). Comparing with traditional CD@Cu-IIP-B synthesized by bulk polymerization, the CD@Cu-IIP-A exhibited better adsorption capacity. The good fluorescence quenching effect on Cu^{2+} by CD@Cu-IIP-A in 80.0–780.0 μM range was obtained, indicating promising prospect in practical applications of Cu^{2+} sensing. The MIPs-based fluorescent sensors are limited to specific detection extent, so, it needs more studies to develop more sensors for heavy metals detection.

7.3.4 Pathogen Contaminants

Several pathogens are present in the environment, mainly viruses and bacteria, enter the human body through the food chain, and can cause diarrhoea or even lead to epidemics of infectious diseases. Therefore, sensitive and rapid detection of environmental pathogens is essential for environmental protection and prevention of infectious diseases outbreak. The most common methods for quantifying pathogens are polymerase chain reaction (PCR) and bacterial culture testing (Skottrup et al. 2008; Parra et al. 2017). However, these methods are complex, require multiple steps, and are labor intensive with long analysis times (>1 day). MIPs combined with sensors and biosensors devices as biometric elements for the detection of environmental have proven to be a thriving measure due to their superior identification specificity, ease of synthesis, low cost, and environment friendly properties (Cui et al. 2020; Srivastava et al. 2020).

Bacteria

The most frequent environmental contamination is caused by Escherichia coli (E. coli) *O157:H7*, Campylobacter spp, Legionella spp, Giardia, *Bacillus cereus*, Cryptosporidium, Salmonella spp, Y ersinia enterocolitica, *Enterococcus faecalis*, and Shigella (Craun et al. 2006; Scallan et al. 2011). The large size of bacteria, usually micron-sized, and their highly susceptible to morphological changes make the development of bacterial-based blotting sensors challenging.

Yeseren Saylan has developed a bionic optical sensor for the detection of *Enterococcus faecalis*, a common fecal contaminant in seawater and tap water (Fig. 7.4a). The author first synthesized gold nanoparticles and then complexed them with functional monomers to facilitate the preparation of imprinted nanomembrane-based optical sensors. The proposed sensor could detect 6.6×10^3 cfu/mL with a linear response from 5×10^4–1×10^8 cfu/mL. Due to the matrix effect and the change in the refractive index of the solution during the detection study, the recovery rate of fecal *E. coli* in tap water and seawater was 92–95%. With its impressive performance, the optic sensor represents a valuable tool for ensuring water safety and advance bacterial detection technologies (Saylan 2023).

Chen developed an electrochemiluminescence (ECL) biosensor for the quantitative detection of *E. coli O157:H7* based on a polydopamine (PDA) surface-imprinted polymer (SIP) and nitrogen-doped graphene quantum dots (N-GQDs) (Chen et al. 2017) (Fig. 7.4b). First, *E. coli O157:H7* was used as a template in the PDA film formed on the electrode surface. After removing the template, the PDA SIP was available. Subsequently, the target bacteria were adsorbed and specifically bound to an N-GQD$_S$-labeled polyclonal antibody (pAb). *E. coli O157:H7* was detected by ECL signal of N-GQDs using $K_2S_2O_8$ as co-reactant. The sensor can specifically capture and selectively identify the target pathogen. *E. coli O157:H7* at $10^1 - 10^7$ cfu mL^{-1} can be detected. Most important, the work was able to successfully identify *E. coli* O157:H7 in water samples.

In addition to using the entire bacteria for the synthesis of imprinting, specific components of the bacterial surface, such as sugar parts, proteins, lipids, etc., can also be extracted for the synthesis of imprinting. For example, Yeseren Saylan extracted bacterial surface proteins from *Enterococcus faecalis* (E. faecalis) and then printed them onto nanoparticles (Saylan et al. 2019) (Fig. 7.4c). The nanoparticles were then used to modify the surface of the sensor to selectively detect waterborne bacteria, resulting in a detection range of 5×10^4–5×10^8 cfu/mL and a detection limit of 3.4×10^4 cfu/mL. The selectivity of the constructed sensor was evaluated using *Bacillus subtilis*, *E. coli* and *Staphylococcus aureus* as interfering bacteria. These results showed that the sensor had good selectivity. Moreover, the sensor can detect different concentrations of Enterococcus faecalis in seawater and tap water samples with a recovery rate of 90–95%.

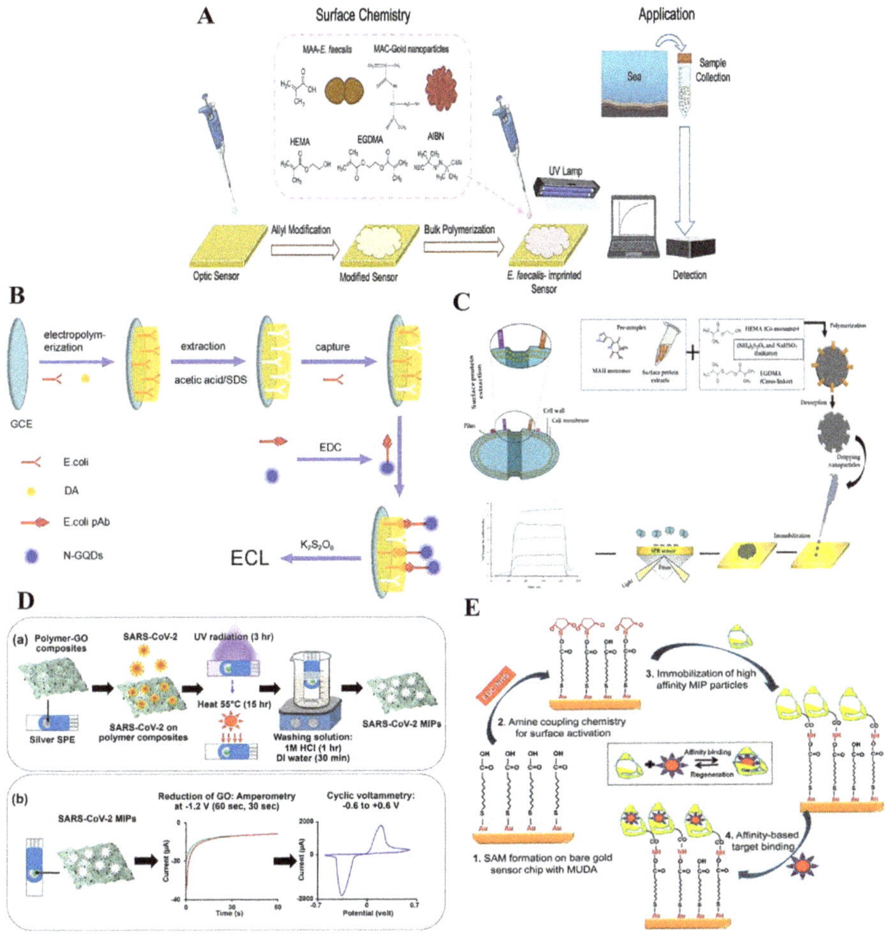

Fig. 7.4 **a** Scheme of optic sensor preparation. Reproduced with permission (Saylan 2023). Copyright 2023, Elsevier. **b** Schematic presentation of the fabrication procedure of the biosensor and the detection process. Reproduced with permission (Chen et al. 2017). Copyright 2017, American Chemical Society. **c** Scheme of nanoparticle and sensor preparation. Reproduced with permission (Saylan et al. 2019). Copyright 2019, Multidisciplinary Digital Publishing Institute. **d** This diagram illustrates the protocol for imprinting SARS-CoV-2 on polymer-GO composites. Reproduced with permission (Sukjee et al. 2022). Copyright 2022, Elsevier. **e** Affinity-based sensor assays for virus detection. Reproduced with permission. Copyright 2015, American Chemical Society (Altintas et al. 2015)

Viruses

Major environmental viruses are known to be hepatitis A and E viruses, norovirus, astrovirus, and rotavirus. Virus imprinting has shown very promising outcomes for the diagnosis of virus-associated diseases (Gall et al. 2015).

The severe acute respiratory syndrome coronavirus (SARS-CoV-2) is the virus responsible for causing the pandemic coronavirus disease 2019 (COVID-19) (Wang et al. 2020c). Identifying the presence of this virus in the environment could potentially improve the effectiveness of disease control measures. A study by Wannisa Sukjee proposed an electrochemical polymer complex biosensor for measuring SARS-CoV-2 in the environment (Sukjee et al. 2022) (Fig. 7.4d). The authors prepared MIPs that can specifically recognize and bind SARS-CoV-2 on the surface of graphene oxide (GO). The results showed that the sensitivity of the total relative signal change from 0.01 to 100 fM was 25%. In buffers and wastewater mixed with SARS-CoV-2, LOD was 0.1fM. This work provides a useful assessment of indicators of overall viral content in aquatic environments for medical risk surveillance.

In another interesting work, Zeynep Altintas et al. used bacteriophage *E. coli* (MS_2) as a model pathogenic virus (Altintas et al. 2015) (Fig. 7.4e). Phages are attached to silica microbeads and used as templates for MIPs synthesis to produce nanoparticles with high affinity binding sites by using a new automated solid phase method. It is integrated with microfluidics on a surface plasmon resonance (SPR-2) sensor chip for automatic capture and analysis of viruses. This method provided a good LOD of 5×10^6 pfu/mL and a linear range of 0.33−27 pmol.

7.3.5 Other Environmental Contaminants

The MIPs-based sensors have been applied to monitor other environment contaminants including endocrine disrupting chemicals (EDCs) and organic dyes due to their highly potential risks to human and environment. Yu et al. (2019) prepared MIPs-based electrochemical sensor for detection of rapid and selective detection of p-nonylphenol based on polypyrrole MIPs modified TiO_2 (NP-PPy@TiO_2 MIPs) (Fig. 7.5a). The PPy layer provides enough micro cavities for specific recognition of p-nonylphenol, showing linear concentration range of 10.0 nM–80.0 μM and LOD of 3.9 nM. Another fluorescent MIPs sensor was designed and fabricated by Liu et al. (2023b) using CDs@MIPs for sensitive and selective analysis of bisphenol A (BPA). After addition of BPA, the fluorescence of CDs was significantly quenched according to the electron transfer from CDs (electron donor) to BPA (electron acceptor). The nice linearity with BPA concentration in range of 10.0 nM–2.0 μM and low LOD of as 1.5 nM can be obtained, realizing practical analysis in aqueous and plastic samples. Furthermore, Li et al. (2017) proposed ratio metric fluorescence sensor combined with mesoporous structured MIPs for highly analysis of 2,4,6-trinitrophenol (TNP). The CdTe QDs were embedded in mesoporous-structured 2-acrylamide-6-methoxybenzothiazole (AAMBT) modified silica (AAMBT&SiO_2) along with the surface imprinting process. In case of ratio metric detection, the fluorescence of AAMBT was quenched in the presence of TNP due to the resonance energy transfer and the fluorescence of QDs can be preserved. Under optimum conditions, the as-fabricated ratio metric fluorescence MIPs sensors achieved a low LOD of 43.0 nM within 3 min, which was verified by TNP analysis in

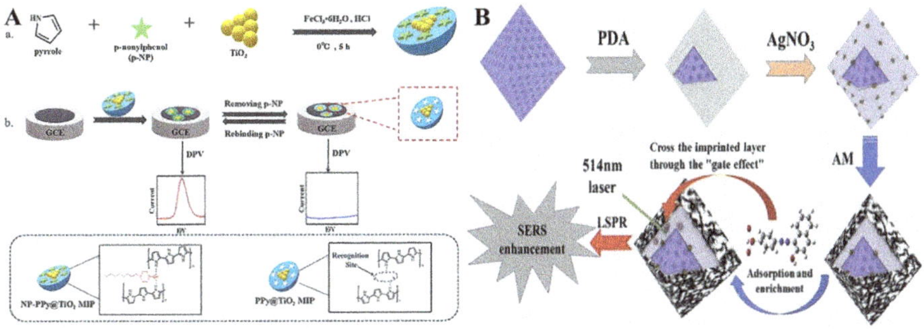

Fig. 7.5 **a** Preparation of PPy@TiO_2 MIP and fabrication of PPy@TiO_2 MIP/Nafion/GCE electrochemical sensor. Reproduced with permission (Yu et al. 2019). Copyright 2019, Elsevier. **b** Synthesis flow chart of Ag@UIO-66(NH_2)/PDA@MIPs (AUP-MIPs) and mechanism of SERS detection for orange II dyes. Reproduced with permission (Xue et al. 2021). Copyright 2021, Elsevier

water sample. Moreover, He et al. (2019) proposed microplate chemiluminescence MIPs sensor based on dummy-template for one-step determination of Sudan dyes. The MIPs layer was coated on the surface of conventional microplate, ensuring only one sample-loading step before signal acquisition. The sensors achieved rapid detection process (10 min), good sensitivity (1.0–5.0 pg/mL), and promising recovery (70.5–92.2%). Li's group prepared SERS based MIPs sensor based on Ag@UIO-66 (NH_2)/PDA-MIPs for orange II dyes detection in water sample (Fig. 7.5b) (Xue et al. 2021). Firstly, the UIO-66 (NH_2) was modified by thin PDA layer, followed by decoration of AgNPs and encapsulation of MIPs shell. The highly analysis of orange II can be realized thanks to the synergistic effect of specific recognition of MIPs high adsorbility of UIO-66 (NH_2) and promising enhancement of AgNPs. The detection linearity in 0.1–10.0 nM range ensured the sensitive analysis of orange II. In addition, Foguel et al. (2015) fabricated evanescent wave fiber optic sensor using MIPs as the recognition element for determination of basic red 9 (BR9) dyes. The MIPs layer was immobilized on the polystyrene optical waveguide with 4-cm long by dip-coating method. The BR9 can be detected as low as at μM level, suggesting good sensitivity in organic dyes determination. Ongoing endeavours to more MIPs-based sensors will significantly promote their widespread applications in EDCs and organic dyes analysis. Table 7.1 lists the research progress in environmental pollutant detection based on MIPs-based sensors and biosensors in recent years.

7.4 Conclusions and Perspectives

During the past years, MIPs-based sensors have gained significant development in monitoring environmental pollution (Mostafiz et al. 2021; Rebelo et al. 2021; Keçili et al. 2023). The MIPs-based sensors exhibit unique advantages in rapidity, sensitivity, selectivity, and convenient portability (Ramanavicius et al. 2021). This chapter

Table 7.1 The research progress in environmental pollutant detection based on MIPs-based sensors and biosensors in recent years

Analyte	Type	Linear range	LOD	Samples	Recovery	References
NOR	LSV	0.1 nM–10 µM	0.012 nM	Milk	99.4–105.2%	Li et al. (2022)
AMO	QCM	2–40 nM	0.2 nM	Water	/	Ayankojo et al. (2018a)
SMZ	DPV	1.0–8.2 nM	0.9 nM	Tap water	/	Ayankojo et al. (2020)
CFX	ET	0–300 µM	1.0 µM	Pharmaceuticals and human urine	/	Wang et al. (2022b)
MFX						
LFX						
ERY	DPV	2–20 nM	1.1–1.6 nM	Tap water	93–108%	Nguyen et al. (2023)
AZI						
CLA						
ENR	FL	63.0 ng/L–60.0 µg/L	8.0 ng/L	Fish	83.7–87.7%	Tang et al. (2017)
NOR	FL	10–100 ng/mL	2.6 ng/mL	Pharmaceutical preparations	98.2–104.4%	Wang et al. (2020a)
	CL	0.1–20 µg/mL	0.1 µg/mL		97.6–99.1%	
DOX	FL	0.5–30 µg/L	0.2 µg/L	Environment water	92.5–105.2%	Li et al. (2023a)
OFL	CL	0.83–40.0 nM	0.25 nM	Water	/	Wang et al. (2022c)
KAN	CL	0.1–25 mg/L	4.7 µg/L	Milk, Penaeu chinensis, and Carassius auratus	60.4–88.7%	Song et al. (2022b)
TC	CL	0.2–200 µM	0.07 µM	River, tap water, milk and soil extract	93.5–104.5%	Liu et al. (2023a)
ENR	SERS	2.5–250 nM	1.5 nM	Water	/	Carrasco et al. (2016)
AMO	SPR	0.1–2.6 nM	73.0 pM	Tap water	93.4–96%	Ayankojo et al. (2018b)
TC	FET	8.0 nM–1.0 µM	7.94 nM	Water	/	Chen et al. (2023b)
CPF	PEC	0.01–100 ng/mL	7.4 pg/mL	Drinking and river water	/	Sun et al. (2017)

(continued)

Table 7.1 (continued)

Analyte	Type	Linear range	LOD	Samples	Recovery	References
CYP	DPV	0.2 pM–1.0 nM	0.07 pM	Soil, mackerel, crayfish and water	96.2–100.4%	Li et al. 2019a)
CYT	DPV	1–500 ng/mL	0.20 ng/mL	Pears	79.6–92.1%	Zhang et al. (2019)
TCF	QCM	0–250 ng/mL	4.63 ng/mL	Iceberg lettuce	92.7–98.1%	Dayal et al. (2019)
DDT	EIS	60.0 pM–1.0 mM	6.0 pM	Radish	89–102%	Miao et al. (2020)
Aminotriazole	ECL	5.0 aM–1.0 pM	1.2 aM	Tap water, river water, pineapple, orange, tea, banana, carrot	96.7–103.9%	Liu et al. (2021)
OMT	FL	50.0–400.0 ng/mL	18.8 ng/mL	Cabbage and soil	81.4–101%	Huang et al. (2018a)
Thiacloprid	FL	0.1–10.0 mg/L	0.03 mg/L	Water	101.1–110.0%	Liu et al. (2018)
CPF	FL	10.0 ng/mL–1.0 μg/mL	3.4 ng/mL	Juice and environmental water	96.4–100.3%	Eskandari et al. (2020)
λ-cyhalothrin	FL	1.0–350.0 μg/g	0.25 μg/g	Broccoli, potatoes, carrots, spinach, and tea	80.5–111.6%	Wei et al. (2020)
3-PBD	CL	0.1–1.0 μg/mL	0.052 μg/mL	River water, fruit juice, and beverage	90.0–98.9%	Ye et al. (2018)
CPF	CL	0.2–10.0 μM	0.2 μM	Apple and carrot	83.0–98.0%	Chen et al. (2021)
AP	CL	2.5–200.0 μg/kg	0.9 μg/kg	Bean sprout, cauliflower, leek, onion, pakchoi and carrot	83.5–116.2%	Qi et al. (2021)
Cu^{2+}	DPV	70.0 nM–0.1 mM	23.0 nM	Waste water	/	Zhihua et al. (2011)
Ag^{+}	DPV	0.05–150 μg/L	15 ng/L	Environment water	97.3–103.0%	Ghanei-Motlagh and Taher (2017)
Gd^{3+}	DPV	6.0 nM–48 μM	4.5 nM	Environment water	93.3–108.7%	Alizadeh et al. (2014)

(continued)

Table 7.1 (continued)

Analyte	Type	Linear range	LOD	Samples	Recovery	References
Hg^{2+}	DPV	0.5 nM–10 nM, 0.08 μM–2 μM	0.1 nM	Waste water	94.7–104.2%	Bahrami et al. (2015)
Be^{2+}	ECL	70.0 pM–8.0 nM	24.3 pM	Well and rain water	95.0–101.0%	Li et al. (2015)
Co^{2+}	ECL	1.0 nM–0.1 μM	0.3 nM	Environmental water, soil samples and agricultural products	87.5–111.3%	Li et al. (2019b)
Li^{+}	FL	0.5–50 μM	0.2 μM	Lake water	102–111.6%	Shao et al. (2017)
Cu^{2+}	FL	0.25–2 mg/L 3–10 mg/L	0.016 mg/L 0.028 mg/L	Tap and river water	99.3–105.4%	Wang et al. (2021)
Cu^{2+}, Cd^{2+}, Hg^{2+}, Pb^{2+}	FL	0.5–5.0 μM	1.0 nM	Water	/	Xia et al. (2021)
Cu^{2+}	FL	80.0–780.0 μM	3.17 μM	Water	/	Bai et al. (2023)
E. faecalis	PEC	5×10^4–10^8 cfu/mL	6.6×10^3 cfu/mL	Seawater, tap water	92–95%	Saylan (2023)
E. coli O157:H7	ECL	10–10^8 cfu/mL	8 cfu/mL	Water	99.4–102%	Chen et al. (2017)
E. faecalis	SPR	5×10^4–5×10^8 cfu/mL	3.4×10^4 cfu/mL	Seawater, tap water	90–95%	Saylan et al. (2019)
SARS-CoV-2	EC	0.1–100 fM	0.1 fM	Wastewater	/	Sukjee et al. (2022)
MS_2	SPR	0.33–27 pmol	5×10^6 pfu/mL	/	/	Altintas et al. (2015)
p-nonylphenol	DPV	10.0 nM–80.0 μM	3.9 nM	Milk powder	98.4–100.7%	Yu et al. (2019)
BPA	FL	10.0 nM–2.0 μM	1.5 nM	Aqueous and plastic samples	97.2–103.6%	Liu et al. (2023b)
TNP	FL	100–700 nM	43.0 nM	Water samples	93.4–102.9%	Li et al. (2017)
Sudan dyes	CL	0.005–100 ng/mL	1.0–5.0 pg/mL	Eggs	70.5–92.2%	He et al. (2019)
Orange II dyes	SERS	0.1–10.0 nM	0.1 nM	Lake water	92.2–114.4%	Xue et al. (2021)

briefly introduce the applications of MIPs-based sensors in monitoring of common environmental pollution including antibiotics, pesticides, heavy metals, pathogens, EDCs and organic dyes, etc. While the MIPs-based sensors provide various advantages and advantages for environmental pollution monitoring, but there are still some limitations for overcoming. The research of sensors and biosensors should be strengthened to improve the stability and reproducibility of sensors, improving the potential of miniaturized devices, expanding application range, and help achieve the United Nations Sustainable Development Goals (SDGs).

Although, MIPs-based sensors have been successfully applied in various environment pollution determination, aiming in improving sensitivity and selectivity. The basic limitations of low stability and reproducibility still need to be overcame to confirm the efficient results (Ayivi et al. 2023). Generally, real samples usually have harsh detection conditions, such as pH, salt ions, temperature, and organic solvents (Gu et al. 2023). Keeping stability in above detection conditions during detection process is a major challenge, indicating the necessity in design and fabrication of stable composites for sensors. In addition, the reproducibility provides promising potential in multiple uses of MIPs-based sensors, improving the usage efficiency. The requirement of enhancement of stability and reproducibility is still under consideration.

The MIPs-based sensors show unique potential in miniaturized devices compared to traditional analytical methods (Ayankojo et al. 2020; Wang et al. 2020a; Huang et al. 2018b). With rapid development of portable instruments including miniaturized electrochemical workstation, portable Raman spectrometer, smartphone with identification ability of color, the integration of above instruments with MIPs technology will be considerable prospect in monitoring environment contaminants. The MIPs-based sensors with feature of miniaturized devices can helpful to control and improve the environmental quality individually, as well as by government i.e. environmental protection organization (EPA). More importantly, it can advance the goal of SDGs. In addition, artificial intelligence (AI) technologies including traditional machine learning (ML) and deep learning (DL) can be embedded in above miniaturized MIPs-based sensors for classification and quantification of environmental pollution with various algorithms (Ayres et al. 2021; Fang et al. 2023). The MIPs-based sensors combined with AI technology could be a powerful tool for detection of environmental contaminants with high analytical performance.

Recently, various analytical tools have been combined with MIPs technology for fabrication of sensors in environmental pollution detection, such as EC, FL, UV–Vis, SERS, SPR, FET (Afsharara et al. 2023; Liu et al. 2023a; Chen et al. 2023b; Kumar and Kim 2022; Gao et al. 2019). With the rapid development in design and fabrication of new MIPs-based sensors, more attention should be given on their practical applications. On the one hand, the main environmental applications of MIPs-based sensors are concentrated on antibiotics, pesticides, heavy metals, pathogens, needing expanding to other contaminants including atmospheric pollutants (formaldehyde, SO_2, CO, etc.), biological contaminants (bioamine, seaweed, parasite, etc.), dyes and microplastics in order to meet the SDGs for providing a low-carbon, energy-efficient, and environmentally friendly living environment. On the other hand, the common

MIPs-based sensors are mostly related to electrochemistry, fluorescence, and colorimetric. However, there are some analytical tools show good potentials in detection of environmental contaminants, such as SERS, SPR, and resonance Rayleigh scattering (RRS). Last but not least, the MIPs-based sensors show catalytic performance provide considerable potential in the field of pollutant degradation.

An ideal MIPs-based sensor shows unique features in terms of promising rapidity, high sensitivity, good specificity, long storage time, simple preparation process, and convenient portability as compared to the common sensors. Combining traditional analytical instruments with MIPs technology can improve the performance of detection, the efficiency of in situ detection, and the expansion of the detection of contaminants in the environment, in particular SDGs 6 (clean water and sanitation), 13 (climate action), 14 (life below water) and 15 (life on land) (Huang et al. 2023). Studies have shown that the sensor's convenient portability reduces carbon emissions brought by transportation, and the sensor's high-throughput analysis saves energy. In addition, MIPs-based sensors have the advantage of simple sample pretreatment, which reduces the use of chemical reagents, in line with SDGs 6.3 and 12.4 (reducing the hazardous waste produced).

MIPs-based sensors with various excellent detection performance offer great potential for commercial applications, but there are three main reasons why MIPs-based sensors have not achieved large-scale commercialization. First of all, it is difficult to prepare MIPs-based nanomaterials with excellent properties, high stability and high reproducibility on a large scale. Secondly, it is difficult to uniformly modify MIPs-based nanomaterials on electrodes, or to combine MIPs-based nanomaterials with analytical equipment. Finally, MIPs-based sensors still need to be improved in the construction of miniaturized devices and the combination of AI technology. The MIPs-based sensor will be a prospective tool for environmental pollution analysis and further commercial production after addressing the above challenges.

Abbreviations

3-PBD	3-Phenoxybenzaldehyde
AMO	Amoxicillin
AP	Acephate
AZI	Azithromycin
BPA	Bisphenol A
CDs	Carbon dots
CFX	Ciprofloxacin
CLA	Clarithromycin
CPE	Carbon paste electrode
CPF	Chlorpyris
CQDs	Carbon quantum dots
CYP	Cypermethrin
CYT	Cyhexatin

DOX	Doxycycline
DPV	Differential pulse voltammetry
ECL	Electrochemiluminescence
EDCs	Endocrine disrupting chemicals
ENR	Enrofloxacin
ERY	Erythromycin
FET	Field-effect transistor
GQDs	Graphene quantum dots
IF	Imprinting factor
IFE	Internal filtration effect
IIP	Ion-imprinted polymer
KAN	Kanamycin
LFX	Levofloxacin
LSV	Linear scan voltammetry
MFX	Moxifloxacin
MIPs	Molecularly imprinted polymers
MPA	Methanephosphonic acid
NOR	Norfloxacin
OFL	Ofloxacin
OMT	Omethoate
PEC	Photoelectrochemical
PET	Photoinduced charge transfer
QCM	Quartz crystal microbalance
QDs	Quantum dots
SERS	Surface-enhanced Raman scattering
SMT	Sulfamethizole
SPE	Screen-printed electrode
SPR	Surface plasmon resonance
TC	Tetracycline
TNP	2,4,6-Trinitrophenol

References

Afsharara H, Asadian E, Mostafiz B, Banan K, Bigdeli SA, Hatamabadi D, Keshavarz A, Hussain CM, Keçili R, Ghorbani-Bidkorpeh F (2023) Molecularly imprinted polymer-modified carbon paste electrodes (MIP-CPE): a review on sensitive electrochemical sensors for pharmaceutical determinations. TrAC Trends Anal Chem 160

Alizadeh T, Ganjali MR, Alizadeh H (2014) Competitive extraction of Gd(III) into a carbon paste electrode impregnated with a nano-sized Gd(III)-imprinted polymer as a new method for its indirect voltammetric determination. Microchim Acta 182(5–6):1205–1212

Alizadeh T, Ganjali MR, Akhoundian M, Norouzi P (2016) Voltammetric determination of ultratrace levels of cerium(III) using a carbon paste electrode modified with nano-sized cerium-imprinted polymer and multiwalled carbon nanotubes. Microchim Acta 183(3):1123–1130

Alizadeh T, Hamidi N, Ganjali MR, Nourozi P (2017) Development of a highly selective and sensitive electrochemical sensor for Bi^{3+} determination based on nano-structured bismuth-imprinted polymer modified carbon/carbon nanotube paste electrode. Sens Actuators B Chem 245:605–614

Altintas Z, Gittens M, Guerreiro A, Thompson K-A, Walker J, Piletsky S, Tothill IE (2015) Detection of waterborne viruses using high affinity molecularly imprinted polymers. Anal Chem 87(13):6801–6807

Amulen DR, Spanoghe P, Houbraken M, Tamale A, de Graaf DC, Cross P, Smagghe G (2017) Environmental contaminants of honeybee products in Uganda detected using LC-MS/MS and GC-ECD. PLoS ONE 12(6):e0178546

Ansari S, Karimi M (2017) Novel developments and trends of analytical methods for drug analysis in biological and environmental samples by molecularly imprinted polymers. TrAC, Trends Anal Chem 89:146–162

Ayankojo AG, Reut J, Boroznjak R, Öpik A, Syritski V (2018a) Molecularly imprinted poly(meta-phenylenediamine) based QCM sensor for detecting Amoxicillin. Sens Actuators, B Chem 258:766–774

Ayankojo AG, Reut J, Opik A, Furchner A, Syritski V (2018b) Hybrid molecularly imprinted polymer for amoxicillin detection. Biosens Bioelectron 118:102–107

Ayankojo AG, Reut J, Öpik A, Syritski V (2020) Sulfamethizole-imprinted polymer on screen-printed electrodes: towards the design of a portable environmental sensor. Sens Actuators B Chem 320

Ayivi RD, Obare SO, Wei J (2023) Molecularly imprinted polymers as chemosensors for organophosphate pesticide detection and environmental applications. TrAC Trends Anal Chem 167

Ayres LB, Gomez FJV, Linton JR, Silva MF, Garcia CD (2021) Taking the leap between analytical chemistry and artificial intelligence: a tutorial review. Anal Chim Acta 1161:338403

Bahrami A, Besharati-Seidani A, Abbaspour A, Shamsipur M (2015) A highly selective voltammetric sensor for nanomolar detection of mercury ions using a carbon ionic liquid paste electrode impregnated with novel ion imprinted polymeric nanobeads. Mater Sci Eng C Mater Biol Appl 48:205–212

Bai Q, Huang C, Ma S, Gong B, Ou J (2023) Rapid adsorption and detection of copper ions in water by dual-functional ion-imprinted polymers doping with carbon dots. Sep Purif Technol 315

Bergwerff AA, Van Knapen F (2006) Surface plasmon resonance biosensors for detection of pathogenicmicroorganisms: strategies to secure food and environmental safety. J AOAC Int 89(3):826–831

Bojdi MK, Behbahani M, Sahragard A, Amin BG, Fakhari A, Bagheri A (2014a) A palladium imprinted polymer for highly selective and sensitive electrochemical determination of ultra-trace of palladium ions. Electrochim Acta 149:108–116

Bojdi MK, Mashhadizadeh MH, Behbahani M, Farahani A, Davarani SSH, Bagheri A (2014b) Synthesis, characterization and application of novel lead imprinted polymer nanoparticles as a high selective electrochemical sensor for ultra-trace determination of lead ions in complex matrixes. Electrochim Acta 136:59–65

Bompart M, De Wilde Y, Haupt K (2010) Chemical nanosensors based on composite molecularly imprinted polymer particles and surface-enhanced Raman scattering. Adv Mater 22(21):2343–2348

Carrasco S, Benito-Peña E, Navarro-Villoslada F, Langer J, Sanz-Ortiz MN, Reguera J, Liz-Marzán LM, Moreno-Bondi MC (2016) Multibranched gold-mesoporous silica nanoparticles coated with a molecularly imprinted polymer for label-free antibiotic surface-enhanced raman scattering analysis. Chem Mater 28(21):7947–7954

Chen L, Xu S, Li J (2011) Recent advances in molecular imprinting technology: current status, challenges and highlighted applications. Chem Soc Rev 40(5):2922–2942

Chen L, Wang X, Lu W, Wu X, Li J (2016) Molecular imprinting: perspectives and applications. Chem Soc Rev 45(8):2137–2211

Chen S, Chen X, Zhang L, Gao J, Ma Q (2017) Electrochemiluminescence detection of Escherichia coli O157: H7 based on a novel polydopamine surface imprinted polymer biosensor. ACS Appl Mater Interfaces 9(6):5430–5436

Chen J, Bai H, Xia J, Liu X, Liu Y, Cao Q (2018) Trace detection of Ce^{3+} by adsorption strip voltammetry at a carbon paste electrode modified with ion imprinted polymers. J Rare Earths 36(10):1121–1126

Chen MJ, Yang HL, Si YM, Tang Q, Chow CF, Gong CB (2021) A hollow visible-light-responsive surface molecularly imprinted polymer for the detection of chlorpyrifos in vegetables and fruits. Food Chem 355:129656

Chen X, Li G, Yue X, Peng C, Wang J (2023) Ratiometric fluorescent detection of carbendazim in foods based on metallic nanoclusters self-assembled nanocomplex. Food Chem 136478

Chen X, Li Q, Yuan T, Ma M, Ye Z, Wei X, Fang X, Mao S (2023b) Highly specific antibiotic detection on water-stable black phosphorus field-effect transistors. ACS Sens 8(2):858–866

Cozzolino D (2016) Near infrared spectroscopy as a tool to monitor contaminants in soil, sediments and water—state of the art, advantages and pitfalls. Trends Environ Anal Chem 9:1–7

Crapnell RD, Hudson A, Foster CW, Eersels K, Grinsven BV, Cleij TJ, Banks CE, Peeters M (2019) Recent advances in electrosynthesized molecularly imprinted polymer sensing platforms for bioanalyte detection. Sensors 19(5):1204

Craun MF, Craun GF, Calderon RL, Beach MJ (2006) Waterborne outbreaks reported in the United States. J Water Health 4(S2):19–30

Cui F, Zhou Z, Zhou HS (2020) Molecularly imprinted polymers and surface imprinted polymers based electrochemical biosensor for infectious diseases. Sensors 20(4):996

Dayal H, Ng WY, Lin XH, Li SFY (2019) Development of a hydrophilic molecularly imprinted polymer for the detection of hydrophilic targets using quartz crystal microbalance. Sens Actuators B Chem 300

Donmez M, Yilmaz MD, Kilbas B (2017) Fluorescent detection of dipicolinic acid as a biomarker of bacterial spores using lanthanide-chelated gold nanoparticles. J Hazard Mater 324:593–598

Eersels K, Lieberzeit P, Wagner P (2016) A review on synthetic receptors for bioparticle detection created by surface-imprinting techniques from principles to applications. ACS Sensors 1(10):1171–1187

Eskandari H, Amirzehni M, Hassanzadeh J, Vahid B (2020) Mesoporous MIP-capped luminescent MOF as specific and sensitive analytical probe: application for chlorpyrifos. Mikrochim Acta 187(12):673

Fang L, Jia M, Zhao H, Kang L, Shi L, Zhou L, Kong W (2021) Molecularly imprinted polymer-based optical sensors for pesticides in foods: recent advances and future trends. Trends Food Sci Technol 116:387–404

Fang B, Yu J, Chen Z, Osman AI, Farghali M, Ihara I, Hamza EH, Rooney DW, Yap PS (2023) Artificial intelligence for waste management in smart cities: a review. Environ Chem Lett 1–31

Farid MM, Goudini L, Piri F, Zamani A, Saadati F (2016) Molecular imprinting method for fabricating novel glucose sensor: Polyvinyl acetate electrode reinforced by MnO_2/CuO loaded on graphene oxide nanoparticles. Food Chem 194:61–67

Farooq S, Nie J, Cheng Y, Yan Z, Li J, Bacha SAS, Mushtaq A, Zhang H (2018) Molecularly imprinted polymers' application in pesticide residue detection. Analyst 143(17):3971–3989

Feng S, Gao F, Chen Z, Grant E, Kitts DD, Wang S, Lu X (2013) Determination of α-tocopherol in vegetable oils using a molecularly imprinted polymers–surface-enhanced Raman spectroscopic biosensor. J Agric Food Chem 61(44):10467–10475

Figueiredo L, Erny G, Santos L, Alves A (2016) Applications of molecularly imprinted polymers to the analysis and removal of personal care products: a review. Talanta 146:754–765

Foguel MV, Ton X-A, Zanoni MVB, Sotomayor MDPT, Haupt K, Tse B, Bui S (2015) A molecularly imprinted polymer-based evanescent wave fiber optic sensor for the detection of basic red 9 dye. Sens Actuators B Chem 218:222–228

Gall AM, Mariñas BJ, Lu Y, Shisler JL (2015) Waterborne viruses: a barrier to safe drinking water. PLoS Pathog 11(6):e1004867

Gao L, Zhuang J, Nie L, Zhang J, Zhang Y, Gu N, Wang T, Feng J, Yang D, Perrett S (2007) Intrinsic peroxidase-like activity of ferromagnetic nanoparticles. Nat Nanotechnol 2(9):577–583
Gao W, Li P, Qin S, Huang Z, Cao Y, Liu X (2019) A highly sensitive tetracycline sensor based on a combination of magnetic molecularly imprinted polymer nanoparticles and surface plasmon resonance detection. Mikrochim Acta 186(9):637
Ghanei-Motlagh M, Taher MA (2017) Magnetic silver(I) ion-imprinted polymeric nanoparticles on a carbon paste electrode for voltammetric determination of silver(I). Microchim Acta 184(6):1691–1699
Ghani SM, Rezaei B, Jamei HR, Ensafi AA (2020) Preparation and comparison of molecularly imprinted polymer fluorimetric nanoprobe based on polymer dots and carbon quantum dots for determination of acetamiprid using response surface method. Microchim Acta 187:1–8
Ghasemi F, Hormozi-Nezhad MR, Mahmoudi M (2015) A colorimetric sensor array for detection and discrimination of biothiols based on aggregation of gold nanoparticles. Anal Chim Acta 882:58–67
Gu Y, Li Y, Qiu H, Yang Y, Wu Q, Fan X, Ding Y, Yi L, Ge K, Shen Y (2023) Recent progress on noble-free substrates for surface-enhanced Raman spectroscopy analysis. Coord Chem Rev 497
Guan W, Liu M, Zhang C (2016) Electrochemiluminescence detection in microfluidic cloth-based analytical devices. Biosens Bioelectron 75:247–253
Gui R, Jin H, Guo H, Wang Z (2018) Recent advances and future prospects in molecularly imprinted polymers-based electrochemical biosensors. Biosens Bioelectron 100:56–70
Güney S, Güney O (2016) A novel electrochemical sensor for selective determination of uranyl ion based on imprinted polymer sol–gel modified carbon paste electrode. Sens Actuators B Chem 231:45–53
Guo Y, Zhang L, Zhang S, Yang Y, Chen X, Zhang M (2015) Fluorescent carbon nanoparticles for the fluorescent detection of metal ions. Biosens Bioelectron 63:61–71
He T, Wang GN, Liu JX, Zhao WL, Huang JJ, Xu MX, Wang JP, Liu J (2019) Dummy molecularly imprinted polymer based microplate chemiluminescence sensor for one-step detection of Sudan dyes in egg. Food Chem 288:347–353
Hoellein L, Holzgrabe U (2014) Development of simplified HPLC methods for the detection of counterfeit antimalarials in resource-restraint environments. J Pharm Biomed Anal 98:434–445
Holdgate M (1980) A perspective of environmental pollution. Cambridge University Press, Great Britain
Hu Y, Liao J, Wang D, Li G (2014) Fabrication of gold nanoparticle-embedded metal–organic framework for highly sensitive surface-enhanced Raman scattering detection. Anal Chem 86(8):3955–3963
Huang S, Guo M, Tan J, Geng Y, Wu J, Tang Y, Su C, Lin CC, Liang Y (2018a) Novel fluorescence sensor based on all-inorganic perovskite quantum dots coated with molecularly imprinted polymers for highly selective and sensitive detection of omethoate. ACS Appl Mater Interfaces 10(45):39056–39063
Huang C, Cheng Y, Gao Z, Zhang H, Wei J (2018b) Portable label-free inverse opal photonic hydrogel particles serve as facile pesticides colorimetric monitoring. Sens Actuators B Chem 273:1705–1712
Huang C-W, Lin C, Nguyen MK, Hussain A, Bui X-T, Ngo HH (2023) A review of biosensor for environmental monitoring: principle, application, and corresponding achievement of sustainable development goals. Bioengineered 14(1):58–80
Hussain C, Kharisov B (2016) Advanced environmental analysis: application of nanomaterials. Royal Society of Chemistry, English, pp 48–75
Hussain CM, Kecili R (2019) Modern environmental analysis techniques for pollutants. Elsevier
Jin Y, Dou M, Zhuo S, Li Q, Wang F, Li J (2022) Advances in microfluidic analysis of residual antibiotics in food. Food Control 136:108885
Jolly P, Tamboli V, Harniman RL, Estrela P, Allender CJ, Bowen JL (2016) Aptamer–MIP hybrid receptor for highly sensitive electrochemical detection of prostate specific antigen. Biosens Bioelectron 75:188–195

Ju J, Chen W (2015) In situ growth of surfactant-free gold nanoparticles on nitrogen-doped graphene quantum dots for electrochemical detection of hydrogen peroxide in biological environments. Anal Chem 87(3):1903–1910

Kamra T, Zhou T, Montelius L, Schnadt J, Ye L (2015) Implementation of molecularly imprinted polymer beads for surface enhanced Raman detection. Anal Chem 87(10):5056–5061

Keçili R, Ghorbani-Bidkorbeh F, Dolak I, Canpolat G, Karabörk M, Hussain CM (2021) Functionalized magnetic nanoparticles as powerful sorbents and stationary phases for the extraction and chromatographic applications. TrAC, Trends Anal Chem 143:116380

Keçili R, Hussain CG, Hussain CM (2023) Fluorescent nanosensors based on green carbon dots (CDs) and molecularly imprinted polymers (MIPs) for environmental pollutants: emerging trends and future prospects. Trends Environ Anal Chem 40

Khanmohammadi A, Jalili Ghazizadeh A, Hashemi P, Afkhami A, Arduini F, Bagheri H (2020) An overview to electrochemical biosensors and sensors for the detection of environmental contaminants. J Iran Chem Soc 17:2429–2447

Kostrewa S, Emgenbroich M, Klockow D, Wulff G (2003) Surface-enhanced Raman scattering on molecularly imprinted polymers in water. Macromol Chem Phys 204(3):481–487

Kumar V, Kim KH (2022) Use of molecular imprinted polymers as sensitive/selective luminescent sensing probes for pesticides/herbicides in water and food samples. Environ Pollut 299:118824

Li J, Ma F, Wei X, Fu C, Pan H (2015) A highly selective molecularly imprinted electrochemiluminescence sensor for ultra-trace beryllium detection. Anal Chim Acta 871:51–56

Li M, Liu H, Ren X (2017) Ratiometric fluorescence and mesoporous structured imprinting nanoparticles for rapid and sensitive detection 2,4,6-trinitrophenol. Biosens Bioelectron 89(Pt 2):899–905

Li S, Li J, Luo J, Xu Z, Ma X (2018) A microfluidic chip containing a molecularly imprinted polymer and a DNA aptamer for voltammetric determination of carbofuran. Microchim Acta 185:1–8

Li Y, Zhang L, Dang Y, Chen Z, Zhang R, Li Y, Ye BC (2019a) A robust electrochemical sensing of molecularly imprinted polymer prepared by using bifunctional monomer and its application in detection of cypermethrin. Biosens Bioelectron 127:207–214

Li S, Li J, Ma X, Pang C, Yin G, Luo J (2019b) Molecularly imprinted electroluminescence switch sensor with a dual recognition effect for determination of ultra-trace levels of cobalt (II). Biosens Bioelectron 139:111321

Li G, Qi X, Wu J, Xu L, Wan X, Liu Y, Chen Y, Li Q (2022) Ultrasensitive, label-free voltammetric determination of norfloxacin based on molecularly imprinted polymers and Au nanoparticle-functionalized black phosphorus nanosheet nanocomposite. J Hazard Mater 436:129107

Li Z, Li S, Li Y, Liu M, Jiang L, Niu J, Zhang Y, Zhou Q (2023a) Highly selective and sensitive determination of doxycycline integrating enrichment with thermosensitive magnetic molecular imprinting nanomaterial and carbon dots based fluorescence probe. Sci Total Environ 896:165257

Li H, Sheng W, Haruna SA, Bei Q, Wei W, Hassan MM, Chen Q (2023) Recent progress in photoelectrochemical sensors to quantify pesticides in foods: theory, photoactive substrate selection, recognition elements and applications. TrAC Trends Anal Chem 164

Liu Y, Cao N, Gui W, Ma Q (2018) Nitrogen-doped graphene quantum dots-based fluorescence molecularly imprinted sensor for thiacloprid detection. Talanta 183:339–344

Liu G, Ling J, Li J (2021) Extremely sensitive molecularly imprinted ECL sensor with multiple probes released from liposomes immobilized by a light-triggered click reaction. ACS Sens 6(11):4185–4192

Liu B, Zhu H, Liu J, Wang M, Pan J, Feng R, Hu P, Niu X (2023a) Alkali-etched imprinted Mn-based Prussian blue analogues with superior oxidase-mimetic activity and precise recognition for tetracycline colorimetric sensing. ACS Appl Mater Interfaces 15(20):24736–24746

Liu Y, Zhang Y, Niu J, Nie L, Huang S, Liu H, Yuan S, Zhou Q (2023b) Selective fluorescent probe for sensitive determination of bisphenol A based on molecularly imprinted polymers decorated carbon dots derived from citric acid and ethylenediamine. Chemosphere 324:138303

Lunau M, Lemke A, Walther K, Martens-Habbena W, Simon M (2005) An improved method for counting bacteria from sediments and turbid environments by epifluorescence microscopy. Environ Microbiol 7(7):961–968

Madikizela LM, Tavengwa NT, Tutu H, Chimuka L (2018) Green aspects in molecular imprinting technology: From design to environmental applications. Trends Environ Anal Chem 17:14–22

Mahmoudpour M, Torbati M, Mousavi M-M, de la Guardia M, Ezzati Nazhad Dolatabadi J (2020) Nanomaterial-based molecularly imprinted polymers for pesticides detection: recent trends and future prospects. TrAC Trends Anal Chem 129

Majdinasab M, Daneshi M, Louis Marty J (2021) Recent developments in non-enzymatic (bio)sensors for detection of pesticide residues: focusing on antibody, aptamer and molecularly imprinted polymer. Talanta 232:122397

Mao Y, Bao Y, Gan S, Li F, Niu L (2011) Electrochemical sensor for dopamine based on a novel graphene-molecular imprinted polymers composite recognition element. Biosens Bioelectron 28(1):291–297

Meece JK, Anderson JL, Klein BS, Sullivan TD, Foley SL, Baumgardner DJ, Brummitt CF, Reed KD (2010) Genetic diversity in *Blastomyces dermatitidis*: implications for PCR detection in clinical and environmental samples. Med Mycol 48(2):285–290

Miao J, Liu A, Wu L, Yu M, Wei W, Liu S (2020) Magnetic ferroferric oxide and polydopamine molecularly imprinted polymer nanocomposites based electrochemical impedance sensor for the selective separation and sensitive determination of dichlorodiphenyltrichloroethane (DDT). Anal Chim Acta 1095:82–92

Moein MM, Nakao R, Amini N, Abdel-Rehim M, Schou M, Halldin C (2019) Sample preparation techniques for radiometabolite analysis of positron emission tomography radioligands; trends, progress, limitations and future prospects. TrAC, Trends Anal Chem 110:1–7

Morsi SM, Abd El-Aziz ME, Mohamed HA (2023) Smart polymers as molecular imprinted polymers for recognition of target molecules. Int J Polym Mater Polym Biomater 72(8):612–635

Mostafiz B, Bigdeli SA, Banan K, Afsharara H, Hatamabadi D, Mousavi P, Hussain CM, Keçili R, Ghorbani-Bidkorbeh F (2021) Molecularly imprinted polymer-carbon paste electrode (MIP-CPE)-based sensors for the sensitive detection of organic and inorganic environmental pollutants: a review. Trends Environ Anal Chem 32

Ndunda EN, Mizaikoff B (2016) Molecularly imprinted polymers for the analysis and removal of polychlorinated aromatic compounds in the environment: a review. Analyst 141(11):3141–3156

Negahdary M (2020) Aptamers in nanostructure-based electrochemical biosensors for cardiac biomarkers and cancer biomarkers: a review. Biosens Bioelectron 152:112018

Nguyen VBC, Ayankojo AG, Reut J, Rappich J, Furchner A, Hinrichs K, Syritski V (2023) Molecularly imprinted co-polymer for class-selective electrochemical detection of macrolide antibiotics in aqueous media. Sens Actuators B Chem 374

Nsibande S, Forbes P (2016) Fluorescence detection of pesticides using quantum dot materials–a review. Anal Chim Acta 945:9–22

Parra E, Segura F, Tijero J, Pons I, Nogueras M-M (2017) Development of a real-time PCR for Bartonella spp. detection, a current emerging microorganism. Mol Cell Probes 32:55–59

Poornima V, Alexandar V, Iswariya S, Perumal PT, Uma TS (2016) Gold nanoparticle-based nanosystems for the colorimetric detection of Hg^{2+} ion contamination in the environment. RSC Adv 6(52):46711–46722

Qi P, Wang J, Wang L, Li Y, Jin J, Su F, Tian Y, Chen J (2010) Molecularly imprinted polymers synthesized via semi-covalent imprinting with sacrificial spacer for imprinting phenols. Polymer 51(23):5417–5423

Qi J-f, Tan D, Wang X-j, Ma H-t, Wan Y-c, Hu A, Li L-h, Xiao B, Lu B (2021) A novel acetylcholinesterase biosensor with dual-recognized strategy based on molecularly imprinted polymer. Sens Actuators B Chem 337

Qiao X, Han Y, Tian D, Yang Z, Li J, Zhao S (2019) MOF matrix doped with rare earth ions to realize ratiometric fluorescent sensing of 2, 4, 6-trinitrophenol: synthesis, characterization and performance. Sens Actuators, B Chem 286:1–8

Ramanavicius S, Jagminas A, Ramanavicius A (2021) Advances in molecularly imprinted polymers based affinity sensors (review). Polymers (Basel) 13(6)

Rao TP, Prasad K, Kala R, Gladis JM (2007) Biomimetic sensors for toxic pesticides and inorganics based on optoelectronic/electrochemical transducers—an overview. Crit Rev Anal Chem 37(3):191–210

Rao H, Chen M, Ge H, Lu Z, Liu X, Zou P, Wang X, He H, Zeng X, Wang Y (2017) A novel electrochemical sensor based on Au@ PANI composites film modified glassy carbon electrode binding molecular imprinting technique for the determination of melamine. Biosens Bioelectron 87:1029–1035

Rebelo P, Costa-Rama E, Seguro I, Pacheco JG, Nouws HPA, Cordeiro M, Delerue-Matos C (2021) Molecularly imprinted polymer-based electrochemical sensors for environmental analysis. Biosens Bioelectron 172:112719

Rezaei B, Rahmanian O, Ensafi AA (2014) An electrochemical sensor based on multiwall carbon nanotubes and molecular imprinting strategy for warfarin recognition and determination. Sens Actuators, B Chem 196:539–545

Roushani M, Zalpour N (2021) Selective detection of Asulam with in-situ dopamine electropolymerization based electrochemical MIP sensor. React Funct Polym 169:105069

Roy E, Patra S, Tiwari A, Madhuri R, Sharma PK (2017) Introduction of selectivity and specificity to graphene using an inimitable combination of molecular imprinting and nanotechnology. Biosens Bioelectron 89:234–248

Samsidar A, Siddiquee S, Shaarani SM (2018) A review of extraction, analytical and advanced methods for determination of pesticides in environment and foodstuffs. Trends Food Sci Technol 71:188–201

Saylan Y (2023) Unveiling the pollution of bacteria in water samples through optic sensor. Microchem J 193:109057

Saylan Y, Erdem Ö, Cihangir N, Denizli A (2019) Detecting fingerprints of waterborne bacteria on a sensor. Chemosensors 7(3):33

Scallan E, Hoekstra RM, Angulo FJ, Tauxe RV, Widdowson M-A, Roy SL, Jones JL, Griffin PM (2011) Foodborne illness acquired in the United States—major pathogens. Emerg Infect Dis 17(1):7

Shamsipur M, Hashemi B, Dehdashtian S, Mohammadi M, Gholivand MB, Garau A, Lippolis V (2014) Silver ion imprinted polymer nanobeads based on a aza-thioether crown containing a 1,10-phenanthroline subunit for solid phase extraction and for voltammetric and potentiometric silver sensors. Anal Chim Acta 852:223–235

Shao H, Li C, Ma C, Sun L, Chen R, Cheng R, Liu Y, Yan Y, Sun Q, Wu C (2017) An ion-imprinted material embedded carbon quantum dots for selective fluorometric determination of lithium ion in water samples. Microchim Acta 184(12):4861–4868

Shi Z, Wang Z, Li K, Wang Y, Li Z, Zhu Z (2023) MXene fibers-based molecularly imprinted disposable electrochemical sensor for sensitive and selective detection of hydrocortisone. Talanta 125100

Skottrup PD, Nicolaisen M, Justesen AF (2008) Towards on-site pathogen detection using antibody-based sensors. Biosens Bioelectron 24(3):339–348

Song Z, Zhai X, Jiang C, Chen R, Ye S, Tong J, Dramou P, He H (2022a) Sensitive and selective detection of carbamazepine in serum samples by bionic double-antibody sandwich method based on cucurbit [7] uril and molecular imprinted polymers. Biosens Bioelectron 203:114037

Song J, He K, Xing B, Pei Y, Wang D, Wang Y, Li S, Li J, Huan W, Zhang Y, Hammock BD (2022b) Rapid measurement of residual kanamycin using highly specific biomimetic recognition paper-based chip. Anal Chem 94(50):17567–17576

Speltini A, Scalabrini A, Maraschi F, Sturini M, Profumo A (2017) Newest applications of molecularly imprinted polymers for extraction of contaminants from environmental and food matrices: a review. Anal Chim Acta 974:1–26

Srivastava KR, Awasthi S, Mishra PK, Srivastava PK (2020) Biosensors/molecular tools for detection of waterborne pathogens. Waterborne Pathogens 237–277

Stoian IA, Iacob BC, Dudas CL, Barbu-Tudoran L, Bogdan D, Marian IO, Bodoki E, Oprean R (2020) Biomimetic electrochemical sensor for the highly selective detection of azithromycin in biological samples. Biosens Bioelectron 155:112098

Sukjee W, Thitithanyanont A, Manopwisedjaroen S, Seetaha S, Thepparit C, Sangma C (2022) Virus MIP-composites for SARS-CoV-2 detection in the aquatic environment. Mater Lett 315:131973

Sun X, Gao C, Zhang L, Yan M, Yu J, Ge S (2017) Photoelectrochemical sensor based on molecularly imprinted film modified hierarchical branched titanium dioxide nanorods for chlorpyrifos detection. Sens Actuators, B Chem 251:1–8

Sun Q, Fu L, Yin C, Wu M, Liu H, Niu N, Chen L (2022) Construction of biomass carbon dots@molecularly imprinted polymer fluorescent sensor array for accurate identification of 5-nitroimidazole antibiotics. Sens Actuators B Chem 373

Tan L, Guo M, Tan J, Geng Y, Huang S, Tang Y, Su C, Lin C, Liang Y (2019) Development of high-luminescence perovskite quantum dots coated with molecularly imprinted polymers for pesticide detection by slowly hydrolysing the organosilicon monomers in situ. Sens Actuators, B Chem 291:226–234

Tang Y, Li M, Gao X, Liu X, Gao J, Ma T, Li J (2017) A NIR-responsive up-conversion nanoparticle probe of the NaYF4:Er, Yb type and coated with a molecularly imprinted polymer for fluorometric determination of enrofloxacin. Microchim Acta 184(9):3469–3475

Thakur N, Das TR, Gupta S, Shukla SK, Patra S, Kumar D (2023) Material selection for the fabrication of MIP-based devices: an overview. Molecularly Imprinted Polymers (MIPs), pp 101–114

Ton X-A, Acha V, Haupt K, Bui BTS (2012) Direct fluorimetric sensing of UV-excited analytes in biological and environmental samples using molecularly imprinted polymer nanoparticles and fluorescence polarization. Biosens Bioelectron 36(1):22–28

Wang L, Chen W, Xu D, Shim BS, Zhu Y, Sun F, Liu L, Peng C, Jin Z, Xu C (2009) Simple, rapid, sensitive, and versatile SWNT– paper sensor for environmental toxin detection competitive with ELISA. Nano Lett 9(12):4147–4152

Wang L, Wang Y, Ye D, Liu Q (2020c) Review of the 2019 novel coronavirus (SARS-CoV-2) based on current evidence. Int J Antimicrob Agents 55(6):105948

Wang W, Gong Z, Yang S, Xiong T, Wang D, Fan M (2020a) Fluorescent and visual detection of norfloxacin in aqueous solutions with a molecularly imprinted polymer coated paper sensor. Talanta 208:120435

Wang W, Wang X, Cheng N, Luo Y, Lin Y, Xu W, Du D (2020) Recent advances in nanomaterials-based electrochemical (bio)sensors for pesticides detection. TrAC Trends Anal Chem 132

Wang H, Xu Y, Xu D, Chen L, Qiu X, Zhu Y (2022a) Graphitic carbon nitride for photoelectrochemical detection of environmental pollutants. ACS ES&T Eng 2(2):140–157

Wang M, Ceto X, Del Valle M (2022b) A sensor array based on molecularly imprinted polymers and machine learning for the analysis of fluoroquinolone antibiotics. ACS Sens 7(11):3318–3325

Wang Y-Q, Fang Z, Min H, Xu X-Y, Li Y (2022c) Sensitive determination of ofloxacin by molecularly imprinted polymers containing ionic liquid functionalized carbon quantum dots and europium ion. ACS Appl Nano Mater 5(6):8467–8474

Wang Z, Zhou C, Wu S, Sun C (2021) Ion-imprinted polymer modified with carbon quantum dots as a highly sensitive copper(II) ion probe. Polymers (Basel) 13(9)

Wei J, Yuan X, Zhang Y, Liu H, Sun B (2020) Ionic liquid-sensitized molecularly imprinted polymers based on heteroatom co-doped quantum dots functionalized graphene for sensitive detection of lambda-cyhalothrin. Anal Chim Acta 1136:9–18

Wloch M, Datta J (2019) Chapter two—synthesis and polymerization techniques of molecularly imprinted polymers. Compr, Anal Chem 86:17–40

Xia X, Yang E, Du X, Cai Y, Chang F, Gao D (2021) Nanostructured shell-layer artificial antibody with fluorescence-tagged recognition sites for the trace detection of heavy metal ions by self-reporting microsensor arrays. ACS Appl Mater Interfaces 13(48):57981–57997

Xiao L, Zhu A, Xu Q, Chen Y, Xu J, Weng J (2017) Colorimetric biosensor for detection of cancer biomarker by Au nanoparticle-decorated Bi_2Se_3 nanosheets. ACS Appl Mater Interfaces 9(8):6931–6940
Xu D-D, Deng Y-L, Li C-Y, Lin Y, Tang H-W (2017) Metal-enhanced fluorescent dye-doped silica nanoparticles and magnetic separation: a sensitive platform for one-step fluorescence detection of prostate specific antigen. Biosens Bioelectron 87:881–887
Xue Y, Shao J, Sui G, Ma Y, Li H (2021) Rapid detection of orange II dyes in water with SERS imprinted sensor based on PDA-modified MOFs@Ag. J Environ Chem Eng 9(6)
Yadav AK, Verma D, Dalal N, Kumar A, Solanki PR (2022) Molecularly imprinted polymer-based nanodiagnostics for clinically pertinent bacteria and virus detection for future pandemics. Biosens Bioelectron X:100257
Yan X, Zhang Z, Zhang R, Yang T, Hao G, Yuan L, Yang X (2021) Rapid detection of dimethoate in soybean samples by microfluidic paper chips based on oil-soluble CdSe quantum dots. Foods 10(11)
Yang Y, Fang G, Liu G, Pan M, Wang X, Kong L, He X, Wang S (2013) Electrochemical sensor based on molecularly imprinted polymer film via sol–gel technology and multi-walled carbon nanotubes-chitosan functional layer for sensitive determination of quinoxaline-2-carboxylic acid. Biosens Bioelectron 47:475–481
Yang Y, Cao Y, Wang X, Fang G, Wang S (2015) Prussian blue mediated amplification combined with signal enhancement of ordered mesoporous carbon for ultrasensitive and specific quantification of metolcarb by a three-dimensional molecularly imprinted electrochemical sensor. Biosens Bioelectron 64:247–254
Yang Y, Chang Y, Guo Y, Yu L, Zhang G, Zhai D, Wang X, Sun X (2019) Fluorometric microplate-based dimethoate assay using CdSe/ZnS quantum dots coated with a molecularly imprinted polymer. Mikrochim Acta 186(8):589
Yang Y, Yan W, Wang X, Yu L, Zhang J, Bai B, Guo C, Fan S (2021a) Development of a molecularly imprinted photoelectrochemical sensing platform based on NH2-MIL-125 (Ti)–TiO_2 composite for the sensitive and selective determination of oxtetracycline. Biosens Bioelectron 177:113000
Yang X, Guo Q, Yang J, Chen S, Hu F, Hu Y, Lin H (2021b) Synergistic effects of layer-by-layer films for highly selective and sensitive electrochemical detection of trans-resveratrol. Food Chem 338:127851
Yang W, Ma Y, Sun H, Huang C, Shen X (2022a) Molecularly imprinted polymers based optical fiber sensors: a review. TrAC, Trends Anal Chem 152:116608
Yang Y, Shi Z, Wang X, Bai B, Qin S, Li J, Jing X, Tian Y, Fang G (2022b) Portable and on-site electrochemical sensor based on surface molecularly imprinted magnetic covalent organic framework for the rapid detection of tetracycline in food. Food Chem 395:133532
Yang L, Wang X, Zhang F, Yu L, Bai B, Zhang J, Zhang B, Tian Y, Qin S, Yang Y (2024) Two birds with one stone: a universal design and application of signal-on labeled fluorescent/electrochemical dual-signal mode biosensor for the detection of tetracycline residues in tap water, milk and chicken. Food Chem 430:136904
Ye J, Chen Y, Liu Z (2014) A boronate affinity sandwich assay: an appealing alternative to immunoassays for the determination of glycoproteins. Angew Chem Int Ed 53(39):10386–10389
Ye T, Yin W, Zhu N, Yuan M, Cao H, Yu J, Gou Z, Wang X, Zhu H, Reyihanguli A, Xu F (2018) Colorimetric detection of pyrethroid metabolite by using surface molecularly imprinted polymer. Sens Actuators B Chem 254:417–423
You M, Yang S, An Y, Zhang F, He P (2020) A novel electrochemical biosensor with molecularly imprinted polymers and aptamer-based sandwich assay for determining amyloid-β oligomer. J Electroanal Chem 862:114017
Yu M, Wu L, Miao J, Wei W, Liu A, Liu S (2019) Titanium dioxide and polypyrrole molecularly imprinted polymer nanocomposites based electrochemical sensor for highly selective detection of p-nonylphenol. Anal Chim Acta 1080:84–94
Zhang C, Zhao F, She Y, Hong S, Cao X, Zheng L, Wang S, Li T, Wang M, Jin M, Jin F, Shao H, Wang J (2019) A disposable molecularly imprinted sensor based on Graphe@AuNPs modified

screen-printed electrode for highly selective and sensitive detection of cyhexatin in pear samples. Sens Actuators, B Chem 284:13–22

Zhao Q, Lu D, Zhang G, Zhang D, Shi X (2021) Recent improvements in enzyme-linked immunosorbent assays based on nanomaterials. Talanta 223:121722

Zhihua W, Xiaole L, Jianming Y, Yaxin Q, Xiaoquan L (2011) Copper(II) determination by using carbon paste electrode modified with molecularly imprinted polymer. Electrochim Acta 58:750–756

Chapter 8 Nano-molecularly Imprinted Polymers for Advanced Environmental Monitoring: Synthesis, Applications, and Performance

Alma Mejri and Abdelmoneim Mars

8.1 Introduction

Molecularly imprinted polymers (MIPs) have gained significant attention in the last twenty years due to their unique preparation methods and broad range of applications (BelBruno 2019). These polymers mimic the 'lock and key' mechanism and exhibit molecular binding characteristics similar to antibodies and biological receptors. They are synthesized by copolymerizing functional monomers and cross-linkers in the presence of a target molecule acting as a molecular template (Liu et al. 2021; Mostafa et al. 2021). The resulting polymer matrix is highly cross-linked with recognition cavities that replicate the template's shape, size, and chemical functionality, enabling selective rebinding in complex mixtures (Mostafa et al. 2021).

MIPs are known for their structural predictability, exceptional sensitivity, and identification capabilities, making them versatile across various domains (Gao et al. 2020). Furthermore, they can be integrated with various sensing techniques, specifically electrochemical sensing, which is gaining traction due to its high sensitivity and practical applications. Due to their theoretical flexibility, customized receptors can be created to target a wide range of analytes, making them suitable for applications in pharmaceuticals, pesticides, nucleic acids, proteins, and other fields (Carballido et al. 2022). Despite their advantages such as cost-effectiveness, durability, and adaptability, challenges remain in scaling up laboratory advancements to large-scale sensor

A. Mejri · A. Mars (✉)
Desalination and Natural Water Valorization Laboratory (LaDVEN), Water Researches and Technologies Center (CERTE), BP 273, 8020 Soliman, Tunisia
e-mail: abdelmoneim.mars@gmail.com

A. Mejri
National Institute of Applied Sciences and Technology (INSAT), University of Carthage, Carthage, Tunisia

S. Patra and M. Sillanpaa (eds.), *Molecularly Imprinted Polymers as Artificial Antibodies for the Environmental Health*, https://doi.org/10.1007/978-3-031-58995-9_8

device manufacturing. To overcome sensitivity limitations in MIP-based electrochemical sensors, an innovative approach involves incorporating nanomaterials into Molecularly Imprinted Polymers (MIPs) (Beluomini et al. 2019). The introduction of nanostructured materials such as magnetic nanoparticles (MNPs), graphene oxides (GO), multi/single-walled carbon nanotubes (MWCNTs/SWCNTs), and carbon dots (CDs) onto the planar surface of MIPs-based electrochemical sensors overcomes the absence of electrochemical catalytic effects that often hinders sensitivity (Fresco-Cala et al. 2020; Keçili and Hussain 2018). This integration significantly improves analytical performance by leveraging the high conductivity of these nanomaterials, resulting in a hybrid material with newfound potential and properties. This approach opens avenues for the development of innovative analytical methods. Consequently, ongoing research emphasizes practicality and real-world applicability in laboratory and industrial settings (Fig. 8.1).

Furthermore, (bio)sensors are devices that can convert (bio)chemical interactions into a signal suitable for analytical purposes. According to the International Union of Pure and Applied Chemistry (IUPAC), such devices consist of two functional units: a (bio)receptor and a physicochemical transducer (Mehrotra 2016). The receptors can be activated or doped surfaces or (macro)molecules, and they interact with the

Fig. 8.1 Schematic representation of the possible incorporation of nanomaterials with molecularly imprinted polymer-based (bio)sensors and the analytical techniques used for monitoring the detection process

analyte. The device is called a biosensor if the receptor has a biological origin, such as DNA, antibodies, or enzymes. The physicochemical transducer translates the signal from the receptor-analyte interaction into an understandable value. This classification extends to both chemical and biosensors, which can be distinguished as catalytic or affinity-based devices. Within this framework, sensors based on Molecularly Imprinted Polymers (MIPs) fall into the category of affinity-based devices (Leibl et al. 2021). The interaction between analytes and the binding sites of MIPs is inherently of a (host–guest interaction nature). Monitoring recognition events is feasible through various methods (e.g., optical, gravimetric, or electrochemical), contingent on the type of transducer employed. A crucial requisite for sensors is to uphold a notable specificity for the targeted analyte amidst potentially interfering chemical species, preventing false-positive outcomes.

This chapter critically examines recent literature to evaluate the sensitive determination of environmental pollutants through nanomaterial-modified molecularly imprinted polymer-based sensing platforms. The assessment encompasses diverse polymerization techniques and the utilization of nano-sized modification agents in sensor development.

8.2 Molecularly Imprinted Polymers: Components and Synthesis Strategies

8.2.1 Molecular Imprinting Process

As per the literature, molecular imprinting refers to creating a highly cross-linked polymer matrix around a template molecule to achieve selective molecular recognition. The polymer matrix has a 3D structure with microcavities that complement those of the template molecule, which is crucial for molecular recognition (Lu et al. 2020). The co-presence of the template, functional monomer, cross-linker, and initiator in a porogenic solvent facilitates the interaction between the functional monomer and the template molecule, which leads to the formation of a stable template-monomer complex (Kloskowski et al. 2009). The functional monomers are copolymerized with cross-linkers to stabilize them. A critical step in this process is the removal of the template molecule, which is often accomplished through solvent extraction. The remaining binding sites in the polymer matrix are complementary in shape to the template molecule, allowing the matrix to recognize and bind the template molecules selectively (Kloskowski et al. 2009). Furthermore, functional monomers can form complex structures with a template molecule through both covalent and non-covalent interactions. Non-covalent imprinting is widely preferred due to its simplicity and effectiveness. However, a new technique known as semi-covalent imprinting has emerged, which combines the strength of covalent imprinting with the ease and speed of template binding and removal (Ndunda 2020).

8.2.2 Components of MIP

The development of molecularly imprinted polymers involves several key components (Fig. 8.2).

One of the most important elements is the use of templates, which help to create a polymer network with a specific affinity for particular template molecules. These molecules need to have inert functional groups during polymerization and be able to form stable template-monomer complexes (Pratama et al. 2020). Templates can be divided into various categories, including ions, organic molecules, biomacromolecules, and cells/viruses (Iskierko et al. 2016). Besides, the choice of functional monomers is critical in molecular imprinting technology, as it allows for interaction with template molecules using non-covalent approaches like hydrogen bonding (Pratama et al. 2020). Common functional monomers include carboxylic acids, sulfonic acids, and heteroaromatic bases, with methacrylic acid being a particularly popular choice (Iskierko et al. 2016). The concentration of functional monomers is also important, as it affects the number of binding sites. Cross-linkers, in turn, are essential in fixing functional monomers around template molecules during polymerization, creating a highly cross-linked and rigid polymer network that is resistant to template removal processes. The type and amount of cross-linker influence various characteristics of MIPs, such as morphology, mechanical performance, selectivity, and binding capacity (Kryscio and Peppas 2012). As regards initiators, they play a pivotal role in free radical polymerization, with azobisisobutyronitrile being a commonly used initiator (Pratama et al. 2020). Porogen agents act as reaction media and pore-forming agents during polymerization and are crucial for creating MIPs with desired properties (Pratama et al. 2020). The choice of solvent type has a significant impact on various factors like imprinting efficiency, interactions between

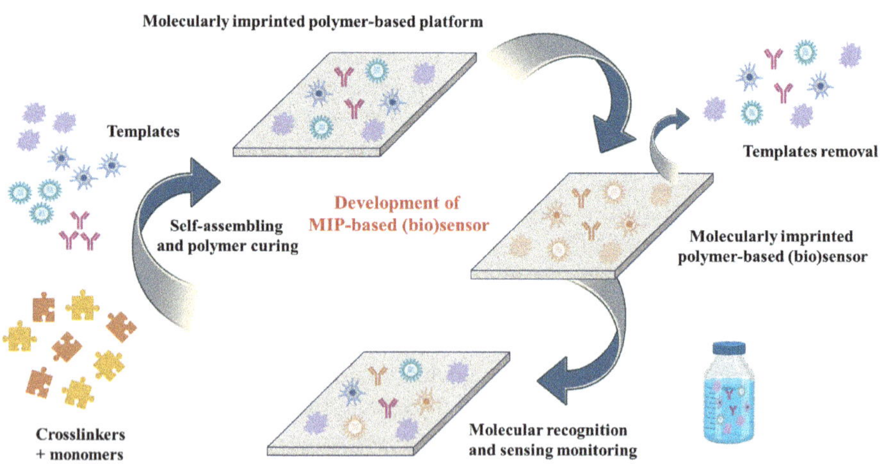

Fig. 8.2 Schematic representation of the development processes of molecularly imprinted polymer based-sensors

template and functional monomers, absorption properties, and polymer morphology. It is important to consider all these components to synthesize molecularly imprinted polymers with desired characteristics successfully.

8.2.3 Synthesis Strategies of MIP

There are several methods available for preparing molecularly imprinted polymers (MIPs), including bulk, suspension, emulsion, precipitation, multi-step, or electrochemical polymerization (Wackerlig and Lieberzeit 2015). In bulk polymerization, a simple process involves utilizing a template, functional monomer, cross-linker, initiator, and porogen (Cheong et al. 2013). However, resulting particles have irregular size and shape, limiting their suitability for sensing applications where mechanical disintegration during grinding and sieving may compromise binding groups, leading to poor reproducibility and selectivity. On the other hand, suspension polymerization, a one-step method, produces spherical particles of relatively large size, typically in the micrometer to millimeter diameter (Wackerlig and Lieberzeit 2015). This method is conducted in a water medium, with perfluorocarbon liquids and mineral oils serving as alternative continuous phases. Despite MIPs obtained through suspension polymerization exhibiting poor recognition due to the dispersing medium, it remains a method applicable on a large scale. In emulsion polymerization, a method with high yield produces monodispersed polymer particles with a size range of tens to hundreds of nanometers (Cheong et al. 2013). Water is typically used as the continuous phase, but its strong polarity and hydrogen bonding capability can impact the imprinting process. Consequently, this technique is considered less attractive for MIP synthesis. In contrast, precipitation polymerization, a single preparative step, results in uniform and spherical particles with diameters typically less than 1 μm (Cheong et al. 2013). This surfactant-free method involves polymerizing monomers in dilute solutions, with resulting polymer particles precipitating from the solution. Careful control of reaction conditions, including solvent polarity, temperature, and stirring speed, is necessary. Furthermore, multi-step swelling polymerization, also known as seed polymerization, involves complex procedures to obtain uniform spherical particles suitable for chromatographic applications (Vasapollo et al. 2011). This method requires the preparation of aqueous emulsions, impacting the imprinting process and selectivity of obtained MIPs. Surface imprinting polymerization, in turn, involves grafting thin MIP layers onto preformed beads, resulting in monodispersed modified surfaces (Cui et al. 2020). While effective, this method is time-consuming and finds applications in separation or sensing through thin imprinted layers. Lastly, electrochemical polymerization proves to be a convenient approach (Sharma et al. 2012). The polymer grows on the working electrode surface without the need for additional immobilization steps. This technique allows for easy control of polymer thickness by managing the current, offering a fast and reproducible process. Notably, electrochemical polymerization is faster than classical bulk polymerization, and the entire process can be monitored on a computer screen. The ability to control film thickness through administered current enhances the reproducibility of electrochemically polymerized MIPs compared to traditional bulk MIPs.

8.2.4 Bulk Imprinted Polymers Versus Nano-imprinted Polymers

The production of Bulk Imprinted Polymers has been revolutionized by molecular imprinting technology, making them highly effective in adsorption applications. The traditional process involves combining templates and functional monomers, followed by cross-linkers and initiators, and system polymerization. However, the challenge of removing templates through crushing and grinding during post-processing can compromise both elution rates and adsorption efficiency, particularly when templates are close to the center. To address this issue, imprinting technology has evolved to focus on establishing active sites on the surface layer of imprinted materials. This innovative approach ensures that Bulk MIPs maintain their efficacy while reducing the need for post-processing. To address these limitations, nano-Molecularly Imprinted Polymers (nano-MIPs) are known to be superior to their bulk counterparts due to several reasons. The smaller size of nano-MIPs means they have a higher density of binding sites, which results in better sensitivity and selectivity. They can recognize and bind to target molecules more quickly due to their nanoscale dimensions, which facilitate rapid diffusion and mass transfer. The small size of nano-MIPs also makes them ideal for precise targeting and detection at the molecular level, making them useful in fields such as medicine and environmental monitoring. Furthermore, nano-Imprinted Polymers offer a more advanced approach that involves a two-step process of carrier modification and polymerization. This technique uses nanomaterials such as silica, metal oxides, metals and carbon nanomaterials to form complexes with templates, which results in recognition sites located on the surface layer. The resulting nano-imprinted polymers exhibit superior selectivity and adsorption capacity compared to traditional bulk molecular imprinting techniques. Notably, a magnetic molecular imprinted polymer utilizing surface-imprinted technology demonstrated exceptional selectivity and adsorption capacity for versatile target analytes compared to its conventional bulk MIP counterpart.

8.3 MIP Based Sensors

The global increase in human population has brought about a multitude of challenges, particularly in the form of water and air pollution, which pose significant threats to human health and well-being. The existence of various hazardous chemical and biological pollutants including heavy metals, pesticides, organic substances, viruses, and antigens in soil, air, and wastewater requires a comprehensive evaluation of related health risks and identification of these pollutants (Selvolini and Marrazza 2017). In this perspective, molecularly imprinted polymers have been integrated into sensing methods, and their performance in terms of selectivity, robustness, and sensitivity has been enhanced by using different nanomaterials.

8.3.1 MIP Based Electrochemical Sensors

Electrochemical sensors have gained popularity in the exploration of various biological, environmental, industrial, and pharmacological substances (Rebelo et al. 2021). Their appeal lies not only in their sustained reliability, heightened sensitivity, and precision but also in their economical nature, swift operation, and convenient scalability (Rebelo et al. 2021). Electrochemical sensing has become widespread due to its versatility in reporting signals, including voltage (potentiometry), current (amperometry), or electrochemical impedance (impedimetry), and the low theoretical detection limits resulting from distinctions in Faradaic and non-Faradaic currents. Furthermore, molecularly imprinted polymer (MIP)-based electrochemical sensors have become increasingly popular over the years due to their durability, easy electrochemical control for template removal and cavity regeneration, and customizable cavity designs that can be tailored to specific geometries for target analytes (Scheller et al. 2019). The MIP approach, combined with the advantages of electrochemical sensing techniques, such as operational simplicity, quick response, and high sensitivity, work together to create a highly efficient sensing platform that can be used for a wide range of applications (Fig. 8.3) (Scheller et al. 2019).

Despite these advantages, they do have limitations that can impact their sensitivity due to a low or even absent electrochemical catalytic effect (Scheller et al. 2019). However, to overcome these limitations and enhance the analytical performance of MIPs-based sensing platforms, integrating innovative functional nanomaterials and advanced analytical technologies holds great promise (Yang et al. 2018). This approach could lead to the development of advanced electrochemical sensors with a wide range of applications, including environmental monitoring. Further, modifying or enhancing the electrode surface with nanomaterials is a particularly noteworthy technique, as it can amplify the electrochemical signal, resulting in sensors with heightened sensitivity and selectivity. In the upcoming discussion, we will explore various examples demonstrating the practical applications of sensors that use molecularly imprinted polymers for environmental monitoring. These examples will highlight how the combination of MIP and nanomaterials can significantly improve the performance of sensor analytics.

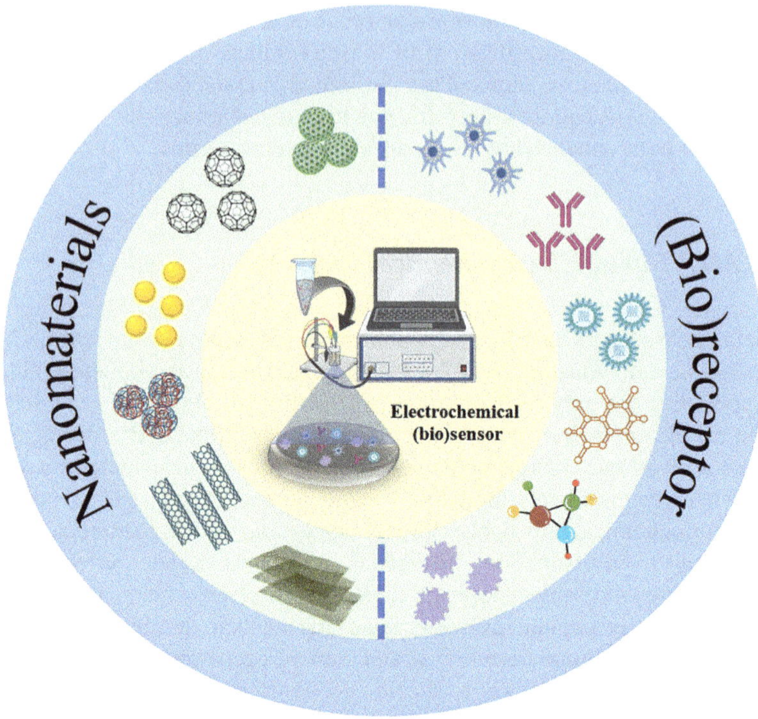

Fig. 8.3 Schematic representation of the (bio)electrochemical sensor and incorporated nanomaterials and (bio)receptors

Sensing of Heavy Metals

Heavy metals have become a major cause of concern due to industrial activities over the past century. The most common heavy metals responsible for human poisonings are mercury, lead, chromium, cadmium, and arsenic, which can enter the body through water, air, or food and cause acute or chronic poisonings (Mitra et al. 2022). Indeed, heavy metal exposure can lead to toxic effects on various organs and tissues, disrupting cellular processes such as growth, proliferation, differentiation, damage-repairing mechanisms, and apoptosis. It is well-established through various studies that heavy metals trigger toxicity through common pathways. These include the generation of reactive oxygen species (ROS), attenuation of antioxidant defense, enzyme inactivation, and the onset of oxidative stress (Mitra et al. 2022). These pathways are implicated in the harmful effects of heavy metals on various body tissues and organs, leading to disruptions in cellular processes such as growth, proliferation, differentiation, damage-repairing mechanisms, and apoptosis. Despite current knowledge about the hazards of heavy metals, incidents of poisoning persist at a significant level, underscoring the need for preventive measures and effective treatments. This emphasizes the importance of monitoring the levels of heavy metals in environmental samples to prevent exposure to these toxic substances. Besides, the ion-imprinted polymer (IIP) is a promising and innovative technique that has emerged from the molecular imprinted polymer strategy, with a specific focus on ion sensing, particularly heavy metals. Its unique ability to recognize ions with heightened specificity and selectivity underscores its potential in this field. The selection of functional monomers in IIPs depends on the molecular structure of the template (Yu et al. 2022). Notable candidates include methacrylic acid (MAA), acrylic acid (AA), acrylamide (AM), and 4-vinylpyridine (4-VP), which are divided into two groups based on their properties. One group has double bonds, while the other has functional groups such as amino and hydroxyl groups, which are adept at chelating metal ions. This strategic selection process further enhances the precision and effectiveness of IIPs in sensing applications. Today, remarkable progress in nanotechnology has given researchers an unprecedented opportunity to use nanomaterials in the creation of IIP-based electrodes (Sala et al. 2022). This has led to the development of ultrasensitive sensors that offer higher selectivity. In this perspective, Carbon-based nanomaterials, including carbon nanotubes, graphene derivatives, and carbon nitride, have garnered extensive utilization in the augmentation of electrochemical responses exhibited by ion-imprinted polymer (IIP)-based electrodes. This application extends to the enhancement of sensitivities across diverse sensing platforms. In the quest for an effective electrochemical sensor tailored for the determination of Cr(VI), S. Wu et al. have developed a sensor based on ion-imprinted chitosan-graphene nanocomposite through a one-step electrodeposition process (Wu et al. 2018). The chitosan-graphene nanocomposite-modified gold electrode was used to intricately synthesize the ion-imprinted polymers. The authors have used graphene to augment the electrochemical response and the active surface area of the sensing electrode. The IIP sensing platform exhibits a wide linear range from 1.0×10^{-9} to 1.0×10^{-5} mol/L, with a low detection limit of 6.4×10^{-10} mol/L (S/N = 3). The sensor is highly selective, particularly in the presence of several ions including Zn(II), Co(II), Cu(II), Ni(II), Mn(II), MnO_4^{-}, $C_2O_4^{2-}$, $S_2O_6^{2-}$, and MoO_4^{2-} ions. It also shows remarkable stability, repeatability, and applicability that have been demonstrated by successfully detecting Cr(VI) ions in tap and river water. Recently, a new electrochemical sensor has been developed for measuring cadmium(II) ions in water. The sensor was made by creating an electropolymerized ion-imprinted poly(ortho-phenylenediamine) P(oPD)/electrochemically reduced graphene (ERGO) composite on a glass carbon electrode (GCE) (Wang et al. 2020). The researchers used ortho-phenylenediamine as the functional monomer and Cd(II) ions as the template. The electrochemical peroxidation method was used to remove the template. The sensor showed excellent selectivity for Cd(II) ions, even in the presence of other heavy metal ions. It had a linear relationship between electrochemical values and Cd(II) concentration in the range of 1 to 50 ng/mL, with a low limit of detection of 0.13 ng/mL (S/N = 3). The proposed electrochemical sensor, imprinted PoPD/ERGO/GCE, was successful in accurately determining trace Cd(II) concentrations in real water samples. As state-of-the-art in the use of carbon nanotubes, a new method for enhancing the sensitivity of ion-imprinted polymer (IIP) sensors has been developed using multi-walled carbon nanotubes (Alizadeh et al. 2017). The technique involves the synthesis of a nano-structured IIP through the copolymerization of itaconic acid-Pb(II) complex and ethylene glycol dimethacrylate via precipitation polymerization. The resulting IIP-nanoparticles were combined with MWCNTs and incorporated into a carbon paste electrode to form a Pb(II)-IIP sensing electrode. The authors used the square wave anodic stripping voltammetry as the determination technique and the results demonstrated a significantly heightened response compared to the electrode embedded with non-imprinted polymer (NIP), indicating the effective creation of recognition sites during the polymerization reaction. The sensor exhibited a linear range from 1.0×10^{-11} to 8.0×10^{-8} mol L^{-1} and an impressive detection limit of 3.8 pmol L^{-1} (S/N = 3), with an outstanding sensitivity of 20,683 A L mol^{-1}. The electrode also showed remarkable selectivity for Pb(II) over common potential interfering heavy metal ions. This new technique shows promise for the trace determination of lead in various samples, including water samples. Furthermore, several studies have shown that carbon nitride, a two-dimensional nanomaterial, can significantly enhance the conductivity of IIP electrode sensors, leading to a significant improvement in the sensing performances of the platforms. For instance, M. R. Ganjali and colleagues have developed a new electrochemical sensor for detecting mercury(II) ions. The sensor utilizes nanoparticles made from an ion-imprinted polymer (Ganjali et al. 2019). The researchers synthesized these nanoparticles using itaconic acid as a functional monomer through a method called precipitation polymerization. They combined these nanoparticles with graphitic carbon nitride (g-C_3N_4) and used them to modify a carbon paste electrode (CPE). This modification transformed the CPE into a highly sensitive platform for Hg(II) anodic stripping analysis. According to the authors, the modified CPE exhibited a higher affinity for Hg(II) ions than an unmodified CPE due to the presence of g-carbon nitride. The sensor showed a suitable linear response in the concentration range of 0.06–25.0 nM, with a low detection limit (S/N = 3) of 18 pM. The researchers validated the performance of the sensor by analyzing Hg(II) in water samples, showing its practical applicability and sensitivity in real applications. On the other hand, incorporating metallic nanomaterials in the IIP platforms is an important aspect of electrochemical analysis. Among these materials, gold nanoparticles (AuNPs) are extensively used due to their unique properties. AuNPs are known for their exceptional electron transfer capabilities, large effective surface area, and high electrical conductivity (Alim et al. 2018; Pingarrón et al. 2008). These properties make them ideal for electrode modification, where they can amplify electrochemical signals and enhance the sensitivity and performance of electrochemical analysis methods. In this scenario, W. Ma and his colleagues have developed an electrochemical sensor for the precise detection of arsenic ions (As^{3+}) in diverse water samples (Ma et al. 2020). The sensor was crafted by incorporating an ion-imprinted polymer (IIP) onto a nanoporous gold (NPG) modified gold electrode (IIP/NPG/GE). This innovative design enables the precise determination of ultra-trace levels of arsenic ions. The NPG was synthesized through an environmentally friendly electrodeposition method, and a layer of IIP was synthesized in situ on the NPG surface via electropolymerization, utilizing As(III) as the template ion and ortho-phenylenediamine as the functional monomer. The sensor exhibited a linear range for As(III) from 2.0×10^{-11} to 9.0×10^{-9} M, with a remarkably lower detection limit of 7.1×10^{-12} M (S/N = 3) demonstrating excellent stability and selectivity. According to the authors, this newly developed sensor provides a reliable and efficient tool for monitoring arsenic levels in diverse water qualities. In the same approach, an efficient electrochemical sensor, combining ion-imprinted polymers and gold nanoparticles, has been

Sensing of Pesticides

Nowadays, it's crucial to accurately detect and quantify pesticides to address water pollution effectively. This helps in designing and implementing efficient removal methods. Although there are different methods available for this purpose, electrochemistry is a promising approach due to its high sensitivity that can reach the femtomolar level and cost-effectiveness (Baranwal et al. 2022; Ramachandran et al. 2019; Scozzari 2008). As a result, electrochemistry could significantly contribute to efficient water remediation strategies. Additionally, sensitivity is a critical feature of sensors, often playing a decisive role in their effectiveness. In this context, bio-inspired molecularly imprinting polymers (MIPs) emerge as exceptional sensing materials. These imprints serve as receptor sites, enabling the recognition of the template during the rebinding step, with remarkable selectivity (Aragay et al. 2012; Munteanu and Apetrei 2022). The increasing utilization of MIP-based electrochemical sensors across various domains, especially in environmental monitoring and pesticide sensing, highlights their efficacy in addressing sensing challenges. Besides, the demand for pesticides continues to persist despite substantial efforts by international and national organizations. The use of organophosphorus pesticides, such as Parathion, Azinphos-methyl, Malathion, Dichlorvos, and Chlorpyrifos, is particularly alarming due to their harmful effects on living beings and their long-lasting presence in the environment (Lakshmi et al. 2020). These pesticides pose severe risks through long-term exposure, presenting dangers of toxicity, mutagenicity, carcinogenicity, and tumorigenicity for humans. In addition, consumers are also at risk of exposure as the frequent use of organophosphorus pesticides on farms can lead to pesticide residues lingering on crops and livestock. As a result, there has been a significant effort to develop rapid, efficient, and accurate sensing strategies to address this issue. To this end, several nanomaterials including graphene derivatives, metal, and metal oxide nanoparticles were combined with molecular imprinted polymers to detect trace levels of a large number of pesticides (Mahmoudpour et al. 2020; Kumar and Kim 2022). Recently, Li and coworkers (2019) have developed an electrochemical sensing platform based on molecularly imprinted polydopamine polymer for the detection of cypermethrin pesticides in water samples. The process detailed in the study employs a silver-modified N co-doped zinc oxide nanocomposite, prepared through a sol–gel method, to enhance the electrochemical signal. Dopamine and resorcinol are utilized as dual monomers, and cypermethrin serves as the template to develop a cypermethrin-imprinted polymer (MIP). This MIP sensor is electropolymerized on a glassy carbon electrode, exhibiting superior selectivity and a heightened response compared to conventional MIPs that use a single monomer.

With the integration of the silver-modified N co-doped zinc oxide nanocomposite into the MIP structure, the sensor demonstrates the capability to detect the target pesticide in a concentration range of 2×10^{-13} to 8×10^{-9} M, boasting a meager detection limit (LOD) of 6.7×10^{-14} M (S/N = 3). The authors propose that this electrochemical sensor holds promise for applications in food safety, drug residue determination, and environmental monitoring, showcasing its potential impact in diverse fields. In another study, Zouaoui et al. (2020) reported the application of an efficient impedimetric MIP sensor for the determination of glyphosate herbicide in various real matrices including water samples. The authors of the study used chitosan as a biopolymer to imprint glyphosate cavities onto the surface of a gold microelectrode. The label-free detection of GLY was carried out using the MIP platform. The results of the Electrochemical Impedance Spectroscopy demonstrated a linear detection range of GLY from 0.31 pg/ml to 50 ng/ml, with a low detection limit of 0.001 pg/ml (S/N = 3). By comparing the MIPs layer's affinity to GLY with non-imprinted CS (NIPs), the study showed a remarkable imprinting factor of 14.5. Additionally, the MIPs' selectivity was demonstrated by detecting various pesticides as potential interferences, resulting in high selectivity factors (SF): 7.9 for glyfosinate, 43.5 for chlorpyrifos, and 14.5 for phosmet. The electrochemical sensor's high sensitivity and impressive selectivity for detecting GLY make it a valuable tool for environmental monitoring. On the other hand, methyl parathion, a widely employed and highly toxic organophosphorus pesticide, poses significant environmental and health concerns as it tends to accumulate in water, soil, and foods due to its low solubility and bioaccumulation properties (Garcia et al. 2003; Senbua et al. 2020). Recently, the electrochemical detection of methyl parathion in environmental samples and foods was achieved by employing a magnetic molecularly imprinted polymer, followed by a readout on a magneto-actuated electrode using square wave voltammetry (Hassan et al. 2018). According to A. Hassan et al., magnetic molecularly imprinted polymer was prepared using methacrylic acid as a functional monomer, selected based on theoretical calculations employing density functional theory (DFT). Further, they examined the binding capacity and selectivity of the polymer towards methyl parathion and compared it with the corresponding magnetic non-imprinted polymer. The magneto-actuated electrochemical sensor developed using the magnetic molecularly imprinted polymer exhibited exceptional analytical performance in detecting methyl parathion with a low detection limit of 1.22×10^{-6} mg L^{-1} and high recovery values. The authors suggested that the magnetic molecularly imprinted polymer sensor can be used for detecting pesticides in environmental control and food safety. In the same context, Wang et al. (2018) applied an innovative electrochemical Sensor for the determination of methyl parathion trace levels in environmental media using electrochemical molecular imprinting polymerization. In the research, the authors developed an electrochemical sensor by using a glassy carbon electrode that was decorated with reduced gold nanoparticles and molecularly imprinted polymer (MIP) on the electrode surface. The sensor was made to detect methyl parathion (MP) as the template molecule. The findings showed that the MIP sensor was effective in quantitatively determining MP in aqueous matrices over a wide dynamic range with an excellent low limit of detection. Moreover, the sensor showed remarkable selectivity towards MP as compared to imidacloprid and fenitrothion, which are structurally similar compounds. The study further highlighted that the current response for MP in different matrix samples did not deviate much, and the recovery rates ranged from 87.7 to 124.8%. Regarding the use of graphene derivatives, He and colleagues (2017) discussed the use of a highly sensitive and selective biomimetic sensor for the direct electrochemical detection of methyl parathion using graphene derivatives. The sensor utilized zinc porphyrin molecularly imprinted polymer microspheres, gold nanoparticles, and carboxyl graphene nanomaterials. To fabricate the sensor, they attached imprinted microspheres onto gold nanoparticles modified graphene nanocomposites. In this study, the imprinted polymer was prepared through free radical polymerization using methyl parathion as the template, zinc porphyrin as the functional monomer, ethylene glycol dimethacrylate as the cross-linking reagent, and azobisisobutyronitrile (AIBN) as the initiator. The electrochemical outputs demonstrated that the introduction of gold-modified graphene nanocomposites increased the effective electrode area and amplified the sensor signal. The biomimetic sensor exhibited a wide linear dynamic range and a low limit of detection in the range of nM. The sensor demonstrated good reproducibility, stability, and selectivity for MP detection, successfully applying it to detect MP in real samples. In the same context, Zhang et al. (2017) reported the incorporation of graphene nanomaterials to improve the electrochemical sensitivity of an imidacloprid sensor. They proposed using para-vinyl benzoic acid (VBA) as a functional monomer that could be immobilized onto graphene in N, N-dimethylformamide (DMF) solvent via π–π interaction. This facilitated the polymerization of uniform MIP layers on the surface of graphene sheets. The synthesis of MIP-modified graphene and the preparation of the MIP-modified graphene-decorated glassy carbon electrode (GCE) were quantitatively controlled, resulting in a sensor with improved stability and specific recognition ability. The results demonstrated that the electrochemical response of the MIP sensor toward the presence of imidacloprid pesticides showed linearity with concentration in the range of 0.5–15 μM, with a low detection limit of 0.10 μM. According to the report, the method can be successfully applied to detect pesticides in several real matrices with satisfactory results. As a last example, Kumar and collaborators (2016) described the application of a superparamagnetic molecularly imprinted polymer for trace-level sensing

8.3.2 MIP-Based Optical Sensors

Environmental pollutants have been traditionally detected through established chromatographic methods like ionic and high-performance liquid chromatography (HPLC), gas chromatography (GC), and mass spectrometry (MS). While these methods offer precise analysis, their reliance on sophisticated equipment and time-consuming sample preparation makes them impractical for urgent scenarios (Zhang et al. 2021; Tajik et al. 2021). In response, researchers have explored alternative approaches, such as optical sensors utilizing molecularly imprinted polymers for swift, sensitive, and accurate detection of hazardous pollutants. Recently, various optical strategies have been combined with nanomaterials, employing the molecularly imprinted polymers approach to enhance sensitivity and provide outstanding selectivity for real-time analysis. This meets the growing demand for more efficient detection methods (Panda and Chakroborty 2022; Long et al. 2013). Within this framework, this section will explore the discussion of MIP-based optical platforms for sensing heavy metals and pesticides, employing techniques such as fluorescence, UV–visible, Raman, optical fiber, and surface plasmon resonance signal variation.

Sensing of Heavy Metals

Fluorescence-based MIP sensors, owing to their high sensitivity and simplicity, stand out as one of the most commonly utilized options in environmental monitoring applications. With the continuous advancement of technologies, various materials have found widespread use in the construction of fluorescence sensing platforms. These materials include semiconductor nanomaterials and carbon materials. Recently, Zhang et al. (2019) reported the development of a novel fluorescence chromium(VI) sensor based on ion-imprinted polymers (IIP) as a sensing strategy and manganese-doped ZnS quantum dots as fluorophores. To achieve this, the QDs were coated with a silica layer and modified with a Cr(VI)-imprinted polymer. The study utilized UV–VIS and fluorescence spectrophotometry to explore the fluorescence quenching mechanism, which revealed that the primary mechanism was the internal filtration effect. The IIP sensor showed significant recognition selectivity for chromium(VI) with a wide dynamic range and a low limit of detection compared to the non-imprinted sensor. The imprinting factor was 2.53, and the sensor demonstrated reusability up to five times. This method was successfully applied to actual water samples, and the results were satisfactory. The IIP sensor is a practical fluorescent sensor for the trace detection of Chromium(VI). Furthermore, Lu and colleagues (2019) have detailed the development of an innovative dual-channel detection system for detecting chromium(III) and lead(II) ions. This system incorporates ion-imprinted fluorescence polymers containing blue and red carbon dots as fluorophores. The blue and red dual-channel dual-emission carbon dots were prepared using a one-step hydrothermal synthesis method within a water-formamide binary system, using citric acid and ethylenediamine as precursors. The choice of both red and blue carbon dots was based on the abilities of chromium(III) and lead(II), to selectively quench the fluorescence of blue and red fluorescence signals, respectively, within the ion-imprinted fluorescence polymers. The study revealed that the dual fluorescence system exhibited sensitive sensing for target ions, with a limit of detection in the nanomolar range. Additionally, the method demonstrated satisfactory recoveries for the detection of Cr(III) and Pb(II) in real water samples, showcasing its potential to significantly simplify the heavy metal ion detection process in complex water environments. Interestingly, the L. Wang group introduced a new microfluidic analytical device that combines a fluorescence sensing cloth-based component with a rotary paper-based device for simultaneous detection of mercury and lead ions (Wang et al. 2022). The cloth-based element utilizes cotton cloth grafted with quantum dots and modified with ion-imprinted polymers to achieve fluorescence sensing. The rotary paper-based device, made of three-layer papers with hydrophilic channels and hydrophobic barriers, facilitates the detection of target ions through clockwise or counterclockwise rotation. The electron transfer fluorescence quenching effect of quantum dots enables the acquisition of fluorescence signals with low detection limits. The authors suggest that this method enhances portability and expands the potential applications of IIP-based nanomaterials for microfluidic devices. Additionally, this device may have promising potential for point-of-care testing of heavy metal ions in environmental monitoring fields. In the same context, an innovative three-dimensional origami ion-imprinted polymer microfluidic paper-based chip device has been developed by Ma and collaborators (2011) for the specific, sensitive, and multiplexed detection of copper(II) and mercury(II) ions. The paper's surface was activated by grafting with cadmium telluride Quantum Dots (CdTe QDs) through amino processing, leading to the formation of Cu(II) or Hg(II) ion-imprinted polymers and CdTe QDs complexes. This complex formation resulted in the fluorescence quenching of quantum dots, as the photoluminescent energy of quantum dots could be transferred to the formed complex. The liquid phase of ion-imprinted polymer-modified quantum dots could be efficiently transferred to the solid glass fiber paper, enhancing the device's portability. The results demonstrated that the developed platform enables the simultaneous detection of both Cu(II) and Hg(II) ions with high selectivity and sensitivity. Based on the report, this device offers convenient quantitative information and shows great potential for extension to the detection of other metal ions in environmental monitoring and food safety applications.

Besides, optical fiber sensors have become increasingly popular as a solution for developing practical sensors. They offer several advantages, including their small size, flexibility, compactness, and cost-effectiveness (Jiao et al. 2020; Joe et al. 2018). Recently, the combination of optic fiber and imprinted polymers has been gaining a lot of attention. Its high sensitivity and wide range of applications in real-world scenarios make it particularly appealing (Yang et al. 2022). In this perspective, Researchers led by Meza López et al. (2021) developed a novel optical sensor based on an ion-imprinted polymer, which has been successfully applied for detecting lead(II) ions. The polymer was prepared using rhodizonate as a selective ligand for Pb(II) and 2-acrylamido-2-methylpropane sulfonic acid as a functional monomer. The polymeric materials were synthesized directly onto a quartz optical fiber, which was then adapted to a bundle of fibers for reflectance measurements. The sensor showed a wide linear response range, low limit of detection, and good analytical performance in reproducibility, repeatability, and selectivity studies. The sensor's feasibility was demonstrated by successfully detecting and quantifying Pb(II) in environmental samples with recovery percentages of almost 100%. Overall, the study suggests that this sensor is highly sensitive and selective for detecting and quantifying Pb(II) in real water samples. More importantly, a novel electrochemical surface plasmon resonance optical fiber sensor was fabricated by the Si group (2019). Based on the report, the sensor incorporates a tilted fiber Bragg grating imprinted in a commercial single-mode fiber, coated with a nanoscale gold film for efficient SPR excitation. The gold-coated optical fiber functions as a working electrode, seamlessly integrating anodic stripping voltammetry and SPR optical sensing. To showcase the sensor's capabilities, the researchers used it to detect trace amounts of lead(II) ions. The results showed that there is a stable and reproducible correlation between real-time ion deposition-stripping cycles and the optical transmission of the optical fiber. The sensor has a remarkable limit of detection and a dynamic range of nearly five orders of magnitude. Additionally, the proposed method is immune to temperature cross-talk, thanks to the temperature-dependent core mode's nature, which is insensitive to the surroundings. The authors suggest that the electrochemical-surface plasmon resonance fiber-optic sensor they developed can be a highly effective tool for environmental monitoring in difficult or isolated settings.

Sensing of Organic Pollutants

Environmental pollution can manifest itself in various ways, one of which is through the presence of hydrophilic/hydrophobic organic compounds. Due to the diverse nature of these pollutants, it is crucial to develop sensors with versatile recognition units that can selectively monitor all types of contaminants. Molecularly imprinted polymers (MIPs) play a pivotal role in this scenario, as they can be synthesized to contain antibody-like molecular recognition sites for any pollutant. This versatility makes MIPs invaluable in the development of sensors that can effectively address the complex and varied nature of environmental pollution (Arabi and Chen 2022). In this sense, Li et al. have developed a new fluorescent nanosensor to determine cyfluthrin, an insecticide (Idris et al. 2023). The sensor is based on a molecularly imprinted silica layer appended to iron selenide quantum dots. The findings revealed that the MIP sensor exhibited high selectivity and sensitivity to cyfluthrin, as evidenced by the fluorescence quenching effect of iron selenide quantum dots. The researchers established a strong linear relationship between fluorescence quenching and increasing cyfluthrin concentration within a wide range, with a high correlation coefficient. The sensing method was also validated for the target insecticide detection in sediment samples, yielding good recoveries and low relative standard deviations. The authors suggested that the reported study successfully introduced a novel and rapid fluorescent nanosensor detection method for the analysis of versatile organic pollutants in environmental samples. In another context, Wei and collaborators (2018) reported the development of a tailored strategy for the fabrication of molecularly imprinted composite sensors designed with highly selective recognition abilities for the determination of sulfanilamides. The approach involves the utilization of quantum dots modified silica spheres coated with molecularly imprinted polymer. Significantly, the developed sensor was successfully applied to determine the target pollutant in real river water, showcasing high selectivity and stability. This approach holds promise as a platform for detecting environmental pollutants. Similarly, the Zhou group (2017) described the synthesis of novel fluorescence molecularly imprinted polymers using manganese-doped zinc sulfide quantum dots and a molecularly imprinted polymers (MIPs) layer anchored on the surface of silica nanoparticles. The synthesis process utilized acrylamide as the functional monomer and ethyl glycol dimethacrylate as the cross-linker through precipitation polymerization. The MIP sensor displayed excellent optical properties and was able to recognize the template molecules effectively. Moreover, the sensor was used to analyze tap water samples, achieving high recovery rates of over 97.80% and a low relative standard deviation (RSD). On the other hand, Zheng and colleagues (2019) investigated the feasibility of using novel core–shell quantum dots (QDs) coated with a molecularly imprinted polymer (MIP) shell for detecting perfluorooctanoic acid, a persistent and bioaccumulating water contaminant. The development of the sensor involved a simple yet effective method. Cadmium sulfide-encapsulated cadmium telluride quantum dots were fabricated and then coated with a molecularly imprinted silica film in a one-pot sol–gel reaction. The creation of the desired MIP-based composite was achieved by attaching the MIP layer to the quantum dots, utilizing 3-aminopropyltriethoxysilane as the functional monomer and tetraethoxysilane as the crosslinker in the presence of aqueous ammonia. The combination of quantum dots and MIP exhibited stable photoluminescence and exceptional selectivity. The presence of perfluorooctanoic acid molecules occupied the templated bond sites, resulting in the quenching of the photoluminescence signal of the MIP-coated quantum dots composite. The findings demonstrated good linearity with a low detection limit in the nanomolar range, showcasing the high efficiency of the developed optic MIP sensor. More importantly, a novel approach was developed using fluorescent quantum dots in conjunction with molecularly imprinted polymers (MIPs) integrated into three-dimensional (3D) origami paper-based microfluidic devices was developed by the Li group (2017). The described innovative method allows for the specific recognition and sensitive detection of phycocyanin. This technique makes it easier to transfer the liquid phase of the Molecularly imprinted polymer conjugated quantum dots to the solid-phase paper base, which makes it highly portable for analysis. The proposed paper MIP sensor microfluidic chip was successfully demonstrated for the sensitive and selective detection of the phycocyanin protein target simply and robustly. The results indicated a dynamic response to phycocyanin in the range of 10–50 mg/L, with a limit of detection of 2 mg/L. The authors suggested that the developed device offers convenient quantitative information and demonstrates the potential for extension to the detection of other proteins or biomarkers in environmental research.

In summary, molecularly Imprinted Polymers (MIPs) have introduced in a new era in environmental monitoring, offering invaluable tools for detecting diverse pollutants. A meticulously detailed summary of the field's progress is presented in Table 8.1. The table highlights four critical parameters that are of particular significance—the structure of the MIP-sensing platform, the limit of detection, linear range, and the nature of sensing approaches employed.

Table 8.1 Summary of MIP sensors reported in the literature for the quantification of environmental pollutants

Structure of MIP-based sensor	Analyte	Type of analytic sensing	Detection limit	Linear range	References
Ion-imprinted chitosan-graphene nanocomposite-modified gold electrode	Cr(VI)	Electrochemical sensors	6.4×10^{-10} M	10^{-9} to 10^{-5} M	Wu et al. (2018)
Ion-imprinted poly(ortho-phenylenediamine)/ reduced graphene modified glass carbon electrode	Cd(II)	Electrochemical sensors	0.13 ng/mL	1–50 ng/mL	Wang et al. (2020)
Ion imprinted polymer nanoparticles modified multi walled carbon nanotubes	Pb(II)	Electrochemical sensors	3.8 pM	1.0×10^{-11} to 8.0×10^{-8} M	Alizadeh et al. (2017)
Ion imprinted polymer/graphitic carbon nitride modified carbon paste electrode	Hg(II)	Electrochemical sensors	18 pM	0.06–25.0 nM	Ganjali et al. (2019)
Ion imprinted polymer/nanoporous gold modified gold electrode	As(II)	Electrochemical sensors	7.1×10^{-12} M	2.0×10^{-11} to 9.0×10^{-9} M	Ma et al. (2020)
Ion imprinted polymers/gold nanoparticles modified glass carbon electrode	Cd(II)	Electrochemical sensors	1.43×10^{10} M	10^{-9} to 10^{-4} M	Huang et al. (2021)
Ion imprinted polymers/silver nanoparticles modified glass carbon electrode	Cd(II)	Electrochemical sensors	1.43×10^{-10} M	10^{-9} to 10^{-4} mol/L	Gong et al. (2021)
Magnetic silver(I) ion-imprinted polymeric nanoparticles modified carbon paste electrode	Ag(I)	Electrochemical sensors	15 ng L^{-1}	0.05–150 μg L^{-1}	Ghanei-Motlagh and Taher (2017)
Ion-imprinted polymer functionalized magnetic nanoparticles (MNPs) modified glass carbon electrode	Co(II)	Electrochemical sensors	0.1 nM	0.5 to 20 and 20 to 500 nM	Torkashvand et al. (2017)
Ion-imprinted polymer/modified magnetic nanoparticles deposited on screen-printed carbon	Ce(IV) Gd(III)	Electrochemical sensors	0.07–0.19 ng mL^{-1}	–	Prasad and Jauhari (2015)
MIP/silver-modified N co-doped zinc oxide nanocomposite/glass carbon electrode	Cypermethrin pesticides	Electrochemical sensors	6.7×10^{-14} M	2×10^{-13} to 8×10^{-9} M	Li et al. (2019)

(continued)

Table 8.1 (continued)

Structure of MIP-based sensor	Analyte	Type of analytic sensing	Detection limit	Linear range	References
MIP/gold microelectrode	Glyphosate herbicide	Electrochemical sensors	0.001 pg/mL	0.31 pg/mL to 50 ng/mL	Zouaoui et al. (2020)
Magnetic molecularly imprinted polymer/glass carbon electrode	Methyl parathion	Electrochemical sensors	1.22×10^{-6} mg L^{-1}	–	Hassan et al. (2018)
MIP-modified graphene-decorated glassy carbon electrode (GCE)	Imidacloprid	Electrochemical sensors	0.10 μM	0.5–15 μM	Zhang et al. (2017)
MIP/superparamagnetic iron oxide nanoparticles/glassy carbon electrode	Mancozeb pesticide	Electrochemical sensors	0.96 μg L^{-1}	5.96–257 μg L^{-1}	Kumar et al. (2016)
Ion-imprinted polymer/Manganese-doped ZnS quantum dots	Cr(VI)	Optical sensor	5.48 μg L^{-1}	20 μg L^{-1} to 1.0 mg L^{-1}	Zhang et al. (2019)
Ion-imprinted fluorescence polymers/blue and red carbon dots	Cr(VI) Pb(II)	Optical sensor	27 nM 34 nM	0.1–5.0 μM	Lu et al. (2019)
Ion-imprinted polymer/Cotton cloth grafted with Quantum dots/rotary paper-based microfluidic analytical device	Pb(II) Hg(II)	Optical sensor	0.07 μg/L 0.18 μg/L	–	Wang et al. (2022)
Ion-imprinted fluorescence polymers/CdTe quantum dots	Hg(II) Cu(II)	Optical sensor	–	0.005–0.5 μM 0.01–1 μM	Ma et al. (2011)
Ion-imprinted fluorescence polymers based on Rhodizonate	Pb(II)	Optical sensor	85 ng L^{-1}	10–100 μg L^{-1}	Meza López et al. (2021)
Ion-imprinted polymer/quantum dots modified silica spheres	Sulfanilamides	Optical sensor	0.17 μM	2.0–30 μM	Wei et al. (2018)
Molecularly imprinted polymers/manganese-doped zinc sulfide quantum dots modified silica nanoparticles	Dibutyl phthalate	Optical sensor	–	5–50 μM	Zhou et al. (2017)

(continued)

Table 8.1 (continued)

Structure of MIP-based sensor	Analyte	Type of analytic sensing	Detection limit	Linear range	References
Molecularly imprinted polymer modified core–shell quantum dots	Phycocyanin	Optical sensor	25 nM	0.25–15.00 μM	Zheng et al. (2019)
Molecularly imprinted polymer/quantum dot based on three-dimensional origami paper microfluidic chip	Phycocyanin	Optical sensor	2 mg/L	10–50 mg/L	Li et al. (2017)

8.4 MIP Based Biosensors

The development of biosensors for environmental monitoring is vital to ensure the sustainability of society. These biosensors are critical in the detection of organic pollutants, potentially toxic elements, and pathogens present in the environment (Huang et al. 2023). By addressing the challenges of environmental pollution, they contribute significantly to safeguarding human health and promoting sustainable development. The continuous research in this field highlights the dedication to creating innovative solutions that enhance our ability to mitigate environmental risks and improve our understanding of them. Besides, molecularly imprinted polymers are becoming increasingly popular in biotechnology due to their rational synthetic strategy and biomimetic design. This renewed interest is driven by the need for analytical platforms that are rapid, accurate, and cost-effective (Wu et al. 2022). MIPs are crosslinked polymer matrices containing molecular recognition sites that are formed during synthesis in the presence of a target template, making them ideal for developing affordable detection modules for target molecules. The attention given to MIPs has spanned decades, underscoring their enduring significance in meeting the evolving needs of analytical science. In the following sections, a summary of the recent and innovative examples of the application of molecularly imprinted polymers for environmental monitoring will be presented. The discussion will focus on the effect of nanomaterials in enhancing the sensing performance of the developed biosensors.

8.4.1 MIP Based Electrochemical Biosensors

The incorporation of nanomaterials as modifiers has brought about functional properties in various sensing applications, particularly in biological sensors. Nanomaterials have proven to be essential in electrochemical analysis, contributing several crucial features such as providing catalytic properties and enhancing the conductive sensing interface (Yola and Atar 2017). Over the last two decades, nanotechnology has become increasingly prevalent in everyday materials. A range of nanostructures are leveraged in molecular imprinting, featuring diverse functionalization and modifications. Additionally, there has been notable progress in improving efficient electrochemical sensors through appropriate electrode modification using Molecularly Imprinted Polymers. This shift from conventional detection methods utilizing bare electrodes has resulted in enhanced selectivity and sensitivity. It has been shown that the sensitivity of electrochemical biosensors based on molecularly imprinted polymers (MIPs) can be improved by increasing the effective sensing areas and modifying the planar sensing surface with electroactive nanomaterials like carbon nanotubes, noble metal nanomaterials, and graphene (Rahman et al. 2022). Notably, MIPs have a dual function in enhancing sensor performance. They accumulate templates on the electrode surface, thus increasing sensitivity, and also aid in the separation of templates from other analytes, leading to improved selectivity. In this perspective, Y. Wang et al. reported the use of multi-walled carbon nanotubes to enhance the sensitivity of the MIP-based electrochemical platform for the sensing of bovine serum albumin protein (Dong et al. 2022). In the study, researchers synthesized a molecularly imprinted polymer film on a glassy carbon electrode surface modified with carboxyl-functionalized multi-walled carbon nanotubes. They utilized an innovative imprinting strategy for proteins in aqueous solutions, utilizing 3-(3-aminopropyl)-1-vinylimidazolium tetrafluoroborate ionic liquid as the functional monomer and N,N′-methylenebisacrylamide as the cross-linker, and ammonium persulfate along with N,N,N′,N′-tetramethylethylenediamine as the initiator. The template for the molecularly imprinted polymer was bovine serum albumin. The electrochemical findings showed a linear correlation between the current difference before and after specific recognition of BSA and its concentration in a wide range of concentrations with a calculated limit of detection in the range of femtomolar. Besides, A research group led by T. S. Anirudhan has developed an efficient potentiometric biosensor for the determination of lindane, an organochlorine pesticide (Anirudhan and Alexander 2015). The sensor utilizes a copper electrode modified with a multi-walled carbon nanotube-based imprinted polymer, which was synthesized using methacrylic acid as the monomer, ethylene glycol dimethacrylate as the cross-linker, α,α'-azobisisobutyronitrile as the initiator, and γ-HCCH as the template. The sensing platform exhibited high sensitivity, low limit of detection, and broad applicability, indicating its potential to determine the levels of a wide range of pesticides. More importantly, Qi and a team of researchers (2021) have introduced a new method for biosensing that can determine the concentration of organophosphorus pesticides. This method uses acetylcholinesterase as the recognition element. The researchers proposed a dual-recognized approach that included molecularly imprinted polymer (MIP) and acetylcholinesterase enzyme, which selectively targeted different moieties of acephate pesticide as simultaneous affinity molecules. This approach was designed to overcome the low selectivity, susceptibility to sample matrix interference, and harsh analytical conditions of conventional biosensors. The sensing strategy relied on the concentration of the pesticide on the molecularly imprinted polymer, and the monitoring of the sensing process was done by controlling the inhibitor activity of the pesticide toward the acetylcholinesterase enzyme. This integrated approach, which combined selective sample pretreatment with acetylcholinesterase inhibition assay in a single sensor, enabled the biosensor to distinguish acephate from other organophosphorus pesticides including carbamate pesticides, and common interferences present in environmental samples. In a similar study, Y. Gu et al. presented an innovative approach that integrates the concept of molecularly imprinted polymers (MIPs) with the utilization of a mimetic enzyme to enhance the catalytic efficiency of MIPs as electrochemical sensors (Qi et al. 2021). The researchers developed an electrochemical method that employed a metronidazole-imprinted polymer with nitroreductase-like activity, which was created using melamine as the functional monomer for MIP and as a component of the mimetic enzyme. During the imprinting process, the redox-active center crucial for catalysis was introduced into the imprinted cavities, resulting in an imprinted polymer with both catalytic centers and recognition sites that exhibited heightened electrocatalytic activity and selectivity. The metronidazole-imprinted biomimetic sensor was evaluated for sensing performance, and the results showed a linear response to metronidazole in a wide range of concentrations with a low detection limit. The authors suggested that the proposed sensor could be used to detect metronidazole in an injection solution, making it feasible for practical applications. This innovative approach not only enhances the catalytic efficiency of MIPs but also extends their utility by incorporating mimetic enzyme properties, showcasing potential advancements in electrochemical sensing technology. On the other hand, a recent study conducted by Jiang and colleagues (2016) presented an innovative magnetic molecularly imprinted polymer nanoparticles-based electrochemical sensor for detecting Gram-negative bacterial quorum signaling molecules, N-acyl-homoserine-lactones. These molecules are known to induce the expression of pathogenic factors. The researchers successfully developed this sensor by incorporating magnetic nanoparticles and molecularly imprinted polymer technology, resulting in magnetic molecularly imprinted polymers that can selectively absorb N-acyl-homoserine-lactones. The synthesis involved surface polymerization, and the particles were deposited onto a magnetic carbon paste electrode. The sensor exhibited good sensing parameters and a notably low limit of detection in the femtomolar range. It represents a valuable tool for the quantitative measurement of Gram-negative bacterial quorum signaling molecules, with potential applications in clinical diagnosis, food analysis, and environmental monitoring. The sensor offers high specificity, excellent reproducibility, robust stability, and real-time detection capability. In a different work, Ji et al. (2015) developed an electrochemical sensor for detecting cholesterol using molecularly imprinted polymer membranes on a glassy carbon electrode modified with multi-walled carbon nanotubes and Au nanoparticles. The team assembled p-aminothiophenol and cholesterol on the modified GCE surface through the formation of Au–S bonds and hydrogen-bonding interactions. Polymer membranes were subsequently formed via electropolymerization in a polymer solution containing p-Aminothiophenol, Tetrachloroauric(III) acid trihydrate, tetrabutylammonium perchlorate, and the template molecule, cholesterol. The MIP sensor demonstrated a linear response in a wide concentration range, with an impressive limit of detection, making it a highly efficient tool for clinical diagnostics of cholesterol. Its features include high-speed real-time detection capability, minimal sample consumption, high sensitivity, low interference, and robust stability, making it ideal for precise and reliable cholesterol detection in practical applications. As a final example, the Ma group (2017) has developed a rapid, simple, and highly sensitive method to determine human immunodeficiency virus p24. They utilized molecularly imprinted polymer-based electrochemical sensors for this purpose. The sensor was constructed using acrylamide as the functional monomer, N,N′-methylenebisacrylamide as the cross-linking agent, and ammonium persulphate as the initiator, and was constructed on the surface of a multi-walled carbon nanotubes (MWCNTs) modified glassy carbon electrode through surface polymerization. The MIPs electrochemical biosensor displayed specific recognition for HIV-p24, with a broad linear detection range and an impressive low detection limit. The authors claim that the performance of the MIP-based sensor surpasses that of most HIV-p24 sensors based on alternative methods. Moreover, the sensor showed excellent selectivity, repeatability, reproducibility, and stability. The accuracy and reliability of the sensor were further confirmed through comparison with enzyme-linked immunosorbent assay (ELISA). This MIPs electrochemical sensor is a significant breakthrough in HIV-p24 detection technology, offering a powerful tool for practical applications with high accuracy and reliability.

8.4.2 MIP Based Optical Biosensors

Optical biosensors have gained significant momentum in the last decade due to their direct, real-time, and label-free detection of various biological and chemical substances (Sharma and Sharma 2023). The detection in optical biosensors relies on the interaction between the optical field and a biorecognition element. Optical biosensors are known for their high specificity, sensitivity, compact size, and cost-effectiveness. Recently, optical sensors integrated with molecularly imprinted polymers have attracted significant interest due to their high selectivity, making them promising for the optical detection of various analytes in real-world applications (Ahmad et al. 2019). As a result, optical MIP-based biosensors have found applications as selective sensors in several domains, including pharmaceuticals and environmental analysis. Several optical MIP sensors incorporate nanoparticles (NPs) and/or nanostructured surfaces, similar to electrochemical sensors have been reported. For instance, Weerathunge et al. (2019) reported the development of a colorimetric MIP-based biosensor that can specifically detect pesticides in river water samples. The biosensor utilizes peroxidase-mimic silver NanoZymes and chlorpyrifos-specific aptamers to monitor the sensing process. The silver nanoparticles, which have tyrosine-capped peroxidase-mimic activity, provide a foundation for rapid and specific detection of chlorpyrifos, an organophosphorus pesticide. The chlorpyrifos aptamer incorporated in the biosensor ensures high specificity, resulting in a colorimetric response only in the presence of chlorpyrifos while remaining unresponsive to other pesticides from organophosphate and non-organophosphate groups. The study demonstrated the biosensor's robustness by directly testing it in environmental samples, specifically river water, and showed excellent recovery rates. The simplicity and rapid response of the biosensor, which takes only two minutes, make it an effective and practical tool for detecting chlorpyrifos and monitoring environmental contaminants. This innovative approach aligns with the need for efficient, specific, and user-friendly methods for environmental monitoring. Furthermore, Geng and colleagues (2018) have proposed an innovative technique to detect kanamycin antibiotics in water samples. They have developed a fluorescent aptamer-functionalized molecularly imprinted polymer (MIP) using surface imprinting in aqueous solutions. The technique used cadmium selenide quantum dots as supports, thiol-modified aptamer, methacrylic acid as functional monomers, and kanamycin as a template. The MIP operates through the double recognition of the aptamer and imprinted cavities, enabling fluorescent sensing of kanamycin. The "thiol-ene" click reaction was used to securely fix the aptamer into the polymer matrix. The results indicated that the MIP has a good linear correlation in a wide concentration range of 0.05–10.0 μg/mL with a low detection limit of 0.013 μg/mL. This technique can be used for the rapid, sensitive, and straightforward sensing of kanamycin in food, water, and biological samples, and has a promising future in various practical applications. In other applications, W. Ming et al. used magnetic molecularly imprinted polymers for the fluorescent detection of trace 17β-estradiol in environmental water. They synthesized a core–shell structured composite, Magnetic Molecularly Imprinted Polymers, which consisted of Fe_3O_4 nanoparticles and molecularly imprinted polymers (Ming et al. 2017). The composite was designed for selective and sensitive fluorescent detection of 17β-estradiol, utilizing the competitive desorption of fluorescein as a fluorescent indicator from the magnetic molecularly imprinted polymer. The study demonstrated a linear correlation with the concentration of 17β-E2 in the range of micromolar, with a low detection limit. The recovery of 17β-E2 from spiked lake and river water samples demonstrated a range of 98.2–103.8%. The authors suggested that the magnetic molecularly imprinted polymer-based method provides a simple, rapid, convenient, cost-effective, and environmentally friendly approach for highly selective and sensitive recognition and detection of non-fluorescent molecules at trace levels. More recently, researchers led by Liu and colleagues (2020) have developed a novel test strip based on molecularly imprinted quantum dots, which has been successfully applied for the fluorescence sensing of propanil herbicide in fish and seawater samples. The molecularly imprinted quantum dots were synthesized through a reversed-phase microemulsion technique, which served as the specific recognition element and signal probe for a fluorescence sensor or test strip to achieve highly sensitive detection of propanil. The MIP-quantum dots demonstrated excellent selectivity and physical–chemical characteristics, exhibiting good linearity within a wide propanil concentration range. The developed MIP-based fluorescence sensor exhibited favorable recoveries ranging, with a relative standard deviation below 6.0% for fish and seawater samples. Furthermore, a fluorescence test strip based on the MIP-quantum dots was developed, which displayed satisfactory recoveries of 90.1–111.1%, with a LOD for propanil in seawater samples at 0.6 μg/L. The proposed fluorescence sensor and test strip were effectively utilized for propanil determination in environmental and aquatic products, highlighting their applicability in real-world scenarios. As regards sensing strategy based on Surface-enhanced Raman scattering, an enzyme-induced molecularly imprinted polymer on SERS substrate for ultrasensitive detection of patulin has been designed by Zhu and colleagues (2020). The MIP-SERS substrate was created to detect patulin without the need for labeling. It combines the selectivity of molecular imprinting polymer and the sensitivity of Surface-Enhanced Raman Scattering (SERS) technology. The substrate was crafted using an aluminum oxide template and gold sputtering to create the SERS substrate. The MIP-SERS substrate was synthesized on the gold substrate by using horseradish peroxidase (HRP) enzyme, incorporating a selective polymer and a high density of SERS "hot spots." The results revealed that the MIP-SERS substrate exhibited exceptional selectivity for patulin mycotoxin, with a wide dynamic concentration range and a low limit of detection. Additionally, it showed acceptable reproducibility, stability, and recoveries ranging from 96.43 to 112.83%. The substrate is user-friendly and does not require complex sample pretreatment, making it a promising candidate for rapid and sensitive detection in food samples. More interestingly, Carneiro et al. (2019) fabricated a dual biorecognition system by combining a molecularly imprinted polymer and antibody in a SERS detection approach. This system was employed for the accurate detection of carcinoembryonic antigen using surface-enhanced Raman spectroscopy. The MIP acted as a pre-concentration scheme for the target protein, while the natural antibody signaled the presence of the target antigen on the MIP platform. Gold-based screen-printed electrodes served as the substrate and gallic acid was utilized for the first time in assembling a MIP film by electropolymerization in the presence of carcinoembryonic antigen. To prevent non-specific binding, the authors suggested coating the SERS layer with a second ultra-thin film of electropolymerized benzoic acid. Gold nanostars served as the metal support for the SERS tag, coupled with 4-aminothiophenol as a Raman reporter and natural antibody for carcinoembryonic antigen as the recognition element. The overall system demonstrated a sensitive response down to 1.0 ng/mL, differentiating it from the blank signal. The proposed method could be extended to other target molecules and has a low cost and time for MIP production due to the electropolymerization approach.

8.5 Outlook and Conclusion

The field of Molecularly Imprinted Polymers (MIPs) seems to be advancing at a fast pace, with significant progress being made and ongoing challenges being addressed. MIPs are cost-effective, robust, and highly selective, which makes them attractive for a wide range of applications. There are various preparation techniques available, and the integration of smart material technology has transformed the landscape of MIPs. This has led to the development of polymers with self-control abilities that can respond to external stimuli, making them highly appealing for diverse applications (Morsi et al. 2023; Thakur et al. 2023). To meet the specific requirements of various stimuli, numerous "smart" systems have been developed, and the success in designing and preparing MIPs with self-control abilities can be attributed to the integration of advanced smart material technology. This integration has attracted considerable attention from both scientific and industrial sectors, leading to the development of enzyme-like catalysts, analytical separations, solid-phase extractions, sensors, and drug delivery systems (Roy et al. 2017). The ability to tailor and monitor smart MIPs within controlled environments with precision is a significant breakthrough. In such environments, smart polymers can exhibit consistent and reliable responses to various external stimuli, marking a significant advancement in this field.

Despite the innovative approaches that researchers have taken to address challenges in the field of MIPs, such as incorporating nanomaterials, smart materials, and refining surface footprints, there are still persistent issues in optimizing physical selectivity, quantifying intermolecular interactions, and validating performance with complex real-world samples. Therefore, concerted efforts are needed to streamline the preparation of intricate surface and target analyte imprinting materials for large-scale production. Additionally, continuous efforts to enhance adsorption capacity and explore the use of imprinting polymers for ions are expected to drive further evolution in the field of MIPs.

References

Ahmad OS, Bedwell TS, Esen C, Garcia-Cruz A, Piletsky SA (2019) Molecularly imprinted polymers in electrochemical and optical sensors. Trends Biotechnol 37(3):294–309. https://doi.org/10.1016/j.tibtech.2018.08.009

Alim S, Vejayan J, Yusoff MM, Kafi AKM (2018) Recent uses of carbon nanotubes & gold nanoparticles in electrochemistry with application in biosensing: a review. Biosens Bioelectron 121:125–136. https://doi.org/10.1016/j.bios.2018.08.051

Alizadeh T, Hamidi N, Ganjali MR, Rafiei F (2017) An extraordinarily sensitive voltammetric sensor with picomolar detection limit for Pb^{2+} determination based on carbon paste electrode impregnated with nano-sized imprinted polymer and multi-walled carbon nanotubes. J Environ Chem Eng 5(5):4327–4336. https://doi.org/10.1016/j.jece.2017.08.009

Anirudhan TS, Alexander S (2015) Design and fabrication of molecularly imprinted polymer-based potentiometric sensor from the surface modified multiwalled carbon nanotube for the determination of lindane (γ-hexachlorocyclohexane), an organochlorine pesticide. Biosens Bioelectron 64:586–593. https://doi.org/10.1016/j.bios.2014.09.074

Arabi M, Chen L (2022) Technical challenges of molecular-imprinting-based optical sensors for environmental pollutants. Langmuir 38(19):5963–5967. https://doi.org/10.1021/acs.langmuir.2c00935

Aragay G, Pino F, Merkoçi A (2012) Nanomaterials for sensing and destroying pesticides. Chem Rev 112(10):5317–5338. https://doi.org/10.1021/cr300020c

Baranwal J, Barse B, Gatto G, Broncova G, Kumar A (2022) Electrochemical sensors and their applications: a review. Chemosensors 10(9):363. https://doi.org/10.3390/chemosensors10090363

BelBruno JJ (2019) Molecularly imprinted polymers. Chem Rev 119(1):94–119. https://doi.org/10.1021/acs.chemrev.8b00171

Beluomini MA, Da Silva JL, De Sá AC, Buffon E, Pereira TC, Stradiotto NR (2019) Electrochemical sensors based on molecularly imprinted polymer on nanostructured carbon materials: a review. J Electroanal Chem 840:343–366. https://doi.org/10.1016/j.jelechem.2019.04.005

Carballido L, Karbowiak T, Cayot P, Gerometta M, Sok N, Bou-Maroun E (2022) Applications of molecularly imprinted polymers and perspectives for their use as food quality trackers. Chem 8(9):2330–2341. https://doi.org/10.1016/j.chempr.2022.05.021

Carneiro MCCG, Sousa-Castillo A, Correa-Duarte MA, Sales MGF (2019) Dual biorecognition by combining molecularly-imprinted polymer and antibody in SERS detection. Application to carcinoembryonic antigen. Biosens Bioelectron 146:111761. https://doi.org/10.1016/j.bios.2019.111761

Cheong WJ, Yang SH, Ali F (2013) Molecular imprinted polymers for separation science: a review of reviews. J Sep Sci 36(3):609–628. https://doi.org/10.1002/jssc.201200784

Cui F, Zhou Z, Zhou HS (2020) Molecularly imprinted polymers and surface imprinted polymers based electrochemical biosensor for infectious diseases. Sensors 20(4):996. https://doi.org/10.3390/s20040996

Dong X, Zhang C, Du X, Zhang Z (2022) Recent advances of nanomaterials-based molecularly imprinted electrochemical sensors. Nanomaterials 12(11):1913. https://doi.org/10.3390/nano12111913

Fresco-Cala B, Batista AD, Cárdenas S (2020) Molecularly imprinted polymer micro- and nanoparticles: a review. Molecules 25(20):4740. https://doi.org/10.3390/molecules25204740

Ganjali MR, Rahmani AR, Shokoohi R, Farmany A, Khazaei M (2019) A highly sensitive and selective electrochemical Mercury(II) sensor based on nanoparticles of Hg(II)-imprinted polymer and graphitic carbon nitride (g-C3N4). Int J Electrochem Sci 14(7):6420–6430. https://doi.org/10.20964/2019.07.80

Gao M, Gao Y, Chen G, Huang X, Xu X, Lv J, Wang J, Xu D, Liu G (2020) Recent advances and future trends in the detection of contaminants by molecularly imprinted polymers in food samples. Front Chem 8:616326. https://doi.org/10.3389/fchem.2020.616326

Garcia S, Abu-Qare A, Meeker-O'Connell W, Borton A, Abou-Donia M (2003) Methyl parathion: a review of health effects. J Toxicol Environ Heal, Part B 6(2):185–210. https://doi.org/10.1080/10937400306471

Geng Y, Guo M, Tan J, Huang S, Tang Y, Tan L, Liang Y (2018) A fluorescent molecularly imprinted polymer using aptamer as a functional monomer for sensing of kanamycin. Sens Actuators, B Chem 268:47–54. https://doi.org/10.1016/j.snb.2018.04.065

Ghanei-Motlagh M, Taher MA (2017) Magnetic silver(I) ion-imprinted polymeric nanoparticles on a carbon paste electrode for voltammetric determination of silver(I). Microchim Acta 184(6):1691–1699. https://doi.org/10.1007/s00604-017-2157-8

Gong H, Hajizadeh S, Liu W, Ye L (2021) Imprinted polymer beads loaded with silver nanoparticles for antibacterial applications. ACS Appl Bio Mater 4(3):2829–2838. https://doi.org/10.1021/acsabm.1c00045

Hassan AHA, Moura SL, Ali FHM, Moselhy WA, Taboada Sotomayor MDP, Pividori MI (2018) Electrochemical sensing of methyl parathion on magnetic molecularly imprinted polymer. Biosens Bioelectron 118:181–187. https://doi.org/10.1016/j.bios.2018.06.052

He B, Mao Y-L, Zhang Y, Yin W, Hou C-J, Huo D-Q, Fa H-B (2017) A porphyrin molecularly imprinted biomimetic electrochemical sensor based on gold nanoparticles and carboxyl graphene composite for the highly efficient detection of methyl parathion. NANO 12(04):1750046. https://doi.org/10.1142/S1793292017500461

Huang W, Liu Y, Wang N, Song G, Yin X, Zhang L, Ni X, Xu W (2021) A sensitive electrochemical sensor based on ion imprinted polymers with gold nanoparticles for high selective detecting Cd(II) ions in real samples. J Inorg Organomet Polym 31(5):2043–2053. https://doi.org/10.1007/s10904-021-01892-8

Huang C-W, Lin C, Nguyen MK, Hussain A, Bui X-T, Ngo HH (2023) A review of biosensor for environmental monitoring: principle, application, and corresponding achievement of sustainable development goals. Bioengineered 14(1):58–80. https://doi.org/10.1080/21655979.2022.2095089

Idris AO, Orimolade B, Dennany L, Mamba B, Azizi S, Kaviyarasu K, Maaza M (2023) A review on monitoring of organic pollutants in wastewater using electrochemical approach. Electrocatalysis 14(5):659–687. https://doi.org/10.1007/s12678-023-00834-x

Iskierko Z, Sharma PS, Bartold K, Pietrzyk-Le A, Noworyta K, Kutner W (2016) Molecularly imprinted polymers for separating and sensing of macromolecular compounds and microorganisms. Biotechnol Adv 34(1):30–46. https://doi.org/10.1016/j.biotechadv.2015.12.002

Ji J, Zhou Z, Zhao X, Sun J, Sun X (2015) Electrochemical sensor based on molecularly imprinted film at Au nanoparticles-carbon nanotubes modified electrode for determination of cholesterol. Biosens Bioelectron 66:590–595. https://doi.org/10.1016/j.bios.2014.12.014

Jiang H, Jiang D, Shao J, Sun X (2016) Magnetic molecularly imprinted polymer nanoparticles based electrochemical sensor for the measurement of Gram-negative bacterial quorum signaling molecules (N-acyl-homoserine-lactones). Biosens Bioelectron 75:411–419. https://doi.org/10.1016/j.bios.2015.07.045

Jiao L, Zhong N, Zhao X, Ma S, Fu X, Dong D (2020) Recent advances in fiber-optic evanescent wave sensors for monitoring organic and inorganic pollutants in water. TrAC, Trends Anal Chem 127:115892. https://doi.org/10.1016/j.trac.2020.115892

Joe H-E, Yun H, Jo S-H, Jun MBG, Min B-K (2018) A review on optical fiber sensors for environmental monitoring. Int J Precis Eng Manuf-Green Tech 5(1):173–191. https://doi.org/10.1007/s40684-018-0017-6

Keçili R, Hussain CM (2018) Recent progress of imprinted nanomaterials in analytical chemistry. Int J Anal Chem 2018:1–18. https://doi.org/10.1155/2018/8503853

Kloskowski A, Pilarczyk M, Przyjazny A, Namieśnik J (2009) Progress in development of molecularly imprinted polymers as sorbents for sample preparation. Crit Rev Anal Chem 39(1):43–58. https://doi.org/10.1080/10408340802570223

Kryscio DR, Peppas NA (2012) Critical review and perspective of macromolecularly imprinted polymers. Acta Biomater 8(2):461–473. https://doi.org/10.1016/j.actbio.2011.11.005

Kumar V, Kim K-H (2022) Use of molecular imprinted polymers as sensitive/selective luminescent sensing probes for pesticides/herbicides in water and food samples. Environ Pollut 299:118824. https://doi.org/10.1016/j.envpol.2022.118824

Kumar S, Karfa P, Patra S, Madhuri R, Sharma PK (2016) Molecularly imprinted star polymer-modified superparamagnetic iron oxide nanoparticle for trace level sensing and separation of mancozeb. RSC Adv 6(43):36751–36760. https://doi.org/10.1039/C6RA03204D

Lakshmi J, Mukhopadhyay K, Ramaswamy P, Mahadevan S (2020) A systematic review on organophosphate pesticide and type II diabetes mellitus. CDR 16(6):586–597. https://doi.org/10.2174/1573399815666190712192844

Leibl N, Haupt K, Gonzato C, Duma L (2021) Molecularly imprinted polymers for chemical sensing: a tutorial review. Chemosensors 9(6):123. https://doi.org/10.3390/chemosensors9060123

Li B, Zhang Z, Qi J, Zhou N, Qin S, Choo J, Chen L (2017) Quantum dot-based molecularly imprinted polymers on three-dimensional origami paper microfluidic chip for fluorescence detection of phycocyanin. ACS Sens 2(2):243–250. https://doi.org/10.1021/acssensors.6b00664

Li Y, Zhang L, Dang Y, Chen Z, Zhang R, Li Y, Ye B-C (2019) A robust electrochemical sensing of molecularly imprinted polymer prepared by using bifunctional monomer and its application in detection of cypermethrin. Biosens Bioelectron 127:207–214. https://doi.org/10.1016/j.bios.2018.12.002

Liu C-X, Zhao J, Zhang R-R, Zhang Z-M, Xu J-J, Sun A-L, Chen J, Shi X-Z (2020) Development and application of fluorescence sensor and test strip based on molecularly imprinted quantum dots for the selective and sensitive detection of propanil in fish and seawater samples. J Hazard Mater 389:121884. https://doi.org/10.1016/j.jhazmat.2019.121884

Liu Z, Xu Z, Wang D, Yang Y, Duan Y, Ma L, Lin T, Liu H (2021) A review on molecularly imprinted polymers preparation by computational simulation-aided methods. Polymers 13(16):2657. https://doi.org/10.3390/polym13162657

Long F, Zhu A, Shi H (2013) Recent advances in optical biosensors for environmental monitoring and early warning. Sensors 13(10):13928–13948. https://doi.org/10.3390/s131013928

Lu H, Xu S, Liu J (2019) One pot generation of blue and red carbon dots in one binary solvent system for dual channel detection of Cr^{3+} and Pb^{2+} based on ion imprinted fluorescence polymers. ACS Sens 4(7):1917–1924. https://doi.org/10.1021/acssensors.9b00886

Lu H, Tian H, Wang C, Xu S (2020) Designing and controlling the morphology of spherical molecularly imprinted polymers. Mater Adv 1(7):2182–2201. https://doi.org/10.1039/D0MA00415D

Ma Q, Ha E, Yang F, Su X (2011) Synchronous determination of mercury(II) and copper(II) based on quantum dots-multilayer film. Anal Chim Acta 701(1):60–65. https://doi.org/10.1016/j.aca.2011.04.051

Ma Y, Shen X-L, Zeng Q, Wang H-S, Wang L-S (2017) A multi-walled carbon nanotubes based molecularly imprinted polymers electrochemical sensor for the sensitive determination of HIV-p24. Talanta 164:121–127. https://doi.org/10.1016/j.talanta.2016.11.043

Ma W, Chang Q, Zhao J, Ye B-C (2020) Novel electrochemical sensing platform based on ion imprinted polymer with nanoporous gold for ultrasensitive and selective determination of As^{3+}. Microchim Acta 187(10):571. https://doi.org/10.1007/s00604-020-04552-9

Mahmoudpour M, Torbati M, Mousavi M-M, De La Guardia M, Ezzati Nazhad Dolatabadi J (2020) Nanomaterial-based molecularly imprinted polymers for pesticides detection: recent trends and future prospects. TrAC, Trends Anal Chem 129:115943. https://doi.org/10.1016/j.trac.2020.115943

Mehrotra P (2016) Biosensors and their applications—a review. J Oral Biol Craniofacial Res 6(2):153–159. https://doi.org/10.1016/j.jobcr.2015.12.002

Meza López FDL, Khan S, Picasso G, Sotomayor MDPT (2021) A novel highly sensitive imprinted polymer-based optical sensor for the detection of Pb(II) in water samples. Environ Nanotechnol Monit Manage 16:100497. https://doi.org/10.1016/j.enmm.2021.100497

Ming W, Wang X, Lu W, Zhang Z, Song X, Li J, Chen L (2017) Magnetic molecularly imprinted polymers for the fluorescent detection of trace 17β-estradiol in environmental water. Sens Actuators, B Chem 238:1309–1315. https://doi.org/10.1016/j.snb.2016.09.111

Mitra S, Chakraborty AJ, Tareq AM, Emran TB, Nainu F, Khusro A, Idris AM, Khandaker MU, Osman H, Alhumaydhi FA, Simal-Gandara J (2022) Impact of heavy metals on the environment and human health: novel therapeutic insights to counter the toxicity. J King Saud Univ—Sci 34(3):101865. https://doi.org/10.1016/j.jksus.2022.101865

Morsi SMM, Abd El-Aziz ME, Mohamed HA (2023) Smart polymers as molecular imprinted polymers for recognition of target molecules. Int J Polym Mater Polym Biomater 72:612–635. https://doi.org/10.1080/00914037.2022.2042287

Mostafa AM, Barton SJ, Wren SP, Barker J (2021) Review on molecularly imprinted polymers with a focus on their application to the analysis of protein biomarkers. TrAC, Trends Anal Chem 144:116431. https://doi.org/10.1016/j.trac.2021.116431

Munteanu IG, Apetrei C (2022) A review on electrochemical sensors and biosensors used in assessing antioxidant activity. Antioxidants 11(3):584. https://doi.org/10.3390/antiox11030584

Ndunda EN (2020) Molecularly imprinted polymers—a closer look at the control polymer used in determining the imprinting effect: a mini review. J Mol Recogn 33(11):e2855. https://doi.org/10.1002/jmr.2855

Panda P, Chakroborty S (2022) Optical sensor technology and its application in detecting environmental effluents: a review. Int J Environ Anal Chem:1–16. https://doi.org/10.1080/03067319.2022.2098480

Pingarrón JM, Yáñez-Sedeño P, González-Cortés A (2008) Gold nanoparticle-based electrochemical biosensors. Electrochim Acta 53(19):5848–5866. https://doi.org/10.1016/j.electacta.2008.03.005

Prasad BB, Jauhari D (2015) Double-ion imprinted polymer @magnetic nanoparticles modified screen printed carbon electrode for simultaneous analysis of cerium and gadolinium ions. Anal Chim Acta 875:83–91. https://doi.org/10.1016/j.aca.2015.02.009

Pratama KF, Manik MER, Rahayu D, Hasanah AN (2020) Effect of the molecularly imprinted polymer component ratio on analytical performance. Chem Pharm Bull 68(11):1013–1024. https://doi.org/10.1248/cpb.c20-00551

Qi J, Tan D, Wang X, Ma H, Wan Y, Hu A, Li L, Xiao B, Lu B (2021) A novel acetylcholinesterase biosensor with dual-recognized strategy based on molecularly imprinted polymer. Sens Actuators, B Chem 337:129760. https://doi.org/10.1016/j.snb.2021.129760

Rahman S, Bozal-Palabiyik B, Unal DN, Erkmen C, Siddiq M, Shah A, Uslu B (2022) Molecularly imprinted polymers (MIPs) combined with nanomaterials as electrochemical sensing applications for environmental pollutants. Trends Environ Anal Chem 36:e00176. https://doi.org/10.1016/j.teac.2022.e00176

Ramachandran R, Chen T-W, Chen S-M, Baskar T, Kannan R, Elumalai P, Raja P, Jeyapragasam T, Dinakaran K, Gnana Kumar GP (2019) A review of the advanced developments of electrochemical sensors for the detection of toxic and bioactive molecules. Inorg Chem Front 6(12):3418–3439. https://doi.org/10.1039/C9QI00602H

Rebelo P, Costa-Rama E, Seguro I, Pacheco JG, Nouws HPA, Cordeiro MNDS, Delerue-Matos C (2021) Molecularly imprinted polymer-based electrochemical sensors for environmental analysis. Biosens Bioelectron 172:112719. https://doi.org/10.1016/j.bios.2020.112719

Roy E, Patra S, Tiwari A, Madhuri R, Sharma PK (2017) Introduction of selectivity and specificity to graphene using an inimitable combination of molecular imprinting and nanotechnology. Biosens Bioelectron 89:234–248. https://doi.org/10.1016/j.bios.2016.02.056
Sala A, Brisset H, Margaillan A, Mullot J-U, Branger C (2022) Electrochemical sensors modified with ion-imprinted polymers for metal ion detection. TrAC, Trends Anal Chem 148:116536. https://doi.org/10.1016/j.trac.2022.116536
Scheller FW, Zhang X, Yarman A, Wollenberger U, Gyurcsányi RE (2019) Molecularly imprinted polymer-based electrochemical sensors for biopolymers. Curr Opin Electrochem 14:53–59. https://doi.org/10.1016/j.coelec.2018.12.005
Scozzari A (2008) Electrochemical sensing methods: a brief review. In: Evangelista V, Barsanti L, Frassanito AM, Passarelli V, Gualtieri P (eds) Algal toxins: nature, occurrence, effect and detection. Springer Netherlands, Dordrecht, pp 335–351. https://doi.org/10.1007/978-1-4020-8480-5_16
Selvolini G, Marrazza G (2017) MIP-based sensors: promising new tools for cancer biomarker determination. Sensors 17(4):718. https://doi.org/10.3390/s17040718
Senbua W, Mearnchu J, Wichitwechkarn J (2020) Easy-to-use and reliable absorbance-based MPH-GST biosensor for the detection of methyl parathion pesticide. Biotechnol Rep 27:e00495. https://doi.org/10.1016/j.btre.2020.e00495
Sharma K, Sharma M (2023) Optical biosensors for environmental monitoring: recent advances and future perspectives in bacterial detection. Environ Res 236:116826. https://doi.org/10.1016/j.envres.2023.116826
Sharma PS, Pietrzyk-Le A, D'souza F, Kutner W (2012) Electrochemically synthesized polymers in molecular imprinting for chemical sensing. Anal Bioanal Chem 402(10):3177–3204. https://doi.org/10.1007/s00216-011-5696-6
Si Y, Lao J, Zhang X, Liu Y, Cai S, Gonzalez-Vila A, Li K, Huang Y, Yuan Y, Caucheteur C, Guo T (2019) Electrochemical plasmonic fiber-optic sensors for ultra-sensitive heavy metal detection. J Lightwave Technol 37(14):3495–3502. https://doi.org/10.1109/JLT.2019.2917329
Tajik S, Beitollahi H, Nejad FG, Dourandish Z, Khalilzadeh MA, Jang HW, Venditti RA, Varma RS, Shokouhimehr M (2021) Recent developments in polymer nanocomposite-based electrochemical sensors for detecting environmental pollutants. Ind Eng Chem Res 60(3):1112–1136. https://doi.org/10.1021/acs.iecr.0c04952
Thakur N, Das TR, Gupta S, Shukla SK, Patra S, Kumar D (2023) Material selection for the fabrication of MIP-based devices: an overview. In: Molecularly imprinted polymers (MIPs). Elsevier, pp 101–114. https://doi.org/10.1016/B978-0-323-91925-8.00006-5
Torkashvand M, Gholivand MB, Azizi R (2017) Synthesis, characterization and application of a novel ion-imprinted polymer based voltammetric sensor for selective extraction and trace determination of cobalt(II) ions. Sens Actuators, B Chem 243:283–291. https://doi.org/10.1016/j.snb.2016.11.094
Vasapollo G, Sole RD, Mergola L, Lazzoi MR, Scardino A, Scorrano S, Mele G (2011) Molecularly imprinted polymers: present and future prospective. IJMS 12(9):5908–5945. https://doi.org/10.3390/ijms12095908
Wackerlig J, Lieberzeit PA (2015) Molecularly imprinted polymer nanoparticles in chemical sensing—synthesis, characterisation and application. Sens Actuators, B Chem 207:144–157. https://doi.org/10.1016/j.snb.2014.09.094
Wang F-R, Lee G-J, Haridharan N, Wu JJ (2018) Electrochemical sensor using molecular imprinting polymerization modified electrodes to detect methyl parathion in environmental media. Electrocatalysis 9(1):1–9. https://doi.org/10.1007/s12678-017-0411-9
Wang J, Hu J, Hu S, Gao G, Song Y (2020) A novel electrochemical sensor based on electropolymerized ion imprinted PoPD/ERGO composite for trace Cd(II) determination in water. Sensors 20(4):1004. https://doi.org/10.3390/s20041004
Wang L, Li B, Wang J, Qi J, Li J, Ma J, Chen L (2022) A rotary multi-positioned cloth/paper hybrid microfluidic device for simultaneous fluorescence sensing of mercury and lead ions

by using ion imprinted technologies. J Hazard Mater 428:128165. https://doi.org/10.1016/j.jhazmat.2021.128165

Weerathunge P, Behera BK, Zihara S, Singh M, Prasad SN, Hashmi S, Mariathomas PRD, Bansal V, Ramanathan R (2019) Dynamic interactions between peroxidase-mimic silver NanoZymes and chlorpyrifos-specific aptamers enable highly-specific pesticide sensing in river water. Anal Chim Acta 1083:157–165. https://doi.org/10.1016/j.aca.2019.07.066

Wei X, Xu G, Gong C, Qin F, Gong X, Li C (2018) Fabrication and evaluation of sulfanilamide-imprinted composite sensors by developing a custom-tailored strategy. Sens Actuators, B Chem 255:2697–2703. https://doi.org/10.1016/j.snb.2017.09.081

Wu S, Dai X, Cheng T, Li S (2018) Highly sensitive and selective ion-imprinted polymers based on one-step electrodeposition of chitosan-graphene nanocomposites for the determination of Cr(VI). Carbohyd Polym 195:199–206. https://doi.org/10.1016/j.carbpol.2018.04.077

Wu L, Li X, Miao H, Xu J, Pan G (2022) State of the art in development of molecularly imprinted biosensors. VIEW 3(3):20200170. https://doi.org/10.1002/VIW.20200170

Yang B, Fu C, Li J, Xu G (2018) Frontiers in highly sensitive molecularly imprinted electrochemical sensors: challenges and strategies. TrAC, Trends Anal Chem 105:52–67. https://doi.org/10.1016/j.trac.2018.04.011

Yang W, Ma Y, Sun H, Huang C, Shen X (2022) Molecularly imprinted polymers based optical fiber sensors: a review. TrAC, Trends Anal Chem 152:116608. https://doi.org/10.1016/j.trac.2022.116608

Yola ML, Atar N (2017) A review: molecularly imprinted electrochemical sensors for determination of biomolecules/drug. Curr Anal Chem 13(1):13–17. https://doi.org/10.2174/1573411012666160601141018

Yu L, Sun L, Zhang Q, Zhou Y, Zhang J, Yang B, Xu B, Xu Q (2022) Nanomaterials-based ion-imprinted electrochemical sensors for heavy metal ions detection: a review. Biosensors 12(12):1096. https://doi.org/10.3390/bios12121096

Zhang M, Zhao HT, Xie TJ, Yang X, Dong AJ, Zhang H, Wang J, Wang ZY (2017) Molecularly imprinted polymer on graphene surface for selective and sensitive electrochemical sensing imidacloprid. Sens Actuators, B Chem 252:991–1002. https://doi.org/10.1016/j.snb.2017.04.159

Zhang MY, Huang RF, Ma XG, Guo LH, Wang Y, Fan YM (2019) Selective fluorescence sensor based on ion-imprinted polymer-modified quantum dots for trace detection of Cr(VI) in aqueous solution. Anal Bioanal Chem 411(27):7165–7175. https://doi.org/10.1007/s00216-019-02100-w

Zhang Y, Zhu Y, Zeng Z, Zeng G, Xiao R, Wang Y, Hu Y, Tang L, Feng C (2021) Sensors for the environmental pollutant detection: are we already there? Coord Chem Rev 431:213681. https://doi.org/10.1016/j.ccr.2020.213681

Zheng L, Zheng Y, Liu Y, Long S, Du L, Liang J, Huang C, Swihart MT, Tan K (2019) Core-shell quantum dots coated with molecularly imprinted polymer for selective photoluminescence sensing of perfluorooctanoic acid. Talanta 194:1–6. https://doi.org/10.1016/j.talanta.2018.09.106

Zhou Z, Li T, Xu W, Huang W, Wang N, Yang W (2017) Synthesis and characterization of fluorescence molecularly imprinted polymers as sensor for highly sensitive detection of dibutyl phthalate from tap water samples. Sens Actuators, B Chem 240:1114–1122. https://doi.org/10.1016/j.snb.2016.09.092

Zhu Y, Wu L, Yan H, Lu Z, Yin W, Han H (2020) Enzyme induced molecularly imprinted polymer on SERS substrate for ultrasensitive detection of patulin. Anal Chim Acta 1101:111–119. https://doi.org/10.1016/j.aca.2019.12.030

Zouaoui F, Bourouina-Bacha S, Bourouina M, Abroa-Nemeir I, Ben Halima H, Gallardo-Gonzalez J, El Alami, El Hassani N, Alcacer A, Bausells J, Jaffrezic-Renault N, Zine N, Errachid A (2020) Electrochemical impedance spectroscopy determination of glyphosate using a molecularly imprinted chitosan. Sens Actuators, B Chem 309:127753. https://doi.org/10.1016/j.snb.2020.127753

Chapter 9
Molecularly Imprinted Polymers as Highly Selective Sorbents in Sample Preparation Techniques for Extracting Emerging Pollutants from Environmental Water

Nur Nadhirah Mohamad Zain, Nor Munira Hashim, Noorfatimah Yahaya, Mazidatulakmam Miskam, Ahmad Husaini Mohamed, Nur Sofiah Abu Kassim, and Muthamilselvi Ponnuchamy

Abbreviation

[VAFMIM]Cl ionic liquid	Monomer-1-vinyl-3-amniformylmethyl imidazolium chloride ionic liquid
AgNP	Silver nanoparticles
AIBN	Azobisisobutyronitrile
APGC-MS/MS	Atmospheric pressure chemical ionization-tandem mass spectrometry
APTES	(3-Aminopropyl) triethoxysilane
ATSs	Amphetamine-type stimulants
AuNS	Gold nanostar

N. N. M. Zain (✉) · N. M. Hashim · N. Yahaya
Department of Toxicology, Advanced Medical and Dental Institute, Universiti Sains Malaysia, 13200 Kepala Batas, Pulau Pinang, Malaysia
e-mail: nurnadhirah@usm.my

M. Miskam
School of Chemical Sciences, Universiti Sains Malaysia, 11800 Pulau Pinang, Malaysia

A. H. Mohamed · N. S. A. Kassim
Faculty of Applied Sciences, School of Chemistry and Environment, Universiti Teknologi MARA, Cawangan Negeri Sembilan, Kampus Kuala Pilah, 72000 Kuala Pilah, Negeri Sembilan, Malaysia

M. Ponnuchamy
Department of Chemical Engineering, School of Bio-engineering, College of Engineering and Technology, SRM Institute of Science and Technology, Kattankulathur, Chennai, Tamil Nadu 603 203, India

S. Patra and M. Sillanpaa (eds.), *Molecularly Imprinted Polymers as Artificial Antibodies for the Environmental Health*, https://doi.org/10.1007/978-3-031-58995-9_9

BDD	Boron doped diamond electrode
BPA	Bisphenol A
CNT-GCE	Carbon nanotube/glassy carbon electrode
CNW/MIP PFOS	Carbon nanowall incorporated into molecularly imprinted polymers sensor for detecting PFOS
CV	Cyclic voltammetry
DBP	Dibutyl phthalate
DCE	α-(2,4-Dichlorophenyl)-1H-imidazole-1-ethanol
DIMSPE	Dummy molecularly imprinted solid phase extraction
DLLE	Dispersive liquid–liquid extraction
DMIPs	Dummy molecularly imprinted polymers
DMPA	Dimethoxy-2-phenylacetophenone
DPV	Differential pulse voltammetry
DSPE	Dispersive solid-phase extraction
EDGMA	Ethylene glycol dimetacrylate
ERY-MIP/SPE	Molecularly imprinted polymers-based sensor for detection of erythromycin combined with screen printed electrode
EtOH	Ethanol
FEN	Fenoxaprop
FT-IR	Fourier transform-infrared spectroscopy
GCE	Glassy carbon electrode
GC-FID	Gas chromatography coupled with flame ionization detector
GC-MS	Gas chromatography coupled with mass spectrometric detection (GC-MS)
GO	Graphene oxide
H_2SO_4	Sulfuric acid
HAL	Haloxyfop
HLB	Hydrophilic-lipophilic balanced
HPEA	Homopiperonylamine
HPLC	High performance liquid chromatography
HPLC-DAD	High performance liquid chromatography coupled with diode array detector
HPLC-PDA	High performance liquid chromatography coupled with photo diode array
HPLC-UV	High performance liquid chromatography coupled ultraviolet detector
IMS	Ion mobility spectrometry
LC-MS/MS	Liquid chromatography with tandem mass spectrometry

LC-qTOF/MS	Liquid Chromatography Quadrupole Time-of-Flight mass spectrometer
LOD	Limit of detection
MAA	Methacrylic acid
MASE-MIP	Membrane assisted solvent extraction and a molecularly imprinted polymer
MCPA	4-Chloro-2-methylphenoxyacetic acid
MCPA	Methylphenoxyacetic acid
MCPB	4-(4-Chloro-2-methylphenoxy)butyric acid
MCPP	Mecoprop
ME	Matrix effects
MeCN	Acetonitrile
MeOH	Methanol
$MgSO_4$	Magnesium sulfate
MHP	Mono hexyl phthalate
MIP@MA/IT-SPME	Molecularly imprinted polymer based-magnetism-assisted-in-tube solid phase microextraction
MIP-AuNS	Molecularly imprinted polymer and gold nanostar modified voltammetric sensor
MIP-CNTs-GCE	Dual template molecularly imprinted polymers-carbon nanotube/glassy carbon electrode
MIP-Co/Fe@CNF/DPV	Molecularly imprinted polymer based on copper/iron alloy driven carbon nanofiber
MIPs	Molecularly imprinted polymers
MIP-SBSE	Molecularly imprinted polymers-stir bar sorptive extraction
MISPE	Molecularly imprinted solid-phase extraction
MMIP-DSPE	Magnetic molecularly imprinted polymer used in dispersive solid phase extraction
MMIPs	Magnetic-based molecularly imprinted polymers
MMISPE	Magnetic molecularly imprinted solid-phase extraction
MPD	M-phenylenediamine
MRLs	Maximum residue limits
MSPE	Magnetic solid-phase extraction
$NaClO_4$	Sodium perchlorate
NaOH	Sodium hydroxide
NTED-MIP	Needle-type extraction device-molecularly imprinted polymers
oPD	O-phenylenediamene

OPFRs	Organophosphate flame retardants
OPPs	Organophosphorus pesticides
PBS	Phosphate buffer saline
PDMS	Polydimethylsiloxane
PFOA	Perfluorooctanoic acid
PFOS	Perfluorooctanesulfonic acid
PSA	Primary secondary amines
PTFE	Polytetrafluoroethylene
QuEChERS	Quick, Easy, Cheap, Effective, Rugged, and Safe
SBSE	Stir bar sorptive extraction
SEM	Scanning electron microscope
SFs	Strobilurin fungicides
SPE	Solid-phase extraction
SPME	Solid-phase microextraction
TEOS	Tetraethyl orthosilicate
TRIM	Trimethylolpropane triacrylate
UHPLC-MS/MS	Ultra-high performance liquid chromatography coupled with tandem mass spectrometry
US EPA	United States Environmental Protection Agency
UV-spectrophotometry	Ultraviolet-spectrophotometer
VSM	Vibrating sample magnetometer
$ZnO/Fe_3O_4@SiO_2–NH_2$/MIP/alginate	Coating zinc oxide and magnetite nanoparticles on a molecularly imprinted polymer embedded in alginate

9.1 Introduction

The emerging pollutants (EPs), as well as their metabolites, are a broad category of compounds that can be found in a wide range of products, including drugs, food additives, plasticizers, and pharmaceuticals (Morin-Crini et al. 2022; Rathi et al. 2021). In recent years, researchers have increasingly focused on these contaminants due to their widespread presence in the environmental water. However, a significant number of these compounds remained unregulated, primarily due to limited knowledge regarding their toxicity, effects and behaviours in the environmental water. According to the United States Environmental Protection Agency (US EPA), EPs are substances that have not been studied properly and pose a threat to human health and the environment (Cantarella et al. 2019; Parlapiano et al. 2021). These substances can be detected in the water supply of wastewater treatment plants and drinking water sources at low concentrations. However, they are not required to be monitored or controlled. Due to the growing awareness about the potential impacts of these contaminants, it is expected that the appropriate regulations and protocols will be issued in the following years. Aqueous solutions are prone to presenting various challenges when it comes to identifying and monitoring contaminants. Therefore, efficient analytical approaches, such as gas chromatography coupled with mass spectrometric detection (GC-MS) and high-performance liquid chromatography (HPLC), are employed to identify and quantify contaminants in environmental water (Sabín López et al. 2020; León-Martínez et al. 2018). However, these methods often fall short in achieving the desired levels of speed and cost-effectiveness due to their requirement for complex and expensive instrumentation, labour-intensive sample pretreatment operations, and their time-consuming nature.

In 2015, the United Nation General Assembly set a goal to ensure that everybody has access to safe and sufficient water by 2030. One of the most challenging factors that the government and private sectors face when it comes to implementing this goal is the development of efficient and effective extraction methods for monitoring environmental water. The use of polymers and other polymeric materials has greatly evolved over the years (Huang et al. 2022). Molecularly imprinted polymers (MIPs) are commonly referred to as analogues to the natural biological antibody systems. They are formed through a lock and key mechanism, which allows them to selectively bind to a specific molecule. Due to the simplicity of the preparation process, molecular imprinting has gained widespread attention over the years (BelBruno 2019). MIPs have recently emerged as a promising extraction tool for developing specific methods incorporated with a detection system to qualify and quantify various types of molecules. The ability to selectively bind to a chemical compound is a characteristic of MIPs. This process is carried out through the use of molecular recognition. It is a method that aims to promote the development of molecular recognition capabilities in synthetic polymers (Khulu et al. 2022a; Dong et al. 2021). Different production methods are utilized, but all of them follow a basic outline: (i) a polymer is produced containing the template or target molecule bond, covalently or non-covalently, to a functional group of the host, (ii) the template molecule is removed from the polymer host, leaving a target-specific cavity available for rebinding, and (iii) the MIPs are then exposed to a target-containing sample. The cavity selectivity allows the uptake of the target molecule from the complex sample (BelBruno 2019). The general protocol for the fabrication of MIPs is illustrated in Fig. 9.1 (Roy et al. 2017).

The process of molecular imprinting involves the assembly of a template molecule and its functional monomers using either reversible or non-covalent bonds. The resulting complexes are then linked together using a cross-linker. After the template molecules have been removed from a cross-linked polymer, their binding cavities are formed. These cavities are designed to have complementary sizes, shapes, and functions to the template molecule. The recognition mechanism of the resulting MIPs is similar to that of the natural ligand-receptor interaction. The fundamental idea of this interaction is the non-covalent binding of two or more molecules. This bonding includes hydrogen bonds, coordinative bonds, hydrophobic forces, π–π interactions, van der Waals forces, and electrostatic and/or electromagnetic effects. In certain instances, the receptor-ligand complex encompasses multiple recognition

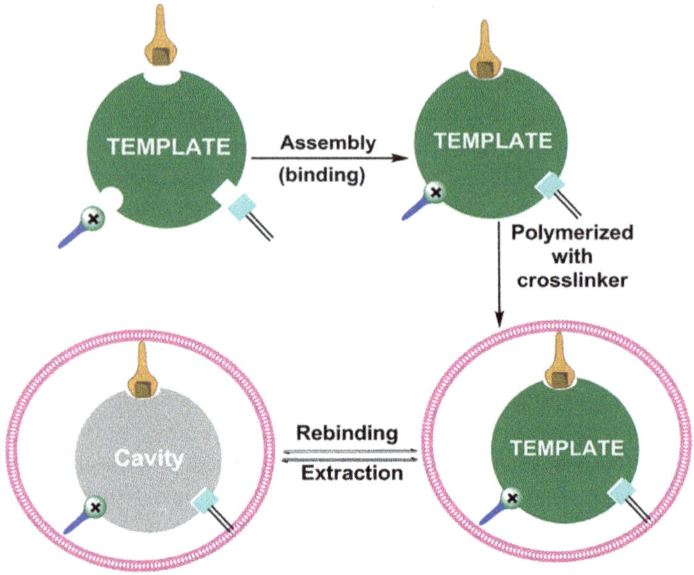

Fig. 9.1 General protocol for fabrication of molecularly imprinted polymers. Reprinted from Roy et al. (2017) with permission from Elsevier

sites, facilitating multivalent binding and enhancing affinity (BelBruno 2019; Pan et al. 2018; Thakur et al. 2023). Currently, typical methods for detecting substances at low concentrations, such as emerging contaminants, include chromatographic separation, gas chromatography, mass spectrometry, and so on. These methods require a lot purification steps, expensive equipment, and skilled personnel. The MIPs' exceptional recognition and selectivity capabilities allow them to be used as sorbents during the sample preparation process (Karrat and Amine 2021). The selection of the functional and template is a critical step in MIPs development. Both molecules should have complementary functional groups and easily dissolved in a compatible solvent. In addition to this, other factors such as the radical initiator and the cross-linker are also taken into account to facilitate the polymerization process (Kaya et al. 2023). Through the formation of specific adsorbing sites within the cavities, a template molecule can be identified and separated from the other molecules in the matrix. This process can be performed repeatedly in order to reclaim the molecule. MIP-based compounds are commonly used as sorbents for various applications as they are inexpensive and reusable (Zhou et al. 2019; Turiel and Martín-Esteban 2019). They are mostly preferred in extraction-based sample preparation technique. The analysis of water sample holds an important place in the application of MIP-based sample preparation to environmental analysis.

The goal of this review is to present the latest trends in the use of MIPs in sample preparation. It also reviews the most commonly used sample preparation techniques. This review serves as an updated source of information for researchers focused on the analytical determination and sample preparation assays. The review also focuses on the various applications of MIPs in the analysis of environmental water samples.

9.2 Fundamentals of Molecular Imprinting Techniques

As synthetic materials with the characteristics of selective binding molecules, MIPs can be templated during synthesis preparation activities through a "lock and key" mechanism. The key features of synthesizing MIPs will further engrave the conditions of the binding sites, such as shape, size, and chemical functionality towards the template molecules, depending on the designation of the template molecule, types of functional monomers, cross-linking agents, and polymerization mechanisms. The synthetic routes of preparation MIPs usually involve surface imprinting, free radical polymerization, nanoscale imprinting, and sol–gel method (Ranjbari et al. 2023). As a tip and tricks strategy for the synthesis of MIPs, the performance of the imprinting process usually depends on three steps as follows (Mabrouk et al. 2023).

i. Pre-polymerization complex formation—involves adding elective template molecules in the porogen, forming complexes with functional monomer molecules (typical bonding involved covalent, non-covalent, and semi-covalent interactions) (Table 9.1).
ii. Polymer preparation—involves the formation of cross-linked polymers with the complexes using cross-linking agents assisted by the initiator. Polymerization methods can be categorized into precipitation polymerization, bulk

Table 9.1 Advantages and disadvantages of covalent and non-covalent molecularly imprinting approaches (Mabrouk et al. 2023)

	Covalent	Non-covalent	Semi-covalent
Advantages	– High selectivity – Homogeneous active binding sites	– Simple and easy operation – Strong binding activity – Easy template removal	– High selectivity – Homogeneous active binding sites
Disadvantages	– Tedious removal steps – Inadequate reversible covalent bonds	– Non-specific active surface binding – Lack of selectivity – Involved complex factors (steric, electronic, and chelate macrocyclic effects, coordination capacity, and solvent effects)	– Poor imprinting efficiency – Inadequate reversible covalent bonds – Tedious removal steps
Approach	– Schiff base, disulphide, ketal, acetal, and boronate esters	– Lewis's acid–base, Van der Waals, hydrophobic, electrostatic, and metal–ligand coordination	– Covalent bonding (during synthesis) and non-covalent bonding (during analysis)

polymerization, surface imprinting polymerization, in-situ polymerization, electropolymerization methods, etc.

iii. Template removal from MIPs—entails removing template molecules attached to the binding of the MIPs to obtain imprinted pores that complement the target analytes (in terms of functional groups and spatial configuration).

Based on the attractive merits offered by the MIPs, molecular imprinting techniques have aroused great interest among researchers which further have been utilized as adsorbents for selective extraction of various compounds. The adaptation of polymerization methods involves selecting the medium to synthesis MIPs with desired characteristics. Table 9.2 highlights the significant advantages and limitations of commonly used polymerization methods adapted to prepare MIPs.

As a new class of adsorbents, MIPs with unique selective three-dimensional cavity typically prepared using dummy templates which pose similar spatial configuration structures and comparable recognition with targeted analytes. This technique allowed accurate quantification as well as prevent low removal of analyte from the template which further leading to environmental issues. There have been many reports of preparation dummy MIPs (DMIPs) as adsorbents for pre-concentrating emerging pollutants in various matrices (Wang et al. 2023b; Surapong et al. 2023; Shahzad et al. 2023; Fu et al. 2022a). To prepare DMIPs, "dummy patterns" which comply with the shape/geometry as well as functional groups is added in the polymerization mixture containing functional monomers groups. For example, Tan and co-workers prepare DMIP using homopiperonylamine (HPEA) as dummy template or class-selective extraction of amphetamine-type stimulants (ATSs). Bulk polymerization method was employed due to the advantages of easily optimized and suitable for most polymerizations with methacrylic acid (MAA) and ethylene glycol dimetacrylate (EDGMA) were used as functional monomers and cross-linking agent, respectively. For template removal, Soxhlet extraction method was selected to detach HPEA dummy template from the prepared DMIP (Tan et al. 2022). In another work, Li et al. prepare magnetic dummy MIPs using sol–gel and surface imprinting polymerization methods with tetraethyl orthosilicate (TEOS), (3-aminopropyl) triethoxysilane (APTES), and mono hexyl phthalate (MHP) were used as cross linker, functional monomer, and dummy template molecule, respectively. Sol–gel and surface imprinting were chosen because the process can be applied at room temperature to avoid any decomposition and reduced embedding as well as promote higher binding rate of binding kinetics (Li et al. 2021a).

Recently, magnetic-based MIPs (MMIPs) have not only exhibited specific selective binding but also demonstrated outstanding magnetic features, attracting significant attention among researchers. For example, Vidal et al. prepared a magnetic and mesoporous MIP using EDGMA as a cross-linking agent and acetonitrile as a porogenic solvent, shown to be an excellent adsorbent for the adsorption process. In the MMIPs, the polymerization of the adsorbent layer occurs on the surface of magnetic nanoparticles, which lessens the incompleteness of template removal, low mass transfer rate, and low binding rates compared to conventional MIPs (Vidal et al. 2023). A study by Hu et al. has established an excellent adsorption capacity, fast

Table 9.2 Commonly used method of polymerization adapted to prepare MIPs

Polymerization method	Advantages	Drawbacks	References
Bulk	– High yield – Easily optimized	– Usage of grinding and screening destroy some part of the binding sites – Low adsorption/desorption efficiency (hindering the mass transfer rate) – Irregular particle size and shape	Muñoz-Bartual et al. (2023), Shah et al. (2023), Mohamed Nageib et al. (2023)
Precipitation	– Particle size, shape structure, and surface morphology can be controlled – No addition of stabilizer during polymerization	– Required optimization for each set of functional monomers	Feng et al. (2021), Surapong et al. (2022)
Surface imprinting	– Reduced embedding and template leakage phenomenon – Facilitate high mass transfer rate, binding kinetics and binding capacity – Easy removal of template molecules – Increase surface area	– Required original template	Wang et al. (2023a), Noor et al. (2023)
Sol–gel	– Mild reaction condition, low cost, – Excellent chemical and mechanical stability – Controllable reaction process – Homogeneity of materials at the molecular surface – Eco-friendly solvent media (water, ethanol)	– Used non-green solvents (toxic)	Bai et al. (2023), Ren and Fan (2023), Yang et al. (2023a)
In-situ	– Strong adhesion to the collector – Controllable structure – No cumbersome processing	– Required optimization of polymerization time (long time produced too-dense polymers, while short time produced less selective binding sites)	Song et al. (2023)

adsorption equilibrium, and excellent reusability of MMIPs adsorbent for enriching and detecting residual strobilurin fungicides in agricultural products. The prepared MMIPs using the surface imprinting technique has successfully maintained the imprinting sites even after several adsorption and desorption mechanisms have been performed (Hu et al. 2023). Other studies related to the development of MMIPs were synthesized by Khongkla and colleagues by developing a coating zinc oxide and magnetite nanoparticles on a molecularly imprinted polymer embedded in alginate ($ZnO/Fe_3O_4@SiO_2$–NH_2/MIP/alginate), forming a composite adsorbent hydrogel bead (Khongkla et al. 2023). Wang et al. obtained a highly selective MMIPs of dibutyl phthalate (DBP) prepared by surface imprinting on SiO_2–$COOH@Fe_3O_4$ magnetic nanoparticles. It was noted that the addition of magnetic microspheres simplified the process of food substrate purification and significantly reduced the pre-concentration time (Wang et al. 2023c).

9.3 Overview of Sample Preparation Techniques

Sample preparation is a crucial step in analytical chemistry that involves the process of extracting and concentrating target analytes from complex sample matrices. Proper sample preparation is essential to obtain accurate and reliable results in various analytical techniques. The sample preparation technique depends on the nature of the analytes, the complexity of the sample matrix, the required detection sensitivity, and the specific analytical method used for subsequent analysis. Approaches to improve and accelerate extraction efficiency and lower solvent consumption include sorbent-based techniques. A recent review evaluates the most recent applications of MIPs in sample preparation techniques (Kaya et al. 2023). This overview section aims to update the state of the art of some cutting-edge sample preparation techniques using MIPs and focuses on relevant applications in emerging pollutants from environmental water (Delinska et al. 2021). Sample preparation techniques such as solid-phase extraction (SPE), solid-phase microextraction (SPME), stir bar sorptive extraction (SBSE), dispersive solid-phase extraction (DSPE), and magnetic solid-phase extraction (MSPE) are highlighted. Each technique has its advantages and limitations, and selecting the most suitable method is essential for obtaining accurate and precise analytical results.

9.3.1 Solid-Phase Extraction (SPE)

SPE is a widely used sample preparation technique in analytical chemistry to isolate, concentrate, and purify analytes from complex sample matrices before analysis. SPE is the most frequently applied extraction technique and is widely used in various fields (environmental analysis, pharmaceuticals, forensics, and food testing) due to its versatility and efficiency (Kraševec and Prosen 2018).

SPE techniques, such as bonded silica phases, polymer-based phases, mixed-mode phases, ion-exchange phases, MIPs, and size-exclusion phases, can be achieved in a one-step extraction, and the methodology is appropriate for isolating trace amounts of chemical compounds from complex matrices. The choice of SPE method and sorbent material depends on the sample complexity, the type of analytes being targeted, and the analytical requirements. The process involves passing the sample through a solid-phase that retains the target analytes while unwanted substances are washed away. MIPs are a type of sorbent that can be used in SPE to isolate and extract specific analytes from complex matrices such as biological and environmental samples. MIPs are prepared by imprinting a template molecule with functional monomers and cross-linkers, resulting in a polymer with selective binding sites for the template molecule.

Sample preparation techniques using SPE with MIPs have been developed for the extraction and isolation of emerging pollutants from environmental water (Pichon et al. 2020). Recent studies have shown that MIPs can be used for the selective extraction and preconcentration of hormones, herbicides, and fluoroquinolones in complex matrices (Mpupa et al. 2021; Margiotta 2019; Shao et al. 2020). In a previous study conducted by Krishnan et al. (2019), MIPs have been used for the solid-phase extraction of andrographolide from plant samples. Recently, MMIPs have been developed for the separation of active compounds from natural products in environmental water (Ariani et al. 2022). The advantages of MMIPs in terms of their selectivity and high specificity are due to their large surface areas and accessible binding sites. MMIPs that have magnetic core shells can be developed further due to their high synthesis yield (Ariani et al. 2022). Figure 9.2 shows a synthesized MMIP applied as a sorbent in magnetic molecularly imprinted solid-phase extraction (MMISPE) for the selective extraction of a mixture of the five herbicides (4-chloro-2-methylphenoxyacetic acid (MCPA), 4-(4-chloro-2- methylphenoxy)butyric acid (MCPB), mecoprop (MCPP), fenoxaprop (FEN) and haloxyfop (HAL) in water samples (Meseguer-Lloret et al. 2022a). In summary, MIPs have been used in SPE for the selective extraction of various analytes from complex matrices. The use of MIPs in SPE has several advantages over traditional sorbents, including high selectivity and performance.

9.3.2 *Solid-Phase Microextraction (SPME)*

SPME technique is a solvent-free method for preparing samples. It involves coating a fiber with a stationary phase and selectively extracting analytes from a sample matrix. This fiber is then inserted into a chromatographic device for analysis and desorption. The SPME technique is widely used for sample preparation for organic compounds as it allows simultaneous extraction of analytes and concentration (Zambonin and Aresta 2021). MIPs have been incorporated into various sample preparation techniques, such as SPME, to produce an efficient sample preparation technique that has expanded its application in the field of environmental sample analysis. The application of MIPs in SPE is called molecularly imprinted solid-phase extraction (MISPE). MISPE has several advantages over traditional sorbents, including high selectivity and performance. A simple solid-phase microextraction technique using MIPs was prepared for the analysis of organophosphate flame retardants, as shown in Fig. 9.3, and bisphenols. The synthesized MIPs were characterized by several techniques such

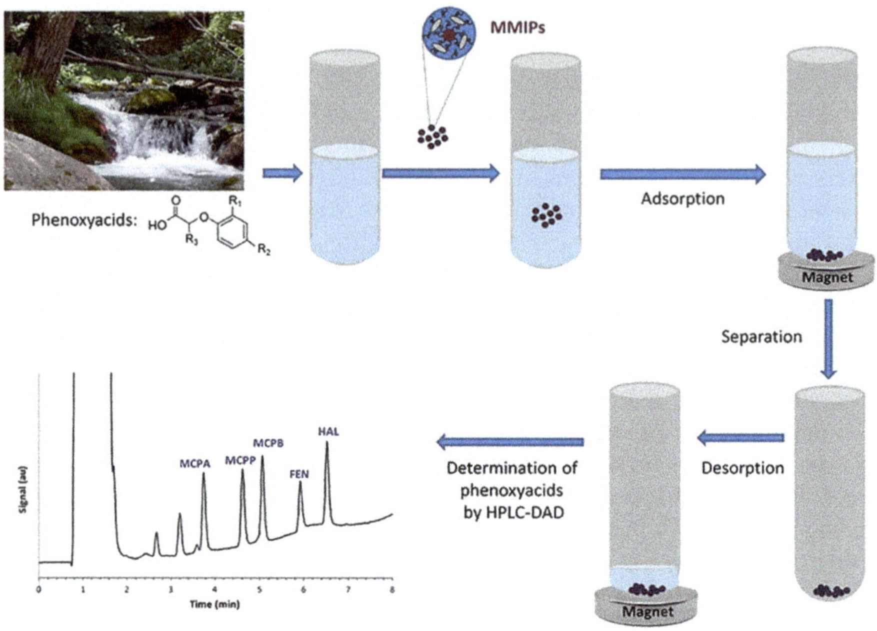

Fig. 9.2 SPE application with magnetic MIPs in environmental waters for phenoxyacids was followed by high performance liquid chromatography coupled with a diode array detector (HPLC-DAD). Reprinted from Meseguer-Lloret et al. (2022a) with permission from Elsevier

as scanning electron microscopy (SEM) and gas chromatography flame photometric detector (Chen et al. 2020a; Zambonin and Aresta 2021; Hu et al. 2015).

9.3.3 Dispersive Solid-Phase Extraction (DSPE)

DSPE is a sample preparation technique that combines the principles of SPE with the simplicity and efficiency of dispersive liquid–liquid extraction (DLLE). DSPE is a sorbent-based technique widely applied in sample preparation for both sample clean-up and analyte preconcentration. DPSE involves the use of a sorbent material, typically a mixture of primary secondary amine (PSA) and magnesium sulfate ($MgSO_4$) or other combinations of sorbents, to extract and clean up analytes from complex sample matrices. The method is particularly popular in multi-residue analysis, where various pesticide residues need to be simultaneously extracted from a single sample. DSPE is also known as QuEChERS (Quick, Easy, Cheap, Effective, Rugged, and Safe), which is a specific version of the DSPE method primarily used for the analysis of pesticide residues in food and agricultural samples.

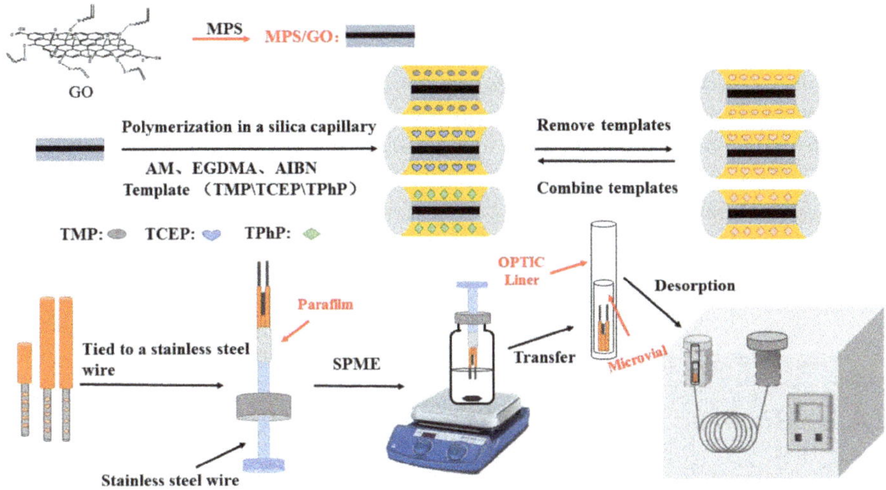

Fig. 9.3 Schematic diagram of for determination of trace level of various organophosphate flame retardants (OPFRs) in environmental matrices. Reprinted from Chen et al. (2020a) with permission from Elsevier

DSPE is a method of sample preparation that involves the use of a sorbent material, such as MIPs, to extract and enrich analytes from water samples. Thus, sample preparation techniques using DSPE with MIPs are becoming increasingly popular for the detection and removal of emerging pollutants from environmental water (Pichon et al. 2020). DSPE is widely used for the analysis of phthalate esters in aqueous samples and triazole in pesticides (Bakhshizadeh Aghdam et al. 2022; Marzi Khosrowshahi et al. 2022). It is known for its simplicity, speed, and cost-effectiveness, making it a preferred technique for laboratories dealing with a large number of samples. A study conducted by Huang et al. (2018) optimized DSPE conditions such as the number of MIPs, the extraction time, the type and volume of desorption solvents to achieve good recoveries and low limits of detection and quantification. Even though DSPE has significance over classic SPE, such as rapidness, easiness, applicability, effectiveness, and affordability, the phase separation step can be difficult (Bakhshizadeh Aghdam et al. 2022).

9.3.4 Stir Bar Sorptive Extraction (SBSE)

SBSE was introduced by Baltussen et al. in 1999 as an alternative to SPME and became one of the most innovative sample preparation techniques used in analytical chemistry to extract and preconcentrate analytes from complex sample matrices. SBSE is a sample preparation technique that uses a small stir bar coated with a sorbent phase to extract analytes from liquid samples. The stir bar is typically made of glass or polytetrafluoroethylene (PTFE) and is coated with a thin layer of sorbent material, such as polydimethylsiloxane (PDMS) or other stationary phases (Baltussen et al. 1999). SBSE is particularly useful for the analysis of volatile and semi-volatile compounds in liquid samples, which can be considered as a limitation (Nogueira 2015). Its applications have been extended to the analysis of non-volatile analytes and solid and liquid samples. This extraction procedure is based on a magnetic stir (0.5–1 mm thickness) coated with MIPs and its application for the analysis of environmental estrogen in water and plastic samples, as well as the extraction of cis-diol-containing catechol (Xu et al. 2014; Chen et al. 2020b). SBSE offers advantages in terms of simplicity, sensitivity, robustness, cost-effectiveness, and environmental friendliness.

9.3.5 Magnetic Solid-Phase Extraction (MSPE)

MSPE is a sample preparation technique that uses magnetic nanoparticles functionalized with specific sorbent materials to extract and preconcentrate analytes from a sample matrix (Majd et al. 2021; Shan et al. 2022). Magnetic nanoparticles can be easily separated from the sample using a magnet, providing an efficient and straightforward way to extract analytes (Jiang et al. 2019). MSPE is considered a greener alternative to traditional extraction methods, as it requires minimal or no organic solvents, reducing the environmental impact of sample preparation. MSPE offers several advantages, including rapid extraction, ease of use, and the ability to handle large sample volumes. It has found applications in various fields, including environmental analysis. These techniques continue to be developed and adopted for various applications in analytical chemistry, benefiting researchers and analysts with their efficiency and sensitivity. The combination of MSPE with MIPs provides a selective and efficient method for detecting and removing lead ion from water samples with humic acid modified magnetic nanoparticles prior to its fame atomic absorption spectrometric detection (Shan et al. 2022). Overall, the use of MSPE with MIPs provides a reliable and efficient method for the detection and removal of emerging pollutants from environmental water, with improved selectivity and high adsorption capacity due to the specific recognition sites of MIPs.

9.4 Applications of MIPs-Based Sample Preparation for Emerging Pollutants Detection

MIPs can be categorized as non-covalent or covalent. The type of bonds that are formed between the template and the monomer determines this distinction. (Kaya et al. 2023). Although fully covalent MIPs can be utilized in analytical chemistry, certain conditions are required to enable the binding of analytes to them covalently

(Alexander et al. 2006). Although semi-covalent solutions can eliminate some of these issues, they may not be able to provide the ideal combination of template and monomer (Chen et al. 2011).

9.4.1 Covalent MIPs

Table 9.3 presents a variety of analytes, including erythromycin, hydroquinone, catechol, perfluorooctanoic acid (PFOA), perfluorooctanesulfonic acid (PFOS), 17 β-estradiol, and methotrexate (Ayankojo et al. 2020; Hu et al. 2022; Wang et al. 2023d; Pierpaoli et al. 2023; Lu et al. 2022; Biyana Regasa and Nyokong 2022; Jara-Cornejo et al. 2023). The analytical methods employed range from differential pulse voltammetry (DPV) to cyclic voltammetry (CV) and ultra-high-performance liquid chromatography-mass spectrometry (UHPLC-MS/MS). These methods reflect the diversity of target compounds and the need for different sensitivity levels and selectivity. The matrices under investigation include lake water, rainwater, dyeing wastewater, river water, wastewater, tap water, treated wastewater, and drinking water. The choice of matrix corresponds to the potential sources of these analytes, and the diversity reflects the practical applicability of MIPs in real-world environmental monitoring.

The methods for creating MIPs vary, with electropolymerization being a common technique. Different potential ranges, solvents, and scan rates are utilized, tailored to the specific analyte and MIP composition. Ayankojo et al. (2020) in their study synthesized a molecularly imprinted polymers-based sensor for the detection of erythromycin combined with a screen-printed electrode (ERY-MIP/SPE) by electropolymerizing erythromycin as the template onto gold working electrode and m-phenylenediamine (MPD) in phosphate buffer saline (PBS) as porogen solvent at a constant potential of 0.63 V. They managed to obtain a low limit of detection (LOD) of 0.1 nM with a higher recovery of erythromycin at 91–102% from lake, rain, and dyeing wastewater (Ayankojo et al. 2020). Meanwhile, Hu and colleagues (2022) utilized a carbon nanotube/glassy carbon electrode (CNT-GCE) as a functional monomer in synthesizing dual template molecularly imprinted polymers-carbon nanotube/glassy carbon electrode (MIP-CNTs-GCE) to detect hydroquinone and catechol from river water. They used 0.5 M sulfuric acid (H_2SO_4) electrolyte to complete the electropolymerization of both hydroquinone and catechol onto CNT-GCE at 50 mV s^{-1} (Hu et al. 2022). In two studies by Pierpaoli et al (2023) and Lu et al. (2022), PFOS was selected as target compound. However, Pierpaoli et al. used carbon nanowall imcorporated into a molecularly imprinted polymers sensor for detecting PFOS (CNW/MIP PFOS) sensor paired with ultra-high performance liquid chromatography coupled with tandem mass spectrometry (UHPLC-MS/MS), while Lu et. al. used MIP and gold nanostar (AuNS) modified voltammetric sensor (MIP-AuNS) in detecting the presence of PFOS (Pierpaoli et al. 2023; Lu et al. 2022).

Wang et al. (2023d) and Jara-Cornejo et al. (2023) utilized pyrrole as a monomer in the synthesis of MIPs to detect perfluorooctanoic acid (PFOA) and methotrexate

Table 9.3 Compilation of studies using covalent and non-covalent MIPs in detecting contaminants from various sources

Covalent MIPs

Analyte	Analytical method	Matrix	Template/monomer/porogen solvent	Polymerization condition (imprinting approach, solvents, temperature/time)	Template eluting (solvents/methods)	LOD/LOQ	Recovery (%)	Imprinting factor	Matrix effect (%)	References
Erythromycin	MIPs-PEC/DPV	Lake, rain and dyeing wastewater	ERY/gold working electrode and MPD/PBS	Electropolymerization at a constant potential of 0.63 V/PBS	Wash with etOH	0.1 nM/ 0.4 nM	91.0–102.0	–	–	Ayankojo et al. (2020)
Hydroquinone and Catechol	MIP-CNTs-GCE/ CV and DPV	River water	Hydroquinone and Catechol/ CNT-GCE/0.5 M H_2SO_4 electrolyte	Elecropolymerization at 50 mV s^{-1}	Immersed in etOH/water solution (1:1, v/v) for 5 min	3.1 μM and 3.5 μM	100.9–102.6 and 103.0–103.6	–	–	Hu et al. (2022)
PFOA	MIP-Co/ Fe@CNF/DPV	Wastewater	PFOA/pyrrole/ MeCN and 0.1 M $NaClO_4$	Electropolymerization at the potential range between – 0.3 V to 0.9 V at 100 mV s^{-1} for 20 cycles	Immersion and elution into NaOH: acetone (1:1, v/v) for 20 min	1.073 × 10^{-9} M	96.0–100.0	–	–	Wang et al. (2023d)
PFOS	CNW/MIP PFOS sensor/ UHPLC-MS/MS	Tap water and treated wastewater	PFOS/BDD and CNW electrode and oPD/PS and MeOH (2:1, v/v)	Electropolymerization between 0 and 0.8 V at a scan rate of 50 mV s^{-1}	Rinsed with distilled water, then kept in a solution of NaOH/MeOH (1:1 v/v) for 20 min, followed by three washes with distilled water	1.2 μg L^{-1}	70.0–110.0	–	–	Pierpaoli et al. (2023)

(continued)

Table 9.3 (continued)

Covalent MIPs										
Analyte	Analytical method	Matrix	Template/monomer/porogen solvent	Polymerization condition (imprinting approach, solvents, temperature/time)	Template eluting (solvents/methods)	LOD/LOQ	Recovery (%)	Imprinting factor	Matrix effect (%)	References
PFOS	MIP and AuNS modified voltammetric sensor	Drinking water	PFOS/AuNS and GCE and oPD/0.1 M acetate buffer: MeOH (2:1, v/v)	Electropolymerization the potential scanning between 0 and 1.0 V versus Ag/AgCl at a scan rate of 50 mV s^{-1}	Rinsed with DI and subsequently immersed into the meOH/water (1:1, v/v) template removal solution for 20 min under mild stirring	0.015 nM/0.041 nM	–	–	–	Lu et al. (2022)
17 β-estradiol	MIP-GO-AgNP	River water	17 β-estradiol/AgNP/GCE modified with GO and imidazole/PBS: acetone (1:1, v/v) containing 0.1 M $NaClO_4$	Electropolymerization between 0.0 and 2.5 V at 200 mVs^{-1} scan rate for 10 cycles	Rinsed with PBS: acetone (4:1, v/v) solution	3.01 fM/10.03 fM	–	–	–	Biyana Regasa and Nyokong (2022)
Methotrexate	Sensor based on MIP, Polypyrrole film and MWCNT/GCE/DPV	River water	Methotrexate/pyrrole and MWCNT/GCE/0.1 mol L^{-1} lithium perchlorate	Electropolymerization between potential range of – 0.4 to 1 V with scan rate of 50 mV s^{-1}, and 25 electropolymerization cycles	Immersed in a sodium phosphate buffer solution of pH 10	2.7×10^{-9} M	102.0–104.0	–	–	Jara-Cornejo et al. (2023)

(continued)

Table 9.3 (continued)

Non-covalent MIPs										
Analytes	Analytical method	Matrix	Template/monomer/porogen solvent/initiator/cross-linker	Polymerization condition (Imprinting approach, solvents, temperature/time)	Template eluting (solvents/methods)	LOD/LOQ	Recovery (%)	Imprinting factor	Matrix effect (%)	References
Diazepam	MIP-SBSE/UHPLC-MS/MS	Natural water	Diazepam/MeCN/AIBN/EGDMA	Bulk polymerization/thermal polymerization in MeCN at 60 ℃ for 24 h	Stirring in 5 mL of 5% (v/v) acetic acid in MeOH for 10 min	15.0 and 50.0 μg L^{-1}	84.0–102.0	19	–	Sorribes-Soriano et al. (2023)
Chlorobenzenes	NTED-MIP/GC-FID	Industrial wastewater, tap water and river water	Mixed chlorobenzenes/[VAFMIM]Cl ionic liquid/AIBN/EGDMA	Precipitation polymerization in MeOH:water at 60 ℃ for 12 h	Soxhlet apparatus using MeOH-acetic acid mixture solvent (9:1) for 36 h	7.3–13.0 ng L^{-1}/0.010–0.030 μg L^{-1}	80.01–97.10	1.79–1.96	–	Yue et al. (2023)
2,4-dinitroaniline	MIP@MA/IT-SPME/HPLC-DAD	River, sewage and reservoir waters were	2,4-dinitroaniline/vinylimidazole/1-propanol and 1,4-butanediol/EGDMA/AIBN	'In-situ' polymerization in 1-propanol/1,4-butanediol at 70 ℃ for 12 h	MeCN/acetic acid (9:1)	0.06 μg L^{-1}/0.19 μg L^{-1}	82.1 –111	3.1	–	Song et al. (2023)

(continued)

Table 9.3 (continued)

Non-covalent MIPs										
Analytes	Analytical method	Matrix	Template/ monomer/porogen solvent/initiator/ cross-linker	Polymerization condition (Imprinting approach, solvents, temperature/time)	Template eluting (solvents/ methods)	LOD/LOQ	Recovery (%)	Imprinting factor	Matrix effect (%)	References
Phenoxy acid herbicides	MMISPE/ HPLC-DAD	Water samples from tap, fountain, irrigation well and marsh	4-chloro-2-methylphenoxy-acetic acid/ 4-vinylpiridine and vinylized magnetite nanoparticles/ MeCN/EGDMA/ AIBN	Surface imprinting polymerization in MeCN at 65 °C for 24 h	Magnetically washed with methanol: acetic acid solution (90:10, v/v)	0.33–0.71 μg L^{-1}/ 1.1–2.4 μg L^{-1}	77–98	1.76	–	Meseguer-Lloret et al. (2022b)
BPA	MMIPSPE/UV-spectrophotometer	Tap water, mineral water, and wastewater	BPA/MAA/ MeCN/water/ EGDMA/ Ammonium persulfate	Radical polymerization in MeCN:water at 70 °C for 2 h	Soxhlet extraction using MeOH/ acetic acid (9:1)	0.03 μg mL^{-1}/ 0.1 μg mL^{-1}	80.0–103.0	1.71	–	Karrat and Amine (2021)
Organophosphorus pesticides	Micro gel thin film MIPs/ LC-MS/MS	Water	2-[diethoxy(sulfanylidene)-l-phosphanyl]amino acetic acid/ 1-octanol/ EGDMA/DMPA	Radical polymerization in 1-octanol under UV light for 30 min	The thin films were washed with MeOH: acetic acid (9:1)	0.001 ng mL^{-1}/ 0.002 ng mL^{-1}	81.9–112	1.5–42.8	–	Azizi et al. (2021)
Pharmaceuticals	MASE-MIP/ LC-qTOF/MS	Surface water	Venlafaxine/ p-vinyl benzoic acid/dimethyl sulfoxide/ EGDMA/AIBN	Radical polymerization in dimethylsulfoxide at 70 °C	Washing with MeOH	0.09 – 0.20 ng mL^{-1}/ 0.31 – 0.69 ng mL^{-1}	38–91	–	–	Khulu et al. (2022a)

(continued)

Table 9.3 (continued)

Non-covalent MIPs										
Analytes	Analytical method	Matrix	Template/monomer/porogen solvent/initiator/cross-linker	Polymerization condition (Imprinting approach, solvents, temperature/time)	Template eluting (solvents/methods)	LOD/LOQ	Recovery (%)	Imprinting factor	Matrix effect (%)	References
BPA	MISPE/HPLC-PDA	River and moat waters, commercial bottled water, waters from a cooling machine and tap water	BPA/isopropanol/3-APTES and phenyl TES	Sol–gel matrix imprinting/isopropanol for 4 h	Washing with MeOH	0.015 ng μL^{-1}/0.045 ng μL^{-1}	85.4–96.1	0.28–1.11	–	Kalogiouri et al. (2020)
Environmental estrogens	Fiber array extraction method based on MIPs/HPLC-DAD	River and lake water	Multi-template- 8 template molecules/acrylamide/MeOH/EGDMA/AIBN	Polymerization in constant temperature at 60 ℃ for 24 h	Washing with MeOH	0.003–0.016 µg L^{-1}	80.8 –114.1	1.46	–	Wang et al. (2020)
Organophosphorus pesticides and their oxon-derivatives	MIP-SPE/HPLC-UV	Tap, river, sewage water, municipal wastewater treatment plant	Different pesticides/MAA/MeCN/EGDMA/AIBN	Polymerization at 60 °C for 18 h	MeCN and MeOH	0.07–0.12 µg L^{-1}/0.23–0.41 µg L^{-1}	79.0–104.0	1.4–2.2	–	Arias et al. (2020)
Polycyclic aromatic hydrocarbons	MMIP-DSPE/APGC-MS/MS	Produced water and river water				1.0–100.0 pg mL^{-1}/2.0–200.0 pg mL^{-1}	4.5–97.0	–	–	Azizi et al. (2020)

(continued)

Table 9.3 (continued)

Non-covalent MIPs										
Analytes	Analytical method	Matrix	Template/ monomer/porogen solvent/initiator/ cross-linker	Polymerization condition (Imprinting approach, solvents, temperature/time)	Template eluting (solvents/ methods)	LOD/LOQ	Recovery (%)	Imprinting factor	Matrix effect (%)	References
Aminoglycoside antibiotics	DMISPE/ LC-MS/MS	River water and reservoirs	Raffinose/MAA/ MeOH/TRIM/ AIBN	Precipitation polymerization/60 °C/ 24 h	Soxhlet extraction with 10% acetic acid in MeOH	0.006–0.6 ng mL^{-1}/ 0.02 – 1.8 ng mL^{-1}	70.8–108.3	–	–	Zhang et al. (2020)
Imidazole fungicides	DMISPE/ HPLC-DAD	River water	DCE/MAA/ EGDMA/MeCN/ AIBN	Bulk polymerization in MeCN at 4 °C for 2 h	Soxhlet extraction in MeOH:acetic acid (9:1)	0.023–0.031 µg L^{-1}/	84.2–95.0	7.0–10.9	–	Sun et al. (2019)

due to high stability and electrical conductivity properties, which resulted in excellent recovery for both PFOA and methotrexate at 96.0–100.0% and 102.0–104.0%, respectively (Wang et al. 2023d; Jara-Cornejo et al. 2023). On the other hand, Biyana Regasa and Nyokong (2022) fully utilized graphene oxide (GO) and silver nanoparticles (AgNP) to introduce additional functional to produce the MIP functionalized with a graphene oxide and silver nanoparticles composite film (MIP-GO-AgNP) for detecting 17 β-estradiol. They added GO to increase the surface area of the composite film, while AgNP acts as a catalyst and charge carrier. Moreover, they also used imidazole, which acts as a p-type-electron acceptor during the adsorption process between the adsorbent and 17 β-estradiol, enabling them to achieve an excellent limit of detection (LOD) for 17 β-estradiol at 3.01 fM (Biyana Regasa and Nyokong 2022).

9.4.2 Non-covalent MIPs

The non-covalent method is commonly used for imprinting. This type of process is advantageous due to the quick removal and relinking of templates. In the non-covalent method, the templates and monomers come together to form complexes through non-covalent interactions such as hydrogen bonding, electrostatic interactions, van der Waals forces, and π–π interactions. Similar interactions are required for analyte rebinding. Non-covalent MIPs are the most popular MIPs in analytical applications despite a time-consuming optimization strategy (Lofgreen and Ozin 2014).

The determination and quantification of targeted pollutants in environmental samples, such as water samples, soils, and air ecosystems are necessary to understand their fate and transport. Generally, the application of MIPs-based sample preparation for EPs includes targeted pharmaceutical compounds, pesticides, and endocrine disruptor compounds. They are present in the environment at very minute concentrations in the environmental matrices, which further challenges the analysis. Furthermore, the maximum residue limits (MRLs) for certain pollutants are at very low ppb levels or even lower. Thus, new selective and specific sorbent materials are required to improve the sensitivity and the detection limits of the targeted pollutants. MIPs-based sample preparation techniques have been developed for the quantification of pharmaceuticals such as diazepam, a group of pharmaceutical substances, and aminoglycoside antibiotics (Khulu et al. 2022a; Sorribes-Soriano et al. 2023; Zhang et al. 2020). Sorribes-Soriano et al. developed a stir bar sorptive extraction combined with ion mobility spectrometry (IMS) or UHPLC-MS/MS. MIP-coated stir bar was developed as a simple, cost-effective, and selective tool to determine diazepam in natural water. In this work, the PTFE-based magnets were treated for surface vinylization and were further attached covalently with the MIP. The MIP was prepared through bulk polymerization and the selectivity demonstrated low extraction efficiency for diazepam's analogues except for medazepam. The LOD obtained from UHPLC-MS/MS (1.2 ng L^{-1}) was more sensitive than the IMS (1.2 μg L^{-1}) (Sorribes-Soriano et al.

2023). Recently, Khulu et al. have proposed a new membrane-assisted solvent extraction and MIP as the two-way technique for the extraction of five pharmaceuticals in surface water. The method was then optimized through a central composite design for selective extraction of the targeted compounds. The approach of using response surface methodology has simplified the overall optimization process. In this study, the MIP was synthesized through radical polymerization in dimethyl sulfoxide using venlafaxine as the template. By using venlafaxine, the MIP was capable of simultaneously extracting the four other pharmaceuticals, while the membrane was used to prevent the co-extraction of bigger molecules. However, low extraction recoveries were obtained for etilefrine and methocarbamol due to the poor recognition by the synthesized MIP and slow mass transfer rate across the membrane bag (Khulu et al. 2022b). The structural analogues of analytes, named the dummy templates were prepared by Zhang et al. (2020) and was employed as an alternative sorbent in SPE method combined with HPLC-DAD for the quantification of aminoglycoside antibiotics in river water and reservoirs. The preparation of dummy templates has overcome the shortcomings in the preparation of traditional MIPs related to the high cost of required standard products and persistent template leakage during the sample preparation that led to inaccurate quantification of trace analytes (Guo et al. 2018). The obtained dummy MIPs exhibited uniform sizes and high selective adsorption capacity for the six aminoglycoside antibiotics that were suitable for the SPE method and avoiding the defect of template leaking during sample pretreatment. LOD as low as 0.006 ng mL^{-1} was obtained using the developed method. The sorbent can be reused up to 15 times, recycling use of the cartridge, and the recoveries of the antibiotics were more than 92%.

Other than pharmaceutical compounds, MIPs have also used in sample preparation methods for the quantification of pesticides such as phenoxy herbicides, organophosphorus pesticides (OPPs), and imidazole fungicides (Arias et al. 2020; Azizi et al. 2021; Sun et al. 2019). LODs in the range of 0.001–0.7 ng mL^{-1} were obtained for all the studied pesticides. The developed methods of MIP-based sample preparation exhibited recoveries of more than 76%. A magnetic SPE has been developed by Meseguer-Lloret et al. to simplify the operations in classical SPE. The development of a magnetic MIP adsorbent as magnetic solid phase extraction (MSPE) in this study not only exhibited excellent selectivity towards phenoxy herbicides but also can be quickly separated and easily recovered from the samples by an external magnet, thus avoiding the blockage issue in SPE (Meseguer-Lloret et al. 2022b; Wang et al. 2017). 4-Chloro-2-methylphenoxy was used as a template in the surface imprinting polymerization, and the adsorbent was characterized comprehensively using scanning electron microscopy (SEM), Fourier transform-infrared spectroscopy (FTIR), and vibrating sample magnetometer (VSM). It is worth noting that the adsorbent can be reused more than 65 times without any significant loss of extraction capacity. A sensitive method based on a thin film MIPs have been fabricated by Azizi et al. using a simple drop-casting technique to produce a low-cost device for the selective, efficient, and reproducible extraction of OPPs. The MIP was optimized with the aid of a pseudo-phase diagram to obtain the micro-gel polymer. The method is applicable for on-site sampling since it avoids preconditioning, carry-over effects, and exhaustive cleaning procedure. The MIP was synthesized through radical polymerization and under UV light for 30 min. An LOD of 0.001 ng mL^{-1} was obtained using the developed method (Azizi et al. 2021). For the determination of endocrine disrupting compounds, MIPs based sample preparation methods were focused on the analysis of bisphenol A (BPA) (Karrat and Amine 2021; Wang et al. 2017) and its analogues, parabens and phthalates compounds (Wang et al. 2020). BPA imprinted sol–gel silica based was synthesized by Kalogiouri et al. via a matrix imprinting approach. Two functional sol–gel precursors were used, namely phenyl triethoxysilane and 3-aminopropyl triethoxysilane. The application of tetraethyl orthosilicate as the cross linker has resulted in sponge-like architecture that enhance the rapid removal of template molecule. Optimization study was conducted using a central composite design with response surface methodology. The imprinting factor of BPA of 6.58 was obtained using the developed adsorbent. The LOD and limit of quantification (LOQ) of 0.015 ng μL^{-1} and 0.045 ng μL^{-1} were obtained, respectively (Arias et al. 2020). A fiber array extraction method based on MIPs as fiber coating was developed for the simultaneous enrichment of bisphenols, parabens, and phthalates from river and lake water (Wang et al. 2020). The prepared MIP fiber array could effectively improve the adsorption, analysis efficiency, and detection sensitivity of the selected analytes. The steps for the preparation of these fibers include (i) eight MIP fibers were prepared using eight template standard molecules in this study; (ii) the fiber array based on MIPs was constructed and consisted of MIP-coated fibers prepared by agitation extraction and ultrasonic desorption in different modes; (iii) the desorbed solution was analyzed by HPLC system. The proposed extraction method coupled with HPLC analysis demonstrated satisfactory recoveries of 80.8–114.1% and good precisions of 0.8–12.4%. The LOD as low as 0.003 $\mu g\ L^{-1}$ was obtained from this work. Other works on MIPs based sample preparation were also developed for other pollutants such as needle-type extraction device based on MIP, magnetism assisted in-tube solid phase microextraction and magnetic MIP based on DSPE were developed for chlorobenzenes, 2,4-dinitroanile, and polycyclic aromatic hydrocarbons, respectively (Wang et al. 2023b; Azizi et al. 2020; Yue et al. 2023). In summary, the research conducted on the applications of MIP-based sample preparation for detecting emerging pollutants predominantly utilized water samples, encompassing tap water, river water, lake water, and wastewater samples. In the creation of MIPs, researchers employed either dummy templates or the specific target analytes pertinent to their studies. In the preparation of MIP adsorbents, MAA, EGDMA, and azobisisobutyronitrile (AIBN) were consistently chosen as the monomer, crosslinker, and initiator, respectively.

9.5 Comparison of Method-Based MIPs to Conventional Method for Emerging Pollutants Detection

Before conducting qualitative and quantitative studies, it is important that the sample preparation process is performed properly to increase the sensitivity and selectivity of the analysis. This step can also help minimize the impact of environmental contaminants on the study. In addition, it is beneficial to use sample preparation methods to pre-concentrate the EPs. With that knowledge in mind, the benefits and goals of sample preparation can be described as follows; (i) pre-concentration of targeted analytes to the obtain the lowest LODs, (ii) removing the matrix's interference effect and improving accuracy and selectivity, (iii) stabilisation of the sample measurement medium through reconstitution with the required solvent and (iv) cleaning the samples to prevent instrument damage (Mohiuddin et al. 2021; Nasiri et al. 2020). SPE, its related techniques including SPME and MSPE, are currently the most often used methods for sample preparation in the determination of the emerging pollutants in environmental water samples (Jafari and Ghani 2023; Mostafa et al. 2023; Sodré and Sampaio 2020; Yang et al. 2023b). Although these standard analytical techniques for the determination of emerging pollutants and monitoring are sensitive, efficient, and possess excellent reproducibility, they are impacted by challenges that affect their use in many applications. Thus, their advantages and disadvantages including encountered challenges as compared to the MIP-based sample preparation techniques were highlighted in the following subsections. The comparisons were carried out based on the adsorbents used, extraction procedures and analytical performances.

9.5.1 Comparison of Conventional Sample Preparation Techniques with MIP-Based Techniques

Adsorbents

The synthesis of adsorbents for SPE and MIP-SPE involves different approaches due to the distinct nature of the extraction techniques. In SPE, the adsorbents used for the extraction can vary depending on the properties of the targeted analytes and the extraction modes. For conventional SPE, the typical adsorbents are made of mesoporous materials, silica-based, carbon-based, polymer-bed or mixtures of few types of the adsorbents (Mostafa et al. 2023). Most of the aforementioned adsorbents are available commercially and synthesized using well-established methods, including the functionalization of silica particles with hydrophobic or polar groups. However, the nature of conventional SPE usually suffers from several drawbacks including high consumption of solvents, labour intensive, co-extracted matrix interferences, low selectivity and compatibility (Li et al. 2021b, 2022).

In order to improve the SPE's mass transfer and feasibility, a modified SPE adsorbent was made using magnetic nanoparticles. This method was then applied to make a type of SPE known as MSPE (Jafari and Ghani 2023; Yang et al. 2023b; Li et al. 2022). The use of magnetic adsorbents enables an easier application without the need for centrifuge or filter steps, relying solely on magnetic interactions. MSPE also overcomes high pressure and blockage issues in packing (El-Deen and Hussain 2023). MSPE is capable of providing rapid and selective extraction, and its ease of magnetic separation makes it a valuable tool for researchers and analysts seeking to extract and concentrate analytes, particularly for the extraction of various emerging pollutants in environmental samples. Although MSPE can be tailored to exhibit selectivity for particular analytes, selectivity remains a major issue in MSPE. Thus, MIPs can also be used to achieve selectivity in MSPE. The development and use of magnetic sorbent materials are two of the most crucial steps. Since they can adsorb analyte and be desorbed using a magnet, the addition of MIPs to the process provides a significant advantage. Due to the unique properties of MIPs, they can be used as both a fast and easy extraction agent (Liu et al. 2022).

The synthesis of adsorbents of MIP-SPE involves different approaches due to the distinct nature of the polymeric adsorbents. The MIP-SPE adsorbents are synthesized when monomers are polymerized in the presence of a template molecule (the target analyte) and useful cross-linking agents. Immediately upon polymerization, the template is eliminated, leaving behind cavities or recognition sites that have a strong affinity for the target analyte (Wang et al. 2017; Sodré and Sampaio 2020). MIP-SPE sorbents are custom-tailored for certain analytes or groups of analytes, allowing selectivity and specificity to be adjusted to suit the needs of a given application. An important step in MIP synthesis is template removal, which guarantees that the sorbent produced is specific for the intended analyte (Fu et al. 2022b; Steinhaeuser et al. 2022). MIP-SPE is a valuable tool for targeted sample preparation because of its versatility and capacity to be tailored for different analytes. The usage of MIPs as sorbents during the preconcentration and clean-up stages of sample preparation is made possible by their exceptional selectivity and recognition capabilities (Fu et al. 2022b). The target analyte can be quickly and easily separated from other molecules in the matrix using reusable MIP-based adsorbents. However, due to partial imprinting of the template, the length of time and solvents required to remove the templates, and the homogeneity of the MIP adsorbents, the synthesis of MIP adsorbents for any sample preparation techniques is laborious.

In conclusion, although MIP-SPE adsorbents are specifically made through a molecular imprinting process targeted to certain analytes, SPE adsorbents are made using pre-existing, commercially accessible sorbents or cartridges. While MIP-SPE requires more complex synthesis steps than SPE, it has the advantage of excellent selectivity and specificity.

Extraction Procedures

For conventional SPE, the extraction procedure involves selecting a suitable SPE cartridge with an adsorbent compatible with the properties of the analytes and sample matrix. The adsorbent is treated with an appropriate solvent before sample loading. After loading, the adsorbent is washed to remove impurities, and the target analytes are washed with the appropriate solution. The eluate containing the isolated analytes is collected for subsequent analysis (Fontanals et al. 2017; Sun et al. 2014). For the MIP-SPE technique, the extraction procedure is similar to the conventional SPE, except for the adsorbents. Thus, it also inherited the same drawbacks, such as clogged cartridges, the use of large amount of solvents, co-eluted matrix, and interference (Meseguer-Lloret et al. 2022b). In order to address these issues, MIPs can be used as a magnetic sorbent material, which provides a significant advantage due to its fast and easy separation. But, their high resistance profile, stability, and selectivity still allow them to be utilized as sorbents for SPE (Fu et al. 2022b; Steinhaeuser et al. 2022).

Meseguer-Lloret et al. (2022a; b) studied the selective extraction of toxic environmental pollutants, specifically phenoxy acid herbicides using magnetic MIPs and SPE from fountain, irrigation, well, and marsh water samples. In addition to their selectivity and fast separation, magnetic MIPs can also be used in sample preparation due to their exceptional binding affinity. They can be recovered from samples by magnets and overcome sorbent blocking. During the polymer synthesis, 4-chloro-2-methylphenoxyacetic acid (MCPA) was used as a template along with 4-vinylpiridine (functional monomer), EGDMA (cross-linker), and acetonitrile (porogenic solvent). The synthesized magnetic MIP and optimized SPE conditions were used for the extraction of selected herbicides from environmental water samples followed by the HPLC-DAD analysis (Fig. 9.4). The linear range was found between 0.003 and 1.0 $\mu g\ L^{-1}$, with LOD values ranging between 0.33 and 0.71 $\mu g\ L^{-1}$. The recovery studies in environmental water samples yielded good recovery percentage values between 79.1 and 96.0%. This study is the first to utilizes magnetic MIPs for the simultaneous and selective determination of phenoxy acid herbicides (Sun et al. 2019).

Extraction Performances

SPE typically relies on commercially available sorbents with predefined properties. While these sorbents can offer selectivity for specific analytes, they may not always meet the requirements for emerging pollutants with diverse chemical structures. In contrast, MIP-SPE is specifically designed to provide exceptional selectivity. In terms of extraction efficiency, MIP-SPE often outperforms conventional SPE techniques. The tailored binding sites in MIPs provide strong and specific interactions with the target analytes, resulting in higher extraction efficiencies and lower detection limits. On the other hand, SPE may capture a broader range of compounds, leading to less efficient extraction and potential interference from co-extracted matrix components. Both SPE and MIP-SPE can be applied to a wide range of emerging pollutants in water samples. However, MIP-SPE's superior selectivity makes it particularly well-suited for challenging analytical tasks where the precise extraction of target compounds is essential. SPE may be preferred for routine analyses with less stringent selectivity requirements or when broad-spectrum extraction is desired.

Fig. 9.4 Representative whole water matrix and SPE procedure. **a** A representative 1 L sample of the representative water matrix, **b** 50 mL of SPE disk eluate from a 1 L sample of representative water matrix, **c** SPE disk eluate after evaporation to 1 mL, **d** SPE disk eluate diluted in 10 mL of H_2O, **e** Diluted SPE disk eluate applied to MISPE column, **f** 3 mL of MISPE eluate of a 0.1 L sample of representative water matrix, **g** HLB-SPE eluate of a 0.1 L sample of representative whole water matrix, **h** 1 mL final sample after MISPE (left) and HLB-SPE (right), respectively (Steinhaeuser et al. 2022)

Fu et al. (2022a, b) conducted a comparative study between SPE and MIP-SPE performance for cathinones (mephedrone, butylone, flephedrone, methedrone and methylone) in river water samples. In terms of the recovery, the synthesized MIPs adsorbent (pseudoephedrine as template) was comparable with Oasis WCX and Oasis MCX ion exchange based-adsorbents. However, concerning matrix effects (ME), lower ME values between − 5.5 and 1.8% were achieved. When the Oasis WCX adsorbent was used to extract a group of synthetic cathinones from river water samples, signal suppression (from − 16 to − 28%) was observed for flephedrone, methylone, methedrone, and butylone despite washing with 5 mL of MeOH (Li et al. 2021b). In another study, signal suppression ranging from − 13 to − 17% was observed for flephedrone, methylone, methedrone, mephedrone and butylone in river water samples when the Oasis MCX sorbent was applied (El-Deen and Hussain 2023). The lower ME values in the present study indicate high selectivity of the MIPs towards cathinones, as the river water samples were efficiently cleaned-up (Fu et al. 2022b).

In other study by Steinhaeuser et al. (2022), a diligent comparison of SPE on a hydrophilic-lipophilic balanced (HLB) phase to SPE on an estrogen-specific MIPs phase for ultra-trace levels of the analytes (1–10 ng L^{-1}) was carried out. Additionally, a two-step procedure combining HLB SPE disks followed by MISPE is also evaluated. In Fig. 9.4, the co-purified dissolved organic carbon after HLB disk and cartridge SPE was clearly visible as a yellowish to deep brown discoloration of the respective eluate. The final sample of the HLB-SPE (Fig. 9.4h) retained a yellowish tint, and small particles of aggregated humic acids were visible on the surface, whereas samples were always perfectly clear after the MISPE and the two-step procedure. While the HLB sorbent retains the analyte almost quantitatively, co-elution of matrix components leads to significant matrix effects that reduce measurement sensitivity and thus lower absolute recoveries. In contrast, the MIP sorbent was far superior in terms of matrix removal, but a considerable portion of analytes was lost. A combined method using an HLB step followed by MISPE for matrix removal proved to be a good compromise, offering decent recovery at very low matrix load. A reduced matrix load not only benefits sensitivity but also extends the life of the HPLC and MS instrument components (Steinhaeuser et al. 2022).

In conclusion, while SPE and MIP-SPE can both be used for the extraction of emerging contaminants from water samples, MIP-SPE offers undisputed advantages in terms of selectivity and extraction efficiency. It is particularly beneficial when dealing with complicated matrices and the requirement for precise target compound extraction. But while choosing amongst these methods, it is essential to take into account the precise needs of the analytical work at hand, taking into consideration variables including selectivity, extraction efficiency, and the type of contaminants being studied.

9.6 Perspective on the Commercialization of MIPs Product

MIPs have emerged significant attention as nano-systems that offer tailored functionality and morphology. Their facile synthesis allows for rapid optimization of design parameters, making MIPs ideal for specific applications. These high-performance MIPs, containing with manipulative properties have been utilized for various technologies applications where affordable and efficient nano-systems are essential for enhancing overall system device performance. Looking ahead, MIPs driven by technology could potentially replace costly and labour-intensive extraction techniques, particularly in the environmental monitoring. In the context of environmental applications, MIPs show great promise in enhancing selectivity and efficiency, owing to their exceptional properties like high selectivity, cost-effectiveness, long-term storage stability, and robust chemical and physical stability in different media. Efforts have been devoted to commercializing and industrially deploying MIPs, resulting in the creation of usable products for both industrial and laboratory applications. Thorough upfront evaluation of risk assessments is vital in the transition of technology from the laboratory to the industry. A part of this systematic approach, including fostering academic-industry partnerships is essential for the successful translation of proof-of-concept MIPs into commercial products. Several notable studies underscore the practical application of MIPs for commercialization and industrial or laboratory use. On this matter, Claude et al. (2017) developed a commercial MIPs dedicated to glyphosate (GLY) and its main metabolite, aminomethylphosphonic acid (AMPA) was assessed as "BPOCIS-like" sampler on natural waters (mineral, ground, and seawaters) as SPE sorbent, before subsequent analysis by UPLC-MS/MS with derivatization. The obtained results were encouraging with 111 and 122 mL day^{-1} as sampling rates for GLY and AMPA, respectively. Mutavdžić Pavlović et al. (2015) employed MIPs in combination with commercial sorbents for the simultaneous extraction of pharmaceuticals from production wastewater. New SPE cartridges were prepared by combination of optimal produced MIPs and Oasis HLB in (400 mg/6 mL) of polypropylene SPE reservoir to improve the performances of the commercial cartridge. The developed method sung newly prepared SPE cartridge has been successfully applied to the analysis of production wastewater samples from pharmaceutical industry. Poma et al. (2013) reported a reusable solid-phase template approach for the synthesis of MIPs nanoparticles, showcasing reliable reusability through prototype automated UV photochemical reactors. The reliable re-use of molecular templates which is demonstrated for more than 30 repeated cycles without significant loss of performance through the immobilized template approach. Berghaus et al. (2014) established a scalable process for producing surface imprinted MIPs with high yield using solid templates, leading to enhanced affinity compared to conventional nanoparticles. The well synthesized involves of core–shell nanoparticles by surface initiated reversible addition fragmentation chain transfer polymerization in presence of chiral template conjugated to magnetic nanoparticles is reported. The approach leads to imprinted nanoparticles featuring enantioselectivity and enhanced affinity compared to nanoparticles prepared using template. The ongoing efforts in the commercialization and industrial application of MIPs, coupled with advancements in technology-driven synthesis and development, underscore their potential as versatile tools in diverse fields, particularly in environmentally sensitive applications. As academic-industry collaborations continue to flourish, the seamless translation of MIPs from laboratory concepts to commercial realities appears promising, with the potential to revolutionize various sectors.

9.7 Challenges and Future Perspective of MIPs in Emerging Pollutants Detection

MIPs stand out as versatile materials with a range of remarkable properties, including easy synthesis, cost-effectiveness, high stability, extended lifespan, and widespread availability. These attributes make MIPs intriguing materials for application across diverse fields, including separation science and pollutant treatment. One of the most distinctive features setting MIPs apart from other materials is their innate selectivity, established during the polymerization process targeting specific analytes. This unique characteristic positions MIPs as an excellent choice for use in complex samples. However, despite their selectivity, MIPs do have certain limitations that constrain their application. Addressing these limitations has become the focus of extensive research aimed at enhancing the utility of MIPs. One crucial aspect of MIPs is their adsorption capacity, particularly significant in applications such as emerging pollutants treatment. To address this issue, MIPs can be combined with different materials like metal–organic frameworks or covalent organic frameworks, which offer high surface areas, increased physical and chemical stability, and porous structures. Another critical consideration with MIPs is their compatibility with water. Given that most emerging pollutants are found in aqueous media, the development of water-compatible MIPs can significantly enhance their performance. Historically, MIPs have predominantly been applied using organic approaches. However, this method has its limitations, including the excessive use of organic solvents and the employment of toxic and hazardous reagents and solvents. These practices not only pose risks but also run counter to the principles of green chemistry. Therefore, synthesizing MIPs trough green and sustainable approaches opens up new possibilities for their synthesis and application in various field.

References

Alexander C, Andersson HS, Andersson LI, Ansell RJ, Kirsch N, Nicholls IA et al (2006) Molecular imprinting science and technology: a survey of the literature for the years up to and including 2003. J Mol Recognit 19(2):106–180. https://doi.org/10.1002/jmr.760

Ariani MD, Zuhrotun A, Manesiotis P, Hasanah AN (2022) Magnetic molecularly imprinted polymers: an update on their use in the separation of active compounds from natural products. Polymers 14(7). https://doi.org/10.3390/polym14071389

Arias PG, Martínez-Pérez-Cejuela H, Combès A, Pichon V, Pereira E, Herrero-Martínez JM et al (2020) Selective solid-phase extraction of organophosphorus pesticides and their oxon-derivatives from water samples using molecularly imprinted polymer followed by high-performance liquid chromatography with UV detection. J Chromatogr A 1626:461346. https://doi.org/10.1016/j.chroma.2020.461346

Ayankojo AG, Reut J, Ciocan V, Opik A, Syritski V (2020) Molecularly imprinted polymer-based sensor for electrochemical detection of erythromycin. Talanta 209:120502. https://doi.org/10.1016/j.talanta.2019.120502

Azizi A, Shahhoseini F, Bottaro CS (2020) Magnetic molecularly imprinted polymers prepared by reversible addition fragmentation chain transfer polymerization for dispersive solid phase extraction of polycyclic aromatic hydrocarbons in water. J Chromatogr A 1610:460534. https://doi.org/10.1016/j.chroma.2019.460534

Azizi A, Shahhoseini F, Langille EA, Akhoondi R, Bottaro CS (2021) Micro-gel thin film molecularly imprinted polymer coating for extraction of organophosphorus pesticides from water and beverage samples. Anal Chim Acta 1187:339135. https://doi.org/10.1016/j.aca.2021.339135

Bai L, Ye T, Si W, Yu J, Yuan M, Cao H et al (2023) Molecular imprinted polymer capped luminescent metal -organic framework for the determination of pyrethroids and its metabolite. Microchem J 193(109047). https://doi.org/10.1016/j.microc.2023.109047

Bakhshizadeh Aghdam M, Farajzadeh MA, Afshar Mogaddam MR (2022) Magnetic dispersive solid phase extraction based on carbonized cellulose–ferromagnetic nanocomposite for screening phthalate esters in aqueous samples. J Chromatogr A 1663:462756. https://doi.org/10.1016/j.chroma.2021.462756

Baltussen E, Sandra P, David F, Cramers C (1999) Stir bar sorptive extraction (SBSE), a novel extraction technique for aqueous samples: theory and principles. J Microcolumn Sep 11(10):737–747. https://doi.org/10.1002/(SICI)1520-667X(1999)11:10%3c737::AID-MCS7%3e3.0.CO;2-4

BelBruno JJ (2019) Molecularly imprinted polymers. Chem Rev 119(1):94–119. https://doi.org/10.1021/acs.chemrev.8b00171

Berghaus M, Mohammadi R, Sellergren B (2014) Productive encounter: molecularly imprinted nanoparticles prepared using magnetic templates. Chem Commun 50(64):8993–8996. https://doi.org/10.1039/C4CC01346H

Biyana Regasa M, Nyokong T (2022) Synergistic recognition and electrochemical sensing of 17β-Estradiol using ordered molecularly imprinted polymer-graphene oxide-silver nanoparticles composite films. J Electroanal Chem 922. https://doi.org/10.1016/j.jelechem.2022.116713

Cantarella M, Carroccio SC, Dattilo S, Avolio R, Castaldo R, Puglisi C et al (2019) Molecularly imprinted polymer for selective adsorption of diclofenac from contaminated water. Chem Eng J 367:180–188. https://doi.org/10.1016/j.cej.2019.02.146

Chen L, Xu S, Li J (2011) Recent advances in molecular imprinting technology: current status, challenges and highlighted applications. Chem Soc Rev 40(5):2922–2942. https://doi.org/10.1039/C0CS00084A

Chen L, Jian Y, Cheng J, Yan L, Huang X (2020a) Preparation and application of graphene oxide-based surface molecularly imprinted polymer for monolithic fiber array solid phase microextraction of organophosphate flame retardants in environmental water. J Chromatogr A 1623:461200. https://doi.org/10.1016/j.chroma.2020.461200

Chen X, Wu F, Tang J, Yang K, Ma Y, Pan J (2020b) Anisotropic emulsion constructed boronate affinity imprinted Janus nanosheets for stir bar sorptive extraction of cis-diol-containing catechol. Chem Eng J 395:124995. https://doi.org/10.1016/j.cej.2020.124995

Claude B, Berho C, Bayoudh S, Amalric L, Coisy E, Nehmé R et al (2017) Preliminary recovery study of a commercial molecularly imprinted polymer for the extraction of glyphosate and AMPA in different environmental waters using MS. Environ Sci Pollut Res 24(13):12293–12300. https://doi.org/10.1007/s11356-017-8844-5

de León-Martínez LD, Rodríguez-Aguilar M, Ocampo-Pérez R, Gutiérrez-Hernández JM, Díaz-Barriga F, Batres-Esquivel L et al (2018) Synthesis and evaluation of a molecularly imprinted polymer for the determination of metronidazole in water samples. Bull Environ Contam Toxicol 100(3):395–401. https://doi.org/10.1007/s00128-017-2249-5

Delinska K, Yavir K, Kloskowski A (2021) Ionic liquids in extraction techniques: determination of pesticides in food and environmental samples. Trends Anal Chem 143(116396). https://doi.org/10.1016/j.trac.2021.116396

Dong C, Shi H, Han Y, Yang Y, Wang R, Men J (2021) Molecularly imprinted polymers by the surface imprinting technique. Eur Polymer J 145:110231. https://doi.org/10.1016/j.eurpolymj.2020.110231

El-Deen AK, Hussain CM (2023) Recent developments in green magnetic nanoparticles for extraction and preconcentration of pollutants from environmental samples. Trends Environ Anal Chem 39:e00211. https://doi.org/10.1016/j.teac.2023.e00211

Feng G, Sun J, Wang M, Wang M, Li Z, Wang S et al (2021) Preparation of molecularly imprinted polymer with class-specific recognition for determination of 29 sulfonylurea herbicides in agro-products. J Chromatogr A 1647(462143). https://doi.org/10.1016/j.chroma.2021.462143

Fontanals N, Marcé RM, Borrull F (2017) Solid-phase extraction followed by liquid chromatography-high resolution mass spectrometry to determine synthetic cathinones in different types of environmental water samples. J Chromatogr A 1524:66–73. https://doi.org/10.1016/j.chroma.2017.10.002

Fu X, Wang X, Xia Z, Huang Y (2022a) Preparation of dummy molecularly imprinted polymers for selective extraction of aromatic amine genotoxic impurities. J Chromatogr A 1685(463617). https://doi.org/10.1016/j.chroma.2022.463617

Fu Y, Pessagno F, Manesiotis P, Borrull F, Fontanals N, Maria Marcé R (2022b) Preparation and evaluation of molecularly imprinted polymers as selective SPE sorbents for the determination of cathinones in river water. Microchem J 175:107100. https://doi.org/10.1016/j.microc.2021.107100

Guo P, Yuan X, Zhang J, Wang B, Sun X, Chen X et al (2018) Dummy-surface molecularly imprinted polymers as a sorbent of micro-solid-phase extraction combined with dispersive liquid–liquid microextraction for determination of five 2-phenylpropionic acid NSAIDs in aquatic environmental samples. Anal Bioanal Chem 410(2):373–389. https://doi.org/10.1007/s00216-017-0727-6

Hu M, Zhang Y, Yang J, Zhou X, Wei Z, Ding X et al (2015) Rapid fabrication of molecularly imprinted polymer fibers for solid phase microextraction of bisphenol A. Se Pu 33(2):123–131. https://doi.org/10.3724/sp.j.1123.2014.10008

Hu C, Huang H, Sun H, Yan Y, Xu F, Liao J (2022) Simultaneous analysis of catechol and hydroquinone by polymelamine/CNT with dual-template molecular imprinting technology. Polymer 242. https://doi.org/10.1016/j.polymer.2022.124593

Hu Y, Liu T, Wang S, Lv M, She Y, Cao B et al (2023) Preparation of magnetic molecularly imprinted polymers for the selective extraction of strobilurin fungicides in agricultural products. J Food Compos Anal 123(105564). https://doi.org/10.1016/j.jfca.2023.105564

Huang S, Xu J, Zheng J, Zhu F, Xie L, Ouyang G (2018) Synthesis and application of magnetic molecularly imprinted polymers in sample preparation. Anal Bioanal Chem 410(17):3991–4014. https://doi.org/10.1007/s00216-018-1013-y

Huang CW, Lin C, Nguyen MK, Hussain A, Bui XT, Ngo HH (2022) A review of biosensor for environmental monitoring: principle, application, and corresponding achievement of sustainable development goals. Bioengineered 14(2165–5987 (Electronic)). https://doi.org/10.1080/21655979.2022.2095089

Jafari Z, Ghani M (2023) Magnetic carbonized cellulose-MIL 101(Fe) composite as a sorbent for magnetic solid phase extraction of selected organophosphorus pesticides combined with high performance liquid chromatography-ultraviolet detection. Talanta Open 7:100212. https://doi.org/10.1016/j.talo.2023.100212

Jara-Cornejo E, Khan S, Vega-Chacon J, Wong A, da Silva Neres LC, Picasso G et al (2023) Biomimetic material for quantification of methotrexate using sensor based on molecularly imprinted polypyrrole film and MWCNT/GCE. Biomimetics (Basel) 8(1). https://doi.org/10.3390/biomimetics8010077

Jiang H-L, Li N, Cui L, Wang X, Zhao R-S (2019) Recent application of magnetic solid phase extraction for food safety analysis. TrAC, Trends Anal Chem 120:115632. https://doi.org/10.1016/j.trac.2019.115632

Kalogiouri NP, Tsalbouris A, Kabir A, Furton KG, Samanidou VF (2020) Synthesis and application of molecularly imprinted polymers using sol–gel matrix imprinting technology for the efficient solid-phase extraction of BPA from water. Microchem J 157:104965. https://doi.org/10.1016/j.microc.2020.104965

Karrat A, Amine A (2021) Solid-phase extraction combined with a spectrophotometric method for determination of Bisphenol-A in water samples using magnetic molecularly imprinted polymer. Microchem J 168. https://doi.org/10.1016/j.microc.2021.106496

Kaya SI, Cetinkaya A, Ozkan SA (2023) Molecularly imprinted polymers as highly selective sorbents in sample preparation techniques and their applications in environmental water analysis. Trends Environ Anal Chem 37(e00193). https://doi.org/10.1016/j.teac.2022.e00193

Khongkla S, Jullakan S, Saeaui S, Nurerk P, Kliangsuwan A, Bunkoed O (2023) A composite zinc oxide and magnetic molecularly imprinted polymer hydrogel adsorbent for the extraction of sulfonamides in milk. Microchem J 191(108865). https://doi.org/10.1016/j.microc.2023.108865

Khulu S, Ncube S, Nuapia Y, Madikizela LM, Tutu H, Richards H et al (2022a) Multivariate optimization of a two-way technique for extraction of pharmaceuticals in surface water using a combination of membrane assisted solvent extraction and a molecularly imprinted polymer. Chemosphere 286(Pt 3):131973. https://doi.org/10.1016/j.chemosphere.2021.131973

Khulu S, Ncube S, Nuapia Y, Madikizela LM, Mavhunga E, Chimuka L (2022b) Development and application of a membrane assisted solvent extraction-molecularly imprinted polymer based passive sampler for monitoring of selected pharmaceuticals in surface water. Water Res 225:119145. https://doi.org/10.1016/j.watres.2022.119145

Kraševec I, Prosen H (2018) Solid-phase extraction of polar benzotriazoles as environmental pollutants: a review. Molecules 23(2501). https://doi.org/10.3390/molecules23102501

Krishnan H, Islam AKM, Ahmad M, Hamzah Z, Nadaraja P (2019) A novel molecular imprint polymer synthesis for solid phase extraction of andrographolide. Indo J Chem 19:219–230. https://doi.org/10.22146/ijc.34369

Li L, Cui Y, Ren L, Li C, Liu Y, Jin Z (2021a) Magnetic dummy molecularly imprinted polymer nanoparticles as sorbent for dispersive solid-phase microextraction of phthalate monoesters in human urine samples. Microchem J 167(106266). https://doi.org/10.1016/j.microc.2021.106266

Li N, Zhang T, Chen G, Xu J, Ouyang G, Zhu F (2021b) Recent advances in sample preparation techniques for quantitative detection of pharmaceuticals in biological samples. TrAC, Trends Anal Chem 142:116318. https://doi.org/10.1016/j.trac.2021.116318

Li S, Ma J, Wu G, Li J, Wang X, Chen L (2022) Magnetic covalent-organic frameworks for the simultaneous extraction of eleven emerging aromatic disinfection byproducts in water samples coupled with UHPLC–MS/MS determination. J Hazard Mater 424:127687. https://doi.org/10.1016/j.jhazmat.2021.127687

Liu Y, Zhang H, Xie D, Lai H, Qiu Q, Ma X (2022) Optimized synthesis of molecularly imprinted polymers coated magnetic UIO-66 MOFs for simultaneous specific removal and determination of multi types of macrolide antibiotics in water. J Environ Chem Eng 10(4). https://doi.org/10.1016/j.jece.2022.108094

Lofgreen JE, Ozin GA (2014) Controlling morphology and porosity to improve performance of molecularly imprinted sol–gel silica. Chem Soc Rev 43(3):911–933. https://doi.org/10.1039/C3CS60276A

Lu D, Zhu DZ, Gan H, Yao Z, Luo J, Yu S et al (2022) An ultra-sensitive molecularly imprinted polymer (MIP) and gold nanostars (AuNS) modified voltammetric sensor for facile detection of perfluorooctance sulfonate (PFOS) in drinking water. Sens Actuators B, Chem 352. https://doi.org/10.1016/j.snb.2021.131055

Mabrouk M, Hammad SF, Abdella AA, Mansour FR (2023) Tips and tricks for successful preparation of molecularly imprinted polymers for analytical applications: a critical review. Microchem J 193(109152). https://doi.org/10.1016/j.microc.2023.109152

Majd M, Yazdanpanah M, Bayatloo MR, Nojavan S (2021) Recent advances and applications of cyclodextrins in magnetic solid phase extraction. Talanta 229:122296. https://doi.org/10.1016/j.talanta.2021.122296

Margiotta NP (2019) Síntesi de polímers amb i sense empremta molecular per a l'extracció en fase sòlida de l'herbicida diclofop-metil. Universitat Politècnica de València, Spain

Marzi Khosrowshahi E, Afshar Mogaddam MR, Javadzadeh Y, Altunay N, Tuzen M, Kaya S et al (2022) Experimental and density functional theoretical modeling of triazole pesticides extraction by Ti2C nanosheets as a sorbent in dispersive solid phase extraction method before HPLC-MS/MS analysis. Microchem J 178:107331. https://doi.org/10.1016/j.microc.2022.107331

Meseguer-Lloret S, Torres-Cartas S, Gómez-Benito C, Herrero-Martínez JM (2022a) Magnetic molecularly imprinted polymer for the simultaneous selective extraction of phenoxy acid herbicides from environmental water samples. Talanta 239:123082. https://doi.org/10.1016/j.talanta.2021.123082

Meseguer-Lloret S, Torres-Cartas S, Gómez-Benito C, Herrero-Martínez JM (2022b) Magnetic molecularly imprinted polymer for the simultaneous selective extraction of phenoxy acid herbicides from environmental water samples. Talanta 239:123082. https://doi.org/10.1016/j.talanta.2021.123082

Mohamed Nageib A, Abdul Halim A, Nordin AN, Ali F (2023) Computational modelling analysis and synthesis of molecularly imprinted polymers (MIPs) with two functional monomers using bulk polymerization. Mater Today: Proc. https://doi.org/10.1016/j.matpr.2023.04.680

Mohiuddin I, Bhogal S, Grover A, Malik AK, Aulakh JS (2021) Simultaneous determination of amitriptyline, nortriptyline, and clomipramine in aqueous samples using selective multi-template molecularly imprinted polymers. Environ Nanotech Monit Manage 16. https://doi.org/10.1016/j.enmm.2021.100527

Morin-Crini N, Lichtfouse E, Fourmentin M, Ribeiro ARL, Noutsopoulos C, Mapelli F et al (2022) Removal of emerging contaminants from wastewater using advanced treatments. A review. Environ Chem Lett 20(2):1333–1375. https://doi.org/10.1007/s10311-021-01379-5

Mostafa A, Shaaban H, Alqarni A, Al-Ansari R, Alrashidi A, Al-Sultan F et al (2023) Multi-class determination of pharmaceuticals as emerging contaminants in wastewater from Eastern

Province, Saudi Arabia using eco-friendly SPE-UHPLC-MS/MS: Occurrence, removal and environmental risk assessment. Microchem J 187:108453. https://doi.org/10.1016/j.microc.2023.108453

Mpupa A, Selahle SK, Mizaikoff B, Nomngongo PN (2021) Recent advances in solid-phase extraction (SPE) based on molecularly imprinted polymers (MIPs) for analysis of hormones. Chemosensors 9(7). https://doi.org/10.3390/chemosensors9070151

Muñoz-Bartual M, Herrero-Martínez JM, Esteve-Turrillas FA (2023) Development of paper-immobilized molecularly imprinted polymers by laser pointer activation for methamphetamine extraction with analysis by ion mobility spectrometry. J Chromatogr A 1692(463852). https://doi.org/10.1016/j.chroma.2023.463852

Mutavdžić Pavlović D, Nikšić K, Livazović S, Brnardić I, Anžlovar A (2015) Preparation and application of sulfaguanidine-imprinted polymer on solid-phase extraction of pharmaceuticals from water. Talanta 131:99–107. https://doi.org/10.1016/j.talanta.2014.06.065

Nasiri M, Ahmadzadeh H, Amiri A (2020) Sample preparation and extraction methods for pesticides in aquatic environments: a review. TrAC, Trends Anal Chem 123:115772. https://doi.org/10.1016/j.trac.2019.115772

Nogueira JMF (2015) Stir-bar sorptive extraction: 15 years making sample preparation more environment-friendly. TrAC, Trends Anal Chem 71:214–223. https://doi.org/10.1016/j.trac.2015.05.002

Noor M, Iqbal KM, Kanwal T, Rehman K, Ali M, Shah MR (2023) Synthesis and characterization of modified silica nanoparticles based molecularly imprinted polymers as a sensor for selective detection of pendimethalin from drinking water. J Mol Struct 1287(135635). https://doi.org/10.1016/j.molstruc.2023.135635

Pan J, Chen W, Ma Y, Pan G (2018) Molecularly imprinted polymers as receptor mimics for selective cell recognition. Chem Soc Rev 47(15):5574–5587. https://doi.org/10.1039/C7CS00854F

Parlapiano M, Akyol Ç, Foglia A, Pisani M, Astolfi P, Eusebi AL et al (2021) Selective removal of contaminants of emerging concern (CECs) from urban water cycle via Molecularly Imprinted Polymers (MIPs): potential of upscaling and enabling reclaimed water reuse. J Environ Chem Eng 9(1):105051. https://doi.org/10.1016/j.jece.2021.105051

Pichon V, Delaunay N, Combès A (2020) Sample preparation using molecularly imprinted polymers. Anal Chem 92(1):16–33. https://doi.org/10.1021/acs.analchem.9b04816

Pierpaoli M, Szopińska M, Olejnik A, Ryl J, Fudala-Ksiażek S, Łuczkiewicz A et al (2023) Engineering boron and nitrogen codoped carbon nanoarchitectures to tailor molecularly imprinted polymers for PFOS determination. J Hazar Mater 458. https://doi.org/10.1016/j.jhazmat.2023.131873

Poma A, Guerreiro A, Whitcombe MJ, Piletska EV, Turner APF, Piletsky SA (2013) Solid-phase synthesis of molecularly imprinted polymer nanoparticles with a reusable template–"plastic antibodies." Adv Func Mater 23(22):2821–2827. https://doi.org/10.1002/adfm.201202397

Ranjbari S, Mohammadinejad A, Johnston TP, Kesharwani P, Oskuee RK, Rezayi M et al (2023) Molecularly-imprinted polymers for the separation and detection of curcumin. Eur Polym J 189(111916). https://doi.org/10.1016/j.eurpolymj.2023.111916

Rathi BS, Kumar PS, Show P-L (2021) A review on effective removal of emerging contaminants from aquatic systems: current trends and scope for further research. J Hazard Mater 409:124413. https://doi.org/10.1016/j.jhazmat.2020.124413

Ren Y, Fan Z (2023) Synthesis of fluorescent probe based on molecularly imprinted polymers on nitrogen-doped carbon dots for determination of tobramycin in milk. Food Chem 416(135792). https://doi.org/10.1016/j.foodchem.2023.135792

Roy E, Patra S, Tiwari A, Madhuri R, Sharma PK (2017) Introduction of selectivity and specificity to graphene using an inimitable combination of molecular imprinting and nanotechnology. Biosens Bioelectron 89:234–248. https://doi.org/10.1016/j.bios.2016.02.056

Sabín López A, Paredes Ramos M, Herrero R, López Vilariño JM (2020) Synthesis of magnetic green nanoparticle—molecular imprinted polymers with emerging contaminants templates. J Environ Chem Eng 8(4):103889. https://doi.org/10.1016/j.jece.2020.103889

Shah SAH, Ramachandran MR, Mansur SA, Md Saleh N, Asman S (2023) Selective recognition of bisphenol a using molecularly imprinted polymer based on choline chloride-methacrylic acid deep eutectic solvent monomer: synthesis characterization and adsorption study. Polymer 283(126279). https://doi.org/10.1016/j.polymer.2023.126279

Shahzad A, Majeed A, Lahiq AA, Alqahtani T, Alqahtani AA, Bashir K et al (2023) Preparation and characterization of dummy template molecularly imprinted polymers coupled with HPLC for selective extraction of spiked cloprostenol from milk samples. Arab J Chem 16. https://doi.org/10.1016/j.arabjc.2023.105045

Shan X, Zhang L, Ye H, Shao J, Shi Y, Tan S et al (2022) Magnetic solid phase extraction of lead ion from water samples with humic acid modified magnetic nanoparticles prior to its fame atomic absorption spectrometric detection. Microchem J 179:107417. https://doi.org/10.1016/j.microc.2022.107417

Shao H, Cherif S, Wang J, Wang Q, Jiang Z (2020) Synthesis and application of Zwitterionic magnetic molecularly imprinted polymer for selective removal of fluoroquinolones from aqueous solution. Curr Anal Chem 16. https://doi.org/10.2174/1573411016999200518084012

Sodré FF, Sampaio TR (2020) Development and application of a SPE-LC-QTOF method for the quantification of micropollutants of emerging concern in drinking waters from the Brazilian capital. Emerging Contaminants 6:72–81. https://doi.org/10.1016/j.emcon.2020.01.001

Song X, Li X, Wang J, Huang X (2023) Adoption of new strategy for molecularly imprinted polymer based in-tube solid phase microextraction to improve specific recognition performance and extraction efficiency. Microchem J 194(109224). https://doi.org/10.1016/j.microc.2023.109224

Sorribes-Soriano A, Albert Esteve-Turrillas F, Armenta S, Manuel Herrero-Martínez J (2023) Molecularly imprinted polymer–stir bar sorptive extraction of diazepam from natural water. Microchem J 186:108354. https://doi.org/10.1016/j.microc.2022.108354

Steinhaeuser LBE, Westphalen T, Kaminski K, Piechotta C (2022) Evaluation, comparison and combination of molecularly imprinted polymer solid phase extraction and classical solid phase extraction for the preconcentration of endocrine disrupting chemicals from representative whole water samples. Talanta Open 6:100163. https://doi.org/10.1016/j.talo.2022.100163

Sun X, Wang J, Li Y, Yang J, Jin J, Shah SM et al (2014) Novel dummy molecularly imprinted polymers for matrix solid-phase dispersion extraction of eight fluoroquinolones from fish samples. J Chromatogr A 1359:1–7. https://doi.org/10.1016/j.chroma.2014.07.007

Sun X, Wang M, Yang L, Wen H, Wang L, Li T et al (2019) Preparation and evaluation of dummy-template molecularly imprinted polymer as a potential sorbent for solid phase extraction of imidazole fungicides from river water. J Chromatogr A 1586:1–8. https://doi.org/10.1016/j.chroma.2018.11.077

Surapong N, Santaladchaiyakit Y, Burakham R (2022) A water-compatible magnetic dual-template molecularly imprinted polymer fabricated from a ternary biobased deep eutectic solvent for the selective enrichment of organophosphorus in fruits and vegetables. Food Chem 384:132475. https://doi.org/10.1016/j.foodchem.2022.132475

Surapong N, Pongpinyo P, Santaladchaiyakit Y, Burakham R (2023) A biobased magnetic dual-dummy-template molecularly imprinted polymer using a deep eutectic solvent as a coporogen for highly selective enrichment of organophosphates. Food Chem 418(136045). https://doi.org/10.1016/j.foodchem.2023.136045

Tan D, Han C, Yu Z, Sun X, Chen J, Dhanjai et al (2022) Dummy molecularly imprinted polymers for class-selective extraction of amphetamine-type stimulants from alcoholic and nonalcoholic beverages. J Chromatogr A 1663(462759). https://doi.org/10.1016/j.chroma.2021.462759

Thakur N, Das TR, Gupta S, Shukla SK, Patra S, Kumar D (2023) Chapter 5—Material selection for the fabrication of MIP-based devices: an overview. In: Singh M (ed) Molecularly imprinted polymers (MIPs). Elsevier, pp 101–114

Turiel E, Martín-Esteban A (2019) Molecularly imprinted polymers-based microextraction techniques. TrAC, Trends Anal Chem 118:574–586. https://doi.org/10.1016/j.trac.2019.06.016

Vidal DF, Silva CF, Borges MMC, Nascimento Jr CS, Borges KB (2023) Magnetic and mesoporous molecularly imprinted polymer synthesized by rational computation design: sample preparation

and analysis of ractopamine. J Chromatogr B 1227(123843). https://doi.org/10.1016/j.jchromb.2023.123843
Wang J, Zhu J, Si L, Du Q, Li H, Bi W et al (2017) High throughput screening of phenoxy carboxylic acids with dispersive solid phase extraction followed by direct analysis in real time mass spectrometry. Anal Chim Acta 996:20–28. https://doi.org/10.1016/j.aca.2017.10.007
Wang D, Xu Z, Liu Y, Liu Y, Li G, Si X et al (2020) Molecularly imprinted polymer-based fiber array extraction of eight estrogens from environmental water samples prior to high-performance liquid chromatography analysis. Microchem J 159:105376. https://doi.org/10.1016/j.microc.2020.105376
Wang J, Yang Y, Shen Q, Shen D, Kang Q (2023a) A smartphone-based long optical path colorimetric turntable for selective determination of malachite green and investigation the specific adsorption behavior of the imprinted cavities within molecularly imprinted polymers. Microchem J 190(108629). https://doi.org/10.1016/j.microc.2023.108629
Wang X, Feng Y, Chen H, Qi Y, Yang J, Cong S et al (2023b) Synthesis of dummy-template molecularly imprinted polymers as solid-phase extraction adsorbents for N-nitrosamines in meat products. Microchem J 185(108271). https://doi.org/10.1016/j.microc.2022.108271
Wang S, Yang J, Sun J, Liu K, Xie X, Hong L et al (2023c) Nanomaterial-based magnetic surface molecularly imprinted polymers for specific extraction and efficient recognition of dibutyl phthalate. Food Chem 426(136621). https://doi.org/10.1016/j.foodchem.2023.136621
Wang Y, Ren R, Chen F, Jing L, Tian Z, Li Z et al (2023d) Molecularly imprinted MOFs-driven carbon nanofiber for sensitive electrochemical detection and targeted electro-Fenton degradation of perfluorooctanoic acid. Sep Purif Technol 310. https://doi.org/10.1016/j.seppur.2023.123257
Xu Z, Yang Z, Liu Z (2014) Development of dual-templates molecularly imprinted stir bar sorptive extraction and its application for the analysis of environmental estrogens in water and plastic samples. J Chromatogr A 1358:52–59. https://doi.org/10.1016/j.chroma.2014.06.093
Yang Y, Liu X, Meng S, Mao S, Tao W, Li Z (2023a) Molecularly imprinted polymers-isolated AuNP-enhanced CdTe QD fluorescence sensor for selective and sensitive oxytetracycline detection in real water samples. J Hazar Mater 458(131941). https://doi.org/10.1016/j.jhazmat.2023.131941
Yang L, Wang S, Xie Z, Xing R, Wang R, Chen X et al (2023b) Deep eutectic solvent - loaded Fe_3O_4@MIL-101(Cr) with core–shell structure for the magnetic solid phase extraction of non-steroidal anti-inflammatory drugs in environmental water samples. Microchem J 184:108150. https://doi.org/10.1016/j.microc.2022.108150
Yue B, Wang X, Lian LL, Wang Y, Gao W, Zhang H et al (2023) A fiber-packed needle-type extraction device with ionic liquid-based molecularly imprinted polymer as coating for extraction of chlorobenzenes in water samples. Microchem J 190. https://doi.org/10.1016/j.microc.2023.108592
Zambonin C, Aresta A (2021) Recent applications of solid phase microextraction coupled to liquid chromatography. Separations 8(3). https://doi.org/10.3390/separations8030034
Zhang Z, Cao X, Zhang Z, Yin J, Wang D, Xu Y et al (2020) Synthesis of dummy-template molecularly imprinted polymer adsorbents for solid phase extraction of aminoglycosides antibiotics from environmental water samples. Talanta 208:120385. https://doi.org/10.1016/j.talanta.2019.120385
Zhou T, Ding L, Che G, Jiang W, Sang L (2019) Recent advances and trends of molecularly imprinted polymers for specific recognition in aqueous matrix: preparation and application in sample pretreatment. TrAC, Trends Anal Chem 114:11–28. https://doi.org/10.1016/j.trac.2019.02.028

Chapter 10
Molecularly Imprinted Polymers-Based Sample Preparation and Preconcentration Methods

Jun Zhang, Shuangxian Yan, Wenxin Liu, Xue Dong, Zhaosheng Liu, and Zehui Wei

10.1 Introduction

The overuse of antibiotics in healthcare, the excessive application of insecticides and pesticides in agriculture or forestry, and the release of industrial wastewater are significant threats to human health and environmental pollution. When analyzing samples of complex matrices such as soil, wastewater, food, and crops, it is challenging to achieve precise and direct analysis, despite the use of highly selective and sensitive chromatographic methods that may also cause instrument contamination or damage. It is crucial to prepare and enrich complex samples before conducting analysis. Therefore, researchers are dedicated to developing highly efficient analytical techniques and new analytical instruments to address these challenges. Sample preparation involves separating and purifying samples, eliminating potential matrices, and converting them into a form that is easier to detect. The process accelerates sample analysis by reducing matrix interference, enhancing analyte purity, boosting selectivity, and decreasing detection limits.

Molecularly imprinted polymers (MIPs) have been used for the separation and purification of samples by virtue of their special cavity structure for specific adsorption of target molecules (Santanu Patra et al. 2016), and new breakthroughs continue to be achieved in the environmental health, food safety, and biological fields. This chapter concentrates on the usage of MIPs as adsorbents in various sample preparation and enrichment techniques, like solid-phase extraction (SPE), solid-phase

J. Zhang · S. Yan · W. Liu · X. Dong · Z. Wei (✉)
School of Pharmacy, Jiangsu Ocean University, Jiangsu, China
e-mail: wei.zehui@163.com

Z. Liu (✉)
School of Pharmacy, Tianjin Medicine University, Tianjin, China
e-mail: liuzhaosheng@tmu.edu.cn

S. Patra and M. Sillanpaa (eds.), *Molecularly Imprinted Polymers as Artificial Antibodies for the Environmental Health*, https://doi.org/10.1007/978-3-031-58995-9_10

microextraction (SPME), magnetic solid-phase extraction (MSPE), matrix solid-phase dispersion (MSPD), membrane-assisted solvent extraction (MASE), stir bar sorption extraction (SBSE), biointerface cell recognition, sensor detection, and several modern instrument integration techniques.

10.2 Molecularly Imprinted Solid Phase Extraction (MI-SPE)

Solid-phase extraction (SPE) is commonly employed in various fields owing to its ability to effectively enrich samples, simplicity, high selectivity, and cost-effectiveness. Its popularity is attributable to these impressive features. Typically, an SPE system comprises a packed bed, such that it is mainly a column, cartridge, or membrane, containing the extraction adsorbent. The selection of an appropriate adsorbent critically impacts SPE efficiency. MIPs demonstrate high selectivity and specific recognition abilities, making them a superior choice for use as SPE adsorbents.

Several papers have reported the use of MI-SPE for enrichment detection of antibiotics and other residues in environmental or food samples. For example, Su et al. (2023) used MIP to selectively extract fluoroquinolones from meat samples such as chicken, fish, shrimp and pork. In addition, Boulanouar et al. (2017) used MIP materials for the selective adsorption of organophosphorus pesticides in vegetable oils, which were injected directly into the chromatographic system after extraction and treatment of the samples to further complete the detection. Chen et al. (2023) produced a distinct activated soybean shell biochar molecularly imprinted polymer (ASBC@MIP) and established an ASBC@MIP solid phase extraction column to separate and enrich carbaryl in rice and maize. A molecularly imprinted polymer with a unique shell and core structure was developed by Shen and coworkers (2023) to effectively enhance the absorption of dibutyl phthalate (DBP) in water. Chapuis et al. (2004) used MIP-SPE for the selective removal of triazine pesticide contaminants from soil extracts and grape juice. For instance, MIPs were employed in the SPE of active compounds for the estrogen receptor in water samples (Yagishita et al. 2019). To conduct the solid-phase extraction, the basic procedure was followed.

- Environmental water pre-treatment: Wastewater samples (1000 mL) were taken from a wastewater treatment plant and filtered through a glass filter to remove macromolecular impurities from the water samples. The solution was pre-enriched for hydrophobic compounds using an adsorbent resin synthesized from styrene–divinylbenzene (SDB). The hydrophobic compounds obtained were eluted using 6.0 mL of methanol (MeOH), followed by drying the eluate under N_2 purging and re-dissolving in 1.0 mL of MeOH.
- Conventional extraction method for sample preparation: The above samples were aliquoted into two portions, and one portion was concentrated to 100 μL as SDB concentrated sample (SDB-Conc), in which the activity of estrogen receptor

contained in the concentrated sample was determined by yeast two-hybrid assay (Y2H), followed by LC-Q-TOFMS to identify its structure.

- MIP-SPE sample preparation: Samples of the other half of the above were diluted to 50 mL with purified water, which were all loaded onto a MIP-SPE column collected the effluent fractions and dried them, with the dried sample re-dissolved in MeOH (100 μL) as the effluent sample (MIP-Pass). Finally, the MIP-SPE column was eluted with 6.0 mL acetonitrile (ACN), drying the eluted fractions and redissolving them in MeOH (100 μL) as MIP adsorption samples (MIP-Conc).

The activity levels of the treated samples were measured by Y2H, and the concentration of E2 served as an index of the activity values. 3.1 ppb and 3.7 ppb were found to be the activity values of SDB-Conch and MIP-Conch, as shown in the results. MIP-conch had a higher activity value than SDB-conch by 1.2 times, despite the absence of any active ingredient in the MIP-Pass. In the chromatogram (Fig. 10.1), the peak value of MIP-conch was much smaller than that of SDB-conch, and most of the inactive substances were detected in MIP-Pass, which indicated that the MIP-conch had a high selectivity for ER. Meanwhile, the large amount of inactive substances present in SDB-Conc samples would interfere with the determination of the activity value, resulting in a lower value than the true value. Therefore, compared with the SDB treatment method, the interfering substances in the samples were basically eliminated after the environmental wastewater was treated with MIP-SPE, and the specific enrichment of target compounds was effective.

10.3 Molecularly Imprinted Solid-Phase Microextraction (MI-SPME)

Solid-phase microextraction (SPME) is the same as SPE, the extraction process includes the adsorption and dissolution desorption of analytes, but its scale is much smaller than SPE. Solid-phase microextraction columns are prepared by encapsulating and grafting MIP onto the surface of carbon nanotubes, mesoporous silica tubes, or polymerized quartz fibers. Microextraction was achieved by submerging the prepared fiber columns directly in the sample solution without the need for additional organic solvents. Combining MIP with SPME provides an ideal adsorbent for trace sample preparation and further improves the sample extraction efficiency.

Feng et al. (2009) used submicrospheres of MIPs as adsorbents using polypropylene membranes as encapsulating materials for the selective extraction of phenolic compounds from aqueous samples. The selective SPME of ochratoxin A from red wines was achieved by synthesizing molecularly imprinted polypyrrole-modified carbon nanotubes, as described by Yu and colleagues (2006). A SPME fiber was made by Chen et al. (2024) using a polymer-coated mesoporous silica tube that was imprinted with tobacco-specific nitrosamines in water. Courtois et al. (2006) synthesized a molecularly imprinted monolithic capillary column for the SPME of local

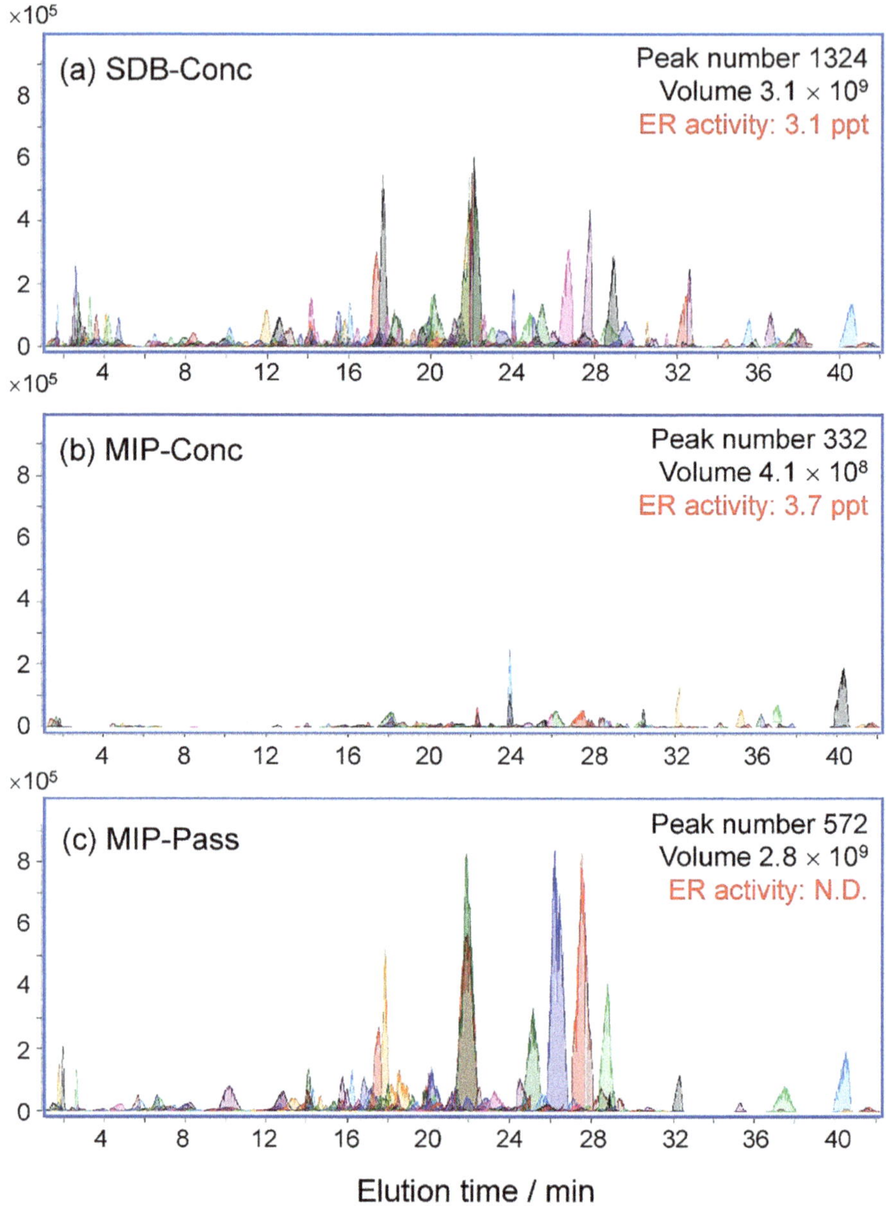

Fig. 10.1 LC-Q-TOFMS chromatograms of the environmental water sample. **a** Concentrated by SDB cartridge, **b** concentrated by MIP, **c** passed through from MIP. Each chromatogram is overwritten with all the peaks of the detected compounds. Reproduced with permission of Yagishita et al. (2019)

anesthetics from samples. Turiel et al. (2007) applied MIP fiber monolithic columns (Fig. 10.2a) for the extraction of triazines from food.

The basic sample preparation procedure was as follows (using the method reported by Turiel et al. (2007) as an example).

- ACN (40 mL) was added to the soil (10 g), which was sonicated for 30 min and mixed thoroughly, followed by filtration of the supernatant (0.45 mm filter) to obtain the filtrate. The filtrate was then purged and dried under N_2 before being dissolved in 1.7 mL of toluene (containing 100 μg/L triazine mixture).

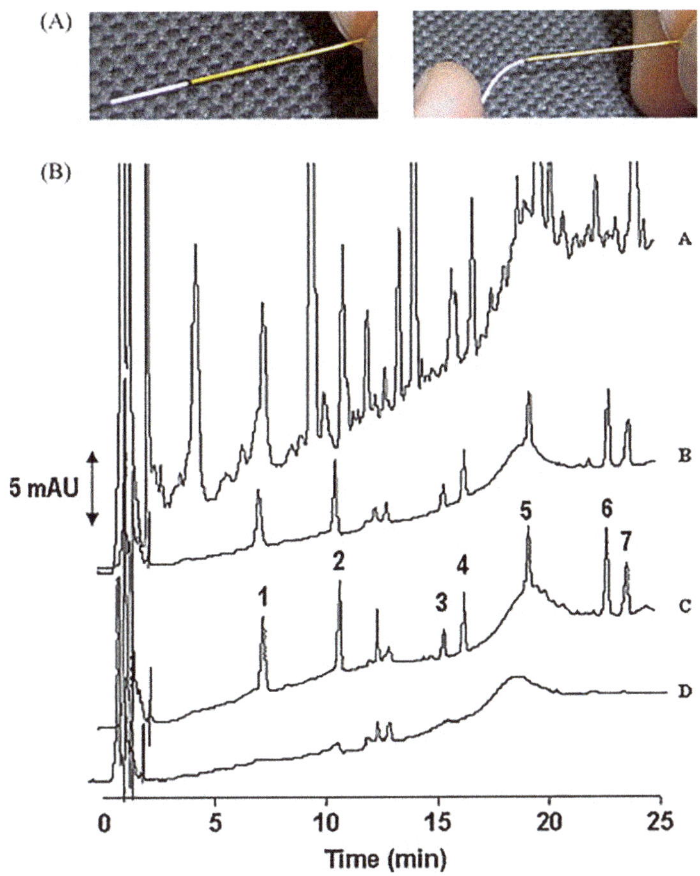

Fig. 10.2 LC-UV chromatograms obtained at 220 nm for **a** a soil sample extract directly injected without any previous cleanup, **b** a soil sample extract enriched with triazines at 0.1 mg L^{-1} concentration level after MI-SPME, **c** a 0.1 mg L^{-1} standard solution of triazines after MI-SPME, and **d** a non-spiked soil sample extract after MI-SPME. Peak numbers: (1) desisopropylatrazine (DIA); (2) Desethylatrazine (DEA); (3) simazine (SIM); (4) cyanazine (CYA); (5) atrazine (ATR); (6) propazine (PPZ); (7) terbutylazine (TER). Reproduced with permission of Turiel et al. (2007)

- Next, 1.7 mL of toluene extract was added into a glass bottle equipped with a 7 × 2 mm Teflon stir bar. The fiber column (1 cm × 0.53 mm) was submerged in the toluene extract under magnetic stirring for MI-SPME (60 min). At the end of the extraction, the fiber column was immersed in toluene for 15 min to remove interfering substances introduced by non-specific interactions.
- Finally, the fiber column was removed and air-dried for 5 min. It was then placed into a 0.4 mL glass vial containing 200 μL of MeOH and sonicated for 5 min to extract the fiber column into MeOH solution. The MeOH extract was collected, and 200 μL of water was added to dilute the sample, which was then spiked by MI-SPME. The same method and fiber column was used to prepare the MI-SPME triazine standard solution and non-spiked soil samples for HPLC analysis.

The findings depicted in Fig. 10.2b. Figure 10.2a represents the results of direct soil sample analysis without extraction column clean-up. The outcome of performing soil sample analysis directly without extraction column purification. Since the soil samples are contaminated with a significant amount of interfering substances, the resulting peaks in the obtained chromatograms are copious and heterogeneous, making it impossible to analyze the target compounds directly. Figure 10.2b represents the result of the spiked soil samples after MI-SPME, which is consistent with the result of the standard solution Fig. 10.2c and eliminates the interferents in Fig. 10.2a. This shows that the soil samples can directly and effectively detect the target compounds with high adsorption selectivity after MI-SPME. Additionally, Fig. 10.2d represents the non-spiked MI-SPME soil sample, which indicated an absence of triazines, attesting to the fiber column's reusability without residue. In summary, MI-SPME is a straightforward, speedy technique that excels in both separating and enriching with high specificity. It proves satisfactory for the enrichment and separation of trace samples.

10.4 Molecularly Imprinted Magnetic Solid-Phase Extraction (MI-MSPE)

Magnetic solid phase extraction (MSPE) is a newly developed and efficient method of SPE and a pre-treatment procedure that utilizes different magnetic materials as adsorbents to enhance the separation of samples by exploiting their magnetic interactions. Various fields have applied MSPE owing to its merits, which include its high efficacy in extraction and selectivity, sensitivity, and ability to save organic solvents. Magnetic adsorbents based on MIPs can be readily manipulated by external magnetic fields to effectively separate, enrich, and elute target analytes. Such adsorbents can be combined with MSPE to facilitate the preparation of complex samples across a broad spectrum of applications.

The use of MSPE for the enrichment and separation of pesticide, fungicide, and antibiotic residues within food products or ambient water has been the subject of numerous investigations in recent years. Huang et al. (2020) used surface imprinting techniques to prepare magnetic molecularly imprinted polymers (MMIPs) and combined them with HPLC for the simultaneous detection of hexestrol (HEX), bisphenol A (BPA), and nonylphenol (NP). Meseguer-Lloret et al. (2022) synthesized MMIPs using 4-chloro-2-methylphenoxyacetic acid (MCPA) as a template for the purification and extraction of herbicides. The processed samples could be injected directly into the HPLC system for detection. Gaho et al. (2022) used the surface molecular imprinting and magnetic separation (NORF-MMIP) technique to selectively remove haloperidol from wastewater and drinking water. Zhao et al. (2020) coupled MIP with magnetic nanoparticles, which can be used as core–shell MMIPs with strong adsorption to selectively identify patulin (PAT) in juice samples, and the magnetic MSPE was used in conjunction with LC-MS coupling for detection. Shan et al. (2022) prepared humic acid-modified magnetic nanoparticles (HA-Fe_3O_4 MNPs) for the enrichment of Pb(II) in water samples using the MSPE technique.

As an example, MMIPs were applied to the MSPE of strobilurin fungicides in agricultural products (Hu et al. 2023), and the sample preparation process was as follows:

- Pre-treatment of agricultural products: The crop samples were first homogenized by taking 10 g of sample with 10 mL of ACN in a centrifuge tube, and then the mixture was vortexed using a vortex mixer to obtain a mixture. After centrifuging the mixture at 8000 rpm, the supernatant was dried at 40 °C (N_2) and then redissolved in a solution of ACN and 20% water.
- MMIPs-MSPE sample preparation: Take 30 mg of MMIP adsorbent into a centrifuge tube and condition with 6 mL of MeOH/water (1:1). Add 4 mL of sample solution and oscillate the adsorbent for 5 min. Separate the adsorbent from the solution with an external magnet and remove the supernatant. The analytes were extracted from the adsorbent by vortexing for 20 s (4 × 1 mL) using MeOH (4 mL) as eluent. Before the solution was desorbed, a 0.22 μm syringe filter was used to filter it. The solution was then added to the HPLC-MS/MS system for analysis. Figure 10.3 shows the schematic diagram of the MSPE extraction process for strobilurin fungicides (SFs).

The results show that MI-MSPE samples have low detection limits and an excellent linear range. Compared with the traditional method, MI-MSPE is simple to operate, and the separation step relies on an external magnet, avoiding the complicated centrifugation step, which is both convenient and efficient.

10.5 Molecularly Imprinted Matrix Solid-Phase Dispersion (MI-MSPD)

Matrix solid-phase dispersion (MSPD) is a kind of adsorption process that involves mixing and grinding the sample and adsorbent to complete the adsorption process, and the extraction process does not require liquid medium, which is suitable for

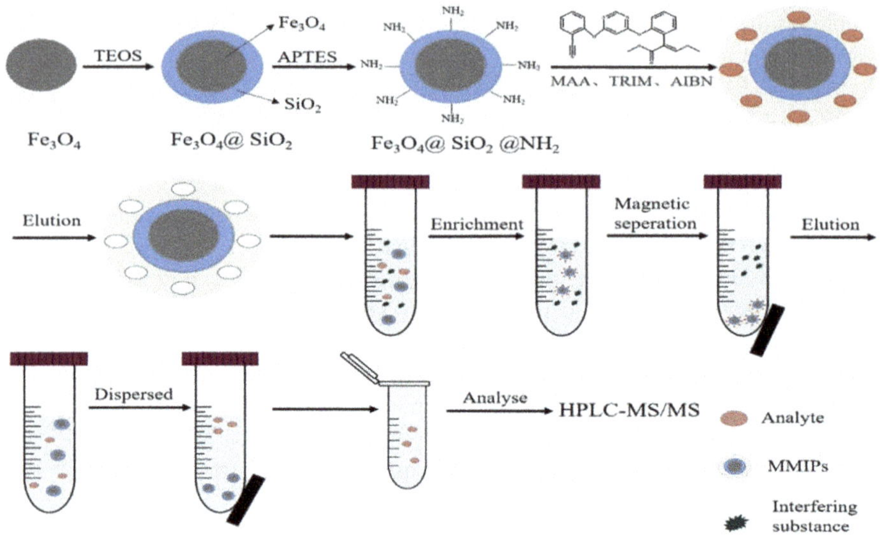

Fig. 10.3 Schematic of the synthesizing process for MMIPs and the MSPE for extracting SFs. Reproduced with permission of Hu (2023)

solid or semi-solid samples. Direct contact between the adsorbent and interfering substances in the sample can affect the extraction results, and MIPs have strong specific adsorption and anti-interference capabilities. Using them as adsorbents for matrix solid-phase dispersion can successfully solve these problem.

Arabi et al. (2020) prepared imprinted nano-microspheres using surface MIT for selective enrichment of the food additive rhodamine B in food samples. Gholami et al. (2019) utilized a one-pot method to synthesize molecularly imprinted biofilms as adsorbents for selective adsorption of parabens in sunscreens by MI-MSPD.

The basic procedure of the MI-MSPD method is shown in Fig. 10.4 (Gholami et al. 2019).

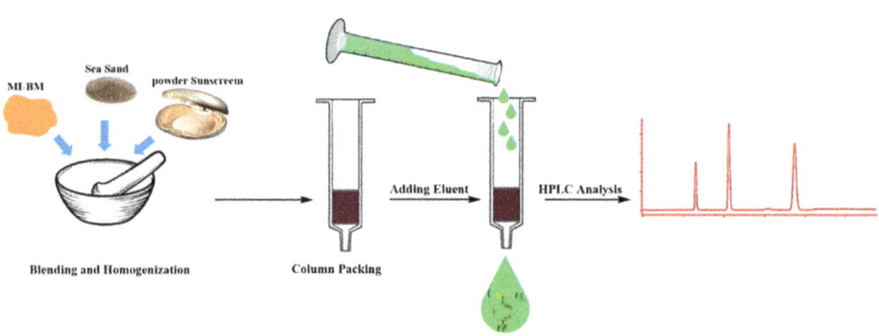

Fig. 10.4 MI-BM-based MSPD procedure for parabens extraction. Reproduced with permission of Gholami et al. (2019)

- Sunscreen powder (0.05 g), molecularly imprinted biofilm (MI-BM) (0.15 g), and sea sand (0.1 g) were ground in a mortar and pestle, which were mixed homogeneously and then transferred to an SPE column.
- The column was washed with hexane (3 mL) to remove the interference of hydrophobic substances, followed by the elution of parabens with 2 mL of MeOH/acetic acid (90/10 v/v) as eluent.
- Drying the eluate using N_2 purge at 35 °C led to its redissolved residue in a mobile phase (50 μL) for HPLC analysis.

MI-MSPD was used to extract the preservative parabens from sunscreens. The results showed that the limit of detection was as low as 11.1 μg/kg, which demonstrated that the method was selective and effective. The preparation of complex samples by MI-MSPD is simple and rapid without pretreatment. As the adsorbent was directly dispersed in the sample, the contact area between the adsorbent and the sample increased, and the adsorption capacity was strong with high extraction efficiency.

10.6 Molecularly Imprinted Membrane-Assisted Solvent Extraction (MI-MASE)

The molecularly imprinted membrane-assisted solvent extraction (MI-MASE) reported in the current literature is based on the addition of a fiber protective membrane to the liquid–liquid microextraction (LLME) method, which is associated with low solvent consumption, good extraction results, and high sensitivity when compared to conventional liquid–liquid extraction (LLE). It can be achieved by using a small-length hollow fiber column and a small amount of solvent. Mhaka et al. (2009) used MI-MASE to isolate triazines from food samples. Khulu et al. (2022) combined membrane-assisted extraction (MAP) with MIPs to extract five anticonvulsant drugs, the cardiotonic drug etilefrine, and other drugs from surface water. Hu et al. (2009) developed a MIP based on fiber-membrane-protected liquid–liquid-solid microextraction (LLSME) sample preparation with specific adsorption of triazines for samples with complex matrices.

The basic operational procedure of MIP-MASE has been described as reference (Hu et al. 2009):

- Collect sludge water from the farmland and homogenize 5 mL of effluent with 15 μL of the standard solution of triazine compounds (1.0 mg/L) in a homogeneous mixture for 3 h. The solution (3 mL) was added to 5 mL glass vials equipped with stirring bars as an aqueous phase for the preparation of spiked samples.
- The MIP-MASE extraction device shown in Fig. 10.5, consists primarily of SPME modules and polypropylene hollow fiber membranes. The membrane prevents contaminants in the sample from coming into direct contact with the extraction column and extends the life of the column. An injection of 6 μL of toluene was

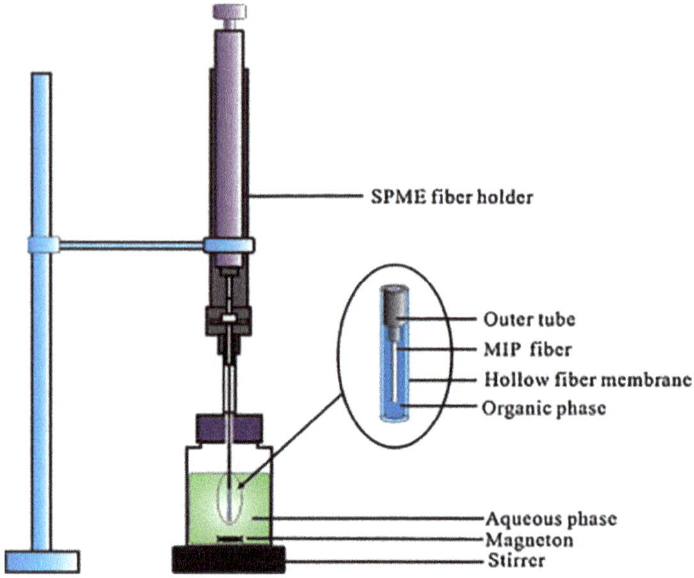

Fig. 10.5 Schematic of the LLSME device. Reproduced with permission of Hu et al. (2009)

made into the fiber membrane's inner lumen. After that, the membrane is sonicated for 20 s to ensure complete penetration. Insert the MIP-coated extraction housing into the membrane chamber. The entire module is then inserted into the aqueous phase and extracted at 1000 rpm for 30 min.

Large molecular interferences and insoluble substances were eliminated from the MIP-MASE process (see Fig. 10.6 for the principle), while the target compounds smoothly entered the interior of the fiber membrane alongside small molecular interferences. Because the fiber column has imprinted cavities of a certain structure, only the desired chemicals are adsorbed selectively on it.

After the extraction was complete, the outer fiber membrane was taken off, and the extraction column's leftover solvent was taken out and put into the SPME-HPLC coupling unit. After 10 min of MeOH elution on the column, the eluate was introduced into the HPLC apparatus for examination.

Before extraction, the water sample was analyzed through chromatography, resulting in significant interference peaks that hindered the direct analysis of the detection outcomes. However, the chromatogram became much clearer, and the analytes were successfully identified after the sample was extracted through the fiber column, which eliminated the interfering substances in the matrix.

Significant advantages of MIP-MASE:

(1) It combines the advantages of LLE and SPME, which is equivalent to a dual concentration procedure and greatly improves sensitivity;

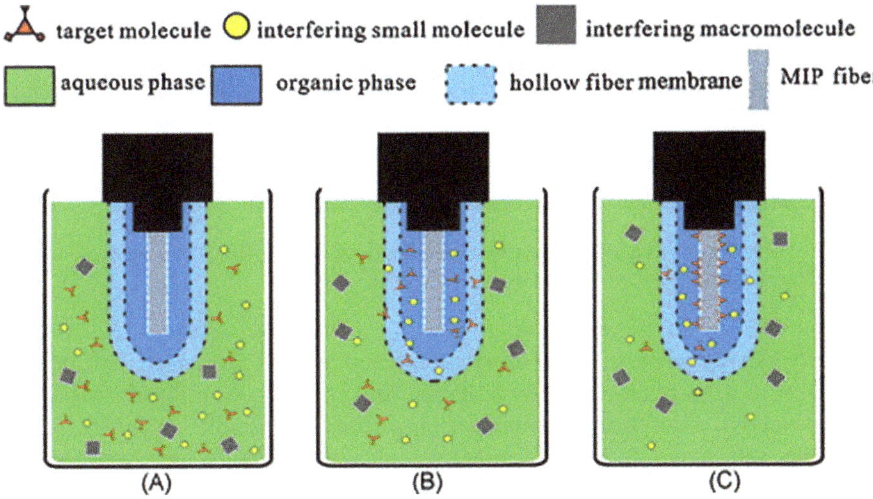

Fig. 10.6 Extraction process of the LLSME technique. **a** Before extraction, **b** during extraction and **c** after extraction. Reproduced with permission of Hu et al. (2009)

(2) The function of the fiber membrane is to protect the MIP adsorbent from contamination of the sample solution and to reduce the interference of low-polarity organic solvents, enhancing the selectivity of aqueous samples.

10.7 Molecularly Imprinted Polymer Stir bar Sorption Extraction (MIP-SBSE)

Molecularly imprinted polymer stir bar sorption extraction (MIP-SBSE) works by coating an adsorbent onto a stir bar and then immersing the bar into the vortexed sample solution, during which the stir bar comes into full contact with the solution and completes the adsorption process. In order to increase the stability of the adsorptive extraction, the stir bar is typically covered with materials such as sol–gel, polymerization, adhesion, etc. Samples that are volatile or semi-volatile can be enriched using MIP-SBSE. Adsorption coating is the main factor affecting extraction efficiency. MIP-SBSE combines the high selectivity of MIP with the high extraction efficiency of SBSE. Therefore, it is commonly used to separate and enrich various complex samples.

Zhu et al. (2015) developed a MIP-SBSE for the enrichment detection of melamine (MA) in milk powder (the appearance of the stir bar is shown in Fig. 10.7). Commercially available polydimethylsiloxane stir bars (PDMS-SB), non-imprinted polymer stir bars (NIP-SB), and MIP-SB were used for MA enrichment detection in milk powder, respectively. The results showed that the signal peaks of MA were almost undetectable in the samples extracted by PDMS-SB and NIP-SB, while there were

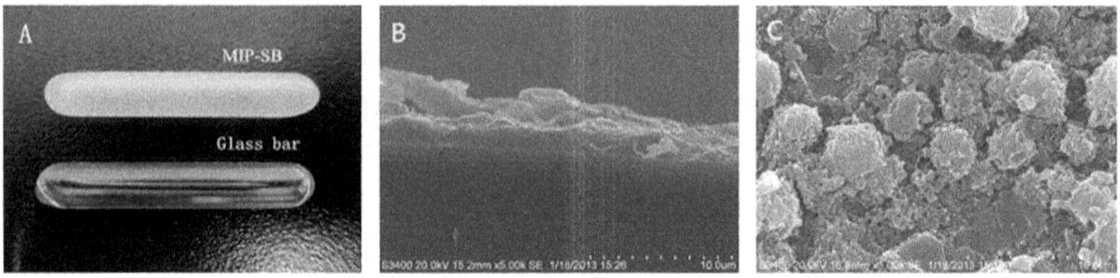

Fig. 10.7 Appearance of a MIP-SB and a bare glass bar (**a**); SEM of MIP-SB (**b**) and cross-section of MIP-SB (**c**). Reproduced with permission of Zhu et al. (2015)

obviously MA signal peaks in the samples enriched by MIP-SB, which indicated the high selective adsorption of MA by MIP-SB. Zhu et al. (2006) utilized a MI-SBSE for the selective adsorption of monocrotophos from soil samples, realizing simple and highly selective adsorption of monocrotophos from complicated samples. Fan et al. (2016) prepared a water-soluble graphene oxide (GO)/MIP coating by wrapping the layer around the surface of a stirring rod for the extraction of propranolol (PRO) from urine. Xu et al. (2016) developed a double-template imprinted stir bar for the treatment of bisphenol A (BPA) in river water. Li et al. (2016) grafted a Fe_3O_4 magnetically imprinted polymer onto a stirring rod for the extraction of clonazepam from Chinese herbal medicine samples by stirring bar adsorption. Chen et al. (2020) used borate affinity-imprinted Janus nanomaterials as an adsorbent coating for the extraction of clonazepam from water by stirring bar adsorption.

MIP-SBSE offers simple operation, low organic solvent usage, and a large specific surface area of the MIP coating with strong specific recognition ability, further improving the extraction efficiency and enrichment effect on analytes.

10.8 Molecularly Imprinted Biointerface Cell Recognition

In addition to the application areas mentioned above, molecular imprinting techniques have recently been developed for cell recognition (Pan et al. 2018a) and isolation. The main imprinting methods for cell recognition consist of whole-cell imprinting and epitope imprinting. Whole-cell imprinting is primarily used for the physical identification and isolation of viruses or bacteria by imprinting intact cells, a method that requires cells of defined shape and relative rigidity. Epitope imprinting is the preparation of MIPs for cellular recognition using specific fragments of recognized epitopes as templates. For example, BelBruno (2019) used peptides (DDDG-GDDD) as epitope templates to prepare MIP biointerface for reversible cell adhesion. This adhesion process is briefly described as follows (Fig. 10.8). The terminal short peptide of a long peptide serves as a template for the production of MIP. The other segment of the adhesion sequence, RGD (Arg-Gly-Asp), is released upon binding of the terminal short peptide to the MIP, allowing the cell to be selectively adsorbed on the MIP substrate and completing the process of cellular recognition. This adhesion

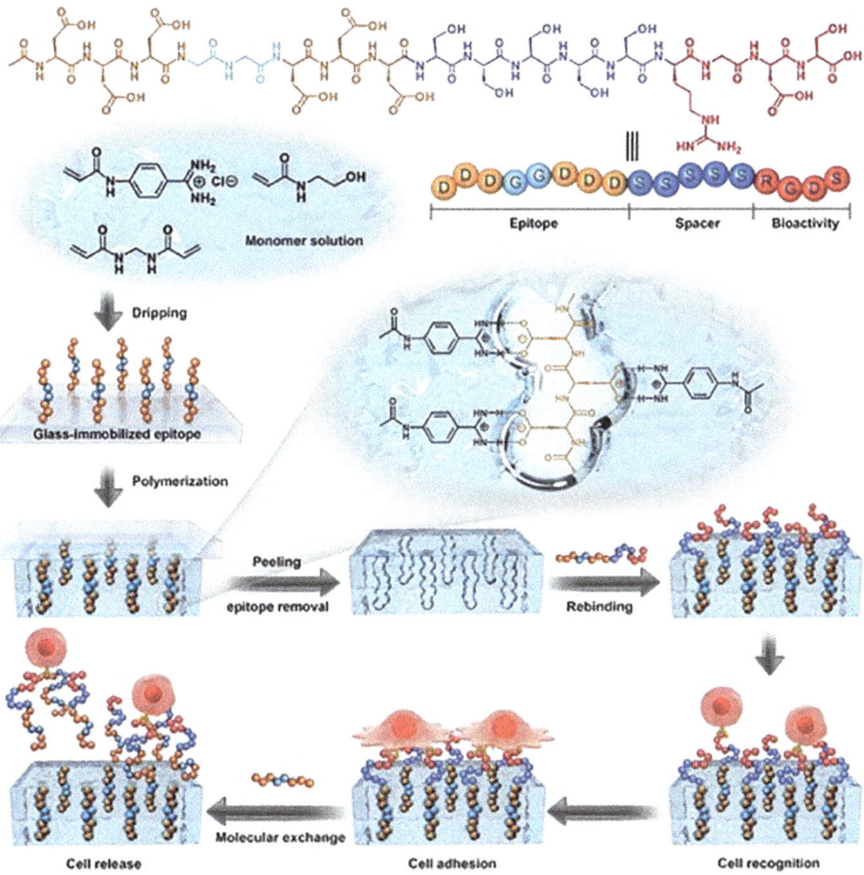

Fig. 10.8 Generation of an epitope-imprinted biointerface (EIB) and dynamic cell adhesion. Reproduced with permission of Pan et al. (2018b)

and release process is reversible and controllable, with a broad application prospect in the fields of disease diagnosis, cell biology, and regenerative medicine.

10.9 Molecularly Imprinted Polymer Sensor Detection

MIPs are used in gas-phase sensors for the detection of toxins, pathogens, and drugs in food and combined with graphene-based nanomaterials for the fabrication of sensing devices, which can greatly improve their selectivity, durability, and stability (Roy et al. 2017). Molecularly imprinted polymer-based materials for device fabrication have potential for intelligence and commercialization (Nitika Thakur et al. 2023). Deng et al. (2014) prepared a MIP sensor using graphene-doped chitosan as

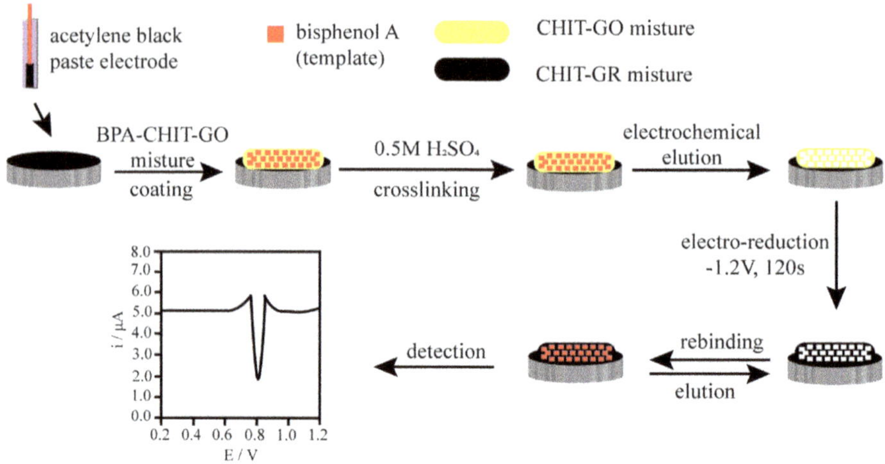

Fig. 10.9 Schematic of the synthesizing process for MMIPs and the MSPE for extracting SFs. Reproduced with permission of Deng et al. (2014)

an electrode for enrichment and detection of bisphenol A in beverages. As shown in Fig. 10.9, bisphenol A was used as a template for the imprinted polymerization reaction with chitosan-graphene oxide on the electrode surface, and the hole imprint of bisphenol A was left on the polymer surface after washing away the excess template. Subsequently, graphene oxide was electrically reduced to graphene in hydrochloric acid solution, and the molecularly imprinted polymer graphene electrode (MIP-GR) was finally obtained. Due to the large specific surface area of graphene, high electrical conductivity, high stability of acetylene black, and strong adsorption capacity, the MIP-GR electrode was able to achieve highly efficient enrichment and selective detection of bisphenol A, with a detection limit of 6.0 nmoL. This imprinted polymer-based sensor recognition mode is expected to be used for rapid trace detection or on-site detection in the field of food hygiene.

10.10 Instrumental Combination Techniques

The on-line process of analyte separation and enrichment was accomplished by connecting the MIP extraction column to the chromatographic system. The high selectivity of MIPs and the automation of the chromatography instrument were utilized to complete the detection and separation of complex matrix samples. Hoshina et al. (2009) synthesized a restricted access medium MIP, which specifically adsorbs barbiturates as antiepileptic drugs, in combination with liquid-quantity coupling to achieve the separation and determination of phenobarbital in river water. Prieto et al.

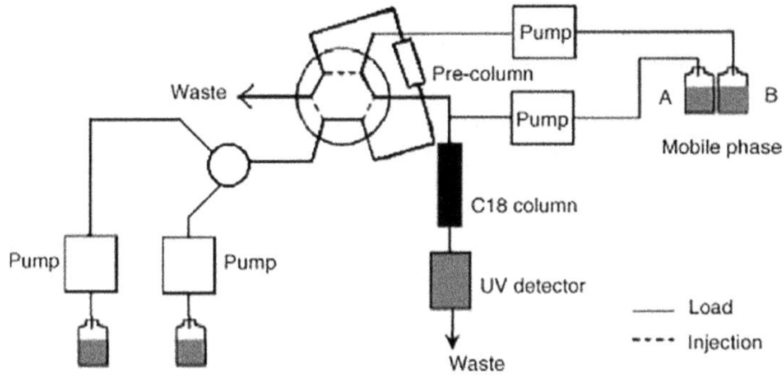

Fig. 10.10 Schematic diagram of an on-line SPE-HPLC system. Reproduced with permission of Ou et al. (2006)

(2011a) used the synthesized MIP as an adsorbent for the extraction and detection of ciprofloxacin in municipal wastewater using filled adsorbent microextraction coupled to LC-MS/MS. Theodoridis et al. (2004) applied MIP to automated sample preparation based on the sequential injection principle, combining the specific adsorption of MIP with the flexibility of HPLC auto-sampling for the extraction of caffeine from water samples. Sun et al. (2009) used a MI-SPE monolithic column coupled with HPLC to efficiently enrich tetracycline in foodstuffs and successfully remove interfering substances from foodstuff samples. Prieto et al. (2011b) combined microextraction with GC-MS to isolate estrogenic compounds from water using MIP as a packed adsorbent. Ou et al. (2006) developed a molecularly imprinted monolithic pre-column conventional reversed-phase chromatographic column for on-line enrichment and detection of bisphenol A in water.

A typical schematic diagram of MIP-SPE coupled with modern instruments shows in Fig. 10.10 (Ou et al. 2006).

- First, the water samples were filtered through a 0.45 μm membrane, and the pH = 3.0 was adjusted. Bisphenol A, along with other phenolic compounds totaling 9 (0.2–1.0 μg/L), were added to the filtrate as the actual samples for reserving.
- The pre-column was sequentially conditioned with ACN and acidified water (pH = 3), followed by the addition of the actual sample to the pre-column, which was washed with 4 mL of ACN and acidified water (80/20 v/v). The flow rate for both steps (1) and (2) was 3.0 mL/min.
- Backflush mode was then used to deliver the eluate to the analytical column for on-line chromatographic analysis. Before reaching the analytical column, mix mobile phase B with mobile phase A, which is water/acetic acid (99/1 v/v).

Molecularly imprinted monolithic columns coupled with HPLC enable efficient enrichment and detection of target compounds. Simple pre-processing of the samples is all that is required to automate the entire process, saving significant labor and time costs.

A summary of relevant studies on the individual methods of sample preparation is detailed in Table 10.1

Each of the above sample preparation methods is highly selective in handling complex matrix samples. Among them, MI-SPE and MI-SPME not only have high recoveries and good stability, but also are user-friendly and without the need for expensive instrumentation. Compared to MI-SPE, MI-SPME treats a much smaller sample volume and has a lower detection limit. MIP-SBSE accomplishes the adsorption and release process in the vortexed sample solution with high extraction efficiency, but there may be a problem of adsorption layer detachment from the stir bar during the sample extraction process. MI-MASE utilizes a fiber membrane that blocks most of the interfering substances in the sample, making it suitable for sample preparation of more complex matrices. The instrument coupling technology combines MIP with sophisticated chromatography instruments that require specialized personnel to operate, with a high degree of automation and extremely low detection limits for trace detection and on-line enrichment of complex matrix samples.

10.11 Outlook and Future Development Trends

Up to now, the methods of MIP in sample preparation and enrichment mainly include SPE, SPME, MSPE, MSPD, SBSE, MASE, etc. MIP-based treatments show excellent extraction efficiency and selectivity in samples with complex matrices compared to traditional extraction methods. Efficient extraction can be achieved with a small amount of solvent and simple steps. MIP is a green, non-toxic, and environmentally friendly material, as well as having a high economic benefit by automating sample preparation in conjunction with chromatography instruments, saving time, labor, and cost. With the occurrence of emerging pollutants and the strengthening people's health awareness, which puts higher demands on sample analysis, sample preparation has become increasingly important in the analytical field. Nevertheless, most of the current applications of MIP in sample preparation have remained in the laboratory stage only, without realizing large-scale commercial applications. In recent years, nanoscale polymer materials, which have a smaller size and a larger specific surface area than micrometers, giving them a higher adsorption capacity than micron-sized materials, and has attracted the focus of many studies. In the future, the combination of molecular imprinting technology with nanomaterials and automated modern analytical instruments will be more widely used in the environmental field, food industry, biosensors, disease diagnosis and drug delivery.

Table 10.1 Summary of enrichment methods

Methods	Analytes	Real samples	Recovery	Limit of detection	References
MI-SPE	Fluoroquinolones	Chicken, fish, shrimp and pork	80.4–110.7%	0.003–0.05 μg/L	Su et al. (2023)
	Organophosphorus	Vegetable oils	73–99%	0.3–2 μg/kg	Boulanouar et al. (2017)
	Carbaryl	Rice and corn	93–101%	3.6 μg/kg	Chen et al. (2023)
	Dibutyl phthalate	River water	83.2%	—	Shen et al. (2023)
	Triazines	Grape juice and soil	88%	—	Chapuis et al. (2004)
	Estrogen receptor	Sewage	—	—	Yagishita et al. (2019)
MI-SPME	Phenolic compounds	Tap water	88.9–102.5%	0.56–4.5 μg/L	Feng et al. (2009)
		River water	80.0–94.0%		
		Sewage water	80.0–90.5%		
	Ochratoxin A	Red wine	—	0.12 μg/L	Yu and Lai (2006)
	Tobacco-specific nitrosamines	Water samples	—	0.1–6.7 μg/L	Chen et al. (2024)
	Anesthetics	Water	—	—	Courtois et al. (2006)
	Triazines	Soil, potato and pea		0.12 μg/L	Turiel et al. (2007)
MI-MSPE	Endocrine disruptor	Lake water and milk	89.9–102.5%	0.1–0.3 μg/L	Huang et al. (2020)
	Herbicides	Water samples	77–98%	0.33–0.71 μg/L	Meseguer-Lloret et al. (2022)
	Patulin	Juice samples	79.4–97.9%	0.1 μg/L	Gaho et al. (2022)
	Norfloxacin	Effluent water and drinking water	75.5–90%	—	Zhao et al. (2020)
	Pb(II)	Water samples	92.4–100.8%	4.9 μg/L	Shan et al. (2022)

(continued)

Table 10.1 (continued)

Methods	Analytes	Real samples	Recovery	Limit of detection	References
	Strobilurin fungicides	Jujube, pear, zucchini and cucumber	71.5–115.8%	0.03–0.49 μg/kg	Hu et al. (2023)
MI-MSPD	Rhodamine B	Chili powder, tomato ketchup and jelly powder	83.6–96.9%	0.14 μg/kg	Arabi et al. (2020)
	Parabens	Sunscreen	88.8–102.8%	11.1 μg/kg	Gholami et al. (2019)
MI-MASE	Triazines herbicides	Lettuce and apple	79–98%	22–38 μg/kg	Mhaka et al. (2009)
	Carbamazepine	Surface water	—	0.09–0.20 μg/L	Khulu et al. (2022)
	Triazine herbicides	Sludge water, watermelon, milk and urine samples	71.1–108.3%	0.006–0.02 μg/L	Hu et al. (2009)
MIP-SBSE	Melamine	Powdered and milk	89.84–100.2%	0.0127 μg/L	Zhu et al. (2015)
	Monocrotophos	Soil	71.5–92.3%	3.5–6.1 μg/kg	Zhu et al. (2006)
	Propranolol	Urine	86.8–106%	0.37 μg/L	Fan et al. (2016)
	Bisphenol A	River water	63–92%	0.003 μg/L	Liu et al. (2016)
		Lake water	89–99%		
	Clonazepam	Herbal health foods	89.8–103.3%	10 μg/L	Li et al. (2016)
	Cis-diol-containing catechol	Water samples	—	—	Chen et al. (2020)
Molecularly imprinted biointerface cell recognition	3T3 cells	Mouse	—	—	Pan et al. (2018b)
Molecularly imprinted polymer sensor detection	bisphenol A	Drinks in plastic bottles or cans	—	6.0 nmoL	Deng et al. (2014)
Instrumental combination techniques	Phenobarbital phenytoin	River water	—	0.0005–0.005 μg/L	Hoshina et al. (2009)

(continued)

Table 10.1 (continued)

Methods	Analytes	Real samples	Recovery	Limit of detection	References
	Fluoroquinolone antibiotics	Waste water	87–115%	—	Prieto et al. (2011a)
	Caffeine	Human urine, coffee and beverages	70%	—	Theodoridis et al. (2004)
	Tetracyclines	Milk and honey	62.6–90.6%	—	Sun et al. (2009)
	Alkylphenols	River and waste water	75–109%	0.087 μg/L	Prieto et al. (2011b)
	Phenolic compounds	River water	95.2–96.0%	0.06 μg/L	Ou et al. (2006)

References

Arabi M et al (2020) Hydrophilic molecularly imprinted nanospheres for the extraction of rhodamine B followed by HPLC analysis: a green approach and hazardous waste elimination. Talanta 215
BelBruno JJ (2019) Molecularly imprinted polymers. Chem Rev 119(1):94–119
Boulanouar S, Combès A, Mezzache S, Pichon V (2017) Synthesis and application of molecularly imprinted polymers for the selective extraction of organophosphorus pesticides from vegetable oils. J Chromatogr A 1513:59–68
Chapuis F, Pichon V, Lanza F, Sellergren B, Hennion MC (2004) Retention mechanism of analytes in the solid-phase extraction process using molecularly imprinted polymers. J Chromatogr B 804(1):93–101
Chen X, Wu F, Tang J, Yang K, Ma Y, Pan J (2020) Anisotropic emulsion constructed boronate affinity imprinted Janus nanosheets for stir bar sorptive extraction of cis-diol-containing catechol. Chem Eng J 395
Chen S et al (2023) Preparation of biochar-based surface molecularly imprinted polymers and evaluation of their selective adsorption and removal of carbaryl from rice and corn. J Chromatogr A 1705
Chen Y et al (2024) Molecularly imprinted polymer sheathed mesoporous silica tube as SPME fiber coating for determination of tobacco-specific nitrosamines in water. Sci Tot Environ 906
Courtois J, Fischer G, Sellergren B, Irgum K (2006) Molecularly imprinted polymers grafted to flow through poly(trimethylolpropane trimethacrylate) monoliths for capillary-based solid-phase extraction. J Chromatogr A 1109(1):92–99
Deng P, Xu Z, Kuang Y (2014) Electrochemical determination of bisphenol A in plastic bottled drinking water and canned beverages using a molecularly imprinted chitosan-graphene composite film modified electrode. Food Chem 157:490–497
Fan W, He M, You L, Zhu X, Chen B, Hu B (2016) Water-compatible graphene oxide/molecularly imprinted polymer coated stir bar sorptive extraction of propranolol from urine samples followed by high performance liquid chromatography-ultraviolet detection. J Chromatogr A 1443:1–9
Feng Q, Zhao L, Lin J-M (2009) Molecularly imprinted polymer as micro-solid phase extraction combined with high performance liquid chromatography to determine phenolic compounds in environmental water samples. Anal Chim Acta 650(1):70–76
Gaho MM et al (2022) Synthesis of novel magnetic molecularly imprinted polymers by solid-phase extraction method for removal of norfloxacin. Chin J Anal Chem 50(6)
Gholami H, Ghaedi M, Arabi M, Ostovan A, Bagheri AR, Mohamedian H (2019) Application of molecularly imprinted biomembrane for advancement of matrix solid-phase dispersion for clean enrichment of parabens from powder sunscreen samples: optimization of chromatographic conditions and green approach. ACS Omega 4(2):3839–3849
Hoshina K, Horiyama S, Matsunaga H, Haginaka J (2009) Molecularly imprinted polymers for simultaneous determination of antiepileptics in river water samples by liquid chromatography–tandem mass spectrometry. J Chromatogr A 1216(25):4957–4962
Hu Y et al (2023) Preparation of magnetic molecularly imprinted polymers for the selective extraction of strobilurin fungicides in agricultural products. J Food Compos Anal 123
Hu Y, Wang Y, Hu Y, Li G (2009) Liquid–liquid–solid microextraction based on membrane-protected molecularly imprinted polymer fiber for trace analysis of triazines in complex aqueous samples. J Chromatogr A 1216(47):8304–8311
Huang XC, Ma JK, Wei SL (2020) Preparation and application of a novel magnetic molecularly imprinted polymer for simultaneous and rapid determination of three trace endocrine disrupting chemicals in lake water and milk samples. Anal Bioanal Chem 412(8):1835–1846
Khulu S et al (2022) Multivariate optimization of a two-way technique for extraction of pharmaceuticals in surface water using a combination of membrane assisted solvent extraction and a molecularly imprinted polymer. Chemosphere 286
Li X et al (2016) Development and application of novel clonazepam molecularly imprinted coatings for stir bar sorptive extraction. J Colloid Interface Sci 468:183–191

Liu R, Feng F, Chen G, Liu Z, Xu Z (2016) Barbell-shaped stir bar sorptive extraction using dummy template molecularly imprinted polymer coatings for analysis of bisphenol A in water. Anal Bioanal Chem 408(19):5329–5335

Meseguer-Lloret S, Torres-Cartas S, Gómez-Benito C, Herrero-Martínez JM (2022) Magnetic molecularly imprinted polymer for the simultaneous selective extraction of phenoxy acid herbicides from environmental water samples. Talanta 239

Mhaka B et al (2009) Selective extraction of triazine herbicides from food samples based on a combination of a liquid membrane and molecularly imprinted polymers. J Chromatogr A 1216(40):6796–6801

Nitika Thakur TRD, Gupta S, Shukla SK, Patra S, Kumar D (2023) Chapter 5—Material selection for the fabrication of MIP-based devices: an overview. Elsevier, pp. 101–114

Ou J, Hu L, Hu L, Li X, Zou H (2006) Determination of phenolic compounds in river water with on-line coupling bisphenol A imprinted monolithic precolumn with high performance liquid chromatography. Talanta 69(4):1001–1006

Pan J, Chen W, Ma Y, Pan G (2018b) Molecularly imprinted polymers as receptor mimics for selective cell recognition. Chem Soc Rev 47(15):5574–5587

Pan J, Chen W, Ma Y, Pan G (2018a) Molecularly imprinted polymers as receptor mimics for selective cell recognition. Chem Soc Rev 47(15):5574–5587

Prieto A et al (2011b) Selective determination of estrogenic compounds in water by microextraction by packed sorbents and a molecularly imprinted polymer coupled with large volume injection-in-port-derivatization gas chromatography–mass spectrometry. Anal Chim Acta 703(1):41–51

Prieto A, Schrader S, Bauer C, Möder M (2011a) Synthesis of a molecularly imprinted polymer and its application for microextraction by packed sorbent for the determination of fluoroquinolone related compounds in water. Anal Chim Acta 685(2):146–152

Roy E, Patra S, Tiwari A, Madhuri R, Sharma PK (2017) Introduction of selectivity and specificity to graphene using an inimitable combination of molecular imprinting and nanotechnology. Biosens Bioelectron 89:234–248

Santanu Patra ER, Madhuri R, Sharma PK (2016) A genuine combination of solvent-free sample preparation technique and molecularly imprinted nanomaterials. Wiley Online Library

Shan X et al (2022) Magnetic solid phase extraction of lead ion from water samples with humic acid modified magnetic nanoparticles prior to its fame atomic absorption spectrometric detection. Microchem J 179

Shen J et al (2023) Preparation and characterisation of photoresponsive molecularly imprinting polymer nanoparticles with hollow structure for extraction and enrichment of dibutyl phthalate. React Func Polym 186

Su L-H, Qian H-L, Yang C, Wang C, Wang Z, Yan X-P (2023) Surface imprinted-covalent organic frameworks for efficient solid-phase extraction of fluoroquinolones in food samples. J Hazar Mater 459

Sun X, He X, Zhang Y, Chen L (2009) Determination of tetracyclines in food samples by molecularly imprinted monolithic column coupling with high performance liquid chromatography. Talanta 79(3):926–934

Theodoridis G, Zacharis CK, Tzanavaras PD, Themelis DG, Economou A (2004) Automated sample preparation based on the sequential injection principle. J Chromatogr A 1030(1–2):69–76

Turiel E, Tadeo JL, Martin-Esteban A (2007) Molecularly imprinted polymeric fibers for solid-phase microextraction. Anal Chem 79(8):3099–3104

Yagishita M et al (2019) Efficient extraction of estrogen receptor–active compounds from environmental surface water via a receptor-mimic adsorbent, a hydrophilic PEG-based molecularly imprinted polymer. Chemosphere 217:204–212

Yu JCC, Lai EPC (2006) Molecularly imprinted polypyrrole modified carbon nanotubes on stainless steel frit for selective micro solid phase pre-concentration of ochratoxin A. React Funct Polym 66(7):702–711

Zhao M et al (2020) Preparation of core-shell magnetic molecularly imprinted polymers for extraction of patulin from juice samples. J Chromatogr A 1615

Zhu X, Cai J, Yang J, Su Q, Gao Y (2006) Films coated with molecular imprinted polymers for the selective stir bar sorption extraction of monocrotophos. J Chromatogr A 1131(1–2):37–44

Zhu L, Xu G, Wei F, Yang J, Hu Q (2015) Determination of melamine in powdered milk by molecularly imprinted stir bar sorptive extraction coupled with HPLC. J Colloid Interface Sci 454:8–13

Chapter 11
Molecularly Imprinted Polymers Based Separation and Purification Techniques

Ilgım Göktürk, Duygu Çimen, Fatma Yılmaz, and Adil Denizli

11.1 Introduction

Although pollutants including antibiotics, medicines, pesticides, dyes, and heavy metal wastes are dangerous, the insufficiency of purifying systems endangers the environment and human health. Regular environmental monitoring and removal of these contaminants are critical for protecting human health and ecosystems. To make reproducible measurements of analytes at very low concentration levels, new analytical procedures must be devised. In recent years, molecularly imprinted polymers (MIPs) have been combined with separation and purification techniques to selectively detect and remove these contaminants. Instead of costly and time-consuming standard analytical procedures, molecularly imprinted polymers may aid in the development of simple, quick, and highly selective analytical methods for detecting and removing harmful pollutants. MIPs are synthetic counterparts of natural biological antibody-antigen complexes that have been investigated since the 1970s. They can selectively bind template molecules during synthesis using a "lock and key" method (Derazshamshir et al. 2020).

In recent years, molecular imprinting technology has become an important strategy for the design and preparation of synthetic materials that mimic the recognition processes found in nature. MIT is a promising approach that can detect biological and chemical components such as nucleotide derivatives, amino acids and proteins, food contaminants and pharmaceuticals (Scorrano et al. 2011). MIT is frequently

I. Göktürk · D. Çimen · A. Denizli (✉)
Department of Chemistry, Hacettepe University, 06800 Ankara, Turkey
e-mail: denizli@hacettepe.edu.tr

F. Yılmaz
Vocational School of Gerede, Department of Chemistry Technology, Bolu Abant Izzet Baysal University, Bolu, Turkey

S. Patra and M. Sillanpaa (eds.), *Molecularly Imprinted Polymers as Artificial Antibodies for the Environmental Health*, https://doi.org/10.1007/978-3-031-58995-9_11

used in different fields such as sensors, separation sciences and purification, catalysis and drug delivery (Çimen et al. 2016, 2018; Kurt et al. 2023; Shama et al. 2023; Zabihi et al. 2023).

Water pollution is accountable for roughly 9 million annual fatalities. Regrettably, there has been no observable alteration in the count of deaths attributed to pollution since 2015 (Fuller et al. 2022). Deaths from pollution, the unintended consequences of industrialization and urbanization, have increased more than 66% since 2000. The combination of increasing urbanization and industrialization, along with ineffective drainage systems, has led to the inadvertent discharge of untreated water, thereby exacerbating the problem of water pollution.

Agricultural waste (AWs), including crop residue, aquaculture waste, and agro-industrial refuse (Koul et al. 2022), as well as industrial waste from heavy metal contamination and plastic production (Briffa et al. 2020), and pharmaceutical waste (Ma et al. 2023; Rozaini et al. 2023) have been identified in aquatic ecosystems. To meet the increasing needs of the fast-growing population, enormous amounts of AWs are created every day on a global basis. Cellulose, the most prevalent biopolymer, is the main component of crop residue and agro-industrial waste, followed by lignin and hemicellulose. AWs and their processing are a global problem. Traditionally, some crop residues have been used for incineration, animal feed, compost, and paper production. At present, the vast majority of these wastes are incinerated or deposited in landfills, leading to contamination of the atmosphere, water, and soil while also contributing to global climate change (Koul et al. 2022). The application of MIPs to improve the efficiency and selectivity of sample preparation and extraction processes can lead to significant advancements in wastewater treatment. This chapter provides a brief discussion of the use of MIPs in the design of materials capable of eliminating environmental contaminants.

11.2 Molecularly Imprinted Polymers

Molecular imprinting technology (MIT) is a synthetic method that uses natural recognition components, like antibodies and biological receptors, to create reliable molecular recognition materials (Ye and Mosbach 2008). During the polymerisation process at MIT, a crosslinker is used to form a three-dimensional polymer network by preparing a complex between a functional monomer and the analyte. After polymerisation, the analyte is removed from the polymer with a suitable desorption solution and recognition sites with the same size, shape and chemical properties as the analyte molecule are formed. In general, molecular recognition involves different intermolecular interactions such as hydrogen bonds, ionic and dipole–dipole interactions between the functional groups and template molecules present in the polymer matrix (Wulff 2013).

MIPs prepared by molecular imprinting technique are synthetic materials designed by selective binding and identification of a template molecule with a specific space structure and durability (He et al. 2021) (Fig. 11.1). MIPs are typically based on

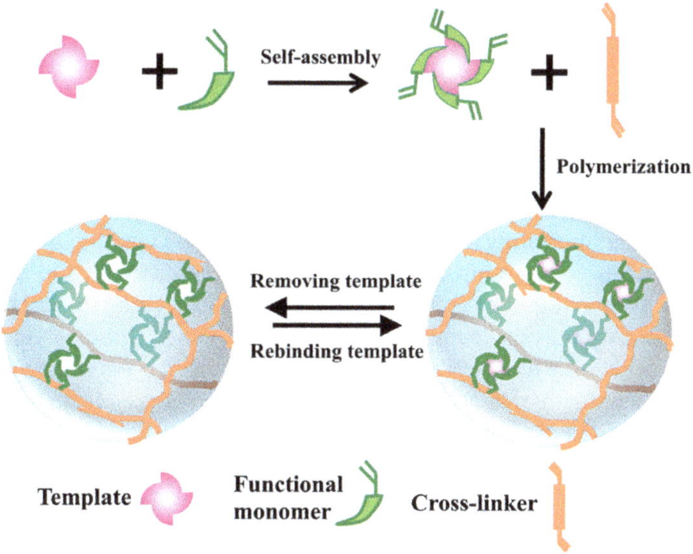

Fig. 11.1 Step-by-step schematic representation of the preparation of MIPs. Reproduced with permission from He et al. (2021)

a polymer matrix, which serves as the support structure for the imprinted sites. Common polymers used in MIPs include acrylics, methacrylates, and polyvinyl polymers. The choice of polymer matrix is crucial, as it affects the physical and chemical properties of the resulting MIP. The template molecule is defined as the target molecule that the MIP is designed to selectively recognise and bind to. It is crucial to choose a molecule with a well-defined structure for the imprinting process (Huang et al. 2017; Belbruno 2019).

Nanoimprinting technology involves creating MIPs with dimensions at the nanoscale (Chen et al. 2020). These nano-MIPs eliminate the need for additional processing, showcasing strong mechanical strength that ensures the longevity of their recognition sites. The diminutive size of MIPs at the nanoscale results in a substantial specific surface area, facilitating nearly complete extraction of the template and optimizing available binding locations. These binding sites are typically situated on or near the surface, enhancing their binding capabilities significantly. Moreover, the accessibility of template molecules to the molecularly imprinted sites is heightened, leading to rapid binding kinetics. Consequently, nanoimprinting technology shows significant promise across various applications and can enhance the stability of molecularly imprinted sensors by generating more resilient nanomaterials (Azimi et al. 2020). In contrast to bulk imprinting, nanoimprinting provides improved accessibility to binding sites, resulting in rapid recognition and reducing the diffusion distance between the target molecule and the binding site, with limited steric hindrance. These advantages are particularly significant when imprinting biomacromolecules (Haupt et al. 2020). Surface-imprinted nanomaterials enable swift template transfer, enhancing both binding capacity and kinetics. Furthermore, they offer the potential to integrate additional functions, providing supplementary features (Kang et al. 2023). Besides, there has been a notable increase in research interest in smart MIPs or stimuli-responsive MIPs, which are meticulously designed MIPs capable of responding to specific external stimuli, thereby altering their affinity for the target molecule and offering the potential for switchable binding or release. The strategy for creating smart MIPs involves integrating a stimuli-responsive material into the molecular imprinting technique, endowing the MIP with the ability to exhibit switchable binding or releasing capacities. Smart MIPs find applications in various fields such as chromatography techniques, antibody mimics, artificial enzymes, drug delivery, chemical catalysis, membranes, sensors, and more. These polymers exhibit distinctive responses when exposed to external stimuli like temperature, incident light, pH, ionic strength, magnetic fields, or the presence of specific molecules (Roy et al. 2017; Thakur et al. 2023; Morsi et al. 2023).

Template molecules, functional monomers, initiators and solvents are the most important parts of MIPs preparation (Algieri et al. 2014; Hasanah et al. 2021):

1. The template or target molecule is the molecule to which the MIP is specifically designed to bind. During the MIP synthesis process, the template molecule is mixed with the monomers and cross-linking agents. It imparts the shape and chemical functionality that the MIP will recognize.
2. Functional monomers are monomers that are chosen to interact with the template molecule through non-covalent interactions like hydrogen bonding, electrostatic interactions, or van der Waals forces. Functional monomers help create specific binding sites in the polymer matrix.
3. Cross-linkers are added to the polymerization mixture to form a three-dimensional network within the MIP. They connect the polymer chains.
4. Solvents are used to dissolve the monomers and initiators, and initiators trigger the polymerization reaction. The choice of solvent and initiator can affect the properties of the resulting MIP.

The choice of molecular imprinting technology depends on the specific application and the nature of the template molecule. The interactions between the template molecule and monomers play a crucial role in determining the specificity and selectivity of the resulting MIPs. There are several molecular imprinting approaches based on these interactions (Wu and Li 2004; Curcio et al. 2009; Dong et al. 2021).

11.2.1 The Interactions Used in Molecular Imprinting

Non-covalent Imprinting Method

Hydrogen Bonding: In this approach, functional monomers form hydrogen bonds with the template molecule. This is a common strategy for creating MIPs for molecules with hydrogen bond-donating or accepting groups.

Ionic Interactions: Functional monomers with charged groups can interact with oppositely charged template molecules. This is useful for imprinting charged species, such as ions or charged biomolecules.

Van der Waals Interactions: Functional monomers with hydrophobic groups can interact with the template molecule through van der Waals forces. This is suitable for hydrophobic templates.

Covalent Imprinting Method

In covalent imprinting, the functional monomers form covalent bonds with the template molecule. This results in highly specific recognition sites. However, it may require additional steps to cleave the covalent bonds for template molecule removal.

Semi-covalent Imprinting Method

This approach involves both covalent and non-covalent approaches. Covalent functional groups are bound to template molecules, leading to a betterment after cleavage of the molecule. The binding sites experience a more uniform distribution. Rebinding which is accomplished using non-covalent interactions, is not affected by any kinetic limitations, apart from diffusion.

Metal-Ion Imprinting Method

In some cases, metal ions can be used as functional monomers. These metal ions can coordinate with specific functional groups on the template molecule. After polymerization, the metal ions can be removed, leaving behind coordination sites that recognize the template.

11.2.2 Methods for Preparation of MIPs

MIPs are created through a process known as molecular imprinting, where a template molecule is used to create binding sites within the polymer matrix. There are several different polymerization methods used in molecular imprinting, each with its advantages and disadvantages (Bie et al. 2015; Lin et al. 2016; Kobayashi et al. 2019; Phan et al. 2019). Here are the main types.

Bulk Polymerization

In bulk polymerization, the monomers, a cross-linking agent, and the template molecule are mixed in a single step and polymerized in a bulk form. After polymerization, the template is removed, leaving behind cavities with complementary shapes and chemical functionality. Bulk polymerisation has the advantage of a straightforward process and the disadvantage of limited control over the size and distribution of binding sites and difficult template removal.

Precipitation Polymerization

The template molecule and monomers are dissolved in a common solvent in precipitation polymerization. Then, a poor solvent is added to induce polymerization. This causes the polymer to precipitate out of the solution, creating imprinted particles. This method has the advantage that well-defined particle morphology can improve separation efficiency. Precipitation conditions must be carefully controlled because the template molecule may not be easily removed.

Suspension Polymerization

In suspension polymerization, the monomers, cross-linking agent, and template are suspended in an aqueous medium. Polymerization is initiated to form imprinted particles suspended in the solution. This method requires control over particle size and distribution, and careful control of suspension conditions while allowing uniform imprinted areas.

Emulsion Polymerization

Emulsion polymerization involves dispersing the monomers, cross-linking agent, and template in an aqueous medium with the help of surfactants. Polymerization occurs within the emulsion droplets. The most important advantage of this method is that it allows the creation of nanoparticles with high surface area and well-defined binding sites. The disadvantage is that the template molecule is difficult to remove.

Surface Imprinting

Surface imprinting is a technique where the template molecule is immobilized on a substrate, and the polymer is then polymerized directly on the surface of the template. After polymerization, the template is removed, leaving behind surface-imprinted sites. While this method has the advantage of creating well-defined binding sites on the surface, it is limited to flat or planar surfaces.

11.3 Applications of Separation and Purification Techniques Using MIPs

The issue of environmental contamination from heavy metals, which is a significant source of global apprehension because of its adverse impacts, is increasingly troublesome. The rapid expansion of agriculture and metal industries, along with fertilizers, improper waste disposal, and pesticides, results in the release of these inorganic pollutants into our soils, waters, and the atmosphere. Efficient elimination of heavy metals and harmful organic compounds from contaminated wastewater offers promise in water treatment, presenting environmental and economic benefits. For example, water in oil (W/O) Pickering HIPEs template stabilized by MSPs as shown in Fig. 11.2 was used to prepare a molecularly imprinted porous adsorbent (MIPA) (Liu et al. 2018).

Soleiman and colleagues created molecularly imprinted polymer-based carbon nanocomposites for the adsorption of Hg(II) ions from an aqueous medium (Abubakar et al. 2023). Sharma and Kandasubramanian investigated the selective detection and extraction of heavy metal ions using molecularly imprinted polymers (Sharma and Kandasubramanian 2020). Profumo et al. investigated novel applications of MIPs for waste and food removal (Speltini et al. 2017). Ejiogu and colleagues investigated the bioremediation of heavy metal-containing tannery effluent (Igiri et al. 2018). Metal ions have been removed using molecularly imprinted polymers as an alternate treatment approach (Erdem et al. 2018). Chen and colleagues (Wang et al. 2019a) investigated green multifunctional monomer-based ion-imprinted polymers to remove copper ions selectively from aqueous medium. Figure 11.3 depicts a schematic representation of the G-HQ-C ion imprinted polymer preparation process.

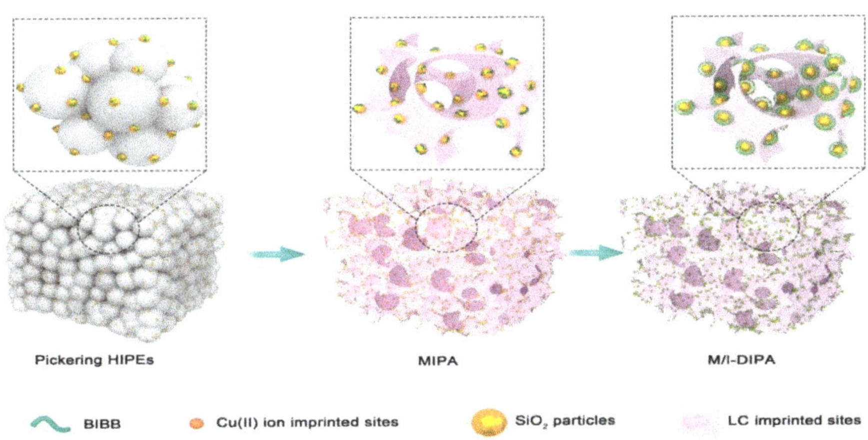

Fig. 11.2 The manufacturing procedure of Pickering HIPEs, MIPA, and M/I-DIPA. Get permission from Liu et al. (2018)

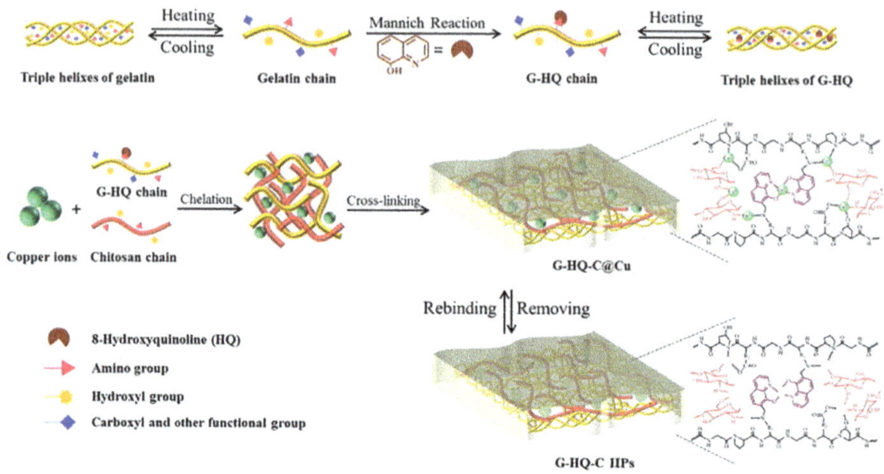

Fig. 11.3 Schematic depiction of the steps involved in getting ready for G-HQ-C ion-imprinted polymers. Get permission from Wang et al. (2019a)

Certain metals can disrupt biological functions and hinder growth, while others can accumulate in various organs, leading to the development of severe illnesses, including cancer (Briffa et al. 2020). Superparamagnetic iron oxide nanoparticles with an alginate coating were employed as nanocomposite adsorbents to remove arsenic from aqueous solutions in environmental applications (Haris et al. 2023). Inexpensive microbial amyloid nanofibers (MAN) have been harnessed for the removal of heavy metal ions. Given the high costs associated with current methods for heavy metal ion removal, it is essential to explore and develop innovative approaches like this one (Shah et al. 2023).

Human and ecological health suffer when active pharmaceutical ingredients (APIs) are exposed to the environment. The most polluted countries are those with low to moderate incomes. It is linked to inadequate infrastructure for managing wastewater and trash. When APIs are produced, used, and disposed of, they emit chemicals into the environment. The three APIs that are most commonly found are metformin, caffeine, and carbamazepine. There is proof that humans' and ecosystems' health suffers when exposed to APIs in the environment. Nur et al. reviewed improved adsorption of non-steroidal anti-inflammatory drugs from environmental waters in the development of offline and online enrichment methods (Rozaini et al. 2023). SupelMIP-SPE is a strongly cross-linked polymer that is one of Sigma Aldrich's commercial SPEs. It is intended for the selective extraction of a set of analytes with comparable structural relationships. Selectivity is added, allowing a template molecule to direct the creation of cavities that are complementary to the target molecule or molecules. This product delivers a high pre-concentration extraction sorbent, increased selectivity, and sensitivity because of the strong chemical interaction between the solid phase and the target molecule. Pharmaceutical waste was found in 1052 samples collected from 104 countries in a prior investigation. Concern

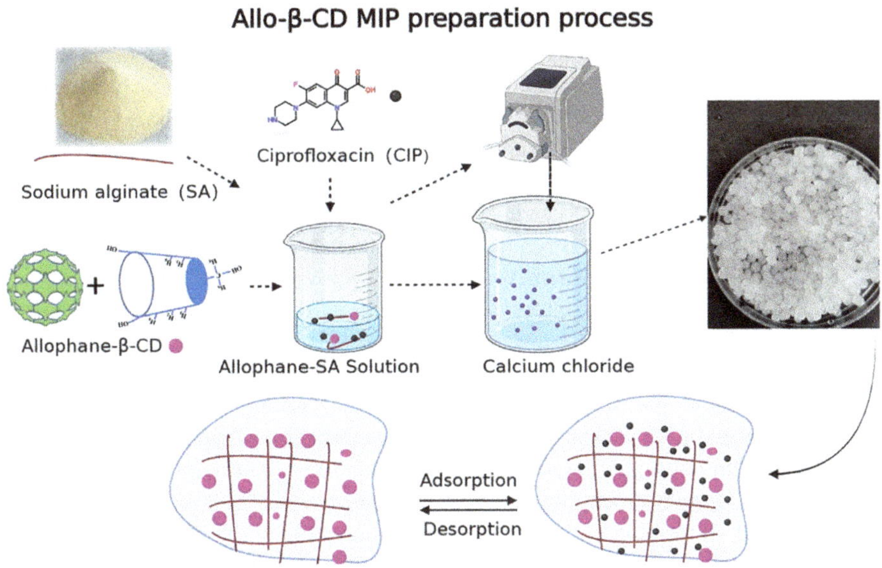

Fig. 11.4 An example of how to prepare Allo-β-CD MIP for effective CIP adsorption. Get permission from Ma et al. (2023)

has been raised about pharmaceutical residues in environmental water (Rozaini et al. 2023). Antibiotics harm the health of people and animals when they are present in water and soil. Allophane-based molecularly imprinted polymers were developed by Yuan and colleagues to efficiently remove antibiotics from aqueous solutions (Ma et al. 2023). Figure 11.4 depicts the preparation procedure of allo-β-CD MIP.

The multi-template imprinting multiplex method using Mt-MIPs has enabled the simultaneous identification of various targets since the 2000s (Shah et al. 2023; Zhao et al. 2023). Mt-MIPs can reduce consumption while saving money, time and labor. Multi-template molecular printing technology simplifies the process by creating multiple recognition sites simultaneously on a single polymeric material using two or more targets as templates, using fewer chemicals and generating less waste (Click 2015; Wang et al. 2023). For the simultaneous purification of multiple targets, such as therapeutic medicines (Ma and Row 2019) and organic contaminants (Ncube et al. 2017), mt-MIPs-based solid-phase extraction (mt-MIPs-SPE) demonstrated enhanced efficiency compared to SPE with inadequate selectivity. In order to create a solid-phase extraction sorbent that could selectively extract and separate 17 triazine herbicides and their metabolites from samples of maize, wheat, and cotton seeds, Wang et al. (2019b) synthesized dual-templates molecularly imprinted polymers utilizing atrazine and pyrometry as the dual templates. Metal–organic frameworks, or MOFs, are novel synthetic materials with large surface areas, porous architectures, and high adsorption affinity that have received a lot of attention nowadays. Mt-MIPs hold significant promise for effectively removing antibiotics simultaneously. In their investigation, Lu et al. used mt-MIPs to extract six phenolic chemicals (Lu et al.

2017). Besides, the polymers used in magnetic solid-phase extraction, or MSPE, are a great way to quickly and very sensitively measure phenolic endocrine disruptors (Bhogal et al. 2021). Burakham and colleagues created molecularly imprinted polymers with two templates in order to isolate eight organophosphorus chemicals quickly from fruit and vegetable samples (Boontongto and Burakham 2021). When it comes to magnetic molecularly imprinted polymers, MSPE is superior to standard SPE because it requires less organic solvents, simplifies operation without requiring centrifugation or filtering, and reduces extraction and separation times. Even after 200 repeated experiments, magnetic molecularly imprinted polymers effectively absorbed three endocrine disruptors in water (Wang et al. 2020a). The manufacturing method of plastic bags and the use of plasticizers like BPA present possible health hazards when food comes into prolonged contact with plastic bags. Consequently, there is widespread attention and interest in the analysis and research methods for BPA and other harmful substances. Because of its exceptional selectivity, molecular imprinting technology has found extensive application in extracting and separating pollutants and frequently employed in conjunction with materials based on MOFs. In a study, Zhu et al. introduced a straightforward, convenient, rapid, and environmentally friendly method for synthesizing molecularly imprinted polymers with magnetic iron MOFs (MCMMIPs) which shows a significant specific surface area (289.4 m^2/g), and an exceptional magnetic separation performance (20 emu/g) (Zhu et al. 2023). Figure 11.5 shows FTIR, XRD, and VSM of MCMMIPs and Fe_3O_4. Compared to traditional industrial effluent treatment techniques, membrane systems are more efficient, boasting lower operating costs and minimal energy requirements. The recent advancements by Ramachandran and Sathishkumar have highlighted the promising outlook of membrane-based techniques for pollutant removal (Ramachandran and Sathishkumar 2023). Table 11.1 outlines the utilization of membrane-based treatment systems for the efficient removal of emerging contaminants (ECs). Because of the primary challenge associated with the use of membrane-based filtration systems is fouling, a visual representation of the various types of fouling is presented in Fig. 11.6.

Bakhshizadeh et al. (2022) conducted studies on the preparation and characterization of mixed matrix membranes, incorporating a PVDF blend and hydrophilic molecularly imprinted MIL-101 (Cr) as a filler (Fig. 11.7). This was aimed at achieving effective and selective removal of dye.

Cyclodextrins (CDs) consist of cyclically linked starch molecules. The multiple hydroxyl groups present in cyclodextrins serve as active sites capable of establishing diverse connections. These can be cross-linked or modified to produce monomers capable of creating branching or linear networks. Additionally, they can engage in inclusion complexes with various guest molecules, encompassing both organic and inorganic compounds, thereby altering the physicochemical properties of the complexes. The monomers derived from cyclodextrins and employed in molecular imprinting exhibit enhanced binding affinity for target molecules. Suah and colleagues emphasized the utility of imprinting technology employing cyclodextrins as a valuable functional monomer in addressing phenolic compound remediation (Syed Yaacob et al. 2023). Mamman and collaborators explored a new magnetic

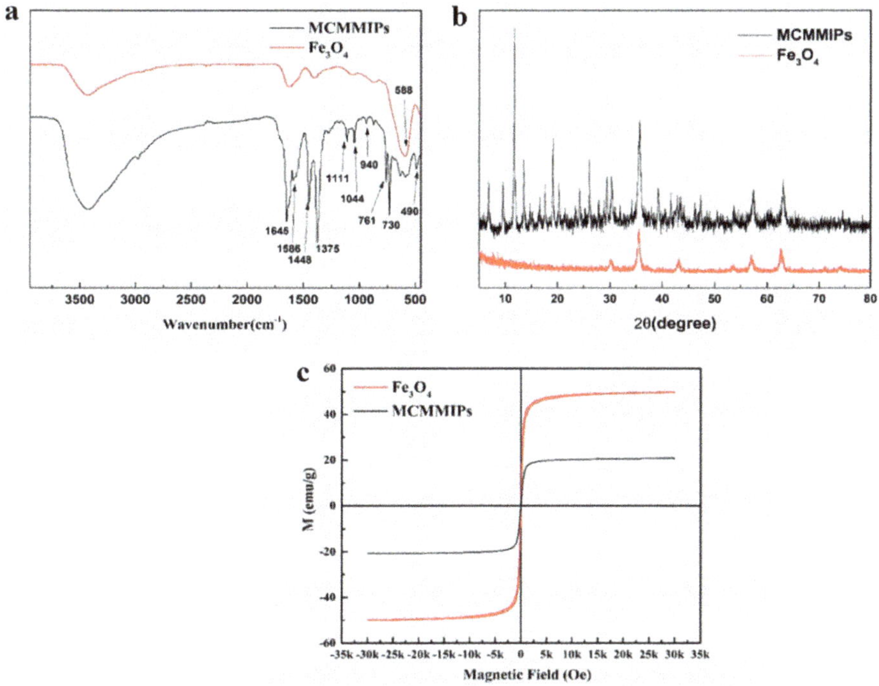

Fig. 11.5 FTIR (**a**), XRD (**b**), and VSM (**c**) of MCMMIPs and Fe_3O_4. Get permission from Zhu et al. (2023)

Table 11.1 Membrane treatment systems: uses across multiple industries (Ramachandran and Sathishkumar 2023)

S. No	Filtration process	Pore size range	Application
1.	Microfiltration	10–1 μm	Colloids, bacteria, cryptosporidium, red blood cells and paint pigment
2.	Ultrafiltration	1–0.01 μm	Organic macromolecules, viruses, gelatin, colloidal silica and proteins
3.	Nanofiltration	0.01–0.001 μm	Organic compounds (pesticides, solvents, dyes, herbicides, endotoxin/pyrogen, divalent dissolved ions and sugar
4.	Reverse osmosis	0.001–0.0001 μm	Monovalent dissolved salts, and metal ions

molecularly imprinted polymer (MMIP) adsorbent for eliminating bisphenol A (BPA) as the template molecule. This involved employing a bulk polymerization process with CDs as a co-monomer alongside MAA (Mamman et al. 2021). The distinctive two-dimensional nanostructure of graphene has garnered significant attention, showing considerable potential for applications in the realm of nanotechnology

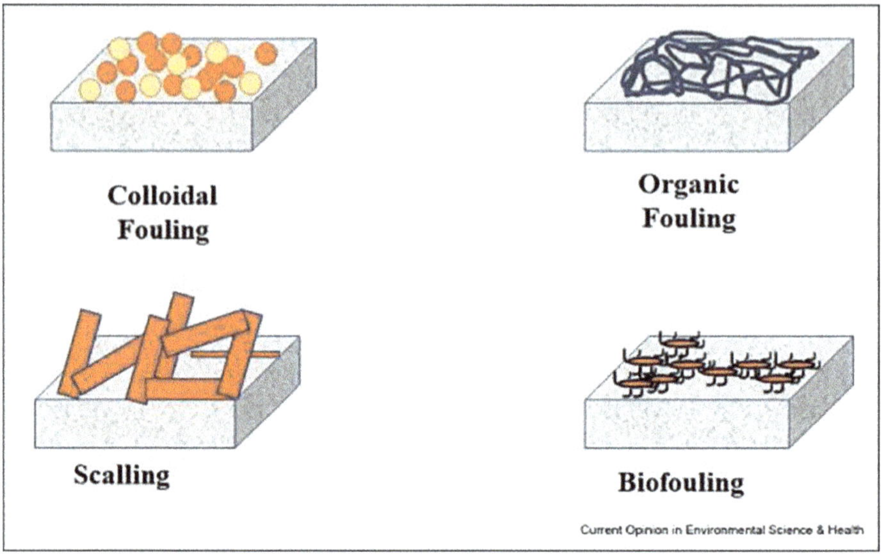

Fig. 11.6 Fouling types. Get permission from Ramachandran and Sathishkumar (2023)

(Çelik et al. 2023). Liu and colleagues synthesized magnetic graphene-based matrix materials modified with zeolitic imidazolate frameworks (MRGO@ZIF-7). This composite was employed for the selective removal of phenolic endocrine-disrupting chemicals from aqueous environments (Li et al. 2023). Small amounts of the target analyte are usually present in environmental samples, such as soil, wastewater, irrigation water, river water, etc., together with a wide variety of other chemicals. That is why excellent selectivity and efficiency are needed in both the extraction and analysis processes to successfully handle this problem. The MIPs were assessed by Kaya et al. as highly selective sorbents that were applied in ambient water analysis sample preparation procedures. Additionally, they provided a summary in Table 11.2 of the most current uses of MIP-based sample preparation methods on environmental samples (Kaya et al. 2023).

11.4 Molecularly Imprinted Bio-Based Materials for Environmental Purification

Utilizing biomass as a primary resource for generating sustainable bio-based materials with high added value presents a mutually beneficial approach, addressing both waste management improvement and environmental preservation. Because they are easily obtainable and abundant, biomass materials present a cost-effective solution with few secondary pollution issues. Furthermore, they are effective functional agents

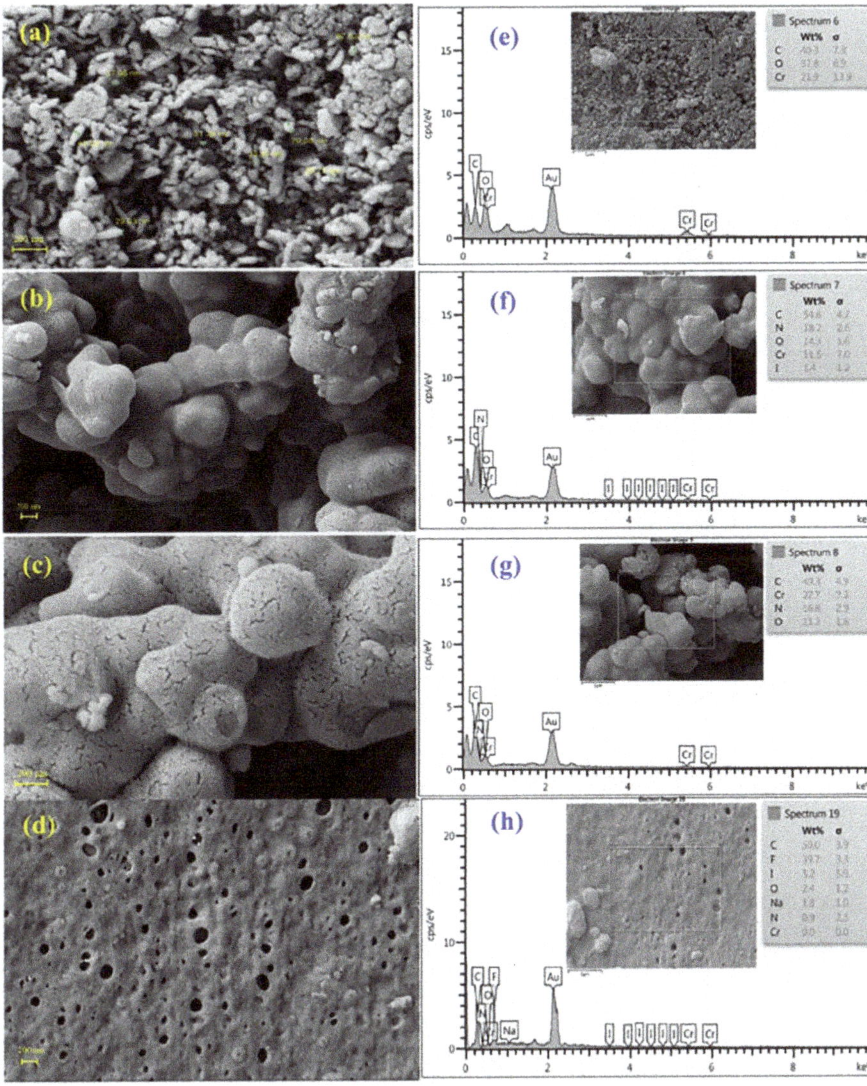

Fig. 11.7 FE-SEM images of MIL-101 (Cr) (**a**), MIRFM (**b**) and MIRFM after washing as filler (**c**) and PVDF/MIRFM membrane (**d**), EDS of MIL-101 (Cr) (**e**), MIRFM (**f**) and MIRFM after washing as filler (**g**) and PVDF/MIRFM membrane (**h**). Get permission from Bakhshizadeh et al. (2022)

for pollutant adsorption and chemical alteration in water bodies due to their abundance of carboxyl, amino, and hydroxyl groups. Apprehensions regarding the increasingly recognized adverse environmental effects of fossil fuels and petrochemicals are on the rise. In order to use cellulose as a biosorbent, its size must be reduced, its specific surface area must be increased, and it must be in pure form with active

Table 11.2 The most recent applications of MIP-based sample preparation techniques to environmental samples (Kaya et al. 2023)

Analyte	MIP type	Sample preparation method	Analysis method	Linear range	LOD/LOQ	Sample	Recovery (%)	Ref
Tetracycline	Cu-CMMIPs	Extraction	HPLC-DAD	10–2500 $\mu g\ L^{-1}$	3.75 $\mu g\ L^{-1}$	Lake water	91.2–101.3	Cui et al. (2023)
Oxazepam	MAA-based MIP	SPE	LC/MS-MS	NA	1.2–4.5 ng L^{-1}	River water	88.0–92.0	Varenne et al. (2022)
Norfloxacin	MMIP	SPE	NA	NA	NA	Wastewater	94.5–95.0	Gaho et al. (2022)
Ciprofloxacin	MIP@ZIF-67 @Fe_3O_4	Magnetic SPE	HPLC	NA	NA	River water	96.52–100.38	Zhou et al. (2022)
Phenoxy acid herbicides	MMIP	SPE	HPLC-DAD	0.033–1 $\mu g\ mL^{-1}$	0.33–0.71 $\mu g\ L^{-1}$	Fountain, irrigation well, and marsh water	79.1–96.0	Meseguer-Lloret et al. (2022)
Phthalate esters	mag-MMIP@MWCNTs	Adsorption	HPLC	NA	NA	Groundwater, surface water, domestic sewage	NA	Deng et al. (2021)
Diclofenac	MIP nanospheres	SPE	NA	NA	NA	Lake and river water	NA	Amaly et al. (2021)
Carbamazepine	Water compatible MIP	SPE	LC-UV LC-MS/MS	0.1–10 $\mu g\ L^{-1}$ 0.1–1 $\mu g\ L^{-1}$	20 ng mL^{-1} 30 ng mL^{-1}	River water	NA	Bosman et al. (2021)
Oxytetracycline	Bulk polymerization-based MIP	SPE	HPLC	NA	0.05 $\mu g\ kg^{-1}$	Soil	80.78–93.75	Cui et al. (2020)
Organophosphorus pesticides	MAA-based MIP	SPE	HPLC	50–1000 $\mu g\ L^{-1}$	0.07–0.12 $\mu g\ L^{-1}$	River and sewage water	80.0–104.0	Arias et al. (2020)
Estrogens	MIP fiber array	Fiber array extraction	HPLC	0.01–200 $\mu g\ L^{-1}$	0.003–0.016 $\mu g.L^{-1}$	River and lake water	80.8–114.1	Wang et al. (2020b)
Aminoglycoside antibiotics	Dummy-template MIP	SPE	HPLC-MS/MS	10–1000 ng mL^{-1}	0.006–0.6 ng.mL^{-1}	Lake and irrigation water	70.8–108.3	Zhang et al. (2020)
17β-estradiol	Hydrophilic magnetic molecularly imprinted nanobeads	Extraction	HPLC-UV	1–100 $\mu g\ L^{-1}$	0.008 $\mu g\ L^{-1}$	Lake water	99.7–102.5	Tian et al. (2020)

(continued)

Table 11.2 (continued)

Analyte	MIP type	Sample preparation method	Analysis method	Linear range	LOD/LOQ	Sample	Recovery (%)	Ref
Fluoroquinolones	dt-MIPs	Dispersive SPE	HPLC	1–200 μg.L^{-1}	0.22 μg L^{-1}	Lake and sea water	80.9–101.0	Lu et al. (2019)
Steroid hormones	Silica gel-based MIPs	SPE	HPLC/DAD	5–100 μg mL^{-1}	0.0182–0.0898 μg mL^{-1}	Wastewater	78.7–101.3	Czarny et al. (2019)
Climbazole	Dummy MIP	SPE	HPLC	0.1–20 μg L^{-1}	0.012 μg L^{-1}	River water	82.3–96.2	Sun et al. (2019)
Ciprofloxacin	Water compatible MIP	SPE	HPLC	0.29–1.47 × 10^5 μg L^{-1}	0.11 μg L^{-1}	Lake and river water, soil	87.33–102.5	Zhu et al. (2019)
Norfloxacin	Precipitation polymerization- based MIP	SPE	HPLC	0.5–10 μg L^{-1}	0.027 μg L^{-1}	Seawater	77.6–86.8	Li et al. (2019)
Fluoroquinolone antibiotics	MIP-hollow fiber	Hollow fiber microextraction	HPLC-UV HPLC-MS/MS	10–500 μg.L^{-1}	0.1–10 μg.L^{-1}	Lake water, groundwater	NA	Barahona et al. (2019)
Nonsteroidal anti-inflammatory drugs residues	Dummy template-imprinted polymers	SPE	UPLC-MS/MS	0.05–100 μg L^{-1}	0.007–0.48 μg L^{-1} 0.03 – 1.58 μg.L^{-1}	River and irrigation water	50.33–127.64	Guo et al. (2018)
BPA	Phenolphthalein-dummy template MIP	SPE	HPLC-DAD	0.02–2.0 mg mL^{-1}	1.3 mg g^{-1}	Sediment	82.1–106.9	Yang et al. (2018)
Triazines	Magnetic MIP	SPE	HPLC-MS/MS	0.25–500 ng L^{-1}	0.02–0.08 ng L^{-1}	River and lake water	84.8–101.7	Zhao et al. (2018)
Chloroacetamide herbicides	Amphiphilic magnetic MIPs	SPE	HPLC	0.1–200 μg L^{-1}	0.03–0.06 μg L^{-1}	Lake and agricultural irrigation	82.1–102.9	Ji et al. (2017)
Fluoxetine	Magnetic chitosan/graphene oxide MIP	SPE	Fiber optic linear array spectrophotometry	0.8–10 μg L^{-1}	0.03 μg L^{-1}	Well and spring water	95.7–104	Barati et al. (2017)
Organophosphorus pesticides	Micro-gel thin film MIP	SPE	LC-MS/MS	0.002–5.0 ng g^{-1}	0.001–0.01 ng g^{-1}	River water	79.7–127	Azizi et al. (2021)
Acesulfame	MAA-based MIP	Paper-based selective disk-type SPE	HPLC-MS-MS	0.1–2.5 μg L^{-1}	0.015 μg L^{-1}	River water	73.0–91.0	Zarejousheghani et al. (2018)

(continued)

Table 11.2 (continued)

Analyte	MIP type	Sample preparation method	Analysis method	Linear range	LOD/LOQ	Sample	Recovery (%)	Ref
Paraquat	hydroxylcucurbit[7]uri--paraquat-based MIP	Stir-bar sorptive extraction	HPLC-UV	100–10,000 ng L^{-1}	8.2 ng L^{-1}	Laboratory tap water and river water	70.0–96.1	Yao et al. (2020)
Piroxicam	Magnetic MIP	Magnetic SPE	Spectrophotometry	1.51×10^{-7}–3.02×10^{-5} M	2.72×10^{-8} M	Tap water and river water	99.0–103.0	Rezaei et al. (2020)
Estrone Estradiol Estriol Testosterone Androstenedione Boldenone	Estrone-based MIP	In–out-tube SPME	HPLC	10.0–1000 μg L^{-1}	0.21–0.80 μg L^{-1}	Sewage plant water	88.5–105.1	Wang et al. (2019c)

binding sites. Furthermore, Bai et al., incorporated cellulose nanocrystals (CNCs) into the polyamide layer, forming a thin-film composite nanofiltration membrane. The objective of this incorporation was to enhance separation performance and introduce antifouling properties by ensuring that the CNCs were in a purified state, with increased specific surface area, reduced size, and active binding sites. Nowadays, bio-based MIPs have gained growing appeal in water treatment due to their numerous advantages, including effective selective adsorption, robust affinity, non-toxicity, straightforward preparation, and cost-effectiveness. In order to produce an intelligent imprinted adsorbent (PR-Cell-MIP) based on cellulose with photo-responsive qualities, Lin et al. used surface-initiated atom transfer radical polymerization. The purpose of this adsorbent is to selectively absorb a pesticide residue (Lin et al. 2020). FT-IR spectra and SEM images of the materials used in the experiment are displayed in Fig. 11.8.

Biosorbents made from lignin have recently gained popularity as viable alternatives to expensive commercial sorbents used in wastewater applications. Through a variety of processes, such as copolymerization, hybridization, grafting, and cross-linking, these biosorbents are engineered (Gao et al. 2021). Ghafuri and colleagues analyzed the current obstacles for lignin-derived (nano)materials for environmental

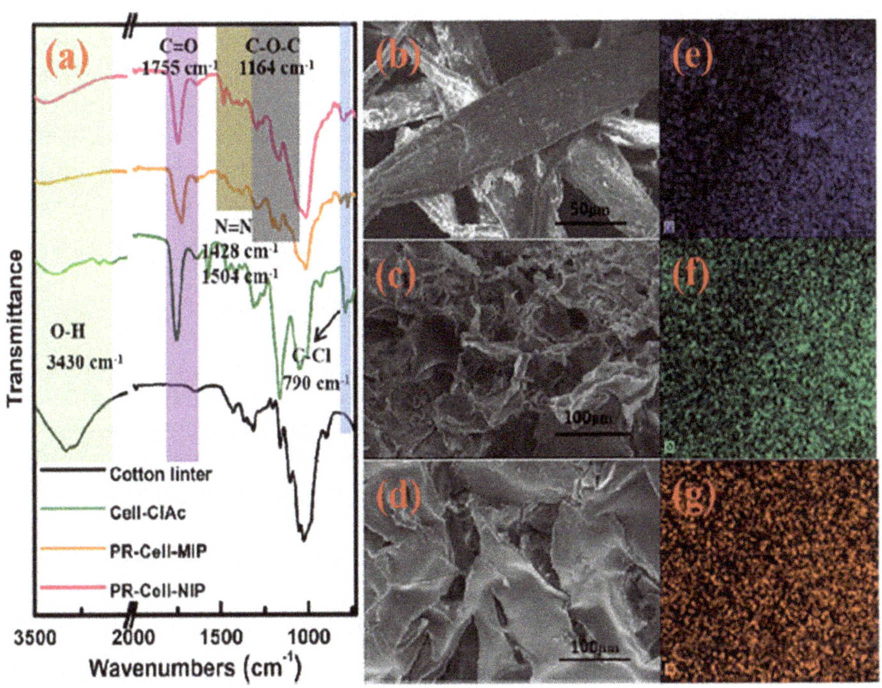

Fig. 11.8 FT-IR spectra and SEM images. **a** FT-IR spectra of the cellulose, Cell-ClAc, PR-Cell-MIP and PR-Cell-NIP. SEM images of **b** cellulose, **c** PR-Cell-MIP and **d** PR-Cell-NIP and **e**–**g** the mapping images of PR-Cell-MIP. Get permission from Lin et al. (2020)

pollution remediation (Sajjadi et al. 2021). Starch, a highly abundant and versatile natural polymer, has attracted considerable attention as a prospective ingredient in the food, chemical, and pharmaceutical sectors. Its advantages encompass affordability, widespread availability, biocompatibility, and biodegradability. However, since natural starch does not have functional groups such as carboxyl, ester and amino, it has a limited adsorption capacity. To overcome this limitation and enhance its applicability, several chemical modification techniques have been utilized to augment the adsorption capability of starch (Hager 2020). Chitosan (CS) stands as the sole naturally existing cationic polysaccharide, predominantly formed through the partial deacetylation of chitin. Boasting numerous amino and hydroxyl groups, CS stands out as one of the extensively employed natural polymers in MIT. Its substantial affinity toward a diverse array of chemicals stems from a multitude of distinctive interactions. Rahangdale and Kumar synthesized acrylamide-functionalized chitosan-based dual surface imprinted polymer by using chitosan as a substrate, epichlorohydrin as a crosslinker, cadmium as a template, and 4-hydroxybenzoic acid as a mimic template. This polymer was created specifically to remove cadmium and salicylic acid from aqueous solutions (Rahangdale and Kumar 2018).

Derived from brown seaweeds, alginate is a renewable natural polysaccharide comprising homopolymeric and heteropolymeric sequences of β-D-mannuronic acid and α-L-guluronic acid. Alginate-based nano-adsorbent materials were created by Qamar et al. as a bioinspired remedy for dangerous environmental contaminants (Qamar et al. 2022). Besides, Dettmer and colleagues conducted a review of the production and environmental applications of gelatin-based composite adsorbents for the removal of contaminants (Rigueto et al. 2021). Gelatin, another widely recognized water-soluble biopolymer, is acquired through the partial hydrolysis of collagen found in various biological connective tissues like cartilage, tendon, bone, and skin.

11.5 Conclusion and Future Outlook

The most serious hazard to the environment and its contents is pollution generated by poisonous metal ions and dyes. Metal plating, paint, and battery engineering industries, which are responsible for the release of metal ions and hazardous dyes into the environment, inflict significant environmental damage. Improper disposal of ECs such as textile dyes, pharmaceutical wastes, toxic metals, and hazardous chemicals, which endanger human health, and pollute the water and the environment. For the effective removal of ECs produced by various industrial processes such as food processing, pharmaceutical manufacture, dyeing, tanning, and so on, several purification approaches (biological, chemical, and physical methods) that comply with environmental regulations are applied.

It is critical to take safeguards against challenges such as population expansion, industrialization, and climate change, which will result in water scarcity in the next decades. There is an urgent need to create cost-effective and energy-efficient techniques for purifying water. Water pollution which has been a major issue since the dawn of civilization, can now be removed because of advances in water technology such as precipitation, coagulation, adsorption, oxidation, and biological metal removal from contaminated groundwater. These techniques, which have been utilized for many years, can still be improved. Many of the above-mentioned strategies can benefit from the development of speedier, fully automated, intelligent platforms that can integrate with detection methods.

In the year 2015, nations embraced the Sustainable Development Goals (SDGs) which aimed to ensure sustainable development in three dimensions: economic, social and environmental, initiating a collective effort to realize the 2030 Agenda (https://sdgs.un.org/2030agenda). This agenda aims to foster prosperity, tackle inequalities, and safeguard the environment. In the 2030 Agenda, UN Environment is tasked with formulating comprehensive strategies for sustainable development. These strategies aim to demonstrate the interconnectedness of enhancing environmental well-being with the realization of social and economic advantages. By striving to diminish environmental risks and enhance the resilience of both societies and the environment, UN Environment's initiatives underscore the environmental aspect of sustainable development, ultimately contributing to socio-economic progress.

A compilation of thirty indicators, spearheaded by UN Environment, succinctly outlines the Sustainable Development Goals and associated targets related to the environment. In areas that are not directly tied to the environment, UN Environment collaborates with other agencies. It is important to emphasize that this collaborative effort is an ongoing process. Within the framework of "GOAL 3: Ensure healthy lives and promote well-being for all at all ages," delves into a comprehensive examination of applications involving molecularly imprinted polymers. The assessment focuses on the purification and elimination of environmental contaminants, contributing to the overarching objective of fostering sustainable and healthy lives for both humans and the environment. By 2030, we plan to reduce the negative effects of harmful chemicals on human health and the environment by environmentally friendly management and safe use of chemicals, reduction and recycling of waste, and more efficient use of water and energy.

MIPs, which are extremely selective and sensitive to template molecules, are a relatively recent strategy in wastewater treatment and are excellent materials for this use. The application of MIPs are unique materials used to increase the efficiency, selectivity and capacity of sample preparation and extraction processes in wastewater application. MIP sorbents also offer reusability in analytical applications, aligning well with green chemistry principles and minimizing solid waste generation (Madikizela et al. 2018; Das et al. 2021; Del Sole et al. 2021; Ostovan et al. 2022). This is particularly crucial for environmental preservation, given the high stability of MIPs against both organic and inorganic solvents, with no degradation at ambient temperatures. The development of new analytical methods and the creation of MIP sorbents for sample preparation should adhere to green chemistry principles. Central to green analysis are key components such as reducing or eliminating chemical substances, minimizing energy consumption, managing analytical waste, and enhancing operator safety. Current emphasis on green chemistry often revolves around waste production, and the introduction of the E factor in different segments of the chemical industry becomes significant. The E factor quantifies the total waste generated per kilogram of product, encompassing not only the desired product but also losses from solvents and chemicals used in the work-up process. In the realm of molecular imprinting technology, the desired product (MIP) is reused in sample preparation, with less attention given to the liquid waste generated during template removal or MIP regeneration. In light of this, the perception of MIP reusability in alignment with green chemistry concepts should be reconsidered, particularly in evaluating its regeneration process.

References

Abubakar L, Yusof NA, Abdullah AH, Mohammad F, Wahid MH, Ismail S, Abdul Rahim Z, Al-Lohedan HA, Soleiman AA (2023) Molecularly imprinted polymer-based nanoporous carbon nanocomposite for effective adsorption of Hg(II) ions from aqueous suspensions. Separations 10:454. https://doi.org/10.3390/separations10080454

Algieri C, Drioli E, Guzzo L, Donato L (2014) Bio-mimetic sensors based on molecularly imprinted membranes. Sensors 14:13863–13912. https://doi.org/10.3390/s140813863

Amaly N, Istamboulie G, El-Moghazy AY, Noguer T (2021) Reusable molecularly imprinted polymeric nanospheres for diclofenac removal from water samples. J Chem Res 45(1–2):102–110. https://doi.org/10.1177/1747519820925998

Arias PG, Martínez-Pérez-Cejuela H, Combès A, Pichon V, Pereira E, Herrero-Martínez JM, Bravo M (2020) Selective solid-phase extraction of organophosphorus pesticides and their oxon-derivatives from water samples using molecularly imprinted polymer followed by high-performance liquid chromatography with UV detection. J Chromatogr A 1626:461346. https://doi.org/10.1016/j.chroma.2020.461346

Azimi M, Ahmadi Golsefidi M, Varasteh Moradi A, Ebadii M, Zafar Mehrabian R (2020) A novel method for extraction of galegine by molecularly imprinted polymer (MIP) technique reinforced with graphene oxide and its evaluation using polarography. J Anal Methods Chem 2020:3646712. https://doi.org/10.1155/2020/3646712

Azizi A, Shahhoseini F, Langille EA, Akhoondi R, Bottaro CS (2021) Micro-gel thin film molecularly imprinted polymer coating for extraction of organophosphorus pesticides from water and beverage samples. Anal Chim Acta 1187:339135. https://doi.org/10.1016/j.aca.2021.339135

Bakhshizadeh S, Sabzehmeidani MM, Ghaedi M, Dashtian K, Abbasi-Asl H (2022) Preparation and characterization of mixed matrix membranes based on PVDF blend and hydrophilic molecularly imprinted MIL-101 (Cr) as filler for efficient selective removal of dye. J Environ Chem Eng 10(6). https://doi.org/10.1016/j.jece.2022.108864

Barahona F, Albero B, Tadeo JL, Martín-Esteban A (2019) Molecularly imprinted polymer-hollow fiber microextraction of hydrophilic fluoroquinolone antibiotics in environmental waters and urine samples. J Chromatogr A 1587:42–49. https://doi.org/10.1016/j.chroma.2018.12.015

Barati A, Kazemi E, Dadfarnia S, Haji Shabani AM (2017) Synthesis/characterization of molecular imprinted polymer based on magnetic chitosan/graphene oxide for selective separation/preconcentration of fluoxetine from environmental and biological samples. J Ind Eng Chem 46:212–221. https://doi.org/10.1016/j.jiec.2016.10.033

Belbruno JJ (2019) Molecularly imprinted polymers. Chem Rev 119(1):94–119. https://doi.org/10.1021/acs.chemrev.8b00171

Bhogal S, Mohiuddin I, Kaur K, Lee J, Brown RJC, Malik AK, Kim KH (2021) Dual-template magnetic molecularly imprinted polymer-based sorbent for simultaneous and selective detection of phenolic endocrine disrupting compounds in foodstuffs. Environ Pollut 275:116613. https://doi.org/10.1016/j.envpol.2021.116613

Bie Z, Chen Y, Ye J, Wang S, Liu Z (2015) Boronate-affinity glycan-oriented surface imprinting: a new strategy to mimic lectins for the recognition of an intact glycoprotein and its characteristic fragments. Angew Chemie Int Ed 54(35):10211–10215. https://doi.org/10.1002/anie.201503066

Boontongto T, Burakham R (2021) Eco-friendly fabrication of a magnetic dual-template molecularly imprinted polymer for the selective enrichment of organophosphorus pesticides for fruits and vegetables. Anal Chim Acta 1186:339128. https://doi.org/10.1016/j.aca.2021.339128

Bosman P, Combès A, Lambert M, Lavison-Bompard G, Pichon V (2021) Development and application of molecularly imprinted polymers for the selective extraction of chlordecone from bovine serum. Separations 8(12):237. https://doi.org/10.3390/separations8120237

Briffa J, Sinagra E, Blundell R (2020) Heavy metal pollution in the environment and their toxicological effects on humans. Heliyon 6(9):e04691. https://doi.org/10.1016/j.heliyon.2020.e04691

Chen Y, Xie Z, Zhang L, Hu X (2020) Effective preparation of magnetic molecularly imprinted polymer nanoparticle for the rapid and selective extraction of cyfluthrin from honeysuckle. J Biomater Sci Polym Ed 31(7):954–968. https://doi.org/10.1080/09205063.2020.1731788

Çelik O, Saylan Y, Göktürk I, Yılmaz F, Denizli A (2023) A surface plasmon resonance sensor with synthetic receptors decorated on graphene oxide for selective detection of benzylpenicillin. Talanta 253:123939. https://doi.org/10.1016/j.talanta.2022.123939

Çimen D, Bereli N, Andaç M, Denizli A (2018) Molecularly imprinted cryogel columns for Concanavalin A purification from jack bean extract. Sep Sci plus 1(6):454–463. https://doi.org/10.1002/sscp.201800039

Çimen D, Göktürk I, Yılmaz F (2016) Removal of iron by chelation with molecularly imprinted supermacroporous cryogel. Artif Cells Nanomed Biotechnol 44(4):1158–1166. https://doi.org/10.3109/21691401.2015.1011810

Click E (2015) Out of space. Lancet Psychiatry 2(9):765. https://doi.org/10.1016/S2215-0366(15)00367-3

Cui S, Cao H, Wang Y, Su Y (2020) A rapid and economical sample pretreatment method using molecularly imprinted polymers for bioavailable content analyses of oxytetracycline in soils. Hum Ecol Risk Assess 26(4):1108–1123. https://doi.org/10.1080/10807039.2018.1558397

Cui Y, Ding J, Su Y, Ding L (2023) Facile construction of magnetic hydrophilic molecularly imprinted polymers with enhanced selectivity based on dynamic non-covalent bonds for detecting tetracycline. Chem Eng J 452:139291. https://doi.org/10.1016/j.cej.2022.139291

Curcio P, Zandanel C, Wagner A, Mioskowski C, Baati R (2009) Semi-covalent surface molecular imprinting of polymers by one-stage mini-emulsion polymerization: Glucopyranoside as a model analyte. Macromol Biosci 9(6):596–604. https://doi.org/10.1002/mabi.200900056

Czarny K, Szczukocki D, Krawczyk B, Juszczak R, Skrzypek S, Gadzała-Kopciuch R (2019) Molecularly imprinted polymer film grafted from porous silica for efficient enrichment of steroid hormones in water samples. J Sep Sci 42(17):2858–2866. https://doi.org/10.1002/jssc.201900281

Das TR, Bhattacherjee P, Patra S, Govender PP, Mangala B, Shukla SK (2021) Metal−organic framework-based artificial antibodies for sensing applications. In: Metal−organic frameworks for environmental sensing. American Chemical Society, pp 155–175 SE–6

Del Sole R, Mele G, Bloise E, Mergola L (2021) Green aspects in molecularly imprinted polymers by biomass waste utilization. Polymers (basel) 13(15):2430. https://doi.org/10.3390/polym13152430

Deng D, He Y, Li M, Huang L, Zhang J (2021) Preparation of multi-walled carbon nanotubes based magnetic multi-template molecularly imprinted polymer for the adsorption of phthalate esters in water samples. Environ Sci Pollut Res 28(5):5966–5977. https://doi.org/10.1007/s11356-020-10970-2

Derazshamshir A, Göktürk I, Tamahkar E, Yılmaz F, Sağlam N, Denizli A (2020) Phenol removal from wastewater by surface imprinted bacterial cellulose nanofibres. Environ Technol (United Kingdom) 41(24):3134–3145. https://doi.org/10.1080/09593330.2019.1600043

Dong C, Shi H, Han Y, Yang Y, Wang R, Men J (2021) Molecularly imprinted polymers by the surface imprinting technique. Eur Polym J 145:110231. https://doi.org/10.1016/j.eurpolymj.2020.110231

Erdem Ö, Saylan Y, Andaç M, Denizli A (2018) Molecularly imprinted polymers for removal of metal ions: an alternative treatment method. Biomimetics 3:38. https://doi.org/10.3390/biomimetics3040038

Fuller R, Landrigan PJ, Balakrishnan K, Bathan G, Bose-O'Reilly S, Brauer M, Caravanos J, Chiles T, Cohen A, Corra L, Cropper M, Ferraro G, Hanna J, Hanrahan D, Hu H, Hunter D, Janata G, Kupka R, Lanphear B, Lichtveld M, Martin K, Mustapha A, Sanchez-Triana E, Sandilya K, Schaefli L, Shaw J, Seddon J, Suk W, Téllez-Rojo MM, Yan C (2022) Pollution and health: a progress update. Lancet Planet Heal 6(6):e535–e547. https://doi.org/10.1016/S2542-5196(22)00090-0

Gaho MM, Memon GZ, Memon J ur R, Arain JB, Arain AJ, Shah A, Samejo MQ (2022) Synthesis of novel magnetic molecularly imprinted polymers by solid-phase extraction method for removal of norfloxacin. Chinese J Anal Chem 50(6):100079. https://doi.org/10.1016/j.cjac.2022.100079

Gao B, Chang Q, Yang H (2021) Selective adsorption of ofloxacin and ciprofloxacin from a binary system using lignin-based adsorbents: quantitative analysis, adsorption mechanisms, and structure-activity relationship. Sci Total Environ 765:144427. https://doi.org/10.1016/j.scitotenv.2020.144427

Guo P, Zhang J, Chen X, Zhao L (2018) Preparation of dummy template-imprinted polymers for the rapid extraction of nonsteroidal anti-inflammatory drugs residues in aquatic environmental samples. Biomed Chromatogr 32(6):1–11. https://doi.org/10.1002/bmc.4193

Hager RA (2020) New trends for removal of water pollutants. In: Robert BH (ed) Prime archives in material science. Vide Leaf, Hyderabad, India

Haris SA, Dabagh S, Mollasalehi H, Ertas YN (2023) Alginate coated superparamagnetic iron oxide nanoparticles as nanocomposite adsorbents for arsenic removal from aqueous solutions. Sep Purif Technol 310:123193. https://doi.org/10.1016/j.seppur.2023.123193

Hasanah AN, Safitri N, Zulfa A, Neli N, Rahayu D (2021) Factors affecting preparation of molecularly imprinted polymer and methods on finding template-monomer interaction as the key of selective properties of the materials. Molecules 26(18). https://doi.org/10.3390/molecules26185612

Haupt K, Medina Rangel PX, Bui BTS (2020) Molecularly imprinted polymers: antibody mimics for bioimaging and therapy. Chem Rev 120(17):9554–9582. https://doi.org/10.1021/acs.chemrev.0c00428

He S, Zhang L, Bai S, Yang H, Cui Z, Zhang X, Li Y (2021) Advances of molecularly imprinted polymers (MIP) and the application in drug delivery. Eur Polym J 143:110179. https://doi.org/10.1016/j.eurpolymj.2020.110179

Huang Z, Zhang P, Yun Y (2017) Preparing molecularly imprinted membranes by phase inversion to separate kaempferol. Polym Adv Technol 28(3):373–378. https://doi.org/10.1002/pat.3898

Igiri BE, Okoduwa SIR, Idoko GO, Akabuogu EP, Adeyi AO, Ejiogu IK (2018) Toxicity and bioremediation of heavy metals contaminated ecosystem from tannery wastewater: a review. J Toxicol 2018:2568038. https://doi.org/10.1155/2018/2568038

Ji W, Sun R, Duan W, Wang X, Wang T, Mu Y, Guo L (2017) Selective solid phase extraction of chloroacetamide herbicides from environmental water samples by amphiphilic magnetic molecularly imprinted polymers. Talanta 170:111–118. https://doi.org/10.1016/j.talanta.2017.04.005

Kang MS, Cho E, Choi HE, Amri C, Lee J-H, Kim KS (2023) Molecularly imprinted polymers (MIPs): emerging biomaterials for cancer theragnostic applications. Biomater Res 27(1):45. https://doi.org/10.1186/s40824-023-00388-5

Kaya SI, Cetinkaya A, Ozkan SA (2023) Molecularly imprinted polymers as highly selective sorbents in sample preparation techniques and their applications in environmental water analysis. Trends Environ Anal Chem 37:1–41. https://doi.org/10.1016/j.teac.2022.e00193

Kobayashi Y, Nakamitsu Y, Zheng Y, Takashima Y, Yamaguchi H, Harada A (2019) Preparation of cyclodextrin-based porous polymeric membrane by bulk polymerization of ethyl acrylate in the presence of cyclodextrin. Polymer (guildf) 177:208–213. https://doi.org/10.1016/j.polymer.2019.06.015

Koul B, Yakoob M, Shah MP (2022) Agricultural waste management strategies for environmental sustainability. Environ Res 206:112285. https://doi.org/10.1016/j.envres.2021.112285

Kurt ZT, Çimen D, Denizli A, Bereli N (2023) Development of optical-based molecularly imprinted nanosensors for adenosine detection. ACS Omega 8(21):18839–18850. https://doi.org/10.1021/acsomega.3c01028

Li H, Chen J, Tan L, Wang J (2019) Solid-phase extraction using a molecularly imprinted polymer for the selective purification and preconcentration of norfloxacin from seawater. Anal Lett 52(18):2896–2913. https://doi.org/10.1080/00032719.2019.1628245

Li Y, Li Y, Ding Z, Wan D, Gao Z, Sun Y, Liu Y (2023) Synthesis of MRGO@ZIF-7-based molecular imprinted polymer by surface polymerization for the fast and selective removal of phenolic endocrine-disrupting chemicals from aqueous environments. Processes 11:1000. https://doi.org/10.3390/pr11041000

Lin ZZ, Zhang HY, Li L, Huang ZY (2016) Application of magnetic molecularly imprinted polymers in the detection of malachite green in fish samples. React Funct Polym 98:24–30. https://doi.org/10.1016/j.reactfunctpolym.2015.11.002

Lin C, Qiu Y, Fan J, Wang M, Ye L, Liu Y, Ye X, Huang X, Lv Y, Liu M (2020) Fabrication of photo-responsive cellulose based intelligent imprinted material and selective adsorption on typical pesticide residue. Chem Eng J 394(2):124841. https://doi.org/10.1016/j.cej.2020.124841

Liu J, Pan J, Ma Y, Liu S, Qiu F, Yan Y (2018) A versatile strategy to fabricate dual-imprinted porous adsorbent for efficient treatment co-contamination of Λ-cyhalothrin and copper(II). Chem Eng J 332:517–527. https://doi.org/10.1016/j.cej.2017.09.079

Lu W, Wang X, Wu X, Liu D, Li J, Chen L, Zhang X (2017) Multi-template imprinted polymers for simultaneous selective solid-phase extraction of six phenolic compounds in water samples followed by determination using capillary electrophoresis. J Chromatogr A 1483:30–39. https://doi.org/10.1016/j.chroma.2016.12.069

Lu W, Liu J, Li J, Wang X, Lv M, Cui R, Chen L (2019) Dual-template molecularly imprinted polymers for dispersive solid-phase extraction of fluoroquinolones in water samples coupled with high performance liquid chromatography. Analyst 144(4):1292–1302. https://doi.org/10.1039/c8an02133c

Ma W, Row KH (2019) Simultaneous determination of levofloxacin and ciprofloxacin in human urine by ionic-liquid-based, dual-template molecularly imprinted coated graphene oxide monolithic solid-phase extraction. J Sep Sci 42(3):642–649. https://doi.org/10.1002/jssc.201800939

Ma Q, Wei Y, Zhao N, Wang S, Zhang B, Liu D, Yuan P (2023) Construction of an allophane-based molecularly imprinted polymer for the efficient removal of antibiotic from aqueous solution. Sci Total Environ 903:166464. https://doi.org/10.1016/j.scitotenv.2023.166464

Madikizela LM, Tavengwa NT, Tutu H, Chimuka L (2018) Green aspects in molecular imprinting technology: From design to environmental applications. Trends Environ Anal Chem 17(October 2017):14–22. https://doi.org/10.1016/j.teac.2018.01.001

Mamman S, Suah FBM, Raaov M, Mehamod FS, Asman S, Zain NNM (2021) Removal of bisphenol A from aqueous media using a highly selective adsorbent of hybridization cyclodextrin with magnetic molecularly imprinted polymer. R Soc Open Sci 8(3). https://doi.org/10.1098/rsos.201604

Meseguer-Lloret S, Torres-Cartas S, Gómez-Benito C, Herrero-Martínez JM (2022) Magnetic molecularly imprinted polymer for the simultaneous selective extraction of phenoxy acid herbicides from environmental water samples. Talanta 239. https://doi.org/10.1016/j.talanta.2021.123082

Morsi SMM, Abd El-Aziz ME, Mohamed HA (2023) Smart polymers as molecular imprinted polymers for recognition of target molecules. Int J Polym Mater Polym Biomater 72(8):612–635. https://doi.org/10.1080/00914037.2022.2042287

Ncube S, Kunene P, Tavengwa NT, Tutu H, Richards H, Cukrowska E, Chimuka L (2017) Synthesis and characterization of a molecularly imprinted polymer for the isolation of the 16 US-EPA priority polycyclic aromatic hydrocarbons (PAHs) in solution. J Environ Manage 199:192–200. https://doi.org/10.1016/j.jenvman.2017.05.041

Ostovan A, Arabi M, Wang Y, Li J, Li B, Wang X, Chen L (2022) Greenificated molecularly imprinted materials for advanced applications. Adv Mater 34(42):1–32. https://doi.org/10.1002/adma.202203154

Phan NHV, Nguyen TKH, Vo TBH, Phan VHN (2019) Synthesis quercetin imprinted polymer by precipitation polymerization method, directed to apply as sorbent for selective solid-phase extraction. Pharm Sci Asia 46(2):108–119. https://doi.org/10.29090/psa.2019.01.017.0038

Qamar SA, Qamar M, Basharat A, Bilal M, Cheng H, Iqbal HMN (2022) Alginate-based nano-adsorbent materials—bioinspired solution to mitigate hazardous environmental pollutants. Chemosphere 288(P3):132618. https://doi.org/10.1016/j.chemosphere.2021.132618

Rahangdale D, Kumar A (2018) Chitosan as a substrate for simultaneous surface imprinting of salicylic acid and cadmium. Carbohydr Polym 202:334–344. https://doi.org/10.1016/j.carbpol.2018.08.129

Ramachandran SK, Sathishkumar P (2023) Membrane-based techniques for pollutants removal: an outlook on recent advancements. Curr Opin Environ Sci Heal 36:100513. https://doi.org/10.1016/j.coesh.2023.100513

Rezaei M, Rajabi HR, Rafiee Z (2020) Selective and rapid extraction of piroxicam from water and plasma samples using magnetic imprinted polymeric nanosorbent: synthesis, characterization and application. Colloids Surf A Physicochem Eng Asp 586:124253. https://doi.org/10.1016/j.colsurfa.2019.124253

Rigueto CVT, Nazari MT, Massuda LÁ, Ostwald BEP, Piccin JS, Dettmer A (2021) Production and environmental applications of gelatin-based composite adsorbents for contaminants removal: a review. Environ Chem Lett 19(3):2465–2486. https://doi.org/10.1007/s10311-021-01184-0

Roy E, Patra S, Tiwari A, Madhuri R, Sharma PK (2017) Introduction of selectivity and specificity to graphene using an inimitable combination of molecular imprinting and nanotechnology. Biosens Bioelectron 89:234–248. https://doi.org/10.1016/j.bios.2016.02.056

Rozaini MNH, Semail NF, Zango ZU, Lim JW, Yahaya N, Setiabudi HD, Tong WY, Shamsuddin R, Chan YJ, Khoo KS, Suliman M, Kiatkittipong W (2023) Advanced adsorptions of non-steroidal anti-inflammatory drugs from environmental waters in improving offline and online preconcentration techniques: an analytical review. J Taiwan Inst Chem Eng. https://doi.org/10.1016/j.jtice.2023.105020

Sajjadi M, Ahmadpoor F, Nasrollahzadeh M, Ghafuri H (2021) Lignin-derived (nano)materials for environmental pollution remediation: current challenges and future perspectives. Int J Biol Macromol 178:394–423. https://doi.org/10.1016/j.ijbiomac.2021.02.165

Scorrano S, Mergola L, del Sole R, Vasapollo G (2011) Synthesis of molecularly imprinted polymers for amino acid derivates by using different functional monomers. Int J Mol Sci 12(3):1735–1743. https://doi.org/10.3390/ijms12031735

Shah SAH, Mansur SA, Sami MH, Asghar R, Asman S (2023) Hybridization and characterization of a deep eutectic solvent molecularly ımprinted polymer membrane for the removal of bisphenol A. Malays J Chem 25(4):232–245. https://doi.org/10.55373/mjchem.v25i4.232

Shama NA, Aşır S, Göktürk I, Yılmaz F, Türkmen D, Denizli A (2023) Electrochemical detection of cortisol by silver nanoparticle-modified molecularly imprinted polymer-coated pencil graphite electrodes. ACS Omega 8(32):29202–29212. https://doi.org/10.1021/acsomega.3c02472

Sharma G, Kandasubramanian B (2020) Molecularly imprinted polymers for selective recognition and extraction of heavy metal ions and toxic dyes. J Chem Eng Data 65(2):396–418. https://doi.org/10.1021/acs.jced.9b00953

Speltini A, Scalabrini A, Maraschi F, Sturini M, Profumo A (2017) Analytica chimica acta newest applications of molecularly imprinted polymers for extraction of contaminants from environmental and food matrices: a review. Anal Chim Acta 974:1–26. https://doi.org/10.1016/j.aca.2017.04.042

Sun X, Wang M, Peng J, Yang L, Wang X, Wang F, Zhang X, Wu Q, Chen R, Chen J (2019) Dummy molecularly imprinted solid phase extraction of climbazole from environmental water samples. Talanta 196:47–53. https://doi.org/10.1016/j.talanta.2018.12.017

Syed Yaacob SFF, Suwaibatu M, Raja Jamil RZ, Mohamed Zain NN, Raoov M, Mohd Suah FB (2023) Review of molecular imprinting polymer: basic characteristics and removal of phenolic contaminants based on the functionalized cyclodextrin monomer. J Chem Technol Biotechnol 98(2):312–330. https://doi.org/10.1002/jctb.7215

Thakur N, Das TR, Gupta S, Shukla SK, Patra S, Kumar D (2023) Chapter 5—material selection for the fabrication of MIP-based devices: an overview. In: Singh MBT-MIP (MIPs) (ed). Elsevier, pp 101–114

Tian X, Song H, Wang Y, Tian X, Tang Y, Gao R, Zhang C (2020) Hydrophilic magnetic molecularly imprinted nanobeads for efficient enrichment and high performance liquid chromatographic detection of 17beta-estradiol in environmental water samples. Talanta 220:121367. https://doi.org/10.1016/j.talanta.2020.121367

Varenne F, Kadhirvel P, Bosman P, Renault L, Combès A, Pichon V (2022) Synthesis and characterization of molecularly imprinted polymers for the selective extraction of oxazepam from complex environmental and biological samples. Anal Bioanal Chem 414(1):451–463. https://doi.org/10.1007/s00216-021-03268-w

Wang L, Li J, Wang J, Guo X, Wang X, Choo J, Chen L (2019a) Green multi-functional monomer based ion imprinted polymers for selective removal of copper ions from aqueous solution. J Colloid Interface Sci 541:376–386. https://doi.org/10.1016/j.jcis.2019.01.081

Wang S, She Y, Hong S, Du X, Yan M, Wang Y, Qi Y, Wang M, Jiang W, Wang J (2019b) Dual-template imprinted polymers for class-selective solid-phase extraction of seventeen triazine herbicides and metabolites in agro-products. J Hazard Mater 367:686–693. https://doi.org/10.1016/j.jhazmat.2018.12.089

Wang X, Huang P, Ma X, Du X, Lu X (2019c) Enhanced in-out-tube solid-phase microextraction by molecularly imprinted polymers-coated capillary followed by HPLC for endocrine disrupting chemicals analysis. Talanta 194:7–13. https://doi.org/10.1016/j.talanta.2018.10.027

Wang D, Liu Y, Xu Z, Zhao D, Liu Y, Liu Z (2020a) Multitemplate molecularly imprinted polymeric solid-phase microextraction fiber coupled with HPLC for endocrine disruptor analysis in water samples. Microchem J 155:104802. https://doi.org/10.1016/j.microc.2020.104802

Wang D, Xu Z, Liu Y, Liu Y, Li G, Si X, Lin T, Liu H, Liu Z (2020b) Molecularly imprinted polymer-based fiber array extraction of eight estrogens from environmental water samples prior to high-performance liquid chromatography analysis. Microchem J 159:105376. https://doi.org/10.1016/j.microc.2020.105376

Wang S, Zhang L, Zeng J, Hu X, Wang X, Yu L, Wang D, Cheng L, Ahmed R, Romanovski V, Li P, Zhang Z (2023) Multi-templates molecularly imprinted polymers for simultaneous recognition of multiple targets: from academy to application. TrAC Trends Anal Chem 166:117173. https://doi.org/10.1016/j.trac.2023.117173

Wu L, Li Y (2004) Metal ion-mediated molecular-imprinting polymer for indirect recognition of formate, acetate and propionate. Anal Chim Acta 517(1–2):145–151. https://doi.org/10.1016/j.aca.2004.05.015

Wulff G (2013) Fourty years of molecular imprinting in synthetic polymers: origin, features and perspectives. Microchim Acta 180(15–16):1359–1370. https://doi.org/10.1007/s00604-013-0992-9

Yang J, Li Y, Huang C, Jiao Y, Chen J (2018) A phenolphthalein-dummy template molecularly imprinted polymer for highly selective extraction and clean-up of bisphenol A in complex biological, environmental and food samples. Polymers (basel) 10(10):1–13. https://doi.org/10.3390/polym10101150

Yao J, Zhang L, Ran J, Wang S, Dong N (2020) Specific recognition of cationic paraquat in environmental water and vegetable samples by molecularly imprinted stir-bar sorptive extraction based on monohydroxylcucurbit[7]uril–paraquat inclusion complex. Microchim Acta 187:1–10. https://doi.org/10.1007/s00604-020-04491-5

Ye L, Mosbach K (2008) Molecular imprinting: synthetic materials as substitutes for biological antibodies and receptors. Chem Mater 20(3):859–868. https://doi.org/10.1021/cm703190w

Zabihi S, Bakhshpour M, Çalışır M, Topçu AA, Denizli A (2023) Preparation of molecular imprinted injectable polymeric micro cryogels for control release of mitomycin C. Polym Bull 80(4):3883–3895. https://doi.org/10.1007/s00289-022-04233-y

Zarejousheghani M, Schrader S, Möder M, Schmidt M, Borsdorf H (2018) A new strategy for accelerated extraction of target compounds using molecularly imprinted polymer particles embedded in a paper-based disk. J Mol Recognit 31(3):1–9. https://doi.org/10.1002/jmr.2629

Zhang Z, Cao X, Zhang Z, Yin J, Wang D, Xu Y, Zheng W, Li X, Zhang Q, Liu L (2020) Synthesis of dummy-template molecularly imprinted polymer adsorbents for solid phase extraction of aminoglycosides antibiotics from environmental water samples. Talanta 208:120385. https://doi.org/10.1016/j.talanta.2019.120385

Zhao Q, Zhou T, Li H, Liu H, Huang N, Xu Y, Ding J, Ding L, Li Y (2018) Preparation of magnetic surface molecularly imprinted polymers for the selective extraction of triazines in environmental water samples. Int J Environ Anal Chem 98(11):1049–1062. https://doi.org/10.1080/03067319.2018.1520226

Zhao D, Peng Z, Fang J, Fang Z, Zhang J (2023) Programmable and low-cost biohybrid membrane for efficient heavy metal removal from water. Sep Purif Technol 306:1–10. https://doi.org/10.1016/j.seppur.2022.122751

Zhou S, Fu J, Zhao P, Tang S, Wu X, Yang Z, Zhang Z (2022) Hollow magnetic molecularly imprinted polymer based on metal-organic framework for capture of ciprofloxacin. Sep Sci plus 5(8):337–348. https://doi.org/10.1002/sscp.202200029

Zhu G, Cheng G, Wang P, Li W, Wang Y, Fan J (2019) Water compatible imprinted polymer prepared in water for selective solid phase extraction and determination of ciprofloxacin in real samples. Talanta 200:307–315. https://doi.org/10.1016/j.talanta.2019.03.070

Zhu Y, Liang M, Liu Y, Zhong M, Zhou B, Huang L, Zhang Z (2023) Magnetic metal-organic framework material synthesized by molecular imprinting technology for analysis of bisphenol A. Chromatographia 86(10):669–676. https://doi.org/10.1007/s10337-023-04277-w

Chapter 12
Adsorption and Removal of Contaminants Using Molecularly Imprinted Polymers

Esra Feyzioğlu-Demir and Nevra Öztürk Atay

12.1 Introduction

About 70 years ago, Dickey developed 25 artificial materials based on silica gel developed in the presence of methyl orange, providing specific recognition of 26 target analytes (Dickey 1955), and with this study he established the concept of "specific adsorption". These artificial polymers were later called "molecular imprinting polymers" (MIP) and were developed with molecular imprinting technology (MIT) (Mabrouk et al. 2023).

MIT is defined as a method of creating molecular locks with binding sites that can recognize template molecules in terms of shape, size and functional groups for the synthesis of MIPs (Chen et al. 2016; Vasapollo et al. 2011).

The molecular imprinting process consists of polymerization of a functional monomer and a molecular template with a crosslinker (Chen et al. 2016). Once the template molecule is eluted, cavities remain within the polymer to which the template molecule or template analogs can rebind (He et al. 2023). A schematic representation of the molecular imprinting process is given in Fig. 12.1.

In the imprinting process, the selectivity and capacity of MIPs are affected by variables such as template molecule, functional monomer and crosslinker interaction, monomer and template molecule concentration and stoichiometry, and solvent and polymerization temperature in the polymerization environment (Batra and Shea 2003).

E. Feyzioğlu-Demir (✉)
Department of Medical Laboratory Technique, Vocational School of Health Services, Izmir University of Economics, 35330 Balçova, Izmir, Turkey
e-mail: esraa.feyzioglu@gmail.com

N. Öztürk Atay
Ege University Central Research Test and Analysis Laboratory Application and Research Center, 35100 Bornova, Izmir, Turkey

Department of Chemistry, Faculty of Science, Ege University, 35100 Bornova, Izmir, Turkey

S. Patra and M. Sillanpaa (eds.), *Molecularly Imprinted Polymers as Artificial Antibodies for the Environmental Health*, https://doi.org/10.1007/978-3-031-58995-9_12

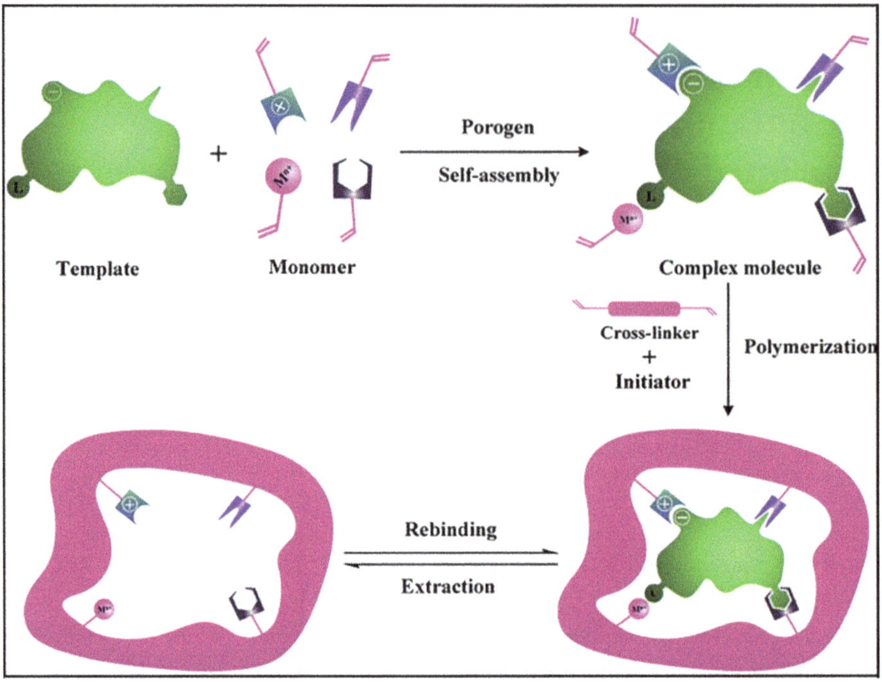

Fig. 12.1 Representative schema of molecular imprinting process. Republished with permission of Elsevier, from Hu et al. (2013); permission conveyed through Copyright Clearance Center

MIPs can be synthesized in different form and morphology, such as blocks or monoliths, thin films, microspheres, nanospheres, nanocomposite membranes, and nanowires according to used synthesized methods and reported in the literature (Vasapollo et al. 2011; Kadhem et al. 2021).

Environmental pollution is increasing rapidly, with the release of pollutants into nature due to factors such as population growth, urbanization and industrial growth (Syed Yaacob et al. 2023). There are many different methods for the detection of contaminants, including liquid chromatography (LC), atomic absorption spectroscopy (AAS), and gas chromatography, however, these techniques have disadvantages such as infrastructure cost and inability to measure on-site. (GC). Our ability to detect and eliminate pollutants must keep pace with the complexity and dynamism of the environment (Veloz Martínez et al. 2022). MIPs have selectivity because they match the analyte as key does to a lock.

The synthesis of molecularly imprinted polymers is convenient and, inexpensive, and the obtained polymers have properties such as high selectivity, excellent mechanical strength, and durability under conditions such as heat and pH (Ávila et al. 2008).

This book chapter will discuss the use of MIPs in environmental applications.

12.2 Adsorption of Pollutant

Adsorption is the process of the mass transfer of pollutants, such as ions and molecules from a phase (usually liquid or gas) to a solid phase surface. Due to this transfer, they accumulate or adhere to surface of the solid substances (Wang and Guo 2020a; Zhang et al. 2023). In recent years, this method has been frequently used instead of methods such as chemical precipitation, ion exchange, electrolysis, and oxidation in wastewater applications because of its low cost, simple design, high performance, and environmentally friendliness (Wang and Guo 2020a, b; Chen et al. 2022a).

The substances adsorbed in the adsorption process are called adsorbates, and the adsorption phase is called adsorbents. Adsorption can occur in two ways: physical adsorption (or physisorption) and chemical adsorption (or chemisorption). Physical adsorption has weak interactions and low enthalpy values, such as van der Waals forces, London forces and polar interactions. Chemical adsorption occurs through covalent and ionic bonds formed by sharing or transferring electrons (complexation, chelation and electrostatic attraction) (Chen et al. 2022a; Wang and Guo 2020b; Wu et al. 2022). In physical adsorption, a multi-layered adsorbate is formed on the adsorbent, and in chemical adsorption, a single-layer adsorbate (Chen et al. 2022a). Chelation or complexation is one of the adsorption mechanism used to describe chemical adsorption. This adsorption mechanism is more stable than physical adsorption. Therefore, chelation is frequently preferred for heavy metal adsorption on MIPs. MIPs are synthesized using different chelating monomers to form chelating bonds with metal ions (Wu et al. 2022).

Several adsorption isotherm and adsorption kinetic models are used to characterize the adsorption mechanism. The adsorption kinetic models give information on the adsorption rate and the performance of the adsorbent, whereas, the adsorption isotherms provide information of adsorption equilibrium mechanism, adsorption capacity, and the inherent properties of the adsorption process (Wang and Guo 2020a; Chen et al. 2022a). In the adsorption kinetic studies, the four most commonly used models are pseudo-first order, pseudo-second order models, Elovich, and intraparticle diffusion (Wang and Guo 2020a). Also, the most commonly used isotherm models are Freundlich, Langmuir, Temkin, Dubinin–Radushkevich, Redlich–Peterson, Koble–Corrigan, and Fritz–Schlünder (Kayranli 2011).

The adsorption performance of MIPs is specified in two ways, first, the interaction between the template molecules and the functional monomers, and, second, the adsorbent's binding capacity. The covalent, semicovalent, or noncovalent binding can occur between template molecule and functional monomer. Figure 12.2 shows the schematic representation of these molecular imprinting approaches.

Covalently imprinted polymers can consist of simply reversible covalent bonds or strong covalent bonds. Semi-covalently imprinted polymers are formed by covalent bonding of the template molecule with one or more polymerizable units; however,

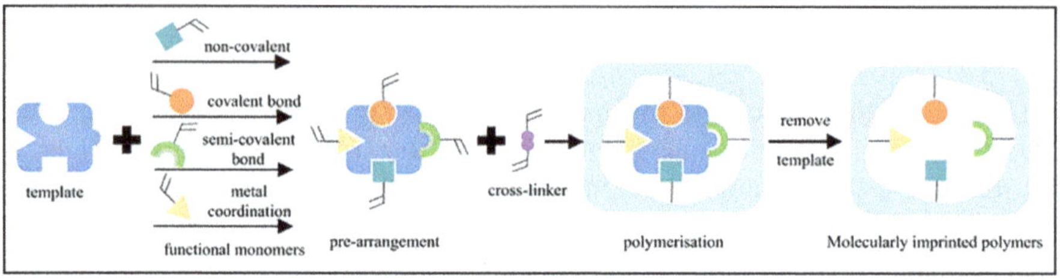

Fig. 12.2 Schematic representation of different molecular imprinting approaches. Republished with permission of Elsevier, from Zhang et al. (2023); permission conveyed through Copyright Clearance Center

the polymer is complex enough to interact with the target molecule through non-covalent bonds. Non-covalent imprinted polymers are formed by hydrogen bonding, van der Waals forces, ionic or Π-Π interaction between the template and functional monomer complex (Zhang et al. 2023; Tarannum et al. 2020) (Fig. 12.3).

The adsorption performances of MIPs are measured by the equilibrium adsorption capacity. Therefore, the adsorption studies are carried out using different adsorption measures, such as contact time, pH, temperature, initial conditions (Zhang et al. 2023). The adsorption capacity of adsorbent is calculated with following equation (Eq. 12.1) (Feyzioğlu Demir and Akgöl 2022):

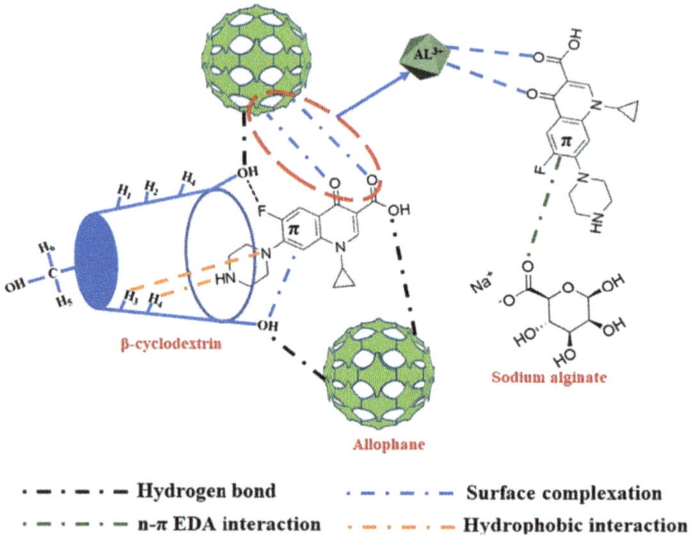

Fig. 12.3 Schematic representation adsorption mechanism of CIP on the Allo-β-CD MIP. Republished with permission of Springer Nature, from Ma et al. (2023) permission conveyed through Copyright Clearance Center

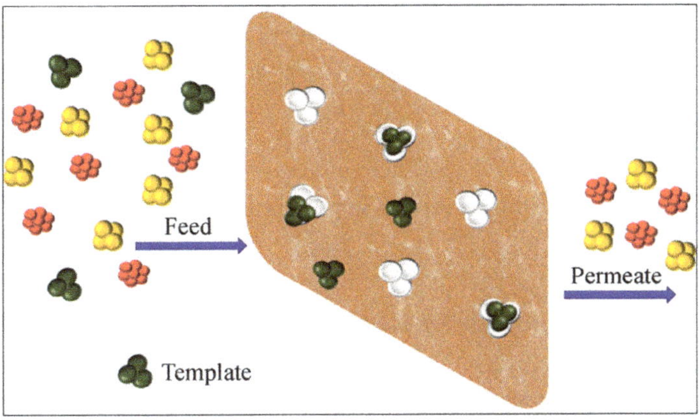

Fig. 12.4 Representation of a molecularly imprinted polymer able to bind selectively the target molecules. Republished with permission of Springer Nature, from Donato and Drioli (2021); permission conveyed through Copyright Clearance Center

$$Q = \frac{(C_i - C_f) \times V}{m} \tag{12.1}$$

In the equation, Q is the adsorption capacity of MIPs (mg/g), C_i and C_f are initial and final concentration of analyte in the solution, V is the total volume of the solution (mL); and m is the dry mass of adsorbent used for adsorption (mg).

Selective adsorption in MIP applications are different from conventional adsorbents, in one important respect; unlike conventional adsorbents, MIPs are aimed to selectively separate the target molecule in a mixed environment. The selective binding of the target molecule to MIP is shown representatively in Fig. 12.4.

The imprinting factor (IF) is calculated to determine the selective adsorption of MIPs using Eq. 12.2:

$$IF = \frac{Q_{MIP}}{Q_{NIP}} \tag{12.2}$$

In the equation, Q_{MIP} and Q_{NIP} are the equilibrium adsorption capacity of the analyte of MIPs and NIPs, respectively. When the value of IF is found as above 1.5, successful imprinting is considered to have occurred (Zhang et al. 2023; Azizi and Bottaro 2020). To compare the affinity of template molecules towards to MIPs, the adsorption of competitive molecules, which have similar molecular structure with the template molecule, is applied and the distribution coefficients (K_d), selectivity coefficients (k), and relative selectivity coefficient (k′) are calculated following equations (Eqs. 12.3, 12.4, and 12.5) (Azizi and Bottaro 2020; Demir et al. 2018):

$$K_d = \frac{Q_e}{C_f} \tag{12.3}$$

$$k = \frac{K_{d(tempale\ molecule)}}{K_{d(competitive\ molecule)}} \tag{12.4}$$

$$k' = \frac{k_{MIP}}{k_{NIP}} \tag{12.5}$$

where, Q_e is the equilibrium adsorption capacity of MIPs.

12.3 MIPs Applications to Removal Pollutants

Environmental pollution has long been considered a threat due to increasing urbanization, changing lifestyles and contemporary levels of industrialization (Rahman et al. 2022). In particular, the release of industrial and agricultural chemicals greatly affects the health of the environmental, including water, soil and air (Azizi and Bottaro 2020). Water pollution has become a major problem due to various chemicals released into water from various industries and agriculture (Huang et al. 2015). The most important are pollutants such as heavy metals, inorganic pollutants, organic pollutants, endocrine disrupting compounds (EDCs), pharmaceuticals, dyes and pesticides (Uygun et al. 2013). All these highly toxic pollutants accumulate biologically and cause great harm to human health and aquatic ecosystems (Huang et al. 2015).

Many pollutants are usually present at very low concentrations such as parts per billion (ppb) and parts per trillion (ppm); therefore, most reported conventional methods do not have sufficient efficiency or sensitivity for their removal. Therefore, using MIPs is an alternative method for removing specific contaminants (Fig. 12.5). Also, they are of great interest due to their easy preparation, environmentally friendly and high selectivity (Metwally et al. 2021).

12.3.1 Removal of Metal Ions

Heavy metals refer to metals with a density greater than 5 g/cm^3 and an atomic weight between 63.5 and 200.6 (Ihsanullah et al. 2016; Fu and Wang 2011). Heavy metals can accumulate in living organisms because they are not biodegradable like organic pollutants. For this reason, toxic metals accumulating in living organisms have caused many neurological, psychological and psychological problems, such as high blood pressure, speech disorders, fatigue, sleep disorders, aggressive behaviors, poor concentration, irritability, mood swings, depression, increased allergic reactions, autoimmune diseases, atherosclerosis and memory loss (Ihsanullah et al. 2016). They are also known as a carcinogenic, mutagenic and toxic (Fu and Wang 2011). The most toxic heavy metals found in industrial wastes are nickel, chromium, lead, cadmium, mercury, arsenic, copper and zinc. The many traditional methods for the removal of heavy metals from wastewater include chemical precipitation,

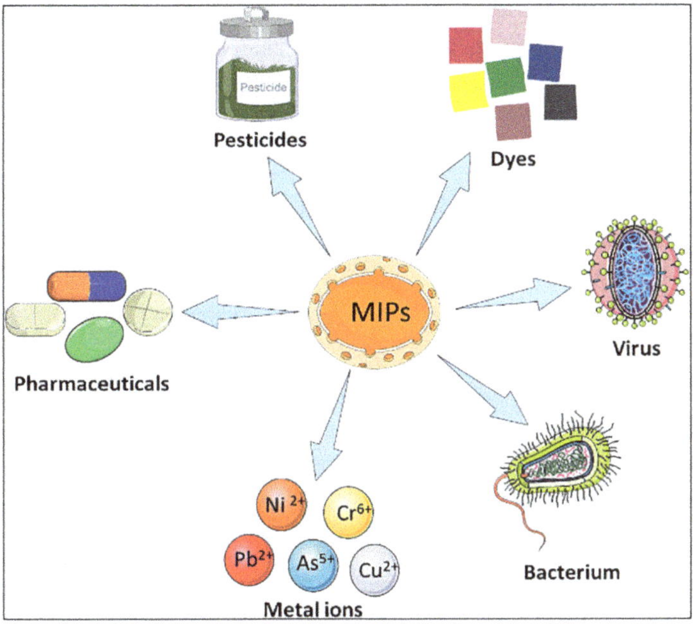

Fig. 12.5 Presentation of MIP applications to removal pollutants. This figure was created using images from Servier Medical Art by Servier is licensed under a Creative Commons Attribution 3.0 Unported License, https://smart.servier.com

flotation, adsorption, ion exchange and electrochemical deposition (Ihsanullah et al. 2016; Barakat 2011). In the recent years, MIPs have been used as an alternative to these methods.

Nickel

Nickel is one of the most abundant non-biodegradable toxic heavy metals in nature. It is also found in many industrial applications such as electroplating, batteries and electronics. Nickel is also one of the micronutrients necessary to increase hormonal activity and strengthen lipid metabolism (Metwally et al. 2021). Excessive nickel exposure causes serious lung problems such as dry cough, bone nose, cyanosis, rapid respiration, shortness of breath, tightness of the chest, chest pain, and pulmonary fibrosis; kidney problems, gastrointestinal distress such as nausea, skin dermatitis, nervous disorders and human carcinogen (Ihsanullah et al. 2016; Fu and Wang 2011; Barakat 2011; Lazar et al. 2023). For these reasons, the removal of nickel from wastewater is of great importance, and several studies on this topic focused on MIPs have been reported in the literature.

In a study, Giove et al. (2023) prepared Ni(II)-imprinted polymer (Ni(II)-IIP) with suspension polymerization method and nearly spherical particles with average particle diameter of 1340 ± 598.72 μm were obtained. 2-(aminomethyl)pyridine was used as a functional monomer. The maximum adsorption capacity of Ni(II)-IIP was found as 169.5 mg/g and 138.9 mg/g at pH 7 and pH 2, respectively and adsorption behavior was also determined to conform with Langmuir model. In the selectivity study of Ni(II)-IIP, Cd(II), Mn(II), Co(II), Mg(II) and Cu(II) were used as selective ions; The selectivity coefficient values were found to be 38.6, 32.9, 25.2, 23.1 and 15.0, respectively, at pH 7, while the values obtained at pH 2 were slightly lower. For the reusability study, adsorption/desorption cycles were carried out five times and regeneration efficiency was determined up to 98.7% for the first cycle and 81.2% for the fifth cycles.

In the another study, Chen et al. (2021a) prepared Ni(II) imprinted magnetic chitosan nanoparticles (MIIPs). The amount of maximum adsorption was determined as about 18.5 mg/g in the 25 mg/L initial Ni(II) concentration. Adsorption isotherms, adsorption kinetics and adsorption thermodynamics were investigated and found to suitable with Langmuir isotherm, pseudo-second order kinetic models and endothermic adsorption respectively. In the selectivity studies, the selectivity coefficient (α) values of MIIPs for Ni(II)/Cu(II) and Ni(II)/Zn(II) were found to be 3.02 and 14.35, respectively. In the reusability study, the adsorption capacity of MIIPs was determined to decrease only about 10% after fifteen adsorption–desorption cycles. In the real wastewaters applications, real nickel plating wastewater was taken from Gaoping Wastewater Treatment Plant. To calculate the removal efficiency of Ni(II), 100 mg MIIP was added to 100 mL of wastewater sample and sorption was carried out for one hour at room temperature. The removal efficiency for Ni(II) was found to reach to 100%. This study was also carried out with non-imprinted (MNIPs) adsorbent and the removal efficiency of MNIPS was determined as only 10.9%.

Chromium

Chromium is abundant in the wastewater of industries such as leather tanning, photography, and metal cleaning. Chromium has two valence form. The trivalent [Cr(III)] chromium is less toxic and stable, whereas the hexavalent [Cr(VI)] chromium is the most toxic, mutagenic, and carcinogenic form. The accumulation of chromium, especially Cr(VI), in the food chain, may lead to various health problems ranging from simple skin irritation to lung cancer. Therefore, many studies focus on their removal from the aquatic environment and drinking water (Wu et al. 2022; Fu and Wang 2011; Erdem et al. 2018; Açikel et al. 2016).

Uygun et al. (2013) synthesized Cr(VI)-imprinted poly (HEMAH) nanoparticles by surfactant-free emulsion polymerization technique to remove Cr(VI) ions (Fig. 12.6). The amount of adsorption capacity was determined as 3,830.58 mg/g in the pH 4. The particle size and specific surface area of nanoparticles were determine as 155.3 nm and 1397.85 m^2/g respectively. Adsorption isotherms and adsorption kinetics were also investigated and found to suitable with Langmuir model and the first order kinetic model respectively. In the selectivity study, K_d, k and k' values

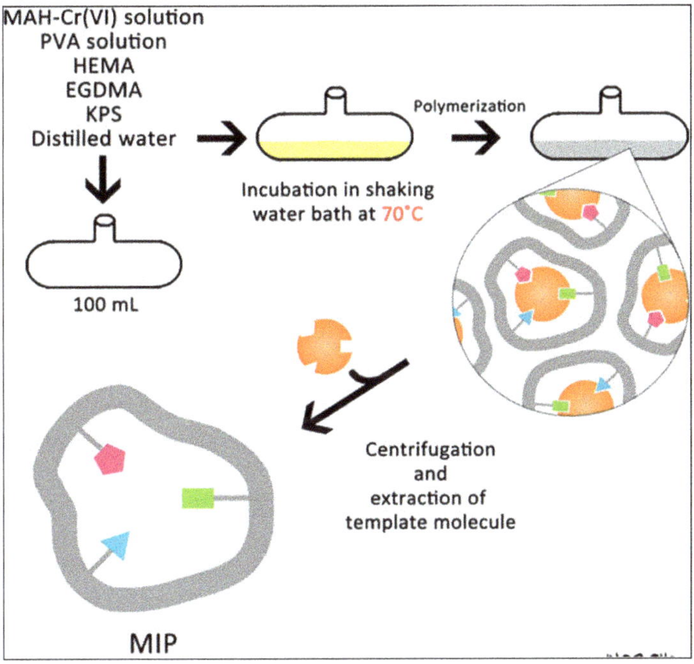

Fig. 12.6 İllustration of the synthesis of Cr(VI)-imprinted poly(HEMAH) nanoparticles. Republished with permission of Springer Nature, from Uygun et al. (2013); permission conveyed through Copyright Clearance Center

of Cr(III) with respect to Cr(VI) ion were calculated and selectivity of nanoparticles towards the Cr(VI) ions 3.51 times greater than Cr(III) ions. In the reusability studies, the fifth adsorption/desorption cycle was carried out and as a result, only 25% decrease in the adsorption capacity of nanoparticles was observed.

In an other study, Dakova et al. (2022) synthesized Cr(III)-imprinted poly(vinyl alcohol)/sodium alginate/AuNPs hydrogel membranes (Cr(III)-IIMs) with the solution blending method. After the characterization, adsorption studies were carried out and the maximum amount of adsorption Cr(III) was determined as 1.175 mg/g. The adsorption kinetics data were found to conform to a pseudo-first-order kinetic model, whereas, adsorption isotherms were also conform to Langmuir isotherm model. In the reusability studies, adoption/desorption experiments were carried out three times. Cr(VI) ions were used to compare with Cr(III) in the selectivity studies. In the real example applications, river water, groundwater, and seawater were diluted with varied concentration ratios of Cr(VI) to Cr(III) and the amounts of adsorbed Cr(III) were determined in the range between 95 and 98%.

Lead

Lead is one of the most toxic heavy metals of all metals. Lead has generally used as alloys in the synthesis of pigments, anticorrosion coatings, batteries, paint, cables, pesticides and pipelines (Erdem et al. 2018; Dahaghin et al. 2020). It can cause various type of damage, such as to the central nervous system, the kidney, the liver and reproductive system, basic cellular processes and brain functions. Also, anemia, insomnia, headache, dizziness, irritability, weakness of muscles, hallucination, and renal damages are examples of its toxic symptoms (Fu and Wang 2011). For this reason, it is very important to remove lead pollution in water, air and soil, and many studies address this subject.

In a study, Jagirani et al. (2022) synthesized lead(II)-imprinted interpenetrating polymer linkage (Pb^{2+}-IP) by the co-precipitation method. After, the characterization experiments, the effect of adsorption conditions on the Pb^{2+} adsorption were also investigate and the maximum amount of Pb^{2+} adsorption was determined as 85.47 mg/g at pH 6, and 10 mg/L initial Pb^{2+} concentration. In the adsorption isotherm and kinetic studies, adsorption behavior was well described by Langmuir isotherm model and pseudo-second order model respectively. In the selectivity experiments, Co^{2+}, Cd^{2+}, Ni^{2+},Cu^{2+}, NO^{3-},Cl^- and $SO_4{}^{2-}$ ions were used as competitive ions and the relative selective factor (K′) of Pb^{2+} ion and competitive ions was greater than 1 because of the imprinting effect. The reproducibility studies of Pb^{2+}-IP was also performed and the adsorption capacity of Pb^{2+}-IP decreased to only 3% after 10 adsorption–desorption cycles.

In the another study, Alnawmasi (2023) prepared Pb(II) imprinted amino-thiol chitosan derivative (Pb-ATCS). In the characterization experiments, fourier transform infrared spectroscopy (FTIR), nuclear magnetic resonance (NMR), scanning electron microscopy (SEM), EDX, and X-ray diffraction (XRD) analysis were carried out. The adsorption condition were also investigated and the optimum adsorption was calculated as 300 mg/g at pH 5. The adsorption behavior of Pb-ATCS was better described by Langmuir model and the pseudo-second order kinetics, respectively, compared to other isotherms and kinetic models. In the selectivity studies, the adsorption amount of Co(II), Hg(II), Cu(II), Zn(II), and Cd(II) ions on the Pb-ATCS were compared with Pb(II) adsorption and the relative selectivity coefficients were determined in the range of 13.95 and 18.87 with good selectivity. In the reuse studies, adsorption capacity of Pb-ATCS decreased from 300.68 to 290.35 mg/g after 5 adsorption–desorption cycles.

Copper

Copper is the third most essential heavy metal and has important roles in the several physiological processes such as oxidation, photosynthesis and carbohydrate, protein metabolism (Rehman et al. 2019; Roushani et al. 2015). Copper must be present in a certain amount in the human body. Its deficiency causes different symptoms such as hypochromic anemia, leucopenia, and osteoporosis (Kuras and Więckowska

2015). On the other hand, excessive amounts of copper cause severe adverse effects, such as liver and kidney damage, immunotoxicity, Wilson disease, insomnia, and developmental toxicity (Barakat 2011; Kuras and Więckowska 2015). Copper is used different industries such as food products, battery manufacturing, metal coating and finishing, metallurgy, and electrical production industries (Chaipuang et al. 2018). The wastes released from these industries cause great harm to the environment. For these reasons, as with other metals, copper removal from wastewater is very important and many studies address this subject.

In a study, Shamsipur and Besharati-Seidani (2011) prepared Cu^{2+}–imprinted polymeric (IIP) nanoparticles by thermal precipitation polymerization technique, using 1,4-dihydroxy-9,10-anthraquinone (quinizarin; QZ) as a ligand. In the characterization experiments, IR spectroscopy, SEM, N_2 adsorption–desorption isotherms and elemental analysis were carried out and the size of nanoparticles were determined as 60–110 nm. In the adsorption studies, the optimum pH value for Cu^{2+} adsorption was found as pH 7.0 and the maximum adsorption capacity was also determined as 346.3 μmol/g. The adsorption behavior was well fitted by Langmuir model. In the selectivity studies, Mn^{2+}, Co^{2+}, Ni^{2+} and Zn^{2+} ions were used as competitive ions and the relative selectivity coefficient (k') were calculated as 21.4, 23.3, 19.7, 21.1 respectively. In the reusability studies, adsorption–desorption cycles were carried out 12 times. In the real example applications, tap water, well water, river water, and spring water were spiked with 10 ng/mL Cu^{2+} ions and the amounts of extracted Cu^{2+} were determined in the range between 95.8 ± 2.8% and 103.9 ± 2.9%.

In an other study, Yu et al. (2022) synthesized Cu(II) ion imprinted bagasse to remove Cu(II) from the leaching solution of electroplating sludge using tetraethylenepentamine as functional monomer. The characterization studies involved SEM, EDS-mapping image, FTIR, zeta potential analysis. In the adsorption studies, the adsorption capacity of imprinted bagasse was defined as 32.0 mg/g and the adsorption process was determined to better fit the Langmuir model and pseudo-second-order kinetics model compared to other isotherms and kinetics models. The Ca(II), Zn(II), Mg(II), Cd(II), and Pb(II) ions were used as competitive ions in the selectivity studies. The values of k' were determined in range of 4.59 and 1.44. In reusability studies, adsorption–desorption cycles were carried out five times with high adsorption capacity. The real example applications involved the leaching solution of electroplating sludge and artificial wastewater, and in these solutions, the amount of Cu(II) adsorption was found as 15.5 mg/g and 17.5 mg/g respectively.

Arsenic

Arsenic pollution originates from two main sources. The first is its natural origin. Arsenic is released from groundwater in geological formations that occur in sedimentary rocks, geothermal water, and weathered volcanic rocks. The second sources is human activities. Human source of arsenics are mining, manufacturing, metallurgy, wood preservation, the use of arsenic in pesticides and herbicides, the combustion of fossil fuels, and arsenic additives (Ihsanullah et al. 2016; Abu Samah et al. 2020).

According to the World Health Organization, the recommended arsenic concentration limit in drinking water is 10 μg/L. Drinking arsenic contaminated water causes diseases such as skin cancers, lungs, bladder and kidney, vascular diseases and diabetes, infant mortality and weight loss of new born babies, hearing loss, hematologic disorders, neurological diseases (Tolkou et al. 2022). Therefore, many studies focus on removing arsenic from the aquatic environment and drinking water.

In their study, Türkmen et al. (2022) synthesized arsenic imprinted magnetic nanoparticles (IIP) using *N*-methacryloyl-l-cysteine (MAC) as a monomer to remove arsenite (As(III)) and arsenate (As(V)) simultaneously. As(III) and As(V) were used as template molecules. FTIR-ATR, Zeta-sizer, SEM, elemental analysis, and surface area techniques were carried out to characterize IIP and NIP magnetic nanoparticles. The average nanoparticle size was found approximately in the range of 33.90–32.30 nm with polydispersity index (PdI) of 0.1–0.05. Optimum adsorption conditions were determined as a pH 5.0, a contact time of 90 min, and 1.0 mg/mL initial arsenic concentration. The maximum adsorption capacity of IIP was as 76.83 mg/g for As(III) and 85.57 mg/g for As(V). The Langmuir isotherm and pseudo-second order model were found to be better description models than others for adsorption behavior of IIP. In the selectivity studies, SO_4^{2-}, NO^{3-}, and PO_4^{3-} ions were used for comparison with arsenic ions. The relative selectivity coefficients (k′) for As(V), PO_4^{3-}, SO_4^{2-}, and NO^{3-} according to IIP-As(III) nanoparticles were calculated as 5.00, 7.27, 9.74, and 9.62, respectively. According to IIP-As(V) nanoparticles, the relative selectivity coefficients (k′) for As(III), PO_4^{3-}, SO_4^{2-}, and NO^{3-} were also calculated as 6.91, 8.29, 9.56, and 9.92, respectively. In the reusability studies, 87.2% and 90% adsorption capacity of IIPs were determined at the end of 10 times adsorption–desorption cycles for As(III) and As(V) respectively. In the real sample applications, the environmental waters, both spiked with As(III)/As(V) ions and collected water from industrial facility were used. The removal efficiency was calculated and found as 72.7% and 90% for As(III) and As(V) respectively.

Another nanomagnetic ion-imprinted polymer (IIP@SiO_2@Fe_3O_4) were prepared by Sadani et al. (2020) for the removal of arsenic (As^{5+}). In this synthesis, 4-vinilpyridine (4-VP) was used as a functional monomer. The IIP@SiO_2@Fe_3O_4 was characterized by SEM, FTIR, XRD, AFM, and Zeta size analysis. The particle diameter was found as 7.54 nm. In the adsorption studies, the maximum removal efficiency of IIP@SiO_2@Fe_3O_4 was defined as 97.46% in the optimum conditions such as at pH 5.26, reaction time 101.93 min, 6.28 mg/L As^{5+} concentration. The adsorption capacity of IIP@SiO_2@Fe_3O_4 was 104.7 mg/g. In the adsorption isotherms studies, the Tempkin model proved a better fit than Freundlich and Langmuir models. The intraparticle diffusion and pseudo-second-order model were the most suitable models to describe adsorption behavior in the adsorption kinetic studies. According to thermodynamic data, the physical and spontaneous endothermic absorption occurred. In reusability studies, adsorption capacity of IIP@SiO_2@Fe_3O_4 was determined 62.50 mg/g with removal efficiency of 59.7% at the end of fourth regeneration cycle. In the toxicology tests, it was determined that IIP@SiO_2@Fe_3O_4 at different dilutions had no growth inhibitory effect on *Escherichia coli*.

Rare Earth Metals

Rare earth elements are 17 metallic elements with atomic numbers between 57 and 71 in Mendeleev's periodic table including lanthanides, yttrium, and scandium (Chen et al. 2022b). In recent years, rare earth elements have become important raw materials due to the developing industry and technology, and more than 100,000 tons of rare earth elements are consumed annually worldwide. These elements are used in a wide range of fields such as battery alloys, petroleum refining, ceramics, phosphors, medical radiation treatment, nuclear energy, medicine (magnetic resonance imaging), fiber optic communications, etc. (Jumadilov et al. 2022; Asadollahzadeh et al. 2021). As a result of such a wide range of uses of these elements, human exposure to these metals is gradually increasing. In recent years, the anthropogenic presence of lanthanides in lake waters and even tap water has been reported (Moussa et al. 2017; Kulaksız and Bau 2011). For this reason, different approaches are reported to remove rare earth elements such as adsorption, solvent extraction, precipitation, fractionation crystallization, membrane separation, and ion-exchange chromatography, etc. (Asadollahzadeh et al. 2021; Dolak et al. 2015). Molecularly imprinted technique is also used as an alternative method for these approaches.

In a study, Yusoff et al. (2017) synthesized lanthanide ion-imprinted polymers (L-IIPs) to separate five lanthanide ions (Pr, Nd, Sm, Eu, and Gd). In the synthesis, 4-vinylpyridine was used as a functional monomer. The L-IIPs were characterized by FT-IR and field emission scanning electron microscopy (FE-SEM) analysis. The maximum adsorption capacity of L-IIPs was determined as 125.3 mg/g, 126.5 mg/g, 127.6 mg/g, 128.2 mg/g, and 129.1 mg/g for Pr, Nd, Sm, Eu, and Gd, respectively at optimum pH of 6. In the adsorption kinetic studies, the pseudo-first-order model was determined to better fit than the pseudo-second-order model. In the selectivity studies, coexisting cations with rare earth elements with close atomic diameters were selected and used as competitive ions. The selectivity coefficient (k) values of L-IIPs were determined in the range of 13.2 and 27.1 with good selectivity. In the reusability studies, the adsorption–desorption cycles were performed 12 times, and about an 8% decrease in adsorption capacity was determined.

In another study, Patra et al. (2015) prepared europium (Eu(III)) imprinted polymer for preconcentration and extraction of Eu(III) from wastewater and coal soil. In the synthesis, graphene oxide and silane-modified magnetic nanoparticles (silane@Fe_3O_4) based dendritic structure was used as a monomer. After the synthesis, the Eu(III) imprinted polymer was modified with silica fiber to obtain a vinyl group. The extraction and detection of Eu(II) from blood samples were performed with a limit of detection = 0.050 pg/mL. Then, the separation and extraction of Eu(III) from soil and water samples in the coal mine area was carried out with high efficiency. In the reusability studies, recovery studies were performed 120 times with 100% efficiency. The storage capacity of this polymer was determined to be stable for 90 days with 98.0% extraction recovery efficiency.

12.3.2 Removal of Pharmaceutics

Pharmaceuticals are defined as specific biological activity molecules because they have many functional groups and physicochemical properties. Pharmaceuticals are generally classified as analgesic (anti-inflammatory and opioid), antibiotic, anti-cancer or cytostatic, antidepressant, antidiabetic, antiseizure, β-blocker, hormone, and lipid regulator, among others (Zare et al. 2022). These pharmaceuticals are used in the treatment of many diseases and the destruction of pathogens in animals as well as humans (Mezzelani et al. 2018). The pharmaceuticals taken into the body of humans and animals are metabolized and the decomposition products formed are excreted with feces. Finally, these compounds enter the sewage system. The leakages of sewage system and discharge of wastewater from treatment facilities into aquatic systems are responsible for pharmaceutical pollution in aquatic environments. In addition, the mixing of wastes formed in chemical synthesis and fermentation in pharmaceutical production factories into water also causes environmental pollution (Zare et al. 2022; Yang et al. 2017). These pollutants filter into the soil through surface waters and reach groundwater, where they are transferred to plants and animals and reach the human body through the food chain. They cause many diseases which can have both acute and long-term consequences in the human body (Zare et al. 2022). For these reasons, it is very important to remove pharmaceutical pollutions from wastewater. MIPs can be used for both specific removal and determination of pharmaceutical pollutants in wastewater. In this part, MIPs applications for pharmaceutical pollutions are discussed.

Antibiotics

Antibiotics are defined as a one of the major pharmaceutical pollutants. They can harm ecosystem survival as they are potentially ecotoxic molecules to non-target species. Frequent exposure to antibiotics, creates bacterial resistance to existing antibiotics (Zare et al. 2022; Mezzelani et al. 2018). For this reason, much research investigates the removal of antibiotics from wastewater using different methods. One of these is molecularly imprinted technology and several studies examine the removal of antibiotics.

Heidari et al. (2023) prepared magnetic molecularly imprinted polymer (MMIP) composite for removal of ciprofloxacin antibiotic. In the polymerization procedure, carboxy methyl dextrin was grafted to poly(aniline-co-meta-phenylenediamine) in the presence of Fe_3O_4/CuO nanoparticles. These MMIPs were characterized by FTIR, XRD, vibrating sample magnetometer (VSM), TGA, EDX, FE-SEM, zeta potential, and BET analyses. In the adsorption studies, adsorption conditions were optimized and the maximum adsorption amount was found to be 1111.1 mg/g (amount of MMIP: 3 mg, initial concentration of ciprofloxacin: 400 mg/L, pH:7, time: 15 min, temperature: 25 °C). The adsorption was well fitted to the Freundlich isotherm model; according to the results of adsorption kinetic data, this model was found to be suitable for the pseudo-second order kinetic model. The major adsorption forces to bond ciprofloxacin onto the MMIP were defined as electrostatic interaction, cation exchange, π-π interactions, and hydrogen bond. In the antibacterial activity of MMIP were also investigated and determined to have an antibacterial effect against *Escherichia coli*, a Gram-negative pathogen (16 mm), and *Staphylococcus aureus*, a Gram-positive pathogen (22 mm). In the reusability studies, after three adsorption–desorption cycles, no significant reduction in adsorption capacity was found.

Hu et al. (2022) synthesized surface molecularly imprinted polymers (SMIPs). Glycidyl methacrylate (GMA)-functionalized poly(maleic anhydride-alt-styrene) (PMS) particles (PMS-GMA) were synthesized in this study. To synthesize NOR-imprinted SMIPs (PMS@MIPs), methacrylic acid and ethylene glycol dimethacrylate were copolymerized in the presence of norfloxacin (NOR) template and PMS-GMA carrier. The major adsorption forces between NOR and PMS@MIPs was non-covalent interactions such as hydrogen bond, electro-static interaction, π-π stacking. In the characterization studies, SEM and FTIR analysis were carried out. In the adsorption studies, the adsorption conditions were investigated and the optimum conditions were determined as 60 min, pH 9 and 300 mg/L initial drug concentration. The maximum adsorption amount was also found as 149.3 mg/g in 60 min. The adsorption behavior was well described by pseudo-second-order and Freundlich model. The selectivity studies, enrofloxacin (ENR), levofloxacin (LEV) and sulfonamides (SA) were used as competitive molecules and the selectivity coefficients (κ) of PMS@MIPs to ENR, LEV and SA were found as 2.66, 2.42 and 3.44, respectively. In the reusability studies, the adsorption capacity of PMS@MIPs reduced slightly to 6.99% after four adsorption–desorption cycles.

Zhang et al. (2020) prepared dummy molecularly imprinted polymers (DMIPs) for aminoglycoside antibiotics (AAs). Streptomycin (STP), paromomycin (PRM), apramycin (APM), kanamycin (KAM), gentamycin (GNT), tobramycin (TOM), spectinomycin (SPC) were used as aminoglycoside antibiotics. DMIPs were synthesized by precipitation polymerization using raffinose (RAF) as template molecule, and methacrylic acid (MAA) as functional monomer. The RAF/ MAA complex was provided via hydrogen bonding interaction between OH groups existing in RAF and COOH groups in MAA. FTIR and SEM analysis were performed to characterize the DMIPs. In the binding studies, kanamycin (KAM) solution was used as model drug and the maximum adsorption amount of KAM was found as 128 mg/g in the 2000 μg/mL initial concentration. DMIPs were packed into SPE cartridge (Agilent) and these DMIPs cartridges were used in the sample treatments. In the specific recognition studies, the water samples spiked at 5 μg/mL of seven AAs were used and the recoveries of DMIPs for AAs were found as 94.7%, 98.0%, 99.0%, 98.5%, 229 95.0%, 99.0%, 51% (STP, KAM, APM, GNT, TOM, PRM and SPC respectively). Environmental water samples spiked with three different drug concentrations (2, 4 and 20 ng/mL or 10, 20 and 100 ng/mL) were obtained from tap water, lake water and irrigation water and the high recoveries were determined in STP, KAM, APM, GNT, TOM and PRM rather than others.

Chen et al. (2021b) synthesized green molecularly imprinted polymer (MIP) for kitasamycin (KITA). They used 1-allyl-3-vinylimidazolium chloride ionic liquid and 2-acrylamide-2-methylpropanesulfonic acid as co-functional monomer on the surface of poly(styrene–divinylbenzene) particles. In the characterization studies, FTIR, SEM and XPS analysis were carried out. In the adsorption studies, the maximum adsorption amount of KITA was determined as 127.06 mg/g in pH = 8–10 at 3 h, 45 °C. The hydrogen bond and π-π interaction were predominated adsorption force in the adsorption mechanism. Also, the absorption of KITA was well fitted by pseudo-second-order kinetic model, and Freundlich isotherm model. A tilmicosin (TIL) and amoxicillin sodium (AMX) were used to compare with KITA adsorption in the selectivity. The relative selectivity coefficient K′ of TIL and AMX were found as 8.52 and 7.59 respectively. In the real sample applications, two cultivated soil samples, three water samples (river water, lake water, sewage effluent), and two muscle samples (pork, chicken) were used with spiked 0.30, 0.90, and 1.50 mg/L KITA. The recovery (%) of KITA from the real environmental samples were determined in range of 92.3% and 108.8%.

Xie et al. (2023) prepared molecularly imprinted polymer (EMIP) using methacrylic acid (MAA) as a functional monomer for removal of erythromycin (ERY). In the characterization studies, SEM, FTIR, XPS, BET analysis and contact angles of absorbents were performed to characterize EMIPs. The specific surface area of EMIP was found as 265.62 m^2/g. In the adsorption studies, the maximum amount of ERY was determined as 4015.51 μg/g in 100 μg/L initial concentration. The adsorption isotherm and adsorption kinetic were found to well descried by Langmuir model and the pseudo-second-order model respectively. The imprinting factor for ERY was calculated and determined as 2.57. The azithromycin (AZM) and oxytetracycline (OTC) were used as interference molecules in the selectivity studies. The adsorption capacity of EMIP was found as 80% after five adsorption/desorption cycles. In the real sample experiments, local wastewater treatment plant was used and the adsorption capacity of EMIP for ERY in the real environment wastewater was determined as 415.4 μg/g.

In the another study, Ma et al. (2023) synthesized 3D porous allophane-β-cyclodextrin molecularly imprinted polymers (Allo-β-CD MIPs) using nanoclay mineral allophane (Allo), β-cyclodextrin (β-CD) as a bifunctional monomer for removal of ciprofloxacin (CIP) (Fig. 12.7). In the characterization studies, SEM, TEM, FTIR, XPS, and zeta potential measurements were carried out. In the adsorption studies, the effects of initial concentration, contact time, pH level, and ion strength on CIP adsorption were investigated and the maximum absorption amount of CIP was determined as 635 mg/g in the optimum adsorption conditions. The adsorption behavior of CIP was well defined by pseudo-second-order kinetic model and Langmuir isotherm models. The compressive forces of CIP adsorption on the Allo-β-CD MIPs were hydrogen bonds, hydrophobic interactions, surface complexation, and n-π EDA interactions. The tetracycline, sulfamethoxazole, and erythromycin were used as competing antibiotics. In the reusability studies, the adsorption capacity of Allo-β-CD MIP was found as 73.39% in the end of five adsorption/desorption cycles.

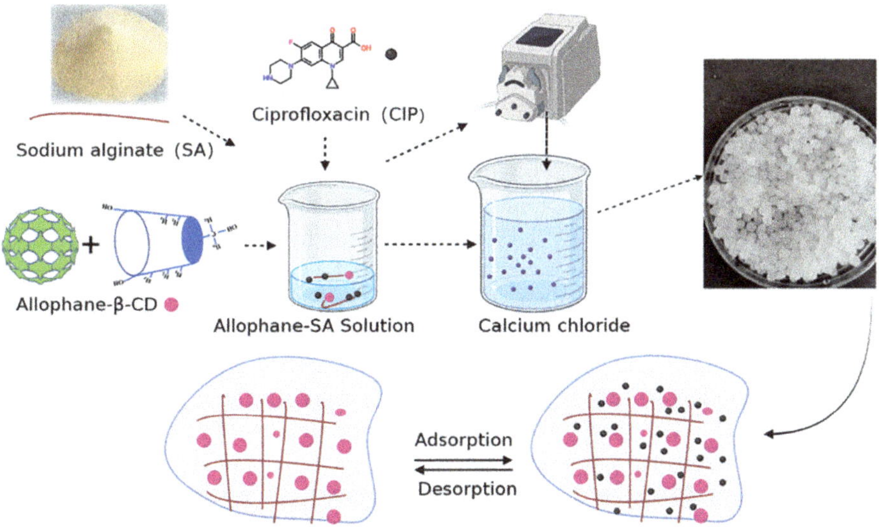

Fig. 12.7 Presentation of preparation process of Allo-β-CD MIP Republished with permission of Elsevier, from Ma et al. (2023); permission conveyed through Copyright Clearance Center

Hormones

Although hormones are important biomolecules for living metabolism, their presence in the environment and food and their involuntary ingestion is a grand problem. They can disrupt the function of the endocrine system and adversely affect growth, development, and reproduction. For this reason, they are also defined as endocrine disruptors (EDs) (Lazarević-Pašti et al. 2023; Oliveira et al. 2020). Removal and determination of hormones from environment waters is very important and many studies are reported on this subject.

In the one study, de Oliveira et al. (2020) prepared mesoporous molecularly imprinted polymer (MMIP) using methacrylic acid (MAA) as a functional monomer for removal of estriol (E3) and estrone (E1: template molecule). The TGA, BET, FTIR, SEM, and surface wettability analysis were performed to characterize the MMIP. In the adsorption experiments, the pH, contact time, initial concentration and temperature were optimized and the maximum adsorption amounts of E3 and E1 were found as 323.1 mg/g and 371.2 mg/g respectively. The adsorption kinetic and adsorption isotherm data were determined to suitable with the pseudo-second order and Langmuir models, respectively. In the reusability studies, the recovery of MMIP for E3 and E1 were defined as 75.4% and 84.6% at the end of five adsorption desorption cycles. In the selectivity experiments, ivermectin, eprinomectin, fipronil, fluazuron, enrofloxacin, norfloxacin, ciprofloxacin, and marbofloxacin drugs were used and the values of k' were calculated in range of 4.72 and 321.08.

Tian et al. (2020) synthesized magnetic molecularly imprinted nanobeads (MMINs) using dopamine as functional monomer for the removal of 17β-estradiol

(E2) in another study. In the characterization experiments, TEM, FTIR, XRD, and vibration sample magnetometry analysis were carried out. The effect of time and initial concentration on the E2 adsorption were investigated and determined as 20 min and 80 μg/mL respectively. Also, the maximum adsorption amount of E2 were found as 41.48 mg/g with 8.07 imprinting factor. The most suitable adsorption kinetic and isotherm models were defined as pseudo-second order and Langmuir models, respectively. In the selectivity studies, four environmental phenolic estrogens estrone (E1), estriol (E3), diethylstilbesterol (DES), and 17α-ethynylestradiol (EE2) were used and the selective factors (SFs) were determined as 4.43, 3.59, 6.07, and 3.43 respectively. In the reusability experiments, the adsorption efficiency was found to be greater than 94.8% after reusing for 10 times. In the real sample experiments, lake water spiked with 0.04, 0.2, and 1.0 μg/L of E2 were used and the recovery of MMINs for E2 were detected as 102.5%, 100.9%, and 99.7% respectively.

Sullivan et al. (2023) prepared molecularly imprinted nanoparticles (NanoMIPs) by one-pot microwave synthesis with suspension polymerization method using methacrylic acid (MAA) as the functional monomer and andarine, RAD-140, estradiol, and testosterone as template molecules. FTIR and SEM analysis were carried out to characterize NanoMIPs. According to SEM images, the particle sizes were determined as 132.7 nm, 143.3 nm, 120.2 nm and 135.5 nm for the andarine, RAD-140, estradiol and testosterone nanoMIPs. In the rebinding studies, the percentage target bound (%) were found as 94.08%, 94.46%, 96.62%, 96.80% for andarine, RAD-140, estradiol and testosterone. In the real sample studies, river water samples spiked with 20 μg/mL the andarine, RAD-140, estradiol and testosterone and percentage bounded (%) was found as 95.12%, 92.27%, 93.03%, and 94.45% respectively.

Non-steroid Anti-inflammatory Drugs

Non-steroid anti-inflammatory drugs (NSAIDs) are one of the most abundant environmental pollutants. One of these is diclofenac. This drug has chronic toxicological effects and is mainly detected in surface water, groundwater and even drinking water. Removal and detection of NSAIDs from environment waters is very important and many studies are reported on this subject. Cantarella et al. (2019) synthesized diclofenac—selective molecularly imprinted polymers (MIPs) by a single-step bulk polymerization process using the methacrylic acid as functional monomer. In the characterization studies, TGA, nuclear magnetic resonance (NMR), BET, and Nonlocal density functional theory (NLDFT) analysis were performed. In the adsorption studies, MIP were removed about 110 μmol/g (∼90% of) diclofenac in the solution in 10 min. In the selectivity studies, acetylsalicylic acid (ASA), trimethoprim (TRIM), and methyl orange (MO) were used as competitive molecules. The proportion of removal was determined as only ∼10% of ASA, less than 3% of TRIM and ∼23% of MO. In the reusability studies, adsorption/desorption cycles were carried out 4 times with high adsorption ratio.

In the other study, Husin et al. (2021) prepared magnetic molecularly imprinted polymer (Fe_3O_4@MIP) by bulk polymerization method using methylacrylate acid and choline chloride-1-butylimidazole (ChCl-BuIM) as functional monomer. Naproxen, non-steroid anti-inflammatory drug, was used as template molecules. FTIR, SEM, elemental analysis, and VSM analysis of Fe_3O_4@MIP were carried out to characterize the polymer. In the adsorption studies, the optimum adsorption conditions were determined to be 15 min, pH 5, 40 mg/L initial concentration of naproxen, 308 K. The adsorption behavior was better defined with the pseudo second order model and Freundlich model. They also determined that naproxen adsorption on Fe_3O_4@MIP was realized by chemisorption mechanism. The amount of adsorption was determined as 178.57 mg/g at 308 K temperature. In the selectivity studies, four different NSAIDs such as ketoprofen, ibuprofen, diclofenac, and fenoprofen were used as competing molecules with a similar structure to naproxen. The relative selectivity coefficients, k' were found as 2.03, 0.92, 2.10, 1.20 respectively. In the reusability studies, the removal percentage was defined as 89.7% at the end of sixth adsorption/desorption cycles. In the real sample experiments, pharmaceutical wastewater samples were obtained from five industrial areas around Sungai Petani, Kedah, Malaysia. These wastewater samples were spiked with 1, 10, and 40 mg/L naproxen. The removal percentages of naproxen from all samples were determined in the range of 93.2–97.1%. The end of the studies, the toxicity of ChCl-BuIM was investigated and MTT assay was performed using the MCF-10A and BEAS-2B cell lines. According to MTT analysis reports, ChCl-BuIM was not toxic in these cell lines. Some MIPs studies used in recent years for the removal of different pharmaceuticals in real samples are given in Table 12.1.

12.3.3 Removal of Dyes

Environmental pollution caused by the wastewater of dye-using industries is a major worldwide problem (Sadia et al. 2022). Industrial areas where natural and synthetic dyes are most used are paper, food, leather, plastic, pharmaceutical, automative, cosmetics and textiles (Malik et al. 2023). Dyes cause many kinds of diseases such as cancer, allergies, dermatitis and skin irritation. Therefore, it is critical to remove dyes from the environment to prevent pollution (Sadia et al. 2023). Adsorption of dye molecules onto a sorbent is one of the most effective and economical methods of color removal. MIPs as sorbents may be a good alternative for the selective separation/ detection of individual dye molecules (Kyzas et al. 2009). In the last few years, molecularly imprinted polymers have been proposed for the removal and/or detection of dyes from contaminated media (Algieri et al. 2014).

Al-Degs et al. (2009) aimed to synthesize MIP-based solid phase extracts with high selectivity for the recognition of reactive red 4, an anionic dye commonly used in the textile industry. The imprinted polymers were tested for the selective extraction and pre-concentration of the dye from water and wastewater. In the study, methacrylic acid and methyl methacrylate were used as functional monomers in the polymerization stage to produce the dye-imprinted polymer. The researchers observed better extraction of reactive red dye after imprinting for both polymers, and

Table 12.1 MIP application in different real sample for pharmaceuticals

Pharmaceuticals	Target pollutant	Molecularly imprinted polymer	Adsorption capacity	Adsorption forces	Synthesis method	Real sample	**Ref**
Antibiotic	Streptomycin, paromomycin, apramycin, kanamycin, gentamycin, tobramycin, spectinomycin	DMIPs	128 mg/g (kanamycin)	Hydrogen bond	Precipitation polymerization	Tap water, lake water and irrigation water	Zhang et al. (2020)
Antibiotic	Kitasamycin	MIP	127.06 mg/g	Hydrogen bond and π-π interaction	Surface imprinting	Two cultivated soil samples, three water samples (river water, lake water, sewage effluent), and two muscle samples (pork, chicken)	Chen et al. (2021b)
Antibiotic	Erythromycin	EMIP	4015.51 μg/g	–	Emulsion polymerization	Local wastewater treatment plant	Xie et al. (2023)
Hormone	17β-estradiol	MMINs	41.48 mg/g	Hydrogen bonding and π-π staking	Surface imprinting	Lake water spiked with 0.04, 0.2, and 1.0 μg/L E2	Tian et al. (2020)
Hormone	Andarine RAD-140 Estradiol Testosterone	NanoMIPs	94.08%, 94.46% 96.62% 96.80%	Hydrogen bonding, Van der Waals and ionic bonding	Suspension polymerization	River water spiked with 20 μg/mL	Sullivan et al. (2023)
Non-steroid anti-inflammatory drugs	Naproxen	Fe_3O_4@MIP	178.57 mg/g	Hydrogen bonding, π-π interaction	Bulk polymerization	Pharmaceutical wastewater spiked with 1, 10, and 40 mg/L	Husin et al. (2021)

DMIPs Dummy-imprinted polymer, ***MIP*** molecularly imprinted polymer, ***EMIP*** erythromycin imprinted polymer, ***MMINs*** magnetic molecularly imprinted nanobeads, ***NanoMIPs*** molecularly imprinted nanoparticles, ***Fe_3O_4@MIP*** magnetic molecularly imprinted polymer, ***E2*** 17β-estradiol they found that the maximum adsorption of the dye to the MIP occurs through the H-bonds between the carboxylic acid groups on the MIP surface and the sulfonated groups of the dye at acidic pH.

Another study on wastewater treatment using MIPs was conducted by Lu et al. (2015). In the study, MIPs with ultra-thin membrane structure were synthesized by ATRP polymerization technique, using Rhodamin B as template molecule. As a result of the kinetic studies carried out to determine the adsorption rate model of the study, the authors emphasized that the adsorption process was a good fit with the pseudo-second order model, and suggested that the rate-limiting step in the adsorption process could be chemisorption which involves valence forces through the sharing or exchange of electrons between the MIP and the template molecule.

A study on the effective removal of acidic dyes from water was investigated with the "synchronously graft-polymerizing and imprinting technique", a new surface printing technique developed by Gao et al. (2015). During polymerization, while Acryloyloxyethyl trimethyl ammoniumchloride (DAC) and *N*,*N*-methylenebisacrylamide (MBA) were produced on the surface of SiO_2 particles, molecular imprinting of template molecule Acid Red (AR) was realized by wrapping AR molecules in cross-linked networks, leading to the development of AR imprinted MIP-PDAC/SiO_2. According to the results of the study, MIP-PDAC/SiO_2 can be considered to have individual recognition selectivity and perfect binding affinity for AR, and it was found that the adsorption of AR onto MIP occurs through electrostatic interactions and pi interactions between the functional monomer DAC and AR.

In an other study, Zhao et al. (2017) focused on removal of methylene blue by using sericin (SS), which is an abundant and renewable resource in nature, with proven efficiency in metal and dye removal. In the study, MB-imprinted sericin/PVA electrospun fiber adsorbent was developed by combining electrospinning technique and molecular imprinting technique. In the pH part of the study, the authors observed that at low pH values, high H+ concentration creates a negative effect due to competitive adsorption between the existing H+, MB and the functional groups of serisine. As pH values increased, more functional groups were deprotonated, and more cationic MB was adsorbed via electrostatic interactions. The study focused on, Langmuir and Freundlich adsorption isotherms, and the authors reported that the adsorption was compatible with the Langmuir adsorption isotherm and that the adsorption between MIPs and methylene blue was monolayer and homogeneous. It was also found that the adsorption was compatible with the pseudo-second order model, which is suitable for the chemiadsorption process.

Another study referring to the textile industry is the development of MIPs for selective adsorption and quantification of Acid green 16 in water samples. In this study by Foguel et al. (2017), 1-vinylimidazole was used as a functional monomer and MIPs were synthesized by bulk method. The authors observed that the acid green 16 dye has a very high binding capacity to the selective binding sites of MIP. Furthermore, MIP showed perfect performance in the extraction of dye in water samples, with a recovery rate close to 100%, which is better than commercial SPE cartridges.

The new amino-functionalized titanium dioxide (TiO_2) surface molecularly imprinted polymer (ATMIP) was developed successfully by Sun et al. (2022) for rapid adsorption of methylene blue (MB). In the study, sodium lignosulfonate (SL) was used as a functional monomer, and they were able to remove methylene blue from the aqueous medium within 10 min with this molecularly imprinted polymer with a surface area of 56.3266 m^2/g. In adsorption kinetics studies, it was found to be compatible with the pesudo second order kinetic model, showing that the adsorption of the template molecule to MIP is a chemical process. In addition, the authors stated that the adsorption occurred in multilayers in accordance with the Freundlich adsorption model.

Another study by Gao et al. (2022) investigated the selective adsorption of methylene blue (MB) by molecularly imprinted sodium alginate/polyethylene oxide nanofiber membranes (SA/PEO-MINP), synthesized by electrospinning method. In the study, it was stated that the adsorption of MB occurs through electrostatic interactions between the negative charges of imprinted regions on SA/PEO-MINP and the cationic MB molecule. Adsorption isotherm experimental data of the study were compatible with the Langmuir isotherm model of adsorption. The kinetic experimental data agreed well with the pseudo-second-order model, indicating that the adsorption of MB on SA/PEO-MINM was chemisorption.

Hybrid MIPs were recently synthesized by Ahmed et al. (2018). In the study, the molecularly imprinted chitosan-TiO_2 nanocomposite (CTNC) material developed with the combination of MIP nanoparticles and chitosan. Synthesized MIP showed a very high selectivity compared to pure chitosan. The study reported that it is compatible with the Langmuir adsorption model, which suppose that the adsorption of RB dye occurs in specific homogeneous regions with a single-layer adsorption, without any interaction between the dye molecules and the solid surface. The study also found that the adsorption of RB dye onto CTNC MIPs was compatible with the pseudo-second-order kinetic model.

In another study, Sedelnikova et al. (2023) synthesized a hybrid model MIPs (MMIP) using nylon 6 and magnetic nanoparticles to capture methylene blue. For the synthesis of the polymer matrix, the transition of nylon-6 from a dissolved form to a solid form was carried out in the presence of template molecules and Fe_3O_4 nanoparticles in the initial solution. To improve the magnetic properties and aggregation ability of MMIPs, MMIPs were synthesized by changing the concentration of the magnetic core. MMIPs were found to have a high adsorption capacity in the range of 20.0–34.8 mg/g and imprinting factors in the range of 4.2 to 5.4. This hybrid polymer showed very high specificity compared to non-imprinted polymer.

Two separate studies published recently by Sadia et al. (2022, 2023) investigated the removal performance of basic blue 3 and acid black 234 from the water environment by molecularly imprinted polymers. In both studies, methacrylic acid, which is a preferred monomer in molecular imprinting procedures, was used as a functional monomer and molecularly imprinted polymers were synthesized by bulk polymerization. According to results, MIPs exhibited higher selectivity for acid black 234 and basic blue 3 than unimprinted polymers. In both studies, the adsorption kinetic model between dye and MIP was found to be compatible with the pseudo second order kinetic model, while the adsorption isotherm model was found to be compatible with the Langmuir adsorption model, which supports the view that the adsorption is monolayered and homogeneous. Table 12.2 gives the important parameters of dye removal studies using some molecular imprinting techniques of recent years.

12.3.4 Removal of Pesticides

Pesticides are a traditional agricultural application used worlwide to neutralize insect damage and increase crop yields and economic returns (Jia et al. 2024; Song et al. 2022). The term pesticide is used for chemicals such as insecticides, molluscicides, fungicides, bactericides, herbicides or rodenticides, depending on the target organism. Besides the target organism, Pesticides are classified according to their chemical structures such as organophosphates, organochlorines, carbamates, pyrethroids, phenyl amides, phenoxy alkanes, trazines, benzoic acid, phthalamides, dipyrides (Majdinasab et al. 2021). More than 100 commercial grades are available today, containing more than 800 active ingredients (Song et al. 2022). These chemicals' use is increasing every year in fields such as agriculture, animal husbandry, aquaculture and forestry (Fang et al. 2021). The excessive and long-term presence of pesticides in various matrices such as water, soil, and vegetables above the recommended doses causes serious social concerns in terms of food safety and environmental protection (Ba Hashwan et al. 2020). Long-term exposure to pesticides can have toxic, mutagenic, carcinogenic and tumorigenic effects in humans (Nageib et al. 2023). Therefore, advanced detection technologies of pesticides are of key importance to prevent pesticide hazard and ensure human health (Mahmoudpour et al. 2020). Continuous monitoring is required to determine whether pesticide residues in soil, food, water and the environment remain within legal limits (Altaf et al. 2021). Molecular imprinting technique is a high-level development in the selective extraction of pesticides from various media.

Liu et al. (2015) developed a new magnetic MIP (MMIP) on the g-C_3N_4–Fe_3O_4 surface for the specific adsorption of atrazine pesticide using Graphite-like carbon nitride (g-C_3N_4). The results of the study showed that g-C_3N_4 is a thin monolayer structure that fixes Fe_3O_4 nanoparticles, and the imprinted polymer is coated on the g-C_3N_4-Fe_3O_4 surface. In the study, it was emphasized that the adsorption mechanism between Atrazine and MIP may be due to hydrogen bonding and non-covalent interactions between atrazine and the monomer (MAA) used during polymerization. According to the Scatchard analysis of the study, MMIP contains high- and low-affinity binding sites.

Masoumi et al. (2016) developed superparamagnetic molecularly imprinted polymers (MMIP) for the selective adsorption of phasolane, diazinone and chlorphyrifos pesticides from aqueous media in 2016. These MMIPs were synthesized by cross-linking methyl methacrylate (MMA) and maleic anhydride (MA) copolymers with triethylene tetramine (TETA), which acts selectively in the adsorption of pesticides. Super paramagnetism was given to the structure by amino-functional

Table 12.2 Some MIP studies for removal of dye

Analyte	Functional monomer	Method/ Material	Binding capacity	Imprinting factor	Selectivity coefficients	Ref.
AB-234	MAA	MIPS/bulk polymerization	83.04 mg/g (298 K)	5.13	AB-234/ AB-1: 6.25 AB-234/ Safranin: 7.89 AB-234/ AY-76: 9.5 AB-234/ BB-3: 11.9	Sadia et al. (2023)
MB	–	MMIPs	110 μmol/g	5.4	MB/PH: 6.1 ± 0.6 MB/TB: 2.1 ± 0.3	Sedelnikova et al. (2023)
BB-3	MAA	MIPs/Bulk polymerization	85.4 mg/g (298 K)	10.73	BB3/ Safranin: 6.18 BB3/ MB:9.66 BB3/BB41: 7.37 BB3/ Thioflavin: 7.29	Sadia et al. (2022)
MB	–	SA/ PEO-MINM	3186.7 mg/ g	–	MB/MO: 7.41	Gao et al. (2022)
AR	DAC	MIP-PDAC/ SiO_2	538 mg/g	–	AR/ALY: 6.46 AR/ AO: 5.32	Gao et al. (2015)
RhB	MAA	MIP Membranes	100.1 mg/g	2.83	RhB/BRhB: 22.57 RhB/Rh6G: 8.79	Lu et al. (2015)

(continued)

Table 12.2 (continued)

Analyte	Functional monomer	Method/ Material	Binding capacity	Imprinting factor	Selectivity coefficients	Ref.
RR	MAA	MIP 1	80.7 mg/g	–	RR/RY: 17.1 RR/RB: 13.9	Al-Degs et al. (2009)
	MMA	MIP 2			RR/RY: 2.65 RR/RB: 2.65	

BB-3 Basic blue 3, ***AB-234*** acid black 234, ***BB41*** basic blue 41, ***MB*** methylene blue, ***AR*** acid red 14, ***RhB*** Rhodamine B, ***RR*** reactive red, ***BR*** basic red, ***RB*** reactive blue, ***RY*** reactive yellow, ***AB-1*** acid black 1, ***AY-76*** acid yellow 76, ***PH*** hydroxyethylphenazine, ***TB*** toluidine blue, ***MO*** methyl orange, ***MAA*** methacrylic acid, ***MMA*** methyl methacrylate, ***MMIP*** magnetic molecularly imprinted polymer, ***SA/PEO-MINM*** sodium alginate/polyethylene oxide nanofibrous MIP membranes, ***MIP*** molecularly imprinted polymer, ***DAC*** acryloyloxyethyltrimethyl ammonium chloride, ***MIP-PDAC/SiO₂*** AR molecule surface-imprinted material MIP-(PDAC-*co*-MBA)/SiO_2, ***ALY*** acid light yellow G, ***AO*** Acid orange II, ***BRhB*** buthyl Rhodamine B, ***Rh 6G*** Rhodamine 6

Fe_3O_4 (Fe_3O_4-NH_2) nanoparticles. Langmuir, Freundlich, Dubkin-Radushkevich and Temkin models were studied to understand the adsorption behavior and it was found to be compatible with the Temkin adsorption model. The authors concluded that due to adsorption-analyte interactions, the adsorption heat decreases linearly with the coverage area, and the binding energies are uniformly distributed up to the maximum binding energy. According to the Langmuir equation, the maximum adsorption capacity for phasolane, diazinone and chlorphyrifos was calculated as 196.07, 192.30 and 172.41 mg/g, respectively. The kinetic model in which the studied pesticides are adsorbed to MMIPs is consistent with the pseudo-second-order model. In the light of this result, the authors emphasized that adsorption is compatible with chemical adsorption, which is a rate control step.

In an other study, So et al. (2018) synthesized a surface molecularly imprinted polymer (SMIP) for the adsorption of CBL (1-naphthyl methyl carbamate), a pesticide frequently used in agriculture. SMIP adsorbent was created by copolymerizing methacrylic acid (MAA) monomer and CBL template molecule with ethylene glycol dimethacrylate (EGDMA) cross-linker on the surface of silica gel particles (Fig. 12.8). The authors emphasized that when the pH value was below 7, the hydrogen bonds between the amide groups and carboxylic groups of CBL became stronger in the imprinted regions, thus achieving better adsorption in acidic conditions. Adsorption isotherms and adsorption kinetics were found to be compatible with the Freundlich equilibrium model and pseudo-second-order kinetic model, respectively. The scatter plot analysis shows that there were two different binding sites in SMIP adsorbents.

Öter and Zorer (2021) synthesized molecularly imprinted polymers by suspension polymerization technique and an adsorbent for selective solid phase extraction

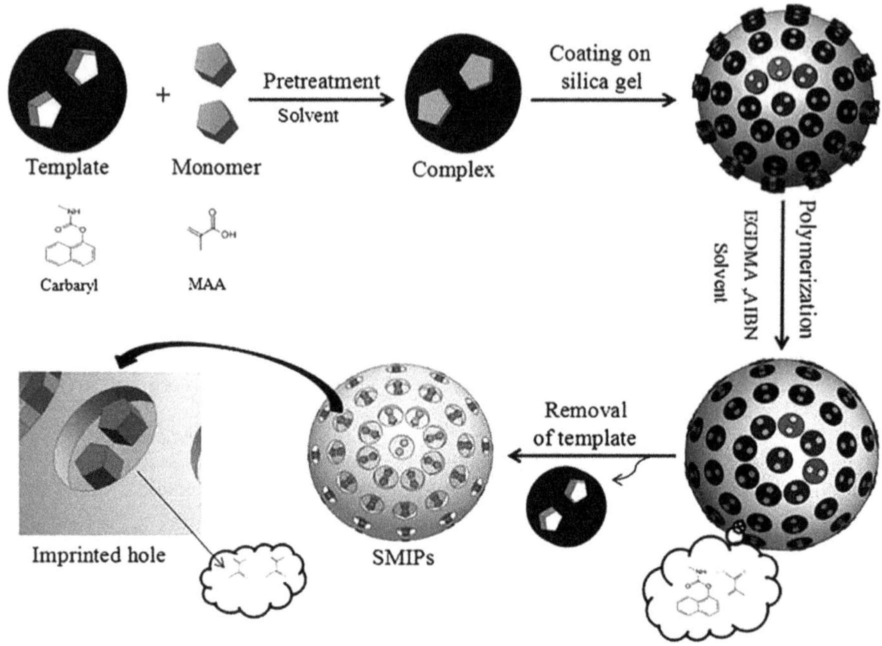

Fig. 12.8 Presentation of synthesis of SMIP presentation of synthesis of SMIP. Republished with permission of Elsevier, from So et al. (2018); permission conveyed through Copyright Clearance Center

was developed for the detection of ziram, a fungicide categorized as a pesticide. In the study, MAA, which can form hydrogen bonds with the template molecule ziram and act as both a hydrogen bond donor and a hydrogen bond acceptor, was used as a functional monomer. Adsorption kinetic studies conducted to determine the ziram binding rate of the adsorbent during the adsorption process were found to be compatible with the pesudo second kinetic adsorption model. Accordingly, the authors emphasized that the adsorption of ziram to MIP was mainly controlled by chemical adsorption. According to the adsorption isotherm results, ziram was observed to be adsorbed to the binding sites distributed homogeneously on polymer particles' surface. The maximum monolayer adsorption value for MIP was found to be 56.98 mg/g.

In a study by Azizi et al. (2021) a thin MIP film prepared on a stainless steel substrate was developed for the enrichment of organophosphate pesticides (OPPs) in water and beverage samples before LC-MS/MS. The study involved the optimization of the composition of MIP according to the relative volume of the porogen using a pseudo-phase diagram to obtain micro-gel polymer, and as a result, it was found that the extraction and desorption were rapid due to the porosity of the microgel. Other factors were optimized, such as the type of template and its proportions with monomer (methacrylic acid) and cross-linker (ethylene glycol dimethacrylate). In the adsorption kinetic study conducted to explain the adsorption behavior, the adsorption kinetics of the other analytes except chlorpyrifos, which has the highest tendency to non-selective hydrophobic interactions in the studied analyte group, were found to be compatible with the pseudo-first-order kinetic model. Based on this, the authors argued that the MIPs they synthesized contained high-affinity imprinted regions, potentially providing better performance for analytes with polar groups. As a result of the study, the authors emphasized that by imprinting, they created new binding sites that could show high affinity for polar functionalities. Investigations of Langmuir and Freundlich adsorption isotherm models, other parameters explaining adsorption behavior, revealed that they were compatible with the Langmuir model, which suggests that the studied OPPs adsorb to MIPs in a monolayer, and homogeneously.

In another study, the MIP based adsorbents were synthesized by Li et al. (2022) to be combined with HPLC-MS/MS for the detection of Organophosphate pesticides from agricultural products. In the study, hybrid composite adsorbents were obtained by synthesizing new core–shell structured zeolitic imidazolate framework@molecularly imprinted polymers (ZIF-8@MIPs) using the surface imprinting technique. These synthesized hybrid composites exhibited specific recognition and rapid adsorption capacity for organophosphate pesticides (Parathion-methyl, malathion, triazophos, fenthion, fenamiphos). It was reported that the dominant interactions between the synthesized hybrid composite and the analyte are π-π interactions and hydrogen bonds. In the study, the binding properties of ZIF-8@MIPs were performed by adsorption equilibrium and Scatchard analysis. According to Scatchard analysis, the binding sites of the analyte in ZIP-8@MIPs were found to be heterogeneous. Qmax values for low-affinity and high-affinity binding sites in MIPs were found to be 39.1 mg/g and 334.57 mg/g, respectively.

In another study, Meseguer-Lloret et al. (2022) performed magnetic molecularly imprinted polymers for the simultaneous selective extraction of phenoxy acid herbicides from environmental water samples. Magnetic molecularly imprinted polymers (MMIP) were synthesized using 4-chloro-2-methylphenoxyacetic acid as template and 4-vinylpyridine as monomer in the presence of vinylated magnetic nanoparticles. It was emphasized in the study that the adsorption of phenoxyacid herbicides to MMIPs occurs through hydrophobic interactions at acidic pH, pi interactions between aromatic groups and electrostatic interactions. The study was carried out at acidic pH because a decrease in the extraction of the analyte was observed at pH values above 6.5, due to the deprotation of the carboxyl group of the compounds and the decrease of the pyridyl groups in the MIP. In the study, the synthesized MMIP exhibited excellent regeneration properties and could be used more than 65 times without a significant loss in extraction capacity.

In an other study, Farooq et al. (2022) carried out a study on the extraction of abamectin (ABM), a highly toxic pesticide that poses a potential risk to the ecological environment. In the study, beta cyclodextrin (β-CD), a preferred monomer in MIP synthesis, which has the ability to form a tight, non-inclusive complex with analytes, was used as a monomer. Within the scope of work, β-CD-based imprinted polymers were synthesized using ionic liquid at room temperature. The synthesized polymers showed high selectivity for abamectin and it was possible to use these polymers repeatedly without loss of performance. A adsorption kinetic study was conducted to explain the binding behavior of the analyte to the polymer in which, a pseudo second order kinetic adsorption model was applied, and the adsorption process of the analyte to the polymer was found to occur via chemisorption. To observe the efficiency of MIPs, Langmuir (5) and Freundlich (6) isotherm models were applied. The data obtained showed that MIPs have single-layer chemical adsorption like the Langmuir model, and the maximum adsorption capacity obtained in the study was found to be 15.08 mg/g.

Another study using MIPs was carried out by Hayat et al. (2022). In the study, MIPs designed in pendant form were synthesized for the extraction of lufenurone, one of the benzoylurea pesticides, from real water samples. In polymerization, lufenurone, which is the template molecule, and vinyl acetate, acrylic acid and acrylonitrile monomers interacted simultaneously and precipitation polymerization was performed. It was reported that monomers have functional groups that can interact with the functional groups of lufernuro through H bonds, dipole–dipole interactions. The experimental studies were performed by batch binding assay and response surface methodology (RSM). In the study, maximum lufenuron adsorption on MIPs was observed at pH 8. The authors emphasized that this is due to hydrogen ions competing with the analyte for binding sites in the acidic environment, while active binding sites are exposed in the basic environment. Since lufenuron was less stable in higher alkaline conditions, adsorption decreased with increasing pH. Lufenuron adsorption on MIP was best described by the pseudo first order kinetic adsorption model. Therefore, the authors emphasized that the rate control step in lufenuron adsorption by MIPs is controlled by a physisorption mechanism.

In another study Silica nanoparticle-based PND imprinted polymers were synthesized by Noor et al. (2023) who investigated their effectiveness in the removal from drinking water of pendimetholine (PND), a synthetic pesticide with carcinogenic effects for humans. In the study, a monomer was synthesized using silica nanoparticles as surface imprinted substrates and MIP synthesis was carried out by providing PND interaction with this monomer. It was emphasizedthat electrostatic interactions were dominant in the re-adsorption of PND to MIPs. The adsorption capacity reached in the study was reported as 532.69 mg/g with PND imprinted polymers. The authors reported that they synthesized polymers that are economical, fast, practical, and have commercial potential for the determination of PND in drinking water.

12.3.5 The Other Contaminants

Mycotoxins are secondary metabolites with toxic properties synthesized by fungi, and mitotoxins are the main cause of contamination of concern in the food and feed industry (Hemmati et al. 2023; Pezeshkpur et al. 2023). The use of MIPs for rapid and selective detection of mycotoxins can be effective.

In a study conducted in 2023 (Pezeshkpur et al. 2023), MIP coupled Fe_3O_4-Au nanocomposite adsorbent, developed with the combination of molecular imprinting technology and nanotechnology, was used for the rapid detection of Aflatoxin B1 (AF B1), a type of mycotoxin, in non-alcoholic berries and barley samples. Methacrilic acid (MAA) was used as a functional monomer and AF B1 was used as a template molecule in the polymerization part of the study. In the study, the adsorption capacity of MIPs was found to be 8.975 mg/g, which is twice the adsorption capacity of unimprinted polymers. The authors concluded that Fe_3O_4-Au@MIP could be an effective and economical alternative adsorbent for the adsorption of AFB1 in food samples, and that the developed method demonstrated satisfactory linearity, low LOD and low LOQ values. In another study, surface molecularly imprinted polymers (FDU-12@MIPs) were synthesized using structural analogues of aflatoxins as template molecules and mesoporous silica FDU-12 as carrier, before an examination of aflatoxin extraction and enrichment performance from food samples (Rui et al. 2019). In the study, 6-Methyl-4-phenylchroman-2-one was used as a pseudo template and a-MAA, together with GMA as a monomer. The pseudo-second-order model was used to describe the adsorption process of MIPs. It was concluded that MIPs synthesized as polymer solid phase extraction combined with high performance liquid chromatography were successful in determining aflatoxin B1 from soy sauce without protein removal process.

Melamine ($C_3H_6N_6$, 1,3,5-triazine-2,4,6-triamine), widely used in the production of adhesives, plastics, melamine resins, kitchenware, and laminates, can cause kidney failure, urolithiasis, bladder cancer, kidney stones, and even death (Hemmati et al. 2023). Therefore, it is very important to identify and extract this contamination potentially present in food and water using selective and effective methods such as MIT. Dursun et al. (2016) combined the benefits of synthesis in monolith format with the advantages of printing technology and synthesized melamine (ME) printed polymeric sorbent in monolith form to be used as a solid phase extraction adsorbent. As a result of the adsorption isotherm study conducted within the scope of adsorption studies, the adsorption process was compatible with the Freundlich adsorption model. Based on this result, the authors emphasized that the adsorbent surface is heterogeneous and has different binding sites with different energies. According to the thermodynamic parameters of the study, it was concluded that the adsorption process was suitable for spontaneous and chemical adsorption. The kinetic model of the study was found to be a pesudo second order kinetic model and it was concluded that the analyte-sorbent relationship was a chemically controlled process. In the study, melamine extraction from a milk sample was also performed and the recycling results and LOD values were found relatively successful. Rolanda et al. (2020) studied melamine extraction with melamine imprinted polymers from blood serum, milk and river water samples. In the polymerization method using MAA as the functional monomer and melamine as the template molecule, the precipitation polymerization method was carried out with a non-covalent approach. The authors reported that the most selective polymer that they had synthesized was able to extract MEL from milk (81.97%), river water (91.52%) and human blood serum (90.66%).

Endocrine disrupting bisphenols, which can cause harmful effects on human health, are a class of chemical compounds widely applied in the production of various plastic and non-plastic products and can penetrate consumables such as food, beverages and dairy products (Hemmati et al. 2023). In recent years, it is possible to encounter studies on the selective and rapid extraction of bisphenol. A using molecular imprinting technology. Shah and Asman (2023) developed deep eutectic solvent molecularly imprinted polymer (DES-MIP) by bulk polymerization via free radical polymerization in a 2023 study. In the polymerization stage, DES was used as the monomer obtained by the combination of choline chloride and methacrylic acid (ChCl-MAA) and bisphenol A (BPA) was used as the template molecule. In the study, the highest adsorption capacity was observed at pH 7, which minimizes H^+ and OH^- concentration, which reduces the adsorption of BPA to the binding sites of DES-MIP, is minimum. Kinetic and isotherm studies fit the pseudo-second-order kinetic model and Langmuir isotherm models, respectively. Thermodynamic studies showed that the adsorption of BPA on DES-MIP was exothermic and spontaneous. Kalogiouria et al. (2020) synthesized molecularly imprinted polymers using sol–gel matrix imprinting technology for the efficient solid-phase extraction of BPA from water. To increase the functional selectivity the polymerization part involved two functional sol–gel precursors, phenyl triethoxysilane (PheTES) and 3-aminopropyl triethoxysilane (3-APTES). According to the results of the study, it was observed that sol–gel MIP had higher specific adsorption and enrichment capacity for BPA compared to the unimprinted polymer. The imprinting factor (IF) was found equal to 6.58 ± 0.23 ($n = 3$), and the sol–gel MIP was found to be sufficiently selective for BPA, providing high adsorption capacity.

There are studies in the literature on the use of MIPs in the isolation of viruses and bacteria that cause contamination in foods, soil and water. Doostmohammadi et al. (2022) performed the synthesis of MIP shells on the surface of polystyrene microparticles (MPs) with binding affinity to *E. coli* OP50 as a bacterial surrogate. In this study, the authors stated that they were able to synthesize uniform monodisperse *E. coli* OP50-imprinted core–shell microparticles (MIP-MPs), which exhibited good binding capacity against target bacterial cells. The findings also suggest the potential discovery of an application in the development of reliable, fast and low-cost biosensors, and will open a new window of interest in the research of new molecular recognition systems. Senehi et al. (2022) synthesized glycoprotein-printed particles to capture the coronavirus. The study's electrostatic potential calculation showed that glycoprotein 43 of human coronavirus-organ culture has positively charged pockets at pH 6 and may facilitate adsorption at lower pH values. This led to the conclusion that adjusting the target virus glycoprotein load through pH adjustment increased adsorption by minimizing repulsive electrostatic interactions.

12.4 Conclusion

With increasing environmental pollution as a result of agricultural and industrial development, various contaminants are released into nature. These contaminants are known to have mutagenic, carcinogenic and toxic effects on humans and animals, and it is imperative that these contaminants are removed from the environment or detected in their current environment. There are many methods used in the detection or removal of contaminants, such as liquid chromatography (LC), atomic absorption spectroscopy (AAS), gas chromatography, (Veloz Martínez et al. 2022) coagulation, bioremediation, chemical precipitation, and oxidation. The adsorption technique is one of these and is thought to be competitive due to its advantages such as simplicity, effectiveness, affordability and sustainability (Wang et al. 2023). MIPs can be used as alternative adsorbents for the selective separation/detection of contaminants to reduce environmental pollution. This chapter provides a comprehensive discussion of the adsorption technique used in the removal/detection of environmental contaminants, the use of MIPs as adsorbents in the selective removal of these contaminants, and the advantages of MIPs over traditional methods. In addition, it provides summaries of selected studies carried out with the combination of molecular imprinting technology and adsorption method for the removal of major contaminants harmful to the ecosystem and human health, such as heavy metals, dyes, pesticides, drugs, bacteria and viruses.

The United Nations Sustainable Development Goals have published 17 directives to ensure health and safety around the world by 2030 (Das et al. 2021). Therefore, the need for more environmentally friendly products has become an urgent need for the emergence of green chemistry (Donato et al. 2022). Beside on this, green molecular imprinting has started to synthesize more ideal MIPs in recent years. There is increasing interest in the use of MIPs developed with greener analytical synthesis protocols using different types of biomass waste derivatives to reduce environmental pollution (Sole et al. 2021). These sustainable materials, called Greenification MIPs (GMIPs), have applications in the fields of biomedicine and therapy, catalysis, nanoparticle detection, energy harvesting and storage, gas sensing and adsorption, and environmental remediation. Environmentally friendly materials such as silica, titanium dioxide, polyethyleneimine and naturally occurring polymers such as bacterial cellulose, chitosan, gelatin and alginates are used in GMIP studies successfully (Ostovan et al. 2022). It is envisaged that green synthesis will be further developed in the future and greenificated MIPs will guide areas such as sustainable environment, sustainable biotechnology, and sustainable health.

References

Abu Samah N, Mat Rosli NA, Abdul Manap AH, Abdul Aziz YF, Mohd Yusoff M (2020) Synthesis & characterization of ion imprinted polymer for arsenic removal from water: a value addition to the groundwater resources. Chem Eng J 394:124900

Açikel SM, Feyzioğlu E, Kuru CI, Türker E, Aslan A, Uygun DA, Akgöl S (2016) Dye incorporated nanopolymers and their application to leather wastewater: a study of removal of Cr(III) ions. J Soc Leather Technol Chem 100

Ahmed MA, Abdelbar NM, Mohamed AA (2018) Molecular imprinted chitosan-TiO2 nanocomposite for the selective removal of Rose Bengal from wastewater. Int J Biol Macromol 107:1046–1053

Al-Degs YS, Abu-Surrah AS, Ibrahim KA (2009) Preparation of highly selective solid-phase extractants for Cibacron reactive dyes using molecularly imprinted polymers. Anal Bioanal Chem 393:1055–1062
Algieri C, Drioli E, Ahmed C, Nasser II, Donato L (2014) Emerging tools for recognition and/or removal of dyes from polluted sites: molecularly imprinted membranes
Alnawmasi JS (2023) Construction of amino-thiol functionalized ion-imprinted chitosan for lead(II) ion removal. Carbohydr Polym 308:120596
Altaf R, Darko DA, Iqbal A, Ullah Z, Darko DA, Shahzeb Khan M, Asif M, Bin S, Al AA (2021) Molecularly imprinted polymers for the detection of chlorpyrifos (an organophosphate pesticide) investigation of kinetic and thermodynamic parameter of sapodilla leaves pyrolysis by using coats redfern method view project synthesis and characterization of aluminium alloy-2024 by Equal channel angular pressing and its applications. View project molecularly imprinted polymers for the detection of chlorpyrifos (an organophosphate pesticide). https://doi.org/10.17509/11154
Asadollahzadeh M, Torkaman R, Torab-Mostaedi M (2021) Extraction and separation of rare earth elements by adsorption approaches: current status and future trends. Sep Purif Rev 50:417–444
Ávila M, Zougagh M, Ríos Á, Escarpa A (2008) Molecularly imprinted polymers for selective piezoelectric sensing of small molecules. TrAC Trends Anal Chem 27:54–65
Azizi A, Bottaro CS (2020) A critical review of molecularly imprinted polymers for the analysis of organic pollutants in environmental water samples. J Chromatogr A 1614:460603
Azizi A, Shahhoseini F, Langille EA, Akhoondi R, Bottaro CS (2021) Micro-gel thin film molecularly imprinted polymer coating for extraction of organophosphorus pesticides from water and beverage samples. Anal Chim Acta 1187:339135
Ba Hashwan SS, Khir MHBM, Al-Douri Y, Ahmed AY (2020) Recent progress in the development of biosensors for chemicals and pesticides detection. IEEE Access 8:82514–82527
Barakat MA (2011) New trends in removing heavy metals from industrial wastewater. Arab J Chem 4:361–377
Batra D, Shea KJ (2003) Combinatorial methods in molecular imprinting. Curr Opin Chem Biol 7:434–442
Cantarella M, Carroccio SC, Dattilo S, Avolio R, Castaldo R, Puglisi C, Privitera V (2019) Molecularly imprinted polymer for selective adsorption of diclofenac from contaminated water. Chem Eng J 367:180–188
Chaipuang A, Phungpanya C, Thongpoon C, Watla-iad K, Inkaew P, Machan T, Suwantong O (2018) Synthesis of copper(II) ion-imprinted polymers via suspension polymerization. Polym Adv Technol 29:3134–3141
Chen L, Wang X, Lu W, Wua X, Lia J (2016) Molecular imprinting: perspectives and applications. Chem Socety Rev 45:2137–2211
Chen Y, Ma X, Peng J (2021a) Highly selective removal and recovery of Ni(II) from aqueous solution using magnetic ion-imprinted chitosan nanoparticles. Carbohydr Polym 271:118435
Chen J, Wang L, Liu Y, Chen L, Li X, Wang X, Zhu G (2021b) Highly selective removal of kitasamycin from the environment by molecularly imprinted polymers: adsorption performance and mechanism. Colloids Surf A Physicochem Eng Asp 625:126926
Chen X, Hossain MF, Duan C, Lu J, Tsang YF, Islam MS, Zhou Y (2022a) Isotherm models for adsorption of heavy metals from water—a review. Chemosphere 307:135545
Chen Z, Li Z, Chen J, Kallem P, Banat F, Qiu H (2022b) Recent advances in selective separation technologies of rare earth elements: a review. J Environ Chem Eng 10:107104
Dahaghin Z, Kilmartin PA, Mousavi HZ (2020) Novel ion imprinted polymer electrochemical sensor for the selective detection of lead(II). Food Chem 303:125374
Dakova I, Vasileva P, Karadjova I (2022) Cr(III) ion-imprinted hydrogel membrane for chromium speciation analysis in water samples. Gels. https://doi.org/10.3390/gels8110757
Das TR, Bhattacherjee P, Patra S, Govender PP, Mangala B, Shukla SK (2021) Metal−organic framework-based artificial antibodies for sensing applications. Met Fram Environ Sens American Chemical Society, pp 155–175 SE–6

Del Sole R, Mele G, Bloise E, Mergola L (2021) Green aspects in molecularly imprinted polymers by biomass waste utilization. Polymers (Basel). https://doi.org/10.3390/polym13152430

Demir EF, Özçalışkan E, Karakaş H, Uygun M, Aktaş Uygun D, Akgöl S, Denizli A (2018) Synthesis and characterization of albumin imprinted polymeric hydrogel membranes for proteomic studies. J Biomater Sci Polym Ed 29:2218–2236

de Oliveira HL, Pires BC, Teixeira LS, Dinali LAF, do Nascimento TA, Borges KB (2020) Mesoporous molecularly imprinted polymer for removal of hormones from aqueous medium. Colloids Surfaces A Physicochem Eng Asp 590:124506

Dickey FH (1955) Specific adsorption. J Phys Chem 59:695–707

Dolak İ, Keçili R, Hür D, Ersöz A, Say R (2015) Ion-imprinted polymers for selective recognition of Neodymium(III) in environmental samples. Ind Eng Chem Res 54:5328–5335

Donato L, Drioli E (2021) Imprinted membranes for sustainable separation processes. Front Chem Sci Eng 15:775–792

Donato L, Nasser II, Majdoub M, Drioli E (2022) Green chemistry and molecularly imprinted membranes. Membranes (Basel). https://doi.org/10.3390/membranes12050472

Doostmohammadi A, Youssef K, Akhtarian S, Tabesh E, Kraft G, Brar SK, Rezai P (2022) Molecularly imprinted polymer (MIP) based core-shell microspheres for bacteria isolation. Polymer (Guildf). https://doi.org/10.1016/j.polymer.2022.124917

Dursun EM, Üzek R, Bereli N, Şenel S, Denizli A (2016) Synthesis of novel monolithic cartridges with specific recognition sites for extraction of melamine. React Funct Polym 109:33–41

Erdem Ö, Saylan Y, Andaç M, Denizli A (2018) Molecularly imprinted polymers for removal of metal ions: an alternative treatment method. Biomimetics. https://doi.org/10.3390/biomimetics3040038

Fang L, Jia M, Zhao H, Kang L, Shi L, Zhou L, Kong W (2021) Molecularly imprinted polymer-based optical sensors for pesticides in foods: recent advances and future trends. Trends Food Sci Technol 116:387–404

Farooq S, Chen B, Ahmad S, Muhammad I, Hussain Q, Wu H (2022) Room-temperature, ionic-liquid-enhanced, beta-cyclodextrin-based, molecularly imprinted polymers for the selective extraction of abamectin. Nanomaterials. https://doi.org/10.3390/nano12061017

Feyzioğlu Demir E, Akgöl S (2022) Synthesis and characterization of double molecular imprinted nanoparticles and investigation to adsorption of respiratory drugs. Polym Technol Mater 61:384–399

Foguel MV, Pedro NTB, Wong A, Khan S, Zanoni MVB, Sotomayor M del PT (2017) Synthesis and evaluation of a molecularly imprinted polymer for selective adsorption and quantification of Acid Green 16 textile dye in water samples. Talanta 170:244–251

Fu F, Wang Q (2011) Removal of heavy metal ions from wastewaters: a review. J Environ Manage 92:407–418

Gao B, Cha X, Chen T, Fan L (2015) Designing and preparing of acid dye surface-imprinted material for effective removal of acid dyes from water. J Environ Chem Eng 3:277–285

Gao T, Guan G, Wang X, Lou T (2022) Electrospun molecularly imprinted sodium alginate/polyethylene oxide nanofibrous membranes for selective adsorption of methylene blue. Int J Biol Macromol 207:62–71

Giove A, El Ouardi Y, Sala A, Ibrahim F, Hietala S, Sievänen E, Branger C, Laatikainen K (2023) Highly selective recovery of Ni(II) in neutral and acidic media using a novel Ni(II)-ion imprinted polymer. J Hazard Mater 444:130453

Hayat M, Raza N, Jamal U, Manzoor S, Abbas N, Khan MI, Lee J, Brown RJC, Kim KH (2022) Targeted extraction of pesticide from agricultural run-off using novel molecularly imprinted polymeric pendants. J Ind Eng Chem 109:202–209

He JY, Li Q, Xu HX, Zheng QY, Zhang QH, Di ZL, Wang CZ, Yuan CS (2023) Recognition and analysis of biomarkers in tumor microenvironments based on promising molecular imprinting strategies with high selectivity. TrAC Trends Anal Chem 162:117033

Heidari G, Afruzi FH, Zare EN (2023) Molecularly imprinted magnetic nanocomposite based on carboxymethyl dextrin for removal of ciprofloxacin antibiotic from contaminated water. Nanomaterials. https://doi.org/10.3390/nano13030489

Hemmati F, Hosseini H, Mostashari P, Aliyeva A, Mousavi Khaneghah A (2023) Application of molecularly imprinted polymers as the sorbent for extraction of chemical contaminants from milk. Int J Environ Health Res. https://doi.org/10.1080/09603123.2023.2207484

Hu Y, Pan J, Zhang K, Lian H, Li G (2013) Novel applications of molecularly-imprinted polymers in sample preparation. TrAC Trends Anal Chem 43:37–52

Hu Y, Zhang C, Jian Z, Chen D, Ma Y, Yang W (2022) Surface-imprinted polymer microsphere based on self-stabilized precipitation polymerization for selective removal of norfloxacin. Appl Surf Sci 574:151706

Huang D-L, Wang R-Z, Liu Y-G, Zeng G-M, Lai C, Xu P, Lu B-A, Xu J-J, Wang C, Huang C (2015) Application of molecularly imprinted polymers in wastewater treatment: a review. Environ Sci Pollut Res 22:963–977

Husin NA, Muhamad M, Yahaya N, Miskam M, Syazni Nik Mohamed Kamal NN, Asman S, Raoov M, Mohamad Zain NN (2021) Application of a new choline-imidazole based deep eutectic solvents in hybrid magnetic molecularly imprinted polymer for efficient and selective removal of naproxen from aqueous samples. Mater Chem Phys 261:124228

Ihsanullah AA, Al-Amer AM, Laoui T, Al-Marri MJ, Nasser MS, Khraisheh M, Atieh MA (2016) Heavy metal removal from aqueous solution by advanced carbon nanotubes: critical review of adsorption applications. Sep Purif Technol 157:141–161

Jagirani MS, Balouch A, Mahesar SA, Alveroğlu E, Kumar A, Tunio A, Abdullah (2022) Selective and sensitive detoxification of toxic lead ions from drinking water using lead(II) ion-imprinted interpenetrating polymer linkage. Polym Bull 79:1887–1909

Jia B, Feng F, Wang X, Song Y, Zhang F (2024) Recent advances in magnetic molecularly imprinted polymers and their application in the food safety analysis. J Futur Foods 4:1–20

Jumadilov T, Kondaurov R, Imangazy A (2022) Application of the remote interaction effect and molecular imprinting in sorption of target ions of rare earth metals. Polymers (Basel). https://doi.org/10.3390/polym14020321

Kadhem AJ, Gentile GJ, Fidalgo de Cortalezzi MM (2021) Molecularly imprinted polymers (MIPs) in sensors for environmental and biomedical applications: a review. Molecules. https://doi.org/10.3390/molecules26206233

Kalogiouri NP, Tsalbouris A, Kabir A, Furton KG, Samanidou VF (2020) Synthesis and application of molecularly imprinted polymers using sol–gel matrix imprinting technology for the efficient solid-phase extraction of BPA from water. Microchem J. https://doi.org/10.1016/j.microc.2020.104965

Kayranli B (2011) Adsorption of textile dyes onto iron based waterworks sludge from aqueous solution; isotherm, kinetic and thermodynamic study. Chem Eng J. https://doi.org/10.1016/j.cej.2011.08.051

Kulaksız S, Bau M (2011) Anthropogenic gadolinium as a microcontaminant in tap water used as drinking water in urban areas and megacities. Appl Geochemistry 26:1877–1885

Kuras MJ, Więckowska E (2015) Synthesis and characterization of a new copper(II) ion-imprinted polymer. Polym Bull 72:3227–3240

Kyzas GZ, Bikiaris DN, Lazaridis NK (2009) Selective separation of basic and reactive dyes by molecularly imprinted polymers (MIPs). Chem Eng J 149:263–272

Lazar MM, Ghiorghita C-A, Dragan ES, Humelnicu D, Dinu MV (2023) Ion-imprinted polymeric materials for selective adsorption of heavy metal ions from aqueous solution. Molecules. https://doi.org/10.3390/molecules28062798

Lazarević-Pašti T, Tasić T, Milanković V, Potkonjak N (2023) Molecularly imprinted plasmonic-based sensors for environmental contaminants—current state and future perspectives. Chemosensors 11:35

Li Y, Li B, Qi Y, Zhang Z, Cong S, She Y, Cao X (2022) Synthesis of metal-organic framework @molecularly imprinted polymer adsorbents for solid phase extraction of organophosphorus pesticides from agricultural products. J Chromatogr B 1188:123081

Liu G, Yang X, Li T et al (2015) Preparation of a magnetic molecularly imprinted polymer using g-C3N4-Fe3O4 for atrazine adsorption. Mater Lett 160:472–475

Lu L, Yue X, Lin F, Huang F, Zhang B, Lin Z (2015) Template-synthesized ultra-thin molecularly imprinted polymers membrane for the selective preconcentration of dyes. J Mater Chem A 3:10959–10968

Ma Q, Wei Y, Zhao N, Wang S, Zhang B, Liu D, Yuan P (2023) Construction of an allophane-based molecularly imprinted polymer for the efficient removal of antibiotic from aqueous solution. Sci Total Environ 903:166464

Mabrouk M, Hammad SF, Abdella AA, Mansour FR (2023) Tips and tricks for successful preparation of molecularly imprinted polymers for analytical applications: a critical review. Microchem J 193:109152

Mahmoudpour M, Torbati M, Mousavi MM, de la Guardia M, Ezzati Nazhad Dolatabadi J (2020) Nanomaterial-based molecularly imprinted polymers for pesticides detection: recent trends and future prospects. TrAC Trends Anal Chem 129:115943

Majdinasab M, Daneshi M, Louis Marty J (2021) Recent developments in non-enzymatic (bio)sensors for detection of pesticide residues: focusing on antibody, aptamer and molecularly imprinted polymer. Talanta 232:122397

Malik S, Khan A, Khan H, Rahman G, Ali N, Khan S, Sotomayor MDPT (2023) Biomimetic electrochemical sensors based on core-shell imprinted polymers for targeted sunset yellow estimation in environmental samples. Biosensors 13:429

Masoumi A, Hemmati K, Ghaemy M (2016) Recognition and selective adsorption of pesticides by superparamagnetic molecularly imprinted polymer nanospheres. RSC Adv 6:49401–49410

Meseguer-Lloret S, Torres-Cartas S, Gómez-Benito C, Herrero-Martínez JM (2022) Magnetic molecularly imprinted polymer for the simultaneous selective extraction of phenoxy acid herbicides from environmental water samples. Talanta 239:123082

Metwally MG, Benhawy AH, Khalifa RM, El Nashar RM, Trojanowicz M (2021) Application of molecularly imprinted polymers in the analysis of waters and wastewaters. Molecules 26:6515

Mezzelani M, Gorbi S, Regoli F (2018) Pharmaceuticals in the aquatic environments: evidence of emerged threat and future challenges for marine organisms. Mar Environ Res 140:41–60

Moussa M, Ndiaye MM, Pinta T, Pichon V, Vercouter T, Delaunay N (2017) Selective solid phase extraction of lanthanides from tap and river waters with ion imprinted polymers. Anal Chim Acta 963:44–52

Nageib AM, Halim AA, Nordin AN, Ali F (2023) Recent applications of molecularly imprinted polymers (MIPs) on screen-printed electrodes for pesticide detection. J Electrochem Sci Technol 14:1–14

Noor M, Iqbal KM, Kanwal T, Rehman K, Ali M, Shah MR (2023) Synthesis and characterization of modified silica nanoparticles based molecularly imprinted polymers as a sensor for selective detection of pendimethalin from drinking water. J Mol Struct 1287:135635

Ostovan A, Arabi M, Wang Y, Li J, Li B, Wang X, Chen L (2022) Greenificated molecularly imprinted materials for advanced applications. Adv Mater 34:2203154

Öter Ç, Zorer ÖS (2021) Molecularly imprinted polymer synthesis and selective solid phase extraction applications for the detection of ziram, a dithiocarbamate fungicide. Chem Eng J Adv 7:100118

Patra S, Roy E, Madhuri R, Sharma PK (2015) Fast and selective preconcentration of europium from wastewater and coal soil by graphene oxide/silane@Fe3O4 dendritic nanostructure. Environ Sci Technol 49:6117–6126

Pezeshkpur M, Tadayon F, Sohrabi MR (2023) A molecularly imprinted polymer based on Fe3O4@Au nanoparticles for detection of Aflatoxin B1 in food samples. Chem Select. https://doi.org/10.1002/slct.202300112

Rahman S, Bozal-Palabiyik B, Unal DN, Erkmen C, Siddiq M, Shah A, Uslu B (2022) Molecularly imprinted polymers (MIPs) combined with nanomaterials as electrochemical sensing applications for environmental pollutants. Trends Environ Anal Chem 36:e00176

Rehman M, Liu L, Wang Q, Saleem MH, Bashir S, Ullah S, Peng D (2019) Copper environmental toxicology, recent advances, and future outlook: a review. Environ Sci Pollut Res 26:18003–18016

Roland RM, Bhawani SA, Wahi R, Ibrahim MNM (2020) Synthesis, characterization, and application of molecular imprinting polymer for extraction of melamine from spiked milk, water, and blood serum. J Liq Chromatogr Relat Technol 43:94–105

Roushani M, Abbasi S, Khani H (2015) Synthesis and application of ion-imprinted polymer nanoparticles for the extraction and preconcentration of copper ions in environmental water samples. Environ Monit Assess 187:219

Rui C, He J, Li Y, Liang Y, You L, He L, Li K, Zhang S (2019) Selective extraction and enrichment of aflatoxins from food samples by mesoporous silica FDU-12 supported aflatoxins imprinted polymers based on surface molecularly imprinting technique. Talanta 201:342–349

Sadani M, Rasolevandi T, Azarpira H, Mahvi AH, Ghaderpoori M, Mohseni SM, Atamaleki A (2020) Arsenic selective adsorption using a nanomagnetic ion imprinted polymer: optimization, equilibrium, and regeneration studies. J Mol Liq 317:114246

Sadia M, Ahmad I, Ali F, Zahoor M, Ullah R, Khan F, Ali E, Sohail A (2022) Selective removal of the emerging dye basic blue 3 via molecularly imprinting technique. Molecules 27:3276

Sadia M, Ahmad I, Ul-Saleheen Z, Zubair M, Zahoor M, Ullah R, Bari A, Zekker I (2023) Synthesis and characterization of MIPs for selective removal of textile dye acid black-234 from wastewater sample. Molecules 28:1555

Sedelnikova A, Poletaeva Y, Golyshev V, Chubarov A, Dmitrienko E (2023) Preparation of magnetic molecularly imprinted polymer for methylene blue capture. Magnetochemistry 9:196

Senehi NL, Ykema MR, Sun R, Verduzco R, Stadler LB, Tao YJ, Alvarez PJJ (2022) Protein-imprinted particles for coronavirus capture from solution. J Sep Sci 45:4318–4326

Shah SAH, Asman S (2023) Evaluation of deep eutectic solvent as a new monomer for molecularly imprinted polymers for removal of bisphenol A. J Polym Res. https://doi.org/10.1007/s10965-023-03581-1

Shamsipur M, Besharati-Seidani A (2011) Synthesis of a novel nanostructured ion-imprinted polymer for very fast and highly selective recognition of copper(II) ions in aqueous media. React Funct Polym 71:131–139

So J, Pang C, Dong H, Jang P, Juhyok U, Ri K, Yun C (2018) Adsorption of 1-naphthyl methyl carbamate in water by utilizing a surface molecularly imprinted polymer. Chem Phys Lett 699:199–207

Song X, Li F, Yan T, Tian F, Ren L, Jiang C, Wang Q, Zhang S (2022) Research progress in the sample pretreatment techniques and advanced quick detection methods of pesticide residues. Process Saf Environ Prot 165:610–622

Sullivan MV, Fletcher C, Armitage R, Blackburn C, Turner NW (2023) A rapid synthesis of molecularly imprinted polymer nanoparticles for the extraction of performance enhancing drugs (PIEDs). Nanoscale Adv

Sun Y, Bai L, Han C, Lv X, Sun X, Wang T (2022) Hybrid amino-functionalized TiO2/sodium lignosulfonate surface molecularly imprinted polymer for effective scavenging of methylene blue from wastewater. J Clean Prod. https://doi.org/10.1016/j.jclepro.2022.130457

Syed Yaacob SFF, Suwaibatu M, Raja Jamil RZ, Mohamed Zain NN, Raoov M, Mohd Suah FB (2023) Review of molecular imprinting polymer: basic characteristics and removal of phenolic contaminants based on the functionalized cyclodextrin monomer. J Chem Technol Biotechnol 98:312–330

Tarannum N, Khatoon S, Dzantiev BB (2020) Perspective and application of molecular imprinting approach for antibiotic detection in food and environmental samples: a critical review. Food Control 118:107381

Tian X, Song H, Wang Y, Tian X, Tang Y, Gao R, Zhang C (2020) Hydrophilic magnetic molecularly imprinted nanobeads for efficient enrichment and high performance liquid chromatographic detection of 17beta-estradiol in environmental water samples. Talanta 220:121367

Tolkou AK, Kyzas GZ, Katsoyiannis IA (2022) Arsenic(III) and Arsenic(V) removal from water sources by molecularly imprinted polymers (MIPs): a mini review of recent developments. Sustainability. https://doi.org/10.3390/su14095222

Türkmen D, Özkaya Türkmen M, Akgönüllü S, Denizli A (2022) Development of ion imprinted based magnetic nanoparticles for selective removal of arsenic (III) and arsenic (V) from wastewater. Sep Sci Technol 57:990–999

Uygun M, Feyzioğlu E, Özçalışkan E, Caka M, Ergen A, Akgöl S, Denizli A (2013) New generation ion-imprinted nanocarrier for removal of Cr(VI) from wastewater. J Nanoparticle Res 15:1–11

Vasapollo G, Del Sole R, Mergola L, Lazzoi MR, Scardino A, Scorrano S, Mele G (2011) Molecularly imprinted polymers: present and future prospective. Int J Mol Sci 12:5908–5945

Veloz Martínez I, Ek JI, Ahn EC, Sustaita AO (2022) Molecularly imprinted polymers via reversible addition-fragmentation chain-transfer synthesis in sensing and environmental applications. RSC Adv 12:9186–9201

Wang J, Guo X (2020a) Adsorption kinetic models: physical meanings, applications, and solving methods. J Hazard Mater 390:122156

Wang J, Guo X (2020b) Adsorption isotherm models: classification, physical meaning, application and solving method. Chemosphere 258:127279

Wang L, Yu J, Wang X, Li J, Chen L (2023) Molecular imprinting-based nanocomposite adsorbents for typical pollutants removal. J Hazard Mater Lett. https://doi.org/10.1016/j.hazl.2022.100073

Wu H, Lin G, Liu C, Chu S, Mo C, Liu X (2022) Progress and challenges in molecularly imprinted polymers for adsorption of heavy metal ions from wastewater. Trends Environ Anal Chem 36:e00178

Xie W-M, Wu Y, Yan W-Y et al (2023) The erythromycin sorption removal at environmentally relevant concentration based on molecular imprinted polymer: performance and mechanism. Environ Pollut 336:122425

Yang Y, Ok YS, Kim K-H, Kwon EE, Tsang YF (2017) Occurrences and removal of pharmaceuticals and personal care products (PPCPs) in drinking water and water/sewage treatment plants: a review. Sci Total Environ 596–597:303–320

Yu J, Li H, Zhou R, Li X, Wu H, Xiao C, Chi R-A (2022) Surface ion imprinted bagasse for selective removal of Cu(II) from the leaching solution of electroplating sludge. Colloids Surf A Physicochem Eng Asp 639:128394

Yusoff MM, Mostapa NRN, Sarkar MS, Biswas TK, Rahman ML, Arshad SE, Sarjadi MS, Kulkarni AD (2017) Synthesis of ion imprinted polymers for selective recognition and separation of rare earth metals. J Rare Earths 35:177–186

Zare EN, Fallah Z, Le VT et al (2022) Remediation of pharmaceuticals from contaminated water by molecularly imprinted polymers: a review. Environ Chem Lett 20:2629–2664

Zhang Y, Zhang X, Wang S (2023) Recent advances in the removal of emerging contaminants from water by novel molecularly imprinted materials in advanced oxidation processes—a review. Sci Total Environ 163702

Zhang Z, Cao X, Zhang Z, Yin J, Wang D, Xu Y, Zheng W, Li X, Zhang Q, Liu L (2020) Synthesis of dummy-template molecularly imprinted polymer adsorbents for solid phase extraction of aminoglycosides antibiotics from environmental water samples. Talanta 208:120385

Zhao R, Li X, Sun B, Li Y, Li Y, Wang C (2017) Preparation of molecularly imprinted sericin/poly(vinyl alcohol) electrospun fibers for selective removal of methylene blue. Chem Res Chinese Univ 33:986–994

Chapter 13
Tetracyclines and an Overview of the Preconcentration/Extractions Method Based on Molecularly Imprinted Polymers

Luana Rianne Rocha and César Ricardo Teixeira Tarley

13.1 Tetracyclines

Tetracycline antibiotics constitute a natural, obtained through fermentation of certain fungi, or semisynthetic group of antibiotics, with a broad antimicrobial activity against infections caused by both gram-positive and gram-negative bacteria (Önal 2011). The first tetracyclines (TCs) were discovered in 1945 by Bejamin Minge Duggar, as a natural product products of the Streptomyces genus actinobacteria found in soils. Em 1948, cultures of Streptomyces aureofaciens gave rise to chlortetracycline (CTC), being prescribed under the trade name Aureomycin®. In 1950, Terramycin (oxytetracycline, OTC) was discovered from cultures of *Streptomyces rimosus*. In 1953, through the catalytic dehalogenation of CTC, the basic structure of tetracycline (TC) was obtained. Synthetic TCs, such as doxycycline (DC), minocycline (MiC) and metacycline (MC) are produced by radical modification of naturally occurring molecules (Chopra and Roberts 2001; Pereira-Maia et al. 2010).

Structurally, all TCs are characterized by having a chemical structure derived from a hydronaphthacene, forming a common linear tetracyclic nucleus with four condensed rings named A, B, C and D. These rings contain functional groups such as a carboxamide group and various other ionizable functional groups such as tricarbonyl amide system, phenolic diketone, and dimethylamine, as shown in Fig. 13.1. These functional groups are responsible for their pKa values and for anionic, cationic, or zwitterionic charges (Dantas et al. 2019; Nelson et al. 2001).

L. R. Rocha · C. R. T. Tarley (✉)
Chemistry Department, Londrina State University (UEL), Londrina, Brazil
e-mail: ctarleyquim@yahoo.com.br

C. R. T. Tarley
Bioanalytical National Institute of Science and Technology (INCTBio), Campinas State University (UNICAMP), Campinas, São Paulo, Brazil

S. Patra and M. Sillanpaa (eds.), *Molecularly Imprinted Polymers as Artificial Antibodies for the Environmental Health*, https://doi.org/10.1007/978-3-031-58995-9_13

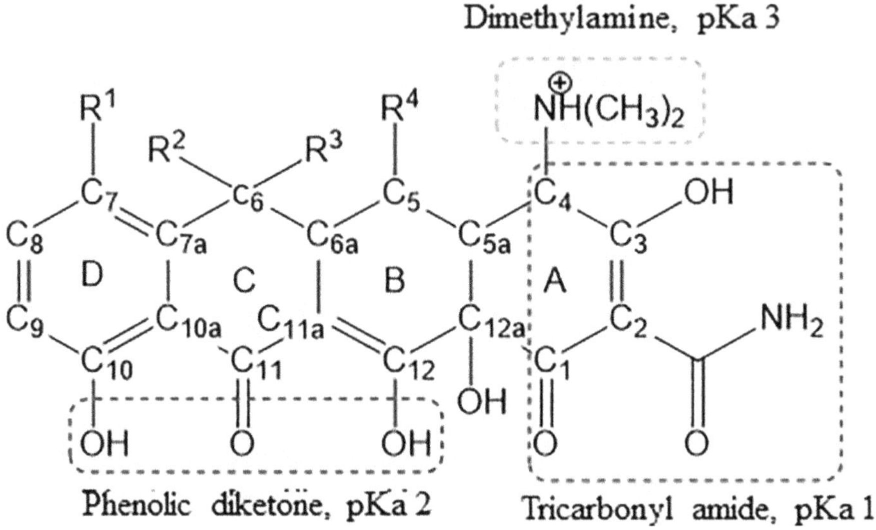

Fig. 13.1 General chemical structure for tetracycline antibiotics and their pKas. Adapted from Sánchez-Polo et al. (2015), Sanli et al. (2009)

The chemical structure and the presence of functional groups are directly related to the biological action of TCs, since the presence of these functional groups can influence the permeability of the molecules through biological membranes, which can lead to the absorption of these antibiotics by bacterial cells (Önal 2011). In its natural form, the configuration of carbon C4 is crucial for the antimicrobial activity, as well as the presence of the amide group at carbon C2, keto-enol system at carbons C10, C11 and C12, in addition to tetracycline nucleus itself (Chopra and Roberts 2001; Pereira-Maia et al. 2010). The 6-deoxy-6-dimethyltetracycline is considered as the minimum pharmacophore since it is the simplest tetracycline with a detectable antibacterial activity, and its structure is shown in Fig. 13.2 (Chopra and Roberts 2001).

TCs are crystalline amphoteric substances with low solubility, usually available in the form of hydrochlorides that exhibit greater solubility, and these solutions beings more stable in acidic solutions (Nelson et al. 2001). The antibiotics penetrate the bacterial cell through active or passive transport, preventing the transfer RNA from attaching to the acceptor ribosome, and thus inhibits protein synthesis and prevents the proliferation of bacteria in the organisms (Chopra and Roberts 2001). In addition to their antibiotic properties, this group exhibits other properties such as anti-inflammatory action and inhibition of lipase, collagenase, apoptosis and angiogenesis, among other properties (Dantas et al. 2019; Nelson et al. 2001; Pereira-Maia et al. 2010). Currently, 8 TCs are marketed for the treatment of bacterial infections

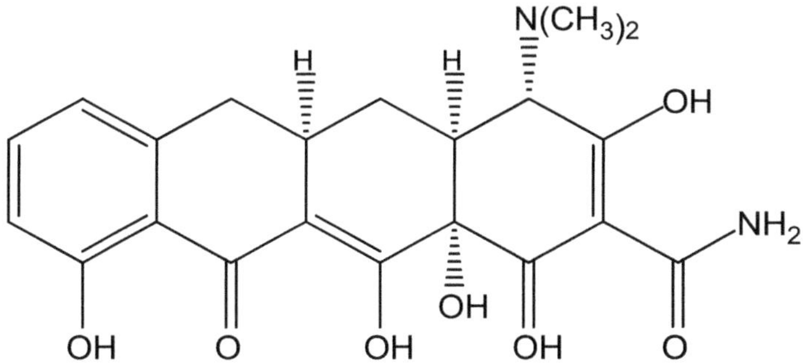

Fig. 13.2 Chemical structure of the 6-deoxy-6-dimethyltetracycline. Adapted from Chopra and Roberts (2001)

in humans and veterinary medicine, and the most common of which is tetracycline (TC), (4*S*,4*aS*,5*aS*,6*S*,12*aR*)-4-(dimethylamino)-1,6,10,11,12*a*-pentahydroxy-6-methyl-3,12-dioxo-4,4*a*,5,5*a*-tetrahydrotetracene-2-carboxamide, in addition to oxytetracycline (OTC), chlortetracycline (CTC) and doxycycline (DC), and their structures are shown in Table 13.1.

One of the mains structural characteristics of TCs is their strong chelation ability, since the formation of a chelate with metal ions directly affects the pharmacological profile of tetracyclines antibiotics (Önal 2011; Pulicharla et al. 2017). The formation of chelates by TCs can occur at various positions of the molecule, due to the presence of electron-donor functional groups such as N and O and the phenolic diketone system that acts as a primary chelation site (Pereira-Maia et al. 2010).

The antibacterial action of TCs occurs due to the formation of a complex between the O at C11 and C12 (Fig. 13.1) with Mg^{2+}, that impair protein synthesis through binding to the 30S subunit of the ribosome, which is responsible for decoding genetic information (Guerra et al. 2016). Interestingly, one the most commonly found mechanism for bacterial resistance to TCs also occurs through metal chelation, and this mechanism involves the formation of metal complex with divalent metal ions, such as Pt^{2+} (Guerra et al. 2016; Lv et al. 2019; Pereira-Maia et al. 2010). Due to this formation of coordination compounds, TCs may potentially be employed in the future of clinical medicine for the treatment of infectious diseases, such as COVID-19, which generated a pandemic and has caused serious consequences throughout the world (Mostafa 2020; Sodhi and Etminan 2020). Studies recently published by Sodhi and Etminan (2020), Mostafa (2020) suggest that the chelating properties of TCs with zinc can cause a limitation in the replication capacity of the novel coronavirus within host cells, which helps to inhibit COVID-19 infection in humans.

Since its discovery in the 1940s and their use in human therapy in the 1950s, it is estimated that global production of TCs antibiotics is around thousands of tons annually (Daghrir and Drogui 2013). Due to their low cost and broad antimicrobial activity, they are widely used in human therapy for the treatment of several types of infections, such as pneumonias, chlamydial infections, acute fever, cholera, acne, for example (Bahrami et al. 2011).

Table 13.1 Structure and chemical properties of the main antimicrobials in the tetracycline group

Antimicrobnial	Systematic name	Chemical structure	Acidity (pKa)	Chemical formula	Molecular mass (g mol^{-1})
Tetracycline (TC)	(4*S*,4*aS*,5*aS*,6*S*,12*aR*)-4-(dimethylamino)-1,6,10,11,12*a*-pentahydroxy-6-methyl-3,12-dioxo-4,4*a*,5,5*a*-tetrahydrotetracene-2-carboxamide	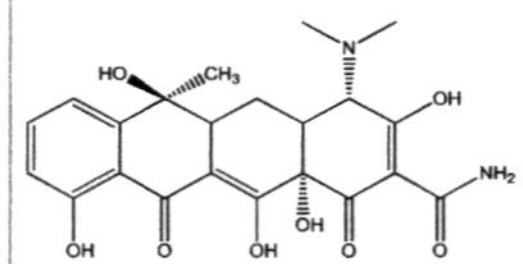	pKa_1 3.3 pKa_2 7.7 pKa_3 9.7	$C_{22}H_{23}N_2O_8$	444.4
Chlortetracycline (CTC)	7-chloro-tetracycline	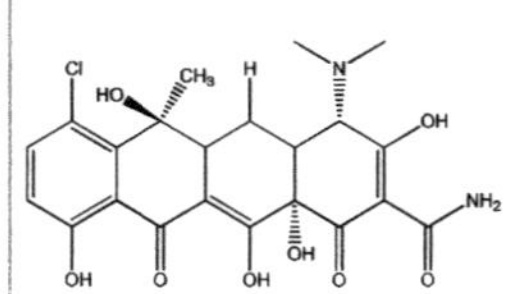	pKa_1 4.5 pKa_2 7.8 pKa_3 9.8	$C_{22}H_{23}ClN_2O_8$	478.8
Doxycycline (DC)	6-deoxy-5-hydroxy-tetracycline	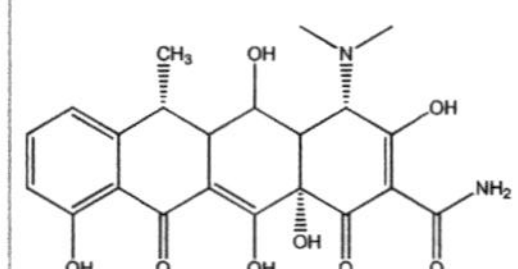	pKa_1 3.0 pKa_2 7.9 pKa_3 9.2	$C_{22}H_{24}N_2O_8$	444.4
Oxytetracycline (OTC)	5-hidroxy-tetracycline	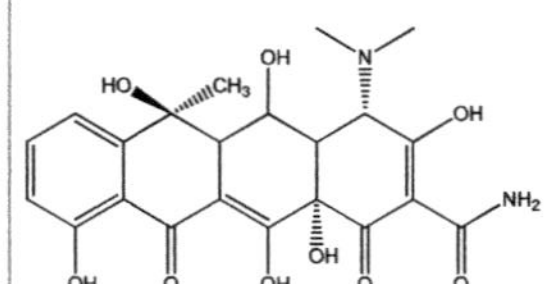	pKa_1 3.0 pKa_2 7.9 pKa_3 9.2	$C_{22}H_{24}N_2O_9$	460.4

Adapted from Pereira-Maia et al. (2010), Pérez-Rodríguez et al. (2016)

In veterinary medicine, TCs are the most widely consumed veterinary antibiotics in the world (Charuaud et al. 2019), as they have been extensively used for the prevention and treatment of respiratory, gastrointestinal, bacterial and other infections, as well a systemic infections and sepsis (Daghrir and Drogui 2013). In the latest study published by the European Medicines Agency (EMA), which assessed 31 European countries, TCs accounted for 30.4% of antimicrobials sold in 2017, approximately 2019 tons, being the best-selling class of antibiotics in these countries (EMA 2019). In the United States of America (USA), according to recent studies released by the U.S. Food and Drug Administration (FDA), TCs were the most sold antimicrobials for use in the production of animal-derived foods, accounting for approximately 66% of sales in the country with more than 3974 tons of TCs (FDA 2019).

TCs antibiotics are also widely used in the agricultural sector as growth promoters for cattle, swine, sheep, poultry, among others, due to their incorporation into animal feed at subtherapeutic doses, aiming to promote rapid weight gain and animal growth (Danner et al. 2019). Despite the existence of data supporting the fact that the use of TCs as growth promoters can lead to bacterial resistance, allergic reactions in human and animals, as well as changes in environmental microflora and bacterial populations, this application for TCs has its positive aspects, such as nutrient absorption efficiency, increased commercial revenue for farmers, among others (Chattopadhyay 2014; Hao et al. 2014).

In recent years, due to the increasing abusive use of TCs antibiotics, the possibility of environmental contamination with such compounds has risen because to the various applications and disposals of this type of substance. The occurrence of these antibiotics in the aquatic environment and in animal-derived foods such as honey, meat, fish, milk, and eggs has received special attention due to their contribution to the increase in infections caused by resistant strains. After the consumption of these substances, approximately 70–90% is excreted and released in its active form into the environment through feces and urine, thus contaminating soils, wastewater, surface waters and, due to leaching, groundwater, resulting in a significant public health issue (Chen et al. 2017; Daghrir and Drogui 2013; Danner et al. 2019; Gaudin 2017). The contamination of drinking water and animal-derived foods can cause health problems for consumers, ranging from possible allergic reactions, liver damage, yellow teeth, gastrointestinal disorders, among others, in addition to potentially increasing bacterial resistance (Rodriguez et al. 2010; Rodríguez et al. 2016).

The presence of TC residues in samples of animal origin, such as milk, for example, has been reported in the literature. Kumar et al. (2022) analyzed samples of fresh and pasteurized (UHT) milk. According to the authors, the presence of OTC was detected in 12 of the 128 fresh milk samples, while for UHT milk samples, 2 of the 45 analyzed samples tested positive. The OTC concentrations found in samples of both types of milk ranged from 31.5 to 150.6 $\mu g\ L^{-1}$, with three fresh milk samples containing OTC residues above the country's tolerance limit (100 $\mu g\ L^{-1}$). Mohebi and et al. (2020) analyzed 14 samples of UHT milk and OTC was found in 3 of them, at concentrations of 15, 30 and 43 $\mu g\ L^{-1}$. Navrátilová et al. (2009) detected OTC residues in 50% of the analyzed milk samples.

Moudgil et al. (2019) analyzed 492 milk samples from dairy farms and divided the study of these samples based on the size of the farms. Thus, out of the 164 milk samples analyzed from large farms, 8 (4.9%) tested positive for OTC residues, with concentrations ranging between 62.7 and 177.0 μg L^{-1}, where 3 samples (1.8%) exceeded the established tolerance limit (100 μg L^{-1}) and 2 (1.2%) for TC with concentrations between 72.6 and 85.1 μg L^{-1}. On the other hand, out of the 164 samples from medium-sized farms, 7 (4.3%) were positive for OTC, quantified in the range 62.7–148.4 μg L^{-1}, with 2 exceeding the tolerance limit and 6 (3 0.7%) for TC, ranging concentrations between 72.6 and 122.6 μg L^{-1}, with 1 sample with residue above the tolerated level. Similary, from small dairy farms, 6 (3.7%) were positive for OTC, with concentrations determined between 55.5 and 170.0 μg L^{-1}, with 3 of these samples above tolerated, while TC was found in 5 (3.0%) samples, with one above the tolerated level. The authors also found OTC residues, with a concentration of 77 μg L^{-1}, in a sample of commercial milk.

The relationship between antibiotic use in livestock farming and milk production, and antimicrobial resistance in humans, has given rise to a significant public health concern, because the administration of such medications to animals can result in the contamination of food products derived from these animals, as well as environmental contamination, which can also arise from alternatives sources. Therefore, due to the significance of monitoring antibiotics, especially TCs, regulatory and government agencies in several countries have established tolerance levels or maximum residue limits (MRL) for the presence of these compounds in different types of matrices. In other words, MRLs represent levels of TCs residues in foods and/or water that can be consumed daily without causing any toxicological harm to human health (Chattopadhyay 2014; Mota et al. 2005).

The Codex Alimentarius, a joint program of the Food and Agriculture Organization (FAO) of the United Nations and the World Health Organization (WHO), is considered a global reference for producers, food processors and consumers, establishing that MRLs for TCs in animal-derived foods are as follows: 600 μg kg^{-1} for poultry, cattle and pig livers; 400 μg kg^{-1} for eggs; 200 μg kg^{-1} for muscle of poultry, cattle, pigs and fish; and 100 μg L^{-1} for cow's milk (Commission 2015).

In European countries like Russia, for instance, the specified limit for TC content in honey should not exceed 0.01 mg kg^{-1}, as recommended by the Technical Regulation of the Customs Union, TR CU 021/2011 (TR CU 021/2011—On Food Safety 2011). The European Union (EU) recommends advocates for safe limits of veterinary drug residues in animal tissues entering the food chain as follows: 100 μg kg^{-1} for muscle, 300 μg kg^{-1} for liver, 600 μg kg^{-1} for kidney, 100 μg kg^{-1} for milk and 200 μg kg^{-1} for eggs (Commission Regulation-No 37/2010 2010).

The WHO establishes that the MRL for TCs in milk is 100 mg L^{-1} (Moreira et al. 2010). In the USA, the Food and Drug Administration (FDA) determines that the maximum TC, OTC, and CTC limits in milk as 80, 30 and 30 mg L^{-1}, respectively, meanwhile the Chinese government (GB 18406.3-2001) allows for a maximum of 100 mg kg^{-1} of TCs in meat (Moreira et al. 2010).

In Brazil, MRL for TCs are regulated in accordance with the National Plan for Residue and Contaminant Control (PNCRC). According to them, 1200 μg kg^{-1} are established for beef, horse, pork, and poultry meat; 200 μg kg^{-1} for fish; 100 μg L^{-1} for bovine milk and 20 μg L^{-1} for honey. These MRLs are adopted for TC, CTC, and OTC individually or for their combination. Additionally, for bovine milk, the PNCRC includes DTC in the combination of these residues (Plano Nacional de Controle de Resíduos e Contaminantes (PNCRC), INSTRUÇÃO NORMATIVA Nº 5, DE 23 DE ABRIL DE 2019 2019). Furthermore, the National Program for the Analysis of Residues of Veterinary Drugs in Food Exposed to Consumption (PAMVet) determines that the maximum acceptable residuelimit for TCs in milk is 100 μg L^{-1} (Lista F2 Da Portaria 344/98 2009; Limites Máximos de Resíduo de Medicamentos Veterinários Em Alimentos de Origem Animal 2018; Instrução Normativa-IN N° 162 2022). It is worth noting that in Brazil, as described in Article 18 of Normative Instruction No. 26 of 2009 from the Ministry of Agriculture, Livestock and Supply (MAPA), TCs "are for exclusive use in veterinary antimicrobial products, and their use as zootechnical additives to improve animal performance or as food preservatives for animals is prohibited." Therefore, the use of these antibiotics as food additives is prohibited in Brazil (Instrução Normativa N°26, de 10 de Julho de 2009 2009).

Based on guidelines and regulations imposed by countries, the assessment and monitoring of tetracycline antibiotics helps ensure a more prudent use and disposal of these medications, which in turn, aids in reducing residues and bacterial resistance. Therefore, the development of new analytical methods with greater selectivity and sensitivity for the detection and quantification of TCs in different types of matrices becomes necessary.

The Association of Official Analytical Chemists (AOAC) recommends the use of High-Performance Liquid Chromatography (HPLC) with ultraviolet (UV) detection as the official method for the analysis of tetracycline residues in animal-derived foods (Multiple Tetracycline Residues in Milk 2012; Chlortetracycline, Oxytetracycline, and Tetracycline in Edible Animal Tissues 2012). Due to their structural characteristics, TCs have several chromophoric groups, exhibiting intense absorption in the UV region between 270 and 360 nm, in both acidic and neutral media. Consequently, the use of UV and diode array detectors (DAD) are quite common for the determination of tetracycline antibiotics (Cherkashina et al. 2018; Pereira-Maia et al. 2010; Pérez-Rodríguez et al. 2018; Yu et al. 2012). Furthermore, the presence of different functional groups in the structures of TCs allows them to be retained in chromatographic systems with reversed-phase columns (C8 and C18) through mechanisms such as ion-pairs formation, competition effects, interaction with silanol groups, among others (Denobile and De Souza Nascimento 2004; Esteban et al. 2007).

Despite advances in the sensitivity of analytical instrumentation for the determination of TCs in complex matrices like food, and the development methods with low limits of detection (LD) (in the range of ng L^{-1} to μg L^{-1}), these results largely depend on proper sample extraction and preparation before instrumental analyses. Nonetheless, many methods may not exhibit sufficient sensitivity to detect residues of the analytes, as they are often found in low concentrations in environmental and food samples. Therefore, it becomes necessary to involve isolation and pre-concentration steps of the analyte, as the failure to remove interfering compounds can hinder its determination and quantification.

When it comes to food samples such as milk, protein precipitation is a crucial step in sample preparation for TCs because they have a tendency to readily bind with proteins. Protein precipitation can be easily achieved by adding precipitating agents or strong acids, and can be done using organic solvents such as acetone (Filik et al. 2016), acetonitrile (ACN) (Aguilera-Luiz et al. 2008; Garrido Frenich et al. 2010), or methanol (MeOH) (Dasenaki and Thomaidis 2015; Liu et al. 2013), or by mixing organic solvents with acidic compounds, such as trifluoroacetic acid (Feng et al. 2016; Nikolaidou et al. 2008), trichloroacetic acid (Wang et al. 2016), formic acid (Piatkowska et al. 2016), acetic acid (Gao et al. 2017; Xue et al. 2016). Alternatively, buffers like EDTA-McIlvaine (Chen et al. 2009; Gao et al. 2020; Jing et al. 2009; Lu et al. 2020; Lv et al. 2012, 2014a, b), citrate buffer (Hu et al. 2008), and sodium succinate buffer (Heller et al. 2006) can also be used for protein precipitation during sample preparation for TCs in food samples. However, protein precipitation only removes the proteins from the matrix, leaving behind other compounds that could also interfere with the analysis. Therefore, more efficient, and selective sample preparation methods are needed to address this issue.

In the literature, there is a wide variety of extraction, clean-up, and pre-concentration methods for the analysis of TCs residues applied to different types of samples. The most commonly used include Solid-Phase Extraction (SPE), solvent extraction (SE), liquid–liquid extraction (LLE), ultrasound-assisted extraction (UAE), magnetic solid-phase extraction (MSPE), dispersive liquid–liquid microextraction (DLLME), solid phase extraction with molecularly imprinted polymer (MISPE), Dispersive Solid-Phase Microextraction (DSPME), QuEChERS, among others (Pérez-Rodríguez et al. 2018).

SPE is the most popular and one of the most important techniques for the extraction and/or pre-concentration of TCs in different types of samples, especially complex ones. This is because SPE allows for sample clean-up, pre-concentration of large sample volumes, reduction of solvent volume, selective analyte extraction, regeneration of the solid phase, and easy combination with different detection techniques (Wen 2019).

SPE involves the solid–liquid partitioning equilibrium, and is based on the separation mechanisms of low-pressure liquid chromatography (Jardim 2010). The typical SPE procedure, illustrated in Fig. 13.3, consists of 4 steps:

- Conditioning (or activation): in this step, the adsorbent material is conditioned or activated using an appropriate solvent, aiming to improve the packing of this material and ensure that it is in suitable physicochemical conditions to facilitate interactions responsible for the retention of the analyte (Jardim 2010; Poole 2002; Wan Ibrahim et al. 2014);
- Sample percolation: in this step, the sample is passed through the adsorbent material under gravity, vacuum, pumping or automated systems. During this stage, the analyte, as well as sometimes other matrix components are retained on the adsorbent material, through mechanisms that include Van Der Walls interactions, hydrogen bonds, dipole–dipole interactions, cationic or anionic exchange and even size exclusion. The analyte is pre-concentrated on the adsorbent during this step,

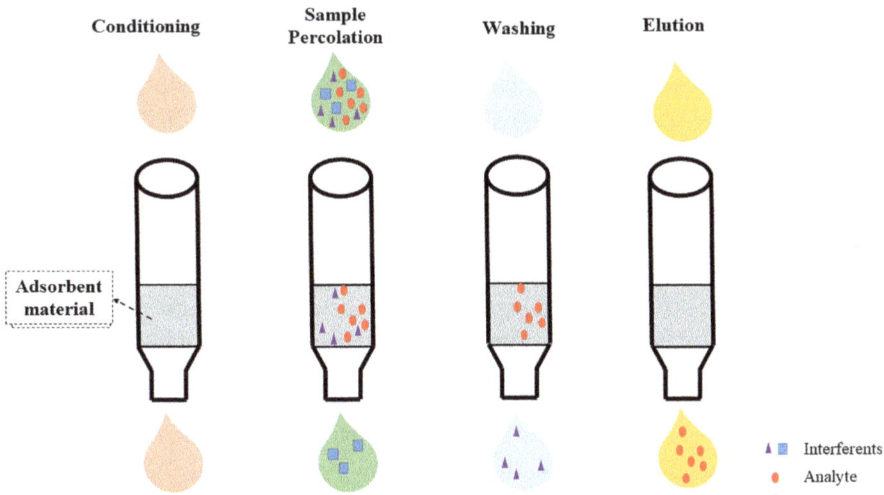

Fig. 13.3 Schematic representation of the steps employed in Solid-Phase Extraction (SPE). Adapted from Luiz et al. (2015)

and depending on the type of sample, different volumes can be percolated. This allows for the passage of large sample volumes, enabling high pre-concentration factors (Jardim 2010; Poole 2002; Wan Ibrahim et al. 2014);

- Cleaning/washing of the column: in this stage, an appropriate solvent is used to wash the column, removing impurities, and enabling purification of the extract. This step helps to eliminate unwanted matrix components and enhance the purity of the analyte (Jardim 2010; Poole 2002);
- Elution: in this step, the analyte is eluted from the adsorbent by using an appropriate solvent that disrupts the analyte-adsorbent interaction. This elution step allows for the release of the analyte from the solid phase so that it can be collected for further analysis (Jardim 2010; Poole 2002; Wan Ibrahim et al. 2014).

To use SPE as a sample preparation step, in addition to considering of the characteristics of the analyte and matrix, parameters such as choice and type of adsorbent material, sample volume and elution solvent should also be considered. The literature contains numerous studies that utilize SPE for extraction/pre-concentration of TCs from different types of samples. Among these, OASIS HLB, a universal polymeric reversed-phase hydrophilic-lipophilic-balanced adsorbent composed of *N*-vinylpyrrolidone and divinylbenzene, is considered the most commonly employed adsorbent Önal (2011). In general, the binding between this adsorbent and these analytes occurs due to a combination of forces, including Van der Waals forces, non-polar interactions, hydrogen bonding, and other relevant interactions (Andrade-Eiroa et al. 2016a, b).

Mamani et al. (2009) developed a method for the extractions of TCs, sulfonamides and chloramphenicols from bovine milk samples for subsequent determination by

HPLC–DAD. The sample preparation involved the precipitation of milk proteins using McIlvaine buffer and trichloroacetic acid in MeOH, followed by SPE using OASIS HLB cartridges. Various authors have also studied the extraction of TCS using SPE cartridges with C18 stationary phases (Fritz and Zuo 2007; Samanidou et al. 2007; Santos et al. 2007). However, the use of HLB cartridges is more recommended for the extraction and pre-concentration of TCs due to their stability in different pH conditions and solvents, as well as their good retention capacity for both polar and hydrophobic compounds (Andrade-Eiroa et al. 2016a, b).

In the recent years, there has been a pursuit of developing SPE adsorbents with specific recognition for TCs, aiming to improve results and achieve greater selectivity, especially due to the complexity of matrices such as food samples, for example. Among these newly developed adsorbents, we can mention metal–organic frameworks (MOF) (Arabsorkhi et al. 2019), zeolite imidazole framework-8 (ZIF-8) (Liu et al. 2017; Yang et al. 2013), carbon/graphene-based adsorbents (de Faria et al. 2017; Sun et al. 2015; Wei et al. 2017; Wu et al. 2017; Xu et al. 2016), magnetic nanoparticles (Castillo-García et al. 2015; Huang et al. 2017; Ibarra et al. 2011), and especially molecularly imprinted polymers (MIP) (Feng et al. 2016; Niu et al. 2016; Wang et al. 2018; Zhang et al. 2020).

13.2 Molecularly Imprinted Polymers for Tetracyclines

MIPs are synthetic polymeric materials that exhibit selective recognition sites that are generated during their polymeric synthesis. These materials have garnered attention due to their advantages such as selectivity, simplicity of preparation, robustness, ease of storage, among other benefits (Chen et al. 2016). Therefore, the search for new adsorbent materials with selective recognition sites has become of great interest to the scientific community at large.

The concept of molecular imprinting began to be studied in 1940s, when Linus Pauling postulated the theory of antibody formation, wherein the polypeptide chain of the antibody molds itself around the antigen, creating a selective configuration for that antigen (Pauling 1940; Tarley et al. 2005). Subsequently, based on selective antigen–antibody interaction, scientists introduced molecular imprinting technology (MIT) to synthesize synthetic materials with a three-dimensional rigid structure around a template molecule, capable of mimicking biological materials while retaining selectivity and specificity. Therefore, MIT was recognized as a method that could provide a specific binding to the template molecule through binding sites in a MIP (Chen et al. 2016).

In general, the synthesis of a MIP can be summarized in three steps, which firstly involves the formation of a complex between template molecule (TM) and functional monomer (FM) through covalent or non-covalent bonds. Next, a cross-linking agent (CA) and a radical initiator (RI) are added to stabilize the binding sites and initiate polymerization. Thus, the template-functional monomer complex is polymerized, creating a polymeric network around the template. Finally, after obtaining

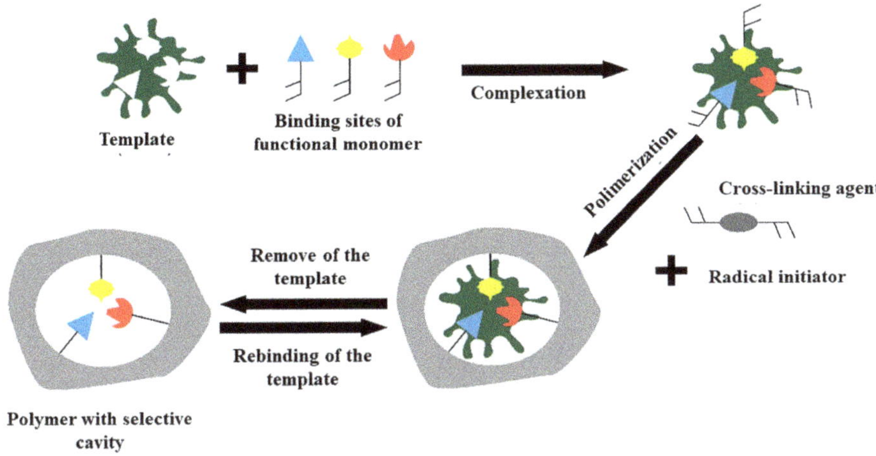

Fig. 13.4 Schematic representation of the general process of MIP synthesis. Adapted from Chen et al. (2016), Tarley et al. (2005), Turiel and Esteban (2019)

the polymer, the template is removed from the polymer matrix, generating three-dimensional cavities with recognition sites in terms of size, shape, and arrangement of functional groups to bind to the chemical species that served as a template (Chen et al. 2016; Turiel and Esteban 2019). Figure 13.4 presents a schematic representation of the general process of MIP synthesis.

The polymerization reaction in the synthesis of MIPs involves the combination of template, functional monomer, cross-linking agent, radical initiator and porogenic solvent, and therefore, the combination of these components affects characteristics such as selectivity, adsorption capacity, as well as their performance for specific applications. The selection and combination of appropriate reagents are crucial steps in the molecular imprinting process to obtain a material with the desired characteristics (Chen et al. 2016; Tarley et al. 2005). Table 13.2 display the reagents and their combinations for the synthesis of various MIPs used in the adsorption of TCs.

The template is the target analyte for which one aims to create a selective cavity, although in some cases, the selective cavity can be formed for a group or family to which the molecule belongs (Barros et al. 2016). While the majority of publications involve only a single molecule as a template, recent studies have reported the utilization of one or more molecules as a template in the synthesis of MIPs, thereby forming selective cavities for two different types of analytes or for a class of substances (Gama and Bottoli 2017; Qu and Zhu 2017; Tang et al. 2016; Xu et al. 2018). Jing et al. (2010) reported the synthesis of a MIP using OTC and CTC as template, resulting in a polymer with high molecular recognition capacity for the class of TCs. The selection of template is based on three main characteristics: not containing groups that engage in or participate in polymerization, being stable under polymerization conditions (high temperatures or exposure to UV light) and, most importantly, containing functional groups capable of interacting with the functional monomer (Tarley et al. 2005; Turiel and Martín-Esteban 2010, 2019).

Table 13.2 Molecularly imprinted polymers reported in the literature for the adsorption of TCs in various samples

Template	Functional monomer	Radical initiator	Cross-linking agent	Porogenic solvent	Other reagents used	Polymerization strategy	Sample/ Sample preparation	Analytical method	Ref.
1 mmol of OTC	3 mmol of MAA	0.15 mmol of AIBN	7 mmol of EGDMA	48 mL of MeOH	7.5 mmol of HEMA + 0.5 mmol of GDMA	Precipitation	Lake water; MISPE	HPLC–DAD	Abrão et al. (2014)
0.1 mmol of TC	0.7 mmol of MAA	80 mg of sodium persulfate	4.9 mmol of EGDMA	30 mL of MeOH		Precipitation	Milk; DSPME	LVSS-CE	Aguilar et al. (2020)
0.511 mmol of TC	6.09 mmol of MAA	27 mg of AIBN	50.56 mmol of EGDMA	15 g of MeOH		Bulk	N/A	UV–VIS	Cai and Gupta (2004)
1.16 mmol of TC	9.28 mmol of MAA	1.02 mmol of AIBN	46.41 mmol of EGDMA	13.33 mL of ACN		Bulk	Pork kidney; MISPE	HPLC–UV	Caro et al. (2005)
1.16 mmol of OTC				13.33 mL of DMSO					
2 mmol of OTC	20.0 mmol of MAA	0.1 g of AIBN	N.I.	100 mL of EtOH:H_2O (9:1, v/v)	120 mmol of styrene + 20 mmol of DVB	Surface imprinting	Egg and chicken tissue; MSPD	LC–MS/MS	Chen et al. (2009)
0.03 mmol of TC	2.4 mmol MAA	0.15 mmol of EBiB, 0.1 of mmol CuCl, 0.1 mmol of PMDETA	14.4 mmol of EGDMA	20 mL of H_2O with 0.3 g of Tweem 20	80 mg of Fe_3O_4-KH570	ATREP	Pork; MSPE	HPLC–UV	Dai et al. (2012)

(continued)

Table 13.2 (continued)

Template	Functional monomer	Radical initiator	Cross-linking agent	Porogenic solvent	Other reagents used	Polymerization strategy	Sample/ Sample preparation	Analytical method	Ref.
0.75 mmol of TC	0.3 mmol of MAA	10 mg of AIBN	1.2 mmol of EGDMA	50 mL of ACN	150 mg of V-MHNTs	Precipitation	N/A	UV–VIS	Dai et al. (2014a, b)
0.5 mmol of TC	1.0 mmol of MAA	15 mg of AIBN	8.0 mmol of EGDMA	60 mL of ACN	100 mg of Fe_3O_4	Precipitation	N/A	HPLC–UV	Dai et al. (2014a, b)
CTC[1]	MAA[1]	40 mg of AIBN	0.3 mmmol of EGDMA	7 mL of Chloroform	N/A	Precipitation	Milk, egg and pork; MISPE	HPLC–DAD	Feng et al. (2016)
1 mmol of TC	8 mmol of MAA	0.1 g of AIBN	20 mmol of EGDMA	100 mL of EtOH with 0.4 g of polyvinylpyrrolidone	0.5 g of Fe_3O_4COOH	Surface imprinting	Milk; MSPE	LC–MS/MS	Gao et al. (2020)
55.6 mg of TC	71.1 mg of AA	12.4 mg of AIBN	1.24 mL of TRIM	11.2 mL of Acetone	Silica fibers	Multiple co-polymerization	Feed and muscle Chicken, milk; SPME	HPLC–UV	Hu et al. (2008)
0.75 mmol of TC	6 mmol of MAA	30 mg of AIBN	3 mL of TRIM	65 mL of MeOH:CAN (30:35, v/v)	N/A	Precipitation	Lobster, duck, honey, and egg; MISPE	LC–MS/MS	Jing et al. (2009)
OTC:CTC (1:1. mol/ mol)	6 mmol of MAA	30 mg of AIBN	3 mL of TRIM	60 mL of MeOH:CAN (30:35, v/v)	N/A	Precipitation	Lobster, honey, and milk; online MISPE	HPLC–UV	Jing et al. (2010)

(continued)

Table 13.2 (continued)

Template	Functional monomer	Radical initiator	Cross-linking agent	Porogenic solvent	Other reagents used	Polymerization strategy	Sample/ Sample preparation	Analytical method	Ref.
TC[2]	MAA[2]	AIBN/ CDB[3]	EDMA[2]	ACN	Polyethylene glicol-1200	Precipitation -RAFT	N/A	HPLC–UV	Liu et al. (2014)
0.25 mmol of TC	3.0 mmol AA	0.25 mmol of BPO	30 mmol of EGDMA	50 mL of MeOH	100 mg of Fe_3O_4@SiO_2-MPS	Surface imprinting	Milk/ MSPE	HPLC–UV and DPV	Pizan-Aquino et al. (2020)
0.4 mmol of TC	1.2 mmol of MAA	120 mg of AIBN	10 mmol of EGDMA	12 mL of ACN + 4 mL of MeOH	0.4 mmol of HEMA/Silica fibers	Bulk-surface imprinting	Milk, chicken and fish/ SPME	HPLC–UV	Lu et al. (2020)
0.580 mmol of TC	4.64 mmol of MAA	N.I.	23.2 mmol of EGDMA	7 mL of DMF	The solution was placed into a Co-60 gamma source	N.I.	N/A	UV–VIS	Akbulut Söylemez (2021)
2.8 mmol of TC	12.2 mmol of MAA	0.7 mmol of AIBN	50 mmol of EGDMA	25 mL of ACN:benzyl alcohol (3:2. v/v)	N/A	N.I.	N/A/SPE	UV–VIS	Suedee et al. (2004)

N.I. Not informed; ***N/A*** not applicable; ***TC*** tetracycline; ***OTC*** oxytetracycline; ***CTC*** chlortetracycline; ***MAA*** methacrylic acid; ***AIBN*** 2,2′-Azobis (isobutyronitrile); ***EGDMA*** ethylene glycol dimethacrylate; ***MeOH*** methanol; ***HEMA*** 2-hydroxyethyl methacrylate; ***GDMA*** glycerol dimethacrylate; ***MISPE*** molecularly imprinted solid-phase extraction; ***HPLC–DAD*** high performance liquid chromatography with diode array ***DSPME*** dispersive solid-phase microextraction; ***LVSS-CE*** large volume sample stacking capillary electrophoresis; ***UV–Vis*** ultraviolet–visible spectroscopy; ***ACN*** acetonitrile; ***DMSO*** dimethyl sulfoxide; ***EtOH*** ethanol; ***MSPD*** matrix solid-phase dispersion; ***LC–MS/MS*** liquid chromatography-tandem mass spectrometry; ***PMDETA*** *N,N,N′N",N"*-pentamethyl diethyenetriamine; ***EBiB*** ethyl-2-bromoisobutryate; ***Fe_3O_4-KH570*** magnetic nanoparticles modified with γ-methacryloxypropylmethoxysilane; ***ATREP*** atom transfer radical emulsion polymerization; ***MSPE*** magnetic solid-phase extraction; ***V-MHNTs*** vinyl-modified magnetic halloysite nanotubes; ***SPME*** solid-phase microextraction; ***TRIM*** 2,2-bis(hydroxymethyl)butanol trimethacrylate; ***CDB*** cumyl dithiobenzoate; ***RAFT*** reversible-addition fragmentation chain-transfer polymerization; *EDMA* ethylene dimethacrylate; ***AA*** acrylic acid; ***BPO*** benzoyl peroxide; ***DVB*** divinylbenzene

[1] Different proportions of Template Molecule (TM):Functional Monomer (FM) were investigated, with the optimal molar ratio determined to be 1:4 (TM:FM)

[2] TC/MAA/EDMA are in the molar ratio 1:8:40, respectively

[3] AIBN/CDB are in the molar ratio 1:2, respectively

The functional monomer, typically composed of a recognition unit and a polymerizable unit, exhibits a dual functionality: it serves the purpose of forming a complex with the template and subsequently incorporating it into the polymer network by participating in the polymerization process through its polymerizable unit. The choice of functional monomer is a critical step for the successful imprinting of template in the polymer matrix, and its choice is closely linked to the characteristics of the template. The strength of the interactions between template and functional monomer influences the characteristics of the MIPs, such as affinity, stability, and selectivity. This is because greater stability in the template-functional monomer complex ensures that the cavities of the formed polymer are minimally or not affected by other species apart from template, imparting a higher uniformity of binding sites and consequently, resulting in greater selectivity of the MIP (Ashley et al. 2017; Chen et al. 2016; Turiel and Martín-Esteban 2019).

As seen in Table 13.2, the MIP synthesized by Hu et al. (2008) was the first MIP reported in the literature employing a functional monomer different from the MAA. In this study, the authors evaluated the MIPs synthesized using functional monomer with acidic, basic, and neutral characteristics, such as methacrylic acid (MAA), 4-vinylpyridine (4-VP) and Acrylamide (AA), respectively. The authors concluded that the MIP using AA as functional monomer exhibited superior homogeneity and morphological structure, as well as the highest adsorptive capacity when compared to the others. Pizan-Aquino and co-authors (Pizan-Aquino et al. 2020) reported the synthesis of a magnetic MIP (M-MIP) for TC, also employing AA as functional monomer, which was chosen through computational analysis in this case. For the purposes of comparison, the authors synthesized a MIP using MAA. The authors concluded that the M-MIP synthesized with functional monomer AA exhibited higher saturation magnetization when compared to the M-MIP made with MAA, as well as greater adsorption of TC onto this material, indicating that M-MIP (AA) provide superior performance.

However, since TC exhibit amphoteric characteristics, the primary and most used functional monomer in the synthesis of MIPs for TCs is MAA. This choice is because MAA possesses both electron-accepting groups, such as the carbonyl, and electro-donating (hydroxyl) groups capable of hydrogen bonding, which can interact with TC that also feature both hydrogen bond acceptor and donor groups. Therefore, it is expected that the interaction between these reagents will create multiple hydrogen bonding sites. Additionally, MAA has the ability to form ionic bonds because it contains the ionizable COOH group, making it a weak acid (pKa = 4.7). This feature allows it to interact with TCs that exhibit varying degrees of ionization in aqueous solution.

Another important factor that affects the characteristics of the MIP is the ratio between template and functional monomer. Because the interactions between template and functional monomer are governed by an equilibrium process, it is typically common to use an excess of functional monomer relation to template to ensure the formation of the maximum possible number of specific binding sites through the template-functional monomer complex. However, it is worth noting that studies on the proportion between these two reagents should be conducted because excessive amounts of functional monomer can increase the number of non-specific binding sites, reducing the selectivity of the MIP (Ashley et al. 2017; Chen et al. 2016; Turiel and Martín-Esteban 2019). Feng et al. (2016) evaluated the molar ratio between template and functional monomer at ratios of 1:4, 1:6 and 1:10 (template:functional monomer) concerning the adsorption capacity of each synthesized material. The authors conclude that there was a minimal effect on the adsorption capacity, and the lower ratio (1:4) was selected. The degree of cross-linking in the binding of TC with the synthesized material was also evaluated by Cai and Gupta (2004), where the authors concluded that with high crosslinking, i.e., a high amount of EGDMA (approximately 80%), there is greater adsorption. They suggested that a rigid structure of the polymer is necessary to maintain the molecular imprinting.

The choice of porogenic solvent used during polymer synthesis plays a fundamental role because it must provide medium solubilization, favoring the polymerization process, while avoiding strong intermolecular interactions with functional monomer and template that could interfere with the formation of the pre-polymerization complex (template-functional monomer). Furthermore, the porogenic solvent also has a direct influence on the morphological characteristics of the material, as it should promote the creation of pores within the polymer. Thus, the proper selection of the porogenic solvent in the pre-polymerization complex favors the formation of a polymer that is either macro-or microporous, as this solvent should not interfere with the template-functional monomer interaction. If interference were to occur, it could result in the formation of small quantities of recognitions sites with low selectivity. Thus, the low affinity between porogenic solvent and functional monomer may lead to the formation of larger pores that facilitate the mass transfer of the analyte to the selective sites, while a solvent whit high affinity for the functional monomer can interfere with the template-functional monomer interaction, resulting in smaller particles and fewer pores (Chen et al. 2016; Cormack and Elorza 2004; Parlapiano et al. 2021; Tarley et al. 2005; Turiel and Martín-Esteban 2019).

Typically, the most commonly used porogenic solvent in MIP synthesis include methanol (MeOH), acetonitrile (ACN), ethanol (EtOH), dimethyl sulfoxide (DMSO), tetrahydrofuran (THF), among others. These solvents can be classified as nonpolar, polar aprotic or polar protic. A polarity of porogens can indeed impact the interaction between the template and the functional monomer, and consequently, the adsorption properties of MIP (Chen et al. 2016). As observed in Table 13.1, the most commonly used porogenic solvent in the synthesis of MIPs for TC are MeOH, and ACN. However, it is worth noting that during the synthesis of MIPs, the use of a solvent mixture is also common, as can be seen in Table 13.1, where the use of mixture such as MeOH:ACN (Jing et al. 2009, 2010) and EtOH:H_2O (Chen et al. 2009) is reported.

Aguilar et al. (2020) assessed the impact of porogenic solvent during the synthesis of a MIP for TC, utilizing ACN as a polar aprotic solvent and MeOH as a polar protic solvent, in various ratios (4:0, 3:1, 2:2, 1:3 and 0:4) ranging from MeOH to ACN. The authors observed that polarity influences the interaction mode of the template-functional monomer complex, and the adsorption of TC decreased with decreasing polarity. The authors attribute these results to the relationship between the imprinting factors of the obtained MIPs and the dielectric constant of the porogenic solvent, given that the dielectric constant of MeOH (33.0) is lower than that of ACN (37.5), thus selecting MeOH as the porogenic solvent due to its superior performance. According to Liu et al. (2014), ACN as the porogenic solvent produces more uniform particles with favorable size, thereby enhancing adsorption. In the study reported by Jing and co-authors (Jing et al. 2009), the authors used a mixture of MeOH:ACN (30:35, v/v) as porogenic solvent. For them, a higher proportion of a less polar solvent promotes the formation of hydrogen bonds between TC and functional monomer (MAA).

The purpose of the crosslinking agent is to control the morphology of the polymeric matrix, stabilize binding sites and provide mechanical stability to the polymer matrix to maintain its molecular recognition capacity. The type and amount of crosslinking agent can indeed affect the selectivity and binding capacity of the MIP. Therefore, the quantify of crosslinking agent in the polymer should be relatively high to maintain a higher degree of crosslinking, thereby ensuring the stability of the recognition sites formed (Chen et al. 2016; Figueiredo et al. 2008; Santos et al. 2012; Turiel and Esteban 2019; Turiel and Martín-Esteban 2019).

Like functional monomer, crosslinking agent also contain one or more vinyl groups in their structure. Mono-vinyl compounds typically form more linear chains, while multi-vinyl compounds (bi-, tri-, etc.) create more branched chains with different molecular arrangements. Among the crosslinking agents, ethylene glycol dimethacrylate (EGDMA) and trimethylolpropane trimethacrylate (TRIM) are the most used in the synthesis of MIPs in general.

Jing et al. (2009) used TRIM and EGDMA in different quantities to evaluate of the influence of crosslinking agent MIPs synthesis. The authors concluded that the MIP synthesized using 3 mL of TRIM showed the best results in terms of imprinting factor and adsorptive capacity. This is because TRIM is a tri-vinyl crosslinking agent, which can favor the formation of porous structures in the polymer more effectively than EGDMA.

The literature review on MIP for TCs has revealed that EGDMA, a bifunctional crosslinking agent, is the most employed crosslinker in the synthesis of MIPs for TCs. This fact is likely associated with its highly stable thermal and mechanical properties, enabling the formation of more stable materials with rapid mass transfer. It is worth noting the high proportion of this reagent used in relation to the functional monomer (Chen et al. 2016; Tarley et al. 2005; Turiel and Esteban 2019). Aguilar et al. (2020) assessed effect of the EGDMA as a crosslinking agent in the synthesis of MIP. Under conditions previously optimized for the template:functional monomer ratio (1:7), the study involving the EGDMA proportion was evaluated under three different conditions: 1:7:25, 1:7:50 and 1:7:75 (template:functional monomer:crosslinking agent). The authors reported that they did not obtain the MIP synthesized in the 1:7:25 ratio due to the low amount of crosslinking agent, which did not facilitate the polymerization process. For the MIP with the highest proportion of EGDMA, 1:7:75 there was a significant decrease in the percentage of TC removal compared to the MIP synthesized in the 1:7:50 ratio. This is attributed to the high degree of polymer crosslinking, which can affect the interaction between template and functional groups, thereby reducing the binding capacity of MIP. Consequently, the MIP synthesized in the 1:7:50 ratio exhibited the best results.

In a polymerization reaction, after the addition of crosslinking agent, a radical initiator is introduced into the reaction medium. The radical initiator, typically featuring weak bonds with low dissociation energy, undergoes bond cleavage within its structure, generating radicals capable of initiating the polymerization reaction. Thus, the role of the radical initiator is to generate free radicals capable of initiating and sustaining the polymerization reaction. Commonly, MIP syntheses involve free radical polymerization reactions, either through photo- or electro polymerization, where the radical initiator can be activated by thermal or photochemical processes. Therefore, the choice of radical initiator should be careful considered, based on the characteristics of the other reagents, as they may be thermo and/or photosensitive. Additionally, it is necessary to remove oxygen from the polymerization synthesis environment, as this leads to an excess of free radicals that can hinder the polymerization reaction. This removal can be achieved by purging with inert gases, such as nitrogen and argon, or using an ultrasound bath. Peroxides and azo compounds are widely used as radical initiator, with 2,2′-azobisisobutyronitrile (AIBN) being the most commonly used in polymeric syntheses via free radicals (Chen et al. 2016; Figueiredo et al. 2008; Santos et al. 2012; Tarley et al. 2005).

The popularity of using AIBN as radical initiaor in MIP syntheses in general is also readily observed for the synthesis of MIPs for TCs, where it serves as the primary and most widely used radical initiator. Pizan-Aquino and co-workers (Pizan-Aquino et al. 2020) consider AIBN an expensive and highly explosive reagent. Consegquently, they were the first to employ benzoyl peroxide (BPO) as a radical initiator in the synthesis of an MIP for TC.

With the advancement of MIT, it has become possible to describe that the interactions occurring between the template and the functional monomer groups during the MIP synthesis are established through covalent, non-covalent and semi-covalent bonds. in the early 1970s, Wuff and Sarhan (1972) introduced the concept of reversible covalent bonding between the template and functional monomer. To achieve this, the authors synthesized an organic polymer for enantiomeric separation of racemic mixtures of glyceric acid, employing 2,3-o-p-vinylphenyl boronic acid as the functional monomer, which contained functional groups capable of selectively binding to template. However, due to the strong covalent bond formed between the template and functional, the removal of template from the polymeric matrix has proven challenging, requiring cleavage of the bond for complete removal (Figueiredo et al. 2008; Wuff and Sarhan 1972).

For MIPs synthesized using a covalent approach, a reversible covalent bond is formed between the template and the functional monomer during the formation of the complex between them. The template is removed by cleaving the formed covalent bonds, and these are reestablished upon rebinding to the template. MIPs formed with this type of approach exhibited more homogeneous and stable binding sites, consequently enhancing their selectivity. However, the use of this approach becomes quite limited due to the covalent binding resulting in slow binding and dissociation, making the extraction of template challenging; in many cases, drastic hydrolysis procedures are required (Ashley et al. 2017; Chen et al. 2016; Tarley et al. 2005; Turiel and Esteban 2019; Turiel and Martín-Esteban 2010).

On the other hand, there is the approach of non-covalent bonding between template and functional monomer, allowing for the removal of template from the polymeric matrix to occur through a simple washing step, involving changes in pH, ionic strength, solvent or some other parameter (Arshady and Mosbach 1981; Figueiredo et al. 2008). The non-covalent approach, based on relatively weak interactions such as Hydrogen bonding, ionic bonds, van Der Walls, among others, between template and functional monomer, features rapid kinetics for binding and removal of the template. The synthesis of MIPs using this approach is the most common due to its simplicity and the availability of a wide range of functional monomers capable of interacting with different molecules that can be used as template. However, MIPs synthesized using this non-covalent approach may exhibit low selectivity due to weak interactions between template and functional monomer when compared to polymers synthesized through the covalent approach. Furthermore, the shows greater heterogeneity of binding sites due to the excess of free functional monomer that can be randomly incorporated into the polymeric matrix. Additionally, it is more influenced by polarity of the porogenic solvent used in polymerization (Ashley et al. 2017; Chen et al. 2016; Tarley et al. 2005; Turiel and Esteban 2019; Vasapollo et al. 2011).

The literature review summarized in Table 13.1 allowed us to observe that all MIPs reported in the literature for TCs were synthesized using a non-covalent approach. This is because, as previously mentioned, the primary functional monomer used in these syntheses is MAA, which interacts with the TCs through hydrogen bonding, meaning a non-covalent bond occurs between the template and the functional monomer.

In order to combine the main advantages presented by covalent and non-covalent approaches, a dual binding mode, known as semi-covalent approach, was introduced by Whitcombe et al. (1995a). They synthesized an imprinted polymer for cholesterol that covalently bonded to functional monomer (4-vinylphenylcarbonate ester) during polymerization. After the removal of the molecule from the polymeric structure through basic hydrolysis cleavage, non-covalent recognition sites were formed, enabling the synthesis of MIPs employing both covalent and non-covalent bonds (Figueiredo et al. 2008; Whitcombe et al. 1995a). Therefore, in this type of synthesis, it is established that polymerization is mediated by covalent bonding, ensuring greater selectivity for the MIP, while the rebinding of the analyte during material usage occurs through non-covalent interactions, which facilitates the elution process of the analyte. One of the main advantages of this approach for MIPs synthesis is that a lower ratio between template and functional monomer is required (usually 1:1 or 1:2) compared to non-covalent synthesis, where rations range from 1: 4 to 1:8. This results in increased selectivity due to the reduction of non-specific binding sites caused by an excess of functional monomer (Beltran et al. 2010; Chen et al. 2016; Turiel and Martín-Esteban 2019).

Isocyanates have been widely employed as the functional monomer in semi-covalent MIP synthesis for various templates. Do Ki et al. (2002) prepared molecularly imprinted spherical silica particles for estrone. For this synthesis, the authors employed estrone as template and 3-(triethoxysilyl)propyl isocyanate (ICPTS) as functional monomer. Thus, the reaction for the formation of the template-functional monomer complex occurred through the reaction between the isocyanate group of functional monomer with the phenol group of the template, resulting in a stable urethane bond at room temperature. However, this bond can easily cleave at high temperatures. For this thermal cleavage, the authors used a mixture of dimethylsulfoxide (DMSO) and water at 180 °C. At high temperature, DMSO promotes the cleavage of the urethane bond and thus, the dissociated isocyanate group is converted into an amino group in the presence of water, as shown in the following scheme. Wang and co-workers also synthesized a MIP for estrone, employing the same synthetic route for the template-functional monomer complex. The difference in this case is that the MIP was synthesized on magnetic nanoparticles (Wang et al. 2009) (Scheme 13.1).

Estrone was also selected by Do Ki and Chang (2006) as the template for the synthesis of molecularly imprinted nanocapsules. In this study, the Estrone-Vinyl complex was synthesized using another isocyanate, 3-isopropenyl-α,α-dimethylbenzyl isocyanate (IPDBMI). In the previously mentioned work by Do Ki, the authors did not provide information regarding the molar ratio between template and the functional monomer. However, in this new study, the authors describe the molar ratio between Estrone and IPDMBI as being 1:1. Isocyanates were also employed in the synthesis of semi-covalent MIPs for Bisphenol A (Güney and Güney 2017), cholesterol (Effting et al. 2022), both utilizing ICPTS as functional monomer and following a similar synthetic route.

In addition to the previously mentioned examples, the semi-covalent approach has also been employed for the synthesis of molecularly imprinted materials for molecules such as polychlorinated aromatic compounds (Lubke et al. 1998), cholesterol (Whitcombe et al. 1995b), propazine (Cacho et al. 2006), phenols (Qi et al. 2010) and Orange II (Zhan et al. 2013).

A common characteristic observed in these studies is the type of template used, which typically consists of large molecules with a limited number of functional groups. Therefore, the semi-covalent synthesis strategy has been used for molecules with few interaction sites with the functional monomer. In this context, the semi-covalent synthesis approach is also referred to as sacrificial chemical imprinting. This is because a as a linking group between the template and the functional monomer serves to attach the template to the functional monomer during polymer formation and acts as a sacrificial spacer between the template and the functionality bound to the polymer. This allows the portion of the molecule that lacks functional groups capable of interacting with the functional monomer to be imprinted in the polymeric matrix due to its occupied volume. Additionally, it helps to prevent steric crowding during the non-covalent rebinding step (Kirsch et al. 2004; Qi et al. 2010).

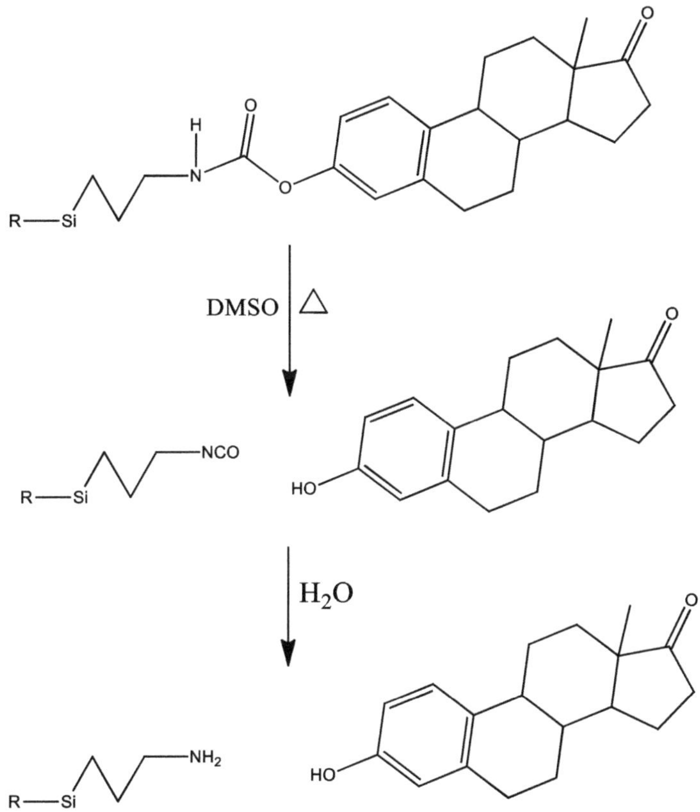

Scheme 13.1 Schematic of the thermal cleavage reaction between a phenol group and an isocyanate group. Adapted from Ki et al. (2002)

It is worth noting that, up to the present moment, the literature does not report MIPs synthesized via semi-covalent methods for polyfunctionalized molecules, such as TCs. This approach presents an alternative to mitigate the formation of heterogeneous sites during synthesis process.

Several strategies have been developed for the synthesis of MIPs. The choice of a specific strategy is influenced by the characteristics it offers, presents depending on the intended purpose for using the MIP in different fields and applications. As mentioned in Table 13.1, the most used methodologies for the synthesis of MIPs for TCs are bulk and precipitation polymerization. These methodologies are widely employed not only for TC-related MIPs but also for MIPs in general. Additionally, other methodologies such as suspension polymerization, surface imprinting, sol–gel, emulsion polymerization, among others, can also be adopted for MIP synthesis (Chen et al. 2016; Tarley et al. 2005; Turiel and Esteban 2019).

Bulk polymerization is indeed the first and most widely employe polymerization method for the synthesis of MIPs. This is primarily due to its simplicity, making it a favored choice for many researchers in the field. Essentially, synthesis using this method takes place in a homogeneous system, involving the mixing of all reagents (template, functional monomer, porogenic solvent, crosslinking agent and radical initiator), and the polymerization reaction is initiated through thermal or photochemical stimuli. At the end of the process, a polymeric mass is obtained, which requires post-treatments including grinding and sieving to obtain smaller particles, followed by solvent washing to remove template. Despites its popularity, this method has some disadvantages such as a limited number of specific recognition sites, resulting in slow binding kinetics and low binding capacity. Furthermore, materials synthesized using this method often have heterogeneous matrix due to the lack of control over the polymerization process. Additionally, the need for grinding can lead to the destruction of binding sites, and the materials may exhibit particles with different sizes and shapes. These drawbacks highlight the importance of considering alternative methods for MIP synthesis, especially when specific recognition and binding properties are critical for the intended application (Chen et al. 2016; Tarley et al. 2005; Turiel and Esteban 2019).

Precipitation polymerization, analogous to the bulk polymerization method, uses higher volumes of the porogenic solvent (often 2 to 10 times more) to prevent the formation of a polymeric mass. This method relies on the growth of polymer chains, which, upon reaching a certain length during the reaction, become insoluble in porogenic solvent and subsequently precipitate. The crosslinking agent AR provides sufficient rigidity to prevent particles coalescence, eliminating the need for additional stabilizing agents. Furthermore, continuous agitation during the process can promotes the formation of more homogeneous particles. As a result, the final product is a material with more uniform particles size, and there is typically a higher yield compared to bulk polymerization. This makes precipitation polymerization an attractive method for the synthesis of MIP, especially when particle size and uniformity are important considerations for the intended application (Chen et al. 2016; Tarley et al. 2005; Vasapollo et al. 2011; Wulff 2013).

The literature review on MIPs for TCs also allowed us to infer that the vast majority of syntheses of these materials occur via free radicals. Indeed, this mechanism is the primary and most widely used method in polymerization synthesis in general, where the cleavage of a double bond in the monomer unit is converted into an active propagation radical (Daoud Attieh et al. 2017). Among the primary advantages of this type of polymerization is the ability to occur under mild reaction conditions, such as room temperature and atmospheric pressure. However, there are also limitations associated with this method, including rapid polymer chain growth and irreversible termination reactions, which can impact the molecular wight distribution of the synthesized material (Beyazit et al. 2016; Cormack and Elorza 2004).

Furthermore, we saw that the numerous MIPs reported for TCs have been synthesized using a non-covalent approach, with MAA being the primary and most used functional monomer. This choice is due to its ability to interact with TCs through hydrogen bonding. However, because TCs have different functional groups, the non-covalent approach can lead to the generation of various interaction sites and, consequently, different conformations of functional monomer and template, resulting in heterogeneous chemical imprinting sites.

In addition to the numerous advancements showcased in recent decades at MIT, the scientific community has been altering the manner in which methods and procedures are devised, aiming to minimize wastage, risks, and hazards to analysts and the environment in accordance with the principles of Green Chemistry (Díaz-Álvarez et al. 2023). Recently, in order to complement the 12 established principles of Green Chemistry, Arabi and collaborators (Arabi et al. 2021) formulated a set of 14 principles, known as "GREENIFICATION" with the aim of promoting the design, preparation, and application of MIPs in a more sustainable and ecological manner. As a result, green MIPs (GMIPs) have emerged as a novel class of synthetic receptors, aligning with sustainability by adhering to the principles of ecologization and minimizing the detrimental impacts conventional MIT may pose to the environment and human health, enabling news applications for MIPs (Donato et al. 2022; Ostovan et al. 2022).

Thus, the principles of "GREENIFICATION" in the synthesis of MIPs for TCs aim to minimize the environmental impacts associated with the process. This involves the adoption of sustainable practices in several stages, such as choosing more eco-friendly solvents, utilizing renewable raw materials, reducing waste, and optimizing processes for energy efficiency. In this way, the goals is to develop more eco-efficient and responsible technologies throughout the entire life cycle of these polymers.

Recently, an approach has been developed for the fabrication of nanocomposite membranes with molecular imprinting of TC, aiming for a more environmentally sustainable profile. In this scenario, biomass-activated carbon nanoparticles were incorporated into the hybrid membrane matrix, composed of porous cellulose acetate and chitosan, using the phase inversion process. These imprinted hybrid membranes demonstrated a remarkable permeation rate of the model molecules compared to the competing structural analogue, OTC (Xing et al. 2020).

13.3 Conclusion and Future Aspects

From the literature review on MIPs for TCs, recent advances in the syntheses of these materials have been observed, particularly emphasizing the extraction of this class of antibiotics from samples with complex matrices for subsequent analysis using analytical techniques. This review underscores that MIPs exhibit greater potential compared to other adsorbent materials, attributable to their high selectivity and other parameters considered important in adsorption processes. However, in the synthesis of these materials, a thorough analysis of the components to be employed is necessary, as studies indicate that synthesis-related parameters influence the synthesized materials and, consequently, their applications. The MIT for the synthesis of new MIPs for TCs in the future can be integrated with other emerging technologies, such as artificial intelligence (AI) and the internet of things (IoT), to establish more intelligent and autonomous systems for monitoring and detection. With the advances in MIT, efforts are being directed towards making detection methods more cost-effective and less complex, thereby enabling their widespread utilization across various scenarios. Furthermore, the integration of ecologization into MIP research for tetracyclines is part of a broader approach to developing sustainable technologies in the field of chemistry and materials. These practices aim to reduce environmental impact and promote sustainability throughout the life cycle of new materials.

Acknowledgements The authors acknowledge the financial support and fellowships of Coordenação de Aperfeiçoamento de Nível Superior (CAPES), Finance Code 001, Conselho Nacional de Desenvolvimento Científico e Tecnológico (CNPq) (Grant No 307432/2017-3, 307505/2021-9, 420097/2021-0, 402387/2020-1, 401256/2020-0) and Instituto Nacional de Ciência e Tecnologia de Bioanalítica (INCT) (FAPESP Grant No 2014/50867-3 and CNPq Grant No 465389/2014-7).

References

Abrão LCC, Maia PP, Figueiredo EC (2014) Determination of tetracyclines by solid-phase extraction with a molecularly imprinted polymer and high-performance liquid chromatography. Anal Lett 47(13):2183–2194. https://doi.org/10.1080/00032719.2014.900780

Aguilar JFF, Miranda JM, Rodriguez JA, Paez-Hernandez ME, Ibarra IS (2020) Selective removal of tetracycline residue in milk samples using a molecularly imprinted polymer. J Polym Res 27(7):176. https://doi.org/10.1007/s10965-020-02139-9

Aguilera-Luiz MM, Vidal JLM, Romero-González R, Frenich AG (2008) Multi-residue determination of veterinary drugs in milk by ultra-high-pressure liquid chromatography-tandem mass spectrometry. J Chromatogr A 1205(1–2):10–16. https://doi.org/10.1016/j.chroma.2008.07.066

Akbulut Söylemez M (2021) Synthesis and characterization of tetracycline-imprinted membranes: a detailed positron annihilation lifetime spectroscopy investigation. J Mol Recognit 34(8):1–10. https://doi.org/10.1002/jmr.2895

Andrade-Eiroa A, Canle M, Leroy-Cancellieri V, Cerdà V (2016a) Solid-phase extraction of organic compounds: a critical review. Part ii. In: TrAC—trends in analytical chemistry, vol 80, pp 655–667). Elsevier B.V. https://doi.org/10.1016/j.trac.2015.08.014

Andrade-Eiroa A, Canle M, Leroy-Cancellieri V, Cerdà V (2016b) Solid-phase extraction of organic compounds: a critical review (Part I). TrAC Trends Anal Chem 80:641–654. https://doi.org/10.1016/j.trac.2015.08.015

Arabi M, Ostovan A, Li J, Wang X, Zhang Z, Choo J, Chen L (2021) Molecular imprinting: green perspectives and strategies. Adv Mater 33(30):1–33. https://doi.org/10.1002/adma.202100543

Arabsorkhi B, Sereshti H, Abbasi A (2019) Electrospun metal-organic framework/polyacrylonitrile composite nanofibrous mat as a microsorbent for the extraction of tetracycline residue in human blood plasma. J Sep Sci 42(8):1500–1508. https://doi.org/10.1002/jssc.201801305

Arshady R, Mosbach K (1981) Synthesis of substrate-selective polymers by host-guest polymerization. Die Makromolekulare Chemie 182(2):687–692. https://doi.org/10.1002/macp.1981.021820240

Ashley J, Shahbazi MA, Kant K, Chidambara VA, Wolff A, Bang DD, Sun Y (2017) Molecularly imprinted polymers for sample preparation and biosensing in food analysis: Progress and perspectives. Biosens Bioelectron 91:606–615. https://doi.org/10.1016/j.bios.2017.01.018

Bahrami FL, Morris D, Pourgholami HM (2011) Tetracyclines: drugs with huge therapeutic potential. Mini Rev Med Chem 12(1):44–52. https://doi.org/10.2174/138955712798868977

Barros LA, Custodio R, Rath S (2016) Design of a new molecularly imprinted polymer selective for hydrochlorothiazide based on theoretical predictions using Gibbs free energy. J Braz Chem Soc 27(12):2300–2311. https://doi.org/10.5935/0103-5053.20160126

Beltran A, Borrull F, Marcé RM, Cormack PAG (2010) Molecularly-imprinted polymers: useful sorbents for selective extractions. TrAC Trends Anal Chem 29(11):1363–1375 (Elsevier). https://doi.org/10.1016/j.trac.2010.07.020

Beyazit S, Tse Sum Bui B, Haupt K, Gonzato C (2016) Molecularly imprinted polymer nanomaterials and nanocomposites by controlled/living radical polymerization. Prog Polym Sci 62:1–21. https://doi.org/10.1016/j.progpolymsci.2016.04.001

Cacho C, Turiel E, Martín-Esteban A, Ayala D, Pérez-Conde C (2006) Semi-covalent imprinted polymer using propazine methacrylate as template molecule for the clean-up of triazines in soil and vegetable samples. J Chromatogr A 1114(2):255–262. https://doi.org/10.1016/j.chroma.2006.02.051

Cai W, Gupta RB (2004) Molecularly-imprinted polymers selective for tetracycline binding. Sep Purif Technol 35(3):215–221. https://doi.org/10.1016/S1383-5866(03)00143-6

Caro E, Marcé RM, Cormack PAG, Sherrington DC, Borrull F (2005) Synthesis and application of an oxytetracycline imprinted polymer for the solid-phase extraction of tetracycline antibiotics. Anal Chim Acta 552(1–2):81–86. https://doi.org/10.1016/J.ACA.2005.07.047

Castillo-García ML, Aguilar-Caballos MP, Gómez-Hens A (2015) A europium- and terbium-coated magnetic nanocomposite as sorbent in dispersive solid phase extraction coupled with ultra-high performance liquid chromatography for antibiotic determination in meat samples. J Chromatogr A 1425:73–80. https://doi.org/10.1016/j.chroma.2015.11.048

Charuaud L, Jarde E, Jaffrezic A, Thomas MF, Le Bot B (2019) Veterinary pharmaceutical residues from natural water to tap water: sales, occurrence and fate. J Hazard Mater 361:169–186. https://doi.org/10.1016/j.jhazmat.2018.08.075

Chattopadhyay MK (2014) Use of antibiotics as feed additives: a burning question. Front Microbiol 5(July):1–3. https://doi.org/10.3389/fmicb.2014.00334

Chen L, Liu J, Zeng Q, Wang H, Yu A, Zhang H, Ding L (2009) Preparation of magnetic molecularly imprinted polymer for the separation of tetracycline antibiotics from egg and tissue samples. J Chromatogr A 1216(18):3710–3719. https://doi.org/10.1016/J.CHROMA.2009.02.044

Chen L, Wang X, Lu W, Wu X, Li J (2016) Molecular imprinting: perspectives and applications. Chem Soc Rev 45(8):2137–2211. https://doi.org/10.1039/c6cs00061d

Chen T, Cheng G, Ahmed S, Wang Y, Wang X, Hao H, Yuan Z (2017) New methodologies in screening of antibiotic residues in animal-derived foods: biosensors. Talanta 175:435–442. https://doi.org/10.1016/j.talanta.2017.07.044

Cherkashina K, Vakh C, Lebedinets S, Pochivalov A, Moskvin L, Lezov A, Bulatov A (2018) An automated salting-out assisted liquid-liquid microextraction approach using 1-octylamine: on-line separation of tetracycline in urine samples followed by HPLC-UV determination. Talanta 184:122–127. https://doi.org/10.1016/j.talanta.2018.02.112

Chlortetracycline, Oxytetracycline, and Tetracycline in Edible Animal Tissues, Pub. L. No. Official Method 995.09, AOAC (2012). http://www.ewenku.net/docs/d5bcb214a7c837

Chopra I, Roberts M (2001) Tetracycline antibiotics: mode of action, applications, molecular biology, and epidemiology of bacterial resistance. Microbiol Mol Biol Rev 65(2):232–260. https://doi.org/10.1128/mmbr.65.2.232-260.2001

Commission CA (2015) Maximum residue limits (MRLs) and risk management recommendations (RMRs) for residues of veterinary drugs in foods CAC/MRL 2-2015 (Issue July)

Commission Regulation-No 37/2010, Pub. L. No. No 37/2010 (2010) Official Journal of the European Union 72. https://ec.europa.eu/health/sites/health/files/files/eudralex/vol-5/reg_2010_37/reg_2010_37_en.pdf

Cormack PAG, Elorza AZ (2004) Molecularly imprinted polymers: synthesis and characterisation. J Chromatogr B Anal Technol Biomed Life Sci 804(1):173–182. https://doi.org/10.1016/j.jchromb.2004.02.013

D'el Rey Dantas FT, Feijó LS, Nogueira CEW, Curcio BDR (2019) Doxiciclina: Uma Revisão Sobre Particularidades E Utilização Clínica Na Espécie Equina. Sci Animal Health 6(2):101. https://doi.org/10.15210/sah.v6i2.12840

Daghrir R, Drogui P (2013) Tetracycline antibiotics in the environment: a review. Environ Chem Lett 11(3):209–227. https://doi.org/10.1007/s10311-013-0404-8

Dai J, Pan J, Xu L, Li X, Zhou Z, Zhang R, Yan Y (2012) Preparation of molecularly imprinted nanoparticles with superparamagnetic susceptibility through atom transfer radical emulsion polymerization for the selective recognition of tetracycline from aqueous medium. J Hazard Mater 205–206:179–188. https://doi.org/10.1016/j.jhazmat.2011.12.056

Dai J, Wei X, Cao Z, Zhou Z, Yu P, Pan J, Zou T, Li C, Yan Y (2014a) Highly-controllable imprinted polymer nanoshell at the surface of magnetic halloysite nanotubes for selective recognition and rapid adsorption of tetracycline. RSC Adv 4(16):7967–7978. https://doi.org/10.1039/c3ra45779f

Dai J, Zhou Z, Zhao C, Wei X, Dai X, Gao L, Cao Z, Yan Y (2014b) Versatile method to obtain homogeneous imprinted polymer thin film at surface of superparamagnetic nanoparticles for tetracycline binding. Ind Eng Chem Res 53(17):7157–7166. https://doi.org/10.1021/ie404140y

Danner MC, Robertson A, Behrends V, Reiss J (2019) Antibiotic pollution in surface fresh waters: Occurrence and effects. Sci Total Environ 664:793–804. https://doi.org/10.1016/j.scitotenv.2019.01.406

Daoud Attieh M, Zhao Y, Elkak A, Falcimaigne-Cordin A, Haupt K (2017) Enzyme-initiated free-radical polymerization of molecularly imprinted polymer nanogels on a solid phase with an immobilized radical source. Angew Chem 129(12):3387–3391. https://doi.org/10.1002/ange.201612667

Dasenaki ME, Thomaidis NS (2015) Multi-residue determination of 115 veterinary drugs and pharmaceutical residues in milk powder, butter, fish tissue and eggs using liquid chromatography-tandem mass spectrometry. Anal Chim Acta 880:103–121. https://doi.org/10.1016/j.aca.2015.04.013

de Faria HD, Rosa MA, Silveira AT, Figueiredo EC (2017) Direct extraction of tetracyclines from bovine milk using restricted access carbon nanotubes in a column switching liquid chromatography system. Food Chem 225:98–106. https://doi.org/10.1016/j.foodchem.2017.01.004

Denobile M, De Souza Nascimento E (2004) Method validation for the determination of the antibiotics residues oxytetracycline, tetracycline, chlortetracycline and doxycycline in milk by high performance liquid chromatography. Revista Brasileira De Ciencias Farmaceuticas/brazilian J Pharm Sci 40(2):209–218. https://doi.org/10.1590/S1516-93322004000200010

Díaz-Álvarez M, Turiel E, Martíns-Esteban A (2023) Recent advances and future trends in molecularly imprinted polymers-based sample preparation. J Sep Sci 46(12):1–19. https://doi.org/10.1002/jssc.202300157

Donato L, Nasser II, Majdoub M, Drioli E (2022) Green chemistry and molecularly imprinted membranes. Membranes 12(5). https://doi.org/10.3390/membranes12050472

Effting L, Prete MC, Urbano A, Effting LM, González MEC, Bail A, Tarley CRT (2022) Preparation of magnetic nanoparticle-cholesterol imprinted polymer using semi-covalent imprinting approach for ultra-effective and highly selective cholesterol adsorption. React Funct Polym 172:105178. https://doi.org/10.1016/j.reactfunctpolym.2022.105178

EMA (2019) European surveillance of veterinary antimicrobial consumption, 2019. 'Sales of veterinary antimicrobial agents in 31 European countries in 2017'. In: Ninth ESVAC report—EMA/294674/2019. https://www.ema.europa.eu/en/documents/report/sales-veterinary-antimicrobial-agents-31-european-countries-2017_en.pdf

Esteban C, Rodrigues CA, Nascimento EDS (2007) Determinação de tetraciclina em líquido sinovial de vacas com doença podal. Revista Brasileira De Ciencias Farmaceuticas/brazilian J Pharm Sci 43(2):205–214. https://doi.org/10.1590/s1516-93322007000200006

FDA (2019) 2018 summary report on antimicrobials sold or distributed for use in food-producing animals

Feng MX, Wang GN, Yang K, Liu HZ, Wang JP (2016) Molecularly imprinted polymer-high performance liquid chromatography for the determination of tetracycline drugs in animal derived foods. Food Control 69:171–176. https://doi.org/10.1016/J.FOODCONT.2016.04.050

Figueiredo EC, Dias ACB, Arruda MAZ (2008) Impressão molecular: Uma estratégia promissora na elaboração de matrizes para a liberação controlada de fármacos. Revista Brasileira De Ciencias Farmaceuticas/brazilian J Pharm Sci 44(3):361–375. https://doi.org/10.1590/S1516-93322008000300005

Filik H, Avan A, Aydar S, Ozyurt D, Demirata B (2016) Determination of tetracycline on the surface of a high-performance graphene modified screen-printed carbon electrode in milk and honey samples. Curr Nanosci 12(4):527–533. https://doi.org/10.2174/1573413712666151214222808

Fritz JW, Zuo Y (2007) Simultaneous determination of tetracycline, oxytetracycline, and 4-epitetracycline in milk by high-performance liquid chromatography. Food Chem 105(3):1297–1301. https://doi.org/10.1016/j.foodchem.2007.03.047

Gama MR, Bottoli CBG (2017) Molecularly imprinted polymers for bioanalytical sample preparation. J Chromatogr B Anal Technol Biomed Life Sci 1043:107–121. https://doi.org/10.1016/j.jchromb.2016.09.045

Gao J, Wang H, Qu J, Wang H, Wang X (2017) Development and optimization of a naphthoic acid-based ionic liquid as a "non-organic solvent microextraction" for the determination of tetracycline antibiotics in milk and chicken eggs. Food Chem 215:138–148. https://doi.org/10.1016/j.foodchem.2016.07.138

Gao W, Li J, Li P, Huang Z, Cao Y, Liu X (2020) Preparation of magnetic molecularly imprinted polymer (MMIP) nanoparticles (NPs) for the selective extraction of tetracycline from milk. Anal Lett 53(7):1097–1112. https://doi.org/10.1080/00032719.2019.1698049

Garrido Frenich A, Aguilera-Luiz del M, Martínez Vidal JL, Romero-González R (2010) Comparison of several extraction techniques for multiclass analysis of veterinary drugs in eggs using ultra-high pressure liquid chromatography-tandem mass spectrometry. Analytica Chimica Acta, 661(2):150–160. https://doi.org/10.1016/j.aca.2009.12.016

Gaudin V (2017) Advances in biosensor development for the screening of antibiotic residues in food products of animal origin—a comprehensive review. Biosens Bioelectron 90(December 2016):363–377. https://doi.org/10.1016/j.bios.2016.12.005

Guerra W, Silva-Caldeira PP, Terenzi H, Pereira-Maia EC (2016) Impact of metal coordination on the antibiotic and non-antibiotic activities of tetracycline-based drugs. Coord Chem Rev 327–328:188–199. https://doi.org/10.1016/j.ccr.2016.04.009

Güney S, Güney O (2017) Development of an electrochemical sensor based on covalent molecular imprinting for selective determination of bisphenol-A. Electroanalysis 29(11):2579–2590. https://doi.org/10.1002/elan.201700300

Hao H, Cheng G, Iqbal Z, Ai X, Hussain HI, Huang L, Dai M, Wang Y, Liu Z, Yuan Z (2014) Benefits and risks of antimicrobial use in food-producing animals. Front Microbiol 5(Jun):1–11. https://doi.org/10.3389/fmicb.2014.00288

Heller DN, Nochetto CB, Rummel NG, Thomas MH (2006) Development of multiclass methods for drug residues in eggs: hydrophilic solid-phase extraction cleanup and liquid chromatography/tandem mass spectrometry analysis of tetracycline, fluoroquinolone, sulfonamide, and β-lactam residues. J Agric Food Chem 54(15):5267–5278. https://doi.org/10.1021/jf0605502

Hu X, Pan J, Hu Y, Huo Y, Li G (2008) Preparation and evaluation of solid-phase microextraction fiber based on molecularly imprinted polymers for trace analysis of tetracyclines in complicated samples. J Chromatogr A 1188(2):97–107. https://doi.org/10.1016/J.CHROMA.2008.02.062

Huang S, Gan N, Liu H, Zhou Y, Chen Y, Cao Y (2017) Simultaneous and specific enrichment of several amphenicol antibiotics residues in food based on novel aptamer functionalized magnetic adsorbents using HPLC-DAD. J Chromatogr, B:anal Technol Biomed Life Sci 1060:247–254. https://doi.org/10.1016/j.jchromb.2017.05.035

Ibarra IS, Rodriguez JA, Miranda JM, Vega M, Barrado E (2011) Magnetic solid phase extraction based on phenyl silica adsorbent for the determination of tetracyclines in milk samples by capillary electrophoresis. J Chromatogr A 1218(16):2196–2202. https://doi.org/10.1016/J.CHROMA.2011.02.046

Instrução Normativa N°26, de 10 de julho de 2009, 9 (2009). http://sistemasweb.agricultura.gov.br/sislegis/action/detalhaAto.do?method=visualizarAtoPortalMapa&chave=1984822284

Instrução Normativa - IN N° 162 (2022)

Jardim ICSF (2010) Extração em Fase Sólida: Fundamentos Teóricos e Novas Estratégias para Preparação de Fases Sólidas Resumo. Scientia Chromatographica 2(1):13–25

Jing T, Gao X, Wang P, Wang Y, Lin Y, Hu X, Hao Q (2009) Determination of trace tetracycline antibiotics in foodstuffs by liquid chromatography—tandem mass spectrometry coupled with selective molecular-imprinted solid-phase extraction. Anal Bioanal Chem 393:2009–2018. https://doi.org/10.1007/s00216-009-2641-z

Jing T, Wang Y, Dai Q, Xia H, Niu J, Hao Q, Mei S, Zhou Y (2010) Preparation of mixed-templates molecularly imprinted polymers and investigation of the recognition ability for tetracycline antibiotics. Biosens Bioelectron 25(10):2218–2224. https://doi.org/10.1016/J.BIOS.2010.02.023

Ki CD, Chang JY (2006) Preparation of a molecularly imprinted polymeric nanocapsule with potential use in delivery applications. Macromolecules 39(9):3415–3419. https://doi.org/10.1021/ma052480q

Ki CD, Oh C, Oh SG, Chang JY (2002) The use of a thermally reversible bond for molecular imprinting of silica spheres. J Am Chem Soc 124(50):14838–14839. https://doi.org/10.1021/ja0277881

Kirsch N, Alexander C, Davies S, Whitcombe MJ (2004) Sacrificial spacer and non-covalent routes toward the molecular imprinting of "poorly-functionalized" N-heterocycles. Anal Chim Acta 504(1):63–71. https://doi.org/10.1016/S0003-2670(03)00510-5

Kumar A, Panda AK, Sharma N (2022) Determination of antibiotic residues in bovine milk by HPLC-DAD and assessment of human health risks in Northwestern Himalayan region, India. J Food Sci Technol 59(1):95–104. https://doi.org/10.1007/s13197-021-04988-8

Limites Máximos de Resíduo de Medicamentos Veterinários em Alimentos de Origem Animal, ANVISA 136 (2018). http://portal.anvisa.gov.br/documents/219201/219401/Med+Vet_Documento+base+discussa~o+18.10/69d161b5-785c-4907-862c-2294b48a79c5

Lista F2 da Portaria 344/98 (2009)

Liu M, Li Y, Han J, Dong X (2014) Synthesis of tetracycline-imprinted polymer microspheres by reversible addition- fragmentation chain-transfer precipitation polymerization using polyethylene glycol as a coporogen. J Sep Sci 37(9–10):1118–1125. https://doi.org/10.1002/jssc.201400034

Liu H, Chen L, Ding J (2017) A core-shell magnetic metal organic framework of type Fe3O4@ZIF-8 for the extraction of tetracycline antibiotics from water samples followed by ultra-HPLC-MS analysis. Microchim Acta 184(10):4091–4098. https://doi.org/10.1007/s00604-017-2442-6

Liu Y, Yang H, Yang S, Hu Q, Cheng H, Liu H, Qiu Y (2013) High-performance liquid chromatography using pressurized liquid extraction for the determination of seven tetracyclines in egg, fish and shrimp. J Chromatogr B Anal Technol Biomed Life Sci 917–918:11–17. https://doi.org/10.1016/j.jchromb.2012.12.036

Lu Y, Lei L, He J, Zhao T (2020) Preparation of hydrophilic molecularly imprinted solid-phase microextraction fiber for the selective removal and extraction of trace tetracyclines residues in animal derived foods. J Sep Sci 43(11):2172–2179. https://doi.org/10.1002/jssc.201901285

Lubke M, Whitcombe MJ, Vulfson EN (1998) A novel approach to the molecular imprinting of polychlorinated aromatic compounds. J Am Chem Soc 120(51):13342–13348. https://doi.org/10.1021/ja9818295

Luiz AL, Maciel EVS, Lanças FM (2015) Miniaturized sample preparation techniques and their application to the determination of residues and contaminants in food samples. Scentia Chromatographica 7(3):157–182. https://doi.org/10.4322/sc.2016.002

Lv YK, Wang LM, Yang L, Zhao CX, Sun HW (2012) Synthesis and application of molecularly imprinted poly(methacrylic acid)-silica hybrid composite material for selective solid-phase extraction and high-performance liquid chromatography determination of oxytetracycline residues in milk. J Chromatogr A 1227:48–53. https://doi.org/10.1016/j.chroma.2011.12.108

Lv Y-K, Zhang J-Q, Guo Z-Y, Wang W-C, Li B-H (2014a) Molecularly imprited solid-phase extraction of tetracycline residue from milk using internal-surface reversed-phase hybrid composite packing materials. Asian J Chem 26(12):3541–3544. https://doi.org/10.14233/ajchem.2014.16198

Lv YK, Zhang JQ, He YD, Zhang J, Sun HW (2014b) Adsorption-controlled preparation of molecularly imprinted hybrid composites for selective extraction of tetracycline residues from honey and milk. New J Chem 38(2):802–808. https://doi.org/10.1039/c3nj00962a

Lv G, Li Z, Elliott L, Schmidt MJ, MacWilliams MP, Zhang B (2019) Impact of tetracycline-clay interactions on bacterial growth. J Hazard Mater 91–97. https://doi.org/10.1016/j.jhazmat.2017.09.029

Mohebi A, Samadi M, Tavakoli HR, Parastouei K (2020) Homogenous liquid–liquid extraction followed by dispersive liquid–liquid microextraction for the extraction of some antibiotics from milk samples before their determination by HPLC. Microchem J 157(April):104988. https://doi.org/10.1016/j.microc.2020.104988

Moreira FTC, Guerreiro JRL, Azevedo VL, Kamel AH, Sales MGF (2010) New biomimetic sensors for the determination of tetracycline in biological samples: Batch and flow mode operations. Anal Methods 2(12):2039. https://doi.org/10.1039/c0ay00511h

Mostafa MA (2020) Doxycycline and minocycline drugs as a treatment proposal for inhibition of ARDS and inflammatory cytokine of SARS-COV-2. Al-Azhar Univ J Virus Res Stud 2(1):1–7. https://doi.org/10.21608/AUJV.2020.106725

Mota RA, Silva KPC, Freitas MFL, Porto WJN, Silva LBG (2005) Utilização indiscriminada de antimicrobianos e sua contribuição a multirresitência bacteriana. Braz J Vet Res Anim Sci 42(6):465–470. https://doi.org/10.11606/issn.1678-4456.bjvras.2005.26406

Moudgil P, Bedi JS, Aulakh RS, Gill JPS (2019) Analysis of antibiotic residues in raw and commercial milk in Punjab, India vis-à-vis human health risk assessment. J Food Saf 39(4):1–8. https://doi.org/10.1111/jfs.12643

Multiple Tetracycline Residues in Milk, Pub. L. No. Official Method 995.04, AOAC (2012). https://wenku.baidu.com/view/0a0b9acca1c7aa00b52acb7d.html?re=view

Navrátilová P, Borkovcová I, Dračková M (2009) Occurrence of tetracycline, chlortetracyclin, and oxytetracycline residues in raw cow's milk. Czech J Food Sci 27(5):379–385

Nelson M, Hillen W, Greenwald RA (2001) Tetracyclines in biology, chemistry and medicine. Birkhäuser Basel

Nikolaidou KI, Samanidou VF, Papadoyannis IN (2008) Development and validation of an HPLC multi-residue method for the determination of seven tetracycline antibiotic residues in bovine liver and kidney according to the European Union Decision 2002/657/EC. J Liq Chromatogr Relat Technol 31(17):2523–2540. https://doi.org/10.1080/10826070802352660

Niu Y, Liu C, Yang J, Ma M, Gong Y, Wang Y, Gong B (2016) Preparation of tetracycline surface molecularly imprinted material for the selective recognition of tetracycline in milk. Food Anal Methods 9(8):2342–2351. https://doi.org/10.1007/s12161-016-0422-9

Önal A (2011) Overview on liquid chromatographic analysis of tetracycline residues in food matrices. Food Chem 127(1):197–203. https://doi.org/10.1016/j.foodchem.2011.01.002

Ostovan A, Arabi M, Wang Y, Li J, Li B, Wang X, Chen L (2022) Greenificated molecularly imprinted materials for advanced applications. Adv Mater 34(42):1–32. https://doi.org/10.1002/adma.202203154

Parlapiano M, Akyol C, Foglia A, Pisani M, Astolfi P, Eusebi AL, Fatone F (2021) Selective removal of contaminants of emerging concern (CECs) from urban water cycle via molecularly imprinted polymers (MIPs): potential of upscaling and enabling reclaimed water reuse. J Environ Chem Eng 9:105051. https://doi.org/10.1016/j.jece.2021.105051

Pauling L (1940) A Theory of the formation of antibodies 2643 [contribution from the gates a theory of the structure and process of formation of antibodies]. J Am Chem Soc 62(10):2643–2657. https://doi.org/10.1021/ja01867a018

Pereira-Maia EC, Silva PP, de Almeida WB, dos Santos HF, Marcial BL, Ruggiero R, Guerra W (2010) Tetraciclinas e glicilciclinas: uma visão geral. Quim Nova 33(3):700–706. https://doi.org/10.1590/S0100-40422010000300038

Pérez-Rodríguez M, Redigolo Pezza H, Pezza L (2016) Ultrasound-assisted dispersive liquid-liquid microextraction of tetracycline drugs from egg supplements before flow injection analysis coupled to a liquid waveguide capillary cell. Anal Bioanal Chem 408:6201–6211. https://doi.org/10.1007/s00216-016-9732-4

Pérez-Rodríguez M, Pellerano RG, Pezza L, Pezza HR (2018) An overview of the main foodstuff sample preparation technologies for tetracycline residue determination. Talanta 182:1–21. https://doi.org/10.1016/J.TALANTA.2018.01.058

Plano Nacional de Controle de Resíduos e Contaminantes (PNCRC), Instrução Normativa N° 5, DE 23 DE ABRIL DE 2019, 2019 21 (2019)

Piatkowska M, Jedziniak P, Zmudzki J (2016) Multiresidue method for the simultaneous determination of veterinary medicinal products, feed additives and illegal dyes in eggs using liquid chromatography-tandem mass spectrometry. Food Chem 197:571–580. https://doi.org/10.1016/j.foodchem.2015.10.076

Pizan-Aquino C, Wong A, Avilés-Félix L, Khan S, Picasso G, Sotomayor MDPT (2020) Evaluation of the performance of selective M-MIP to tetracycline using electrochemical and HPLC-UV method. Mater Chem Phys 245:122777. https://doi.org/10.1016/j.matchemphys.2020.122777

Poole CF (2002) Principles and practice of solid-phase extraction. In: Comprehensive analytical chemistry. Extraction technology fundamental, vol 37, pp 341–387. Elsevier. https://doi.org/10.1016/S0166-526X(02)80049-6

Pulicharla R, Hegde K, Brar SK, Surampalli RY (2017) Tetracyclines metal complexation: significance and fate of mutual existence in the environment. Environ Pollut 221(2016):1–14. https://doi.org/10.1016/j.envpol.2016.12.017

Qi P, Wang J, Wang L, Li Y, Jin J, Su F, Tian Y, Chen J (2010) Molecularly imprinted polymers synthesized via semi-covalent imprinting with sacrificial spacer for imprinting phenols. Polymer 51(23):5417–5423. https://doi.org/10.1016/j.polymer.2010.09.037

Qu Q, Zhu T (2017) Preparation of hybrid-monomer, double-template molecularly imprinted polymers for the purification of green tea extracts. Anal Methods 9(46):6525–6533. https://doi.org/10.1039/c7ay01842h

Rodriguez JA, Espinosa J, Aguilar-Arteaga K, Ibarra IS, Miranda JM (2010) Determination of tetracyclines in milk samples by magnetic solid phase extraction flow injection analysis. Microchim Acta 171(3):407–413. https://doi.org/10.1007/s00604-010-0428-8

Rodríguez MP, Pezza HR, Pezza L (2016) Simple and clean determination of tetracyclines by flow injection analysis. Spectrochimica Acta Part A Mol Biomol Spectros 153:386–392. https://doi.org/10.1016/j.saa.2015.08.048

Samanidou VF, Nikolaidou KI, Papadoyannis IN (2007) Development and validation of an HPLC confirmatory method for the determination of seven tetracycline antibiotics residues in milk according to the European Union Decision 2002/657/EC. J Sep Sci 30(15):2430–2439. https://doi.org/10.1002/jssc.200700057

Sánchez-Polo M, Velo-Gala I, López-Peñalver JJ, Rivera-Utrilla J (2015) Molecular imprinted polymer to remove tetracycline from aqueous solutions. Microporous Mesoporous Mater 203:32–40. https://doi.org/10.1016/j.micromeso.2014.10.022

Sanli S, Sanli N, Alsancak G (2009) Determination of protonation constants of some tetracycline antibiotics by potentiometry and LC methods in water and acetonitrile-water binary mixtures. J Braz Chem Soc 20(5):939–946. https://doi.org/10.1590/S0103-50532009000500020

Santos SM, Henriques M, Duarte AC, Esteves VI (2007) Development and application of a capillary electrophoresis based method for the simultaneous screening of six antibiotics in spiked milk samples. Talanta 71(2):731–737. https://doi.org/10.1016/j.talanta.2006.05.049

Santos MG, de Abrão LC, de Freitas LAS, de Moraes GOI, de Lima MM, Figueiredo EC (2012) Emprego de polímeros de impressão molecular em preparo de amostras para análise de compostos orgânicos: aplicações e tendências. Sci Chromatographica 4(3):161–195. https://doi.org/10.4322/sc.2012.012

Sodhi M, Etminan M (2020) Therapeutic potential for tetracyclines in the treatment of COVID-19. Pharmacother J Hum Pharmacol Drug Ther. https://doi.org/10.1525/jps.2004.33.4.221

Suedee R, Srichana T, Chuchome T, Kongmark U (2004) Use of molecularly imprinted polymers from a mixture of tetracycline and its degradation products to produce affinity membranes for the removal of tetracycline from water. J Chromatogr B 811(2):191–200. https://doi.org/10.1016/j.jchromb.2004.08.044

Sun Y, Tian J, Wang L, Yan H, Qiao F, Qiao X (2015) One pot synthesis of magnetic graphene/carbon nanotube composites as magnetic dispersive solid-phase extraction adsorbent for rapid determination of oxytetracycline in sewage water. J Chromatogr A 1422:53–59. https://doi.org/10.1016/j.chroma.2015.10.035

Tang W, Li G, Row KH, Zhu T (2016) Preparation of hybrid molecularly imprinted polymer with double-templates for rapid simultaneous purification of theophylline and chlorogenic acid in green tea. Talanta 152:1–8. https://doi.org/10.1016/j.talanta.2016.01.046

Tarley CRT, Sotomayor MDPT, Kubota LT (2005) Polímeros biomiméticos em química analítica. Parte 1: Preparo e Aplicações de MIP ("Molecularly Imprinted Polymers") em técnicas de extração e separação. Quimica Nova 28(6):1076–1086. https://doi.org/10.1590/S0100-40422005000600024

TR CU 021/2011-On Food Safety, Pub. L. No. TR CU 021/2011, 212 (2011). http://www.eurexcert.com/TRCUpdf/TRCU-0021-On-food-safety.pdf

Turiel E, Esteban AM (2019) Molecularly imprinted polymers. In: Solid-phase extraction, pp 215–233. Elsevier Inc. https://doi.org/10.1016/B978-0-12-816906-3.00008-X

Turiel E, Martín-Esteban A (2010) Molecularly imprinted polymers for sample preparation: a review. Anal Chim Acta 668(2):87–99. https://doi.org/10.1016/j.aca.2010.04.019

Turiel E, Martín-Esteban A (2019) Molecularly imprinted polymers-based microextraction techniques. TrAC Trends Anal Chem 118:574–586. https://doi.org/10.1016/j.trac.2019.06.016

Vargas Mamani MC, Reyes Reyes FG, Rath S (2009) Multiresidue determination of tetracyclines, sulphonamides and chloramphenicol in bovine milk using HPLC-DAD. Food Chem 117(3):545–552. https://doi.org/10.1016/j.foodchem.2009.04.032

Vasapollo G, Sole RD, Mergola L, Lazzoi MR, Scardino A, Scorrano S, Mele G (2011) Molecularly imprinted polymers: present and future prospective. Int J Mol Sci 12(9):5908–5945. https://doi.org/10.3390/ijms12095908

Wan Ibrahim WA, Abd Ali LI, Sulaiman A, Sanagi MM, Aboul-Enein HY (2014) Application of solid-phase extraction for trace elements in environmental and biological samples: a review. Crit Rev Anal Chem 44(3):233–254. https://doi.org/10.1080/10408347.2013.855607

Wang X, Wang L, He X, Zhang Y, Chen L (2009) A molecularly imprinted polymer-coated nanocomposite of magnetic nanoparticles for estrone recognition. Talanta 78(2):327–332. https://doi.org/10.1016/j.talanta.2008.11.024

Wang Q, Lv Z, Tang Q, Gong C-B, Lam MHW, Ma X-B, Chow C-F (2016) Photoresponsive molecularly imprinted hydrogel casting membrane for the determination of trace tetracycline in milk. J Mol Recognit 29(3):123–130. https://doi.org/10.1002/jmr.2461

Wang S, Zhang J, Li C, Chen L (2018) Analysis of tetracyclines from milk powder by molecularly imprinted solid-phase dispersion based on a metal–organic framework followed by ultra high performance liquid chromatography with tandem mass spectrometry. J Sep Sci 41(12):2604–2612. https://doi.org/10.1002/jssc.201701514

Wei D, Wu S, Zhu Y (2017) Magnetic solid phase extraction based on graphene oxide/nanoscale zero-valent iron for the determination of tetracyclines in water and milk by using HPLC-MS/MS. RSC Adv 7(70):44578–44586. https://doi.org/10.1039/c7ra08203g

Wen Y (2019) Recent advances in solid-phase extraction techniques with nanomaterials. In: Handbook of nanomaterials in analytical chemistry: modern trends in analysis, pp 57–73. Elsevier. https://doi.org/10.1016/B978-0-12-816699-4.00004-9

Whitcombe MJ, Rodriguez ME, Villar P, Vulfson EN (1995a) A new method for the introduction of recognition site functionality into polymers prepared by molecular imprinting: synthesis and characterization of polymeric receptors for cholesterol. J Am Chem Soc 117:7105–7111. https://pubs.acs.org/sharingguidelines

Whitcombe MJ, Rodriguez ME, Villar P, Vulfson EN (1995b) A new method for the introduction of recognition site functionality into polymers prepared by molecular imprinting: synthesis and characterization of polymeric receptors for cholesterol. J Am Chem Soc 117(27):7105–7111. https://doi.org/10.1021/ja00132a010

Wu XJ, Wang GN, Yang K, Liu HZ, Wang JP (2017) Determination of tetracyclines in milk by graphene-based solid-phase extraction and high-performance liquid chromatography. Anal Lett 50(4):641–650. https://doi.org/10.1080/00032719.2016.1194853

Wulff G (2013) Fourty years of molecular imprinting in synthetic polymers: origin, features and perspectives. Microchim Acta 180(15–16):1359–1370. https://doi.org/10.1007/s00604-013-0992-9

Wuff G, Sarhan A (1972) Use of Polymers with enzyme-analogous structures for the resolution of racemates. Angew Chem Int Ed 11:341–345

Xing W, Wu Y, Lu J, Lin X, Yu C, Dong Z, Yan Y, Li C (2020) Biomass-based synthesis of green and biodegradable molecularly imprinted membranes for selective recognition and separation of tetracycline. NANO 15(1):1–18. https://doi.org/10.1142/S1793292020500046

Xu JJ, An M, Yang R, Tan Z, Hao J, Cao J, Peng LQ, Cao W (2016) Determination of tetracycline antibiotic residues in honey and milk by miniaturized solid phase extraction using chitosan-modified graphitized multiwalled carbon nanotubes. J Agric Food Chem 64(12):2647–2654. https://doi.org/10.1021/acs.jafc.6b00748

Xu Y, Li J, Jiang L, Li Z, Li Y, Ding L (2018) Simultaneous determination of sulfonamides and fluoroquinolones from environmental water based on magnetic double-template molecularly imprinting technique. Environ Sci Pollut Res 25(16):16121–16134. https://doi.org/10.1007/s11356-018-1581-6

Xue J, Liu J, Wang C, Tian Y, Zhou N (2016) Simultaneous electrochemical detection of multiple antibiotic residues in milk based on aptamers and quantum dots. Anal Methods 8(9):1981–1988. https://doi.org/10.1039/c5ay03136b

Yang XQ, Yang CX, Yan XP (2013) Zeolite imidazolate framework-8 as sorbent for on-line solid-phase extraction coupled with high-performance liquid chromatography for the determination of tetracyclines in water and milk samples. J Chromatogr A 1304:28–33. https://doi.org/10.1016/j.chroma.2013.06.064

Yu H, Mu H, Hu YM (2012) Determination of fluoroquinolones, sulfonamides, and tetracyclines multiresidues simultaneously in porcine tissue by MSPD and HPLC-DAD. J Pharm Anal 2(1):76–81. https://doi.org/10.1016/j.jpha.2011.09.007

Zhan J, Fang G, Yan Z, Pan M, Liu C, Wang S (2013) Preparation of a semicovalent, molecularly surface imprinted polymer for the rapid determination of trace acid orange II in food and environmental samples. Anal Bioanal Chem 405(19):6353–6363. https://doi.org/10.1007/s00216-013-7036-5

Zhang L, Wang J, Fang G, Deng J, Wang S (2020) A molecularly imprinted polymer capped nitrogen-doped graphene quantum dots system for sensitive determination of tetracycline in animal-derived food. Chem Select 5(2):839–846. https://doi.org/10.1002/slct.201903868

Chapter 14
Limitations and Challenges in the Practical Implementation of MIPs

Akriti Srivastava and Meenakshi Singh

Molecular imprinting (MIP) technology is quite young in comparison to other newer technologies but still it has shown tremendous potential. This is one of the most relevant technologies being capable of creating artificial molecular receptors which are closest to biological receptors till date. It has magnificent capability to replace many materials in industry. Despite of many advantages and prospects, this technology has not been able to reach the market till now. Here in this chapter, an attempt will be made to identify the lacuna and recognize the hurdles, barriers and challenges preventing their practical applications.

14.1 Brief Introduction of Molecularly Imprinted Polymers

Over the past ten years, the development of commercial biosensor technology has accelerated, A valuation of biosensor industry at 18.6 billion USD in 2018 says a lot about the growth of this sector (Ugalmugale and Swain 2018). Although, contemporary research and development are floating numerous new products, still glucose biosensors have occupied major share of biosensor market worldwide (Sadana et al. 2018). Following accounts for their limited commercial uses:

1. Enzyme and antibody stability issues in real-world applications
2. Inadequate enzymes for respective substrates and analytes
3. Inadequacy of a suitable and universal method for transducing binding events into a perceptible electrical signal.

Evolution has provided biology with many fascinating examples of molecular recognition, including those involved in interactions between a ligand and a receptor

A. Srivastava · M. Singh (✉)
Department of Chemistry, MMV Banaras Hindu University, Varanasi 221005, India
e-mail: meenakshi@bhu.ac.in

S. Patra and M. Sillanpaa (eds.), *Molecularly Imprinted Polymers as Artificial Antibodies for the Environmental Health*, https://doi.org/10.1007/978-3-031-58995-9_14

(such as substrate and enzyme, antigen and antibody), and in transport and signal transduction processes. Studies of these molecules have been dependent on our ability to selectively bind these molecules from biological mixtures. Base-pair complementarity provides a robust and powerful tool for selectively isolating and purifying DNA and RNA molecules with desired sequences. This tool will remain instrumental in virtually all aspects of molecular biology research. Antibodies have been widely used for selective protein capture and thus are applied for industrial protein purification, basic biomedical research, and clinical diagnostics. However, antibodies exhibit characteristics that limit their applications. These proteins are large complex molecules that need to be stored carefully. As antibodies are produced by living cells, it is sometimes difficult to control their quality. An alternative to biological sensors comprising of nonbiological components are demand of the day. An ideal molecular recognition agent should have high specificity and be composed of a stable, robust, nonbiological material. The development of synthetic receptors with comparable specificity and sensitivity to the natural antibody-antigen behavior is warranted in modern sensor research. Such molecular recognition holds the possibility of selective as well as sensitive sensors capable of detecting and monitoring targets in a non-invasive approach when combined with new techniques to monitor changes in the recognition assets. Synthetic receptors based on molecular imprinted polymers (MIPs) has shown the capability to serve the purpose. Many instances of MIPs for a variety of biological applications have been reported in the past few decades. MIPs are best described as synthetic analogs to the natural, biological antibody-antigen systems (BelBruno 2018). The idea behind MIPs is to create synthetic, custom receptors for molecular recognition. In most successful situations, they combine selectivity for the target analyte that mimics that of the natural receptors with high chemical and mechanical strength characteristic of such cross-linked compounds. The work of the K. Mosbach led research team almost 25 years ago showed the enormous potential of analytical techniques concerning these materials (Abass and Rzaij 2020).

According to a search of Chemical Abstracts, K. Mosbach and B. Sellergren from Lund published the first work that expressly claimed an 'imprinted polymer' in 1984. Although G. Wulff had been publishing papers in a series entitled 'Enzyme Analog Built Polymers' since 1973, the term 'imprinted polymer' first appeared in one of Wulff's articles in 1985. Multiple aspects of science were produced by these pioneering scholars. While Wulff tended to establish the imprint through covalent binding, the Mosbach group focused on using noncovalent interactions between the host and target (BelBruno 2018). The chemistry required to 'take off' the templating molecules from the MIP is where the two approaches most clearly diverge. The covalent synthesis is significant and is anticipated, Wulff (Liu and Wulff 2004; Wulff 2013; Wulff et al. 2006; Wulff and Knorr 2001; Wulff and Liu 2012) extensively used imprinted polymers in catalytic reactions, as evinced from the title of his early series of papers, while Mosbach (Mosbach and Haupt 1998; Ye and Mosbach 2008; Yilmaz et al. 2005; Zhang et al. 2006) and Sellergren focused on separation and sensing materials. K. Shea working with similar approach, came up with newer biological applications like plastic antibody inserted into living mice (Chen et al. 2016; Hart and Shea 2014; Shea 1994; Hoshino and Shea 2011). The phrase 'molecularly imprinted polymer sensor' originally appears to have been used by S. Piletsky in 1992, while articles by this author explaining the creation of MIPs in general may be found in Russian literature as early as 1989. These groups' research efforts, which range from the late 1990s to more recent years, have sparked work on several applications of this technology.

14.2 MIPs in Sensors

Epidemics such as COVID-19 pose a serious threat to human life, hence increased awareness is imperative. Maintaining a standard fitness level and assessing one's physical strength are turning out to require routine and early-phase monitoring. Many cutting-edge technologies have been applied in practical situations, although they have limited selectivity in intricate matrix systems. Artificial antibodies, such as molecularly imprinted polymers (MIPs), have been investigated as a means to enhance sensor selectivity (Das et al. 2021). In 1990s, the concept of integrating MIPs into conventional biosensor readout technologies serving as transducers, transforming the binding events of molecule(s) to a measurable signal utilizing MIPs as selective recognition elements in the sensing devices and assays first gained attention. Over the last 20 years, improvements in MIP synthesis techniques, microelectronics, and material sciences have stimulated this adaptation, leading to the emergence of several novel sensor and assay platforms. Recently, there has been a lot of interest in research on smart MIPs, also known as stimuli-responsive MIPs. A smart, or stimuli-responsive, MIP has been intelligently developed to sense a particular environmental input (changes in temperature (Qin et al. 2011), incident light (Minoura et al. 2004), pH strength (Zhao et al. 2009), ionic strength (Wu et al. 2011) or even the presence of a specific molecule (Shiraki et al. 2015; Nash et al. 2013)) by changing its affinity for the target molecule and offering the option of capture or release. The general approach to build smart MIP is to add a stimuli-responsive material into the molecular imprinting elements. This stimuli-responsive material provides the MIP a switchable binding or releasing capacity (Morsi et al. 2023). Chemical sensors, purification, analytical separations/extraction, and their determination, as well as enzyme-like catalysis, are all being aided by MIPs (Lowdon et al. 2020).

MIP-based sensors are synthesized incorporating recognition components and target systems. Recognition components use a polymeric matrix to simulate the receptor site. In the presence of functional monomer, cross-linker, and target molecule(s), polymeric matrices are created. MIPs benefit from the stability, sensitivity, selectivity, and resilience of the synthetic polymeric matrix. As a result, MIPs are widely used in sensing methods such as optical, electrochemical, and photoelectrochemical techniques (Fig. 14.1).

Chromatographic systems, lab-on-a-chip systems, and sensor systems are common terms for many of the MIP-integrated systems. Furthermore, point-of-care (POC) sensors are fabricated with MIP-integrated devices. As a result, it improves a target (bio)molecule's selectivity, provides a quick evaluation, low-cost analysis,

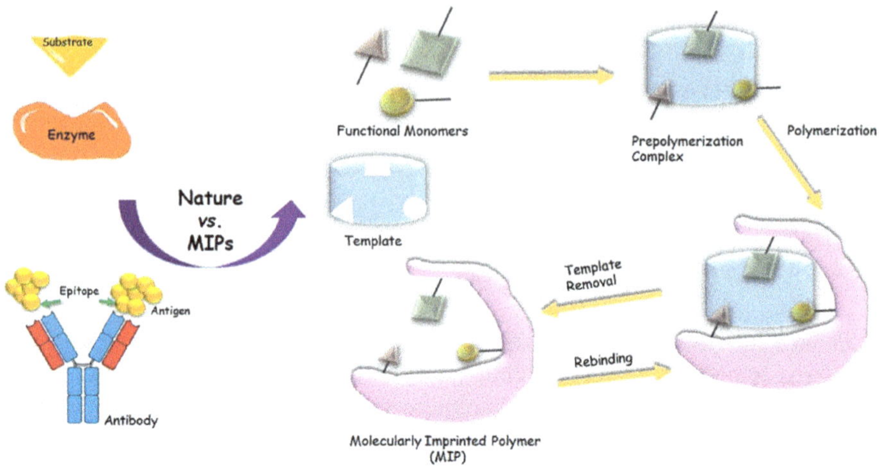

Fig. 14.1 Schematic presentation of molecularly imprinted polymers (MIPs) describing the analogy with antibodies and enzymes with the summarized workflow of MIP-synthesis. *Courtesy* Yarman et al. (2021)

and a biosensing application alternative to conventional approaches (Saylan et al. 2020; Ozcelikay et al. 2022).

14.2.1 Chemiresistors and Capacitance Sensors

Although such sensors also need to be produced with electrodes and/or leads on a dielectric substrate, these electrical procedures constitute the most fundamental measuring techniques for sensor implementation using MIP identification polymers. Conductometric methods are the collective name for such procedures. The requirement for a conducting component is often satisfied by employing a conductive receptive layer over the array of electrodes or by creating a composite MIP with conductive components such as carbon nanotubes, metallic nanoparticles, or conductive copolymers. The majority of conductometric applications use samples in the gaseous form.

Impedance sensors encompass capacitive sensors in the same category. Capacitive sensors are being popular due to their superior resilience, simplicity, and sensitivity. Capacitive sensors with good sensitivity, selectivity, and low sample quantities were used to detect various analytes, particularly in biological applications (El-Akaad et al. 2020). Conceptually, capacitive sensors are relatively simple. A specific type uses a conductive layer comprising of a lattice-like top electrode for sampling, a dielectric MIP, and a solid conducting layer on an insulating substrate. A representation of such a sensor and its corresponding circuit are shown in Fig. 14.2. The two electrodes are linked to a suitable electronic monitor, which might be a simple meter or one of the

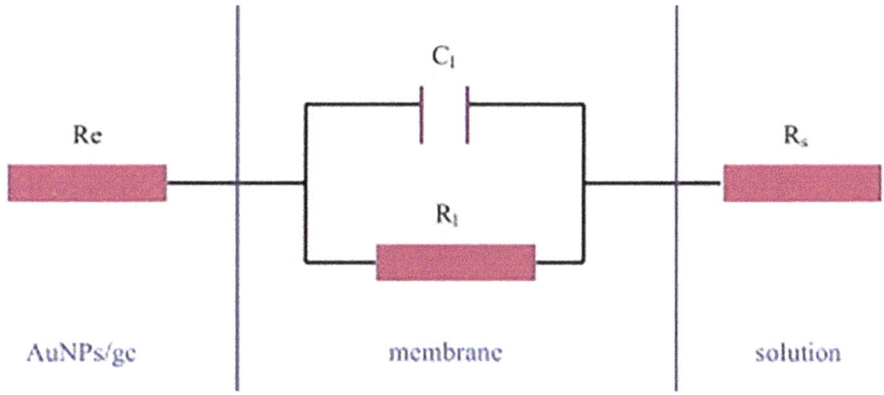

Fig. 14.2 An equivalent circuit of a capacitance sensor designed to detect methyl parathion pesticide. *Courtesy* Li et al. (2012)

multiple specialized capacitance/impedance measurement chips offered by major manufacturers. The capacitance and dissipation factor of the sensor are changed by the adsorption of the template onto the MIP.

Relative to a capacitive sensor, a chemiresistor requires simpler steps to electrode preparation. Only one electrode layer is required, and this can be either a simple interdigitated electrode (IDE) conception with almost 160 pairs of electrodes having MIP coverage throughout the IDE, or it can be a prepared in a more complex electrode pair with the polymer packing the interstitial space between the electrodes. A conductive film integrating with MIP is a crucial part of the chemiresistor. This can be achieved by integrating a conductive polymer, such as polyaniline, polypyrrole, or poly[3,4-(ethylenedioxy)thiophene], (PEDOT), as the imprinted layer, by using a composite material consisting of the MIP and another conductor moiety such as carbon nanotubes (CNTs) or gold nanoparticles, or devising the two layers one by one on the sensor, a conductive and another adlayer-MIP layer coated over that film. It is simple to implement the electronics for monitoring resistance changes brought on by the binding of the target molecule(s). To find the variable resistance, one can use an easy-to-use ohmmeter, a more adaptable Wheatstone bridge, or a constant current applied with a concurrent voltage measurement (BelBruno 2018).

14.2.2 Optical Sensors

Many different MIP-based sensing devices and tests have made substantial use of optical sensing principles as transducers. The detection systems in this group are highly diverse, making it challenging to distinguish between sensors and assays. A lateral flow assay coupled to an SPR analyzer would be an example functioning at the intersection of both. Detectors known as optical sensors monitor a material's optical characteristics and transform these changes—such as light in the form of photons into an electrical signal i.e. electrons. During a binding event, MIP sensors provide a discernible optical change that is then transduced. Optical MIP sensors come in two different varieties. The first, MIP-affinity sensors, may be used to find analytes with built-in optical characteristics, such as fluorescence, optical absorbance, or refractive index. The binding event transduces these optical changes in the analyte, such as fluorescence quenching. Monitoring changes in refractive index brought on by the binding of proteins or other high molecular weight analytes to MIPs mounted on the surface of a surface plasmon sensor is a typical example of this kind of optical MIP sensing.

Optoelectronic MIP sensors make up the second group in this category. Reporter monomers, which have optical characteristics and can sense changes in their environment and respond to the presence of the analyte, are the basis of their functioning mechanism (Ahmad et al. 2019; Altintas et al. 2015). Their photochemical and thermal stability with higher absorption coefficients/ higher quantum yields are basic requisites for any optoelectronic MIP devices. Additionally, molecular events must generate an optical response proportionate to the concentration of the target analyte concentration without entering into any kind of non-specific reactions (Fig. 14.3).

14.2.3 Electrochemical Sensors

Electrochemical sensors are able to transform the outcome of an electrochemical interaction into an operable signal. They offer variations in current, potential, conductivity, and impedance on molecular interactions occurring on interfaces of electrode. Based on the parameter chosen for monitoring these electrochemical events, potentiometric, voltammetric, and conductive/capacitive sensors are generally the most common electrochemical sensors. The changes in the electrochemical property conveyed by molecular level events are used to calculate the concentrations of analyte molecules (Fig. 14.4).

Molecularly imprinted electrochemical sensors combine the high sensitivity of electrochemical detection with the unique identification capability of molecularly imprinted technology. They are easy to mechanically automate and cost-effective to miniaturize while in use. MIP is utilized as a recognition factor for these electrochemical sensors. They can be hyphenated with general analyte characteristics, or modifications to one or more physicochemical system parameters for detection following analyte binding events are taken into account. The design and production of these electrochemical sensors heavily rely on the structure and characteristics of MIP. The use of molecularly imprinted electrochemical sensors in chemical and biological experiments has been the subject of several papers (Yang et al. 2018; Chen et al. 2011; Janata 2010).

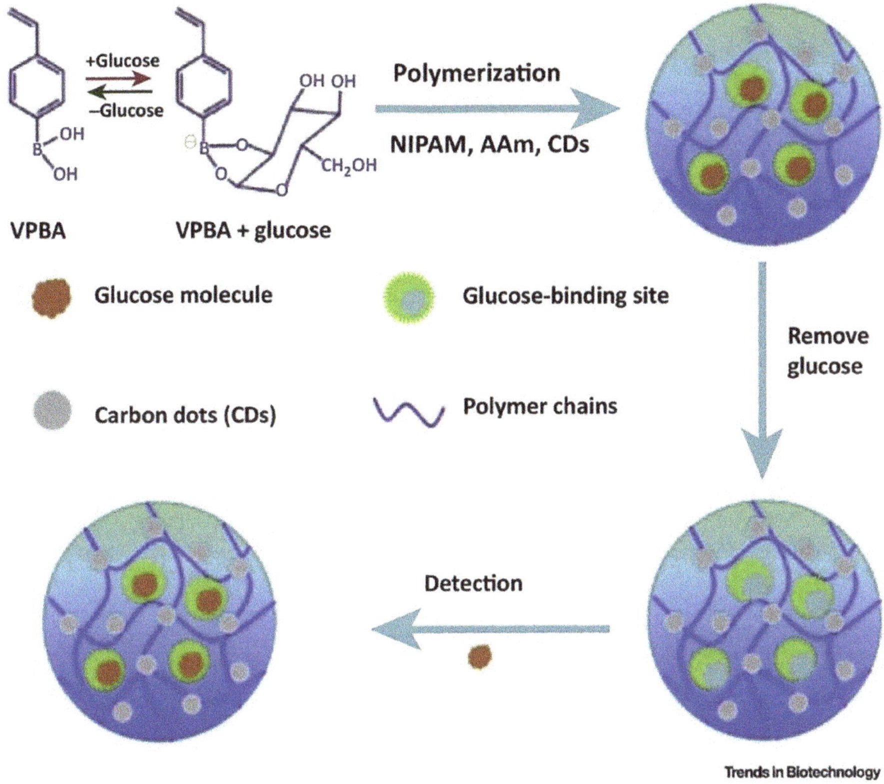

Fig. 14.3 Preparation of molecularly imprinted polymer (MIP) hybrid microgels employed in glucose determination. *Courtesy* Wang et al. (2015)

14.2.4 Quartz Crystal Microbalance (QCM) Sensors

Quartz crystal microbalance (QCM) sensors depend on the piezoelectric behaviour of quartz crystals. Any mass loading on quartz crystal induces a change in its resonating frequency, which is detected by a device that generates a bulk thickness-shear-mode acoustic wave (Akgönüllü et al. 2022; Montagut et al. 2011; Özgür et al. 2017; Marx 2003). A mechanical oscillation with the characteristic frequency, f, is generated in the quartz crystal through an alternating electric field across the quartz crystal through upper and lower metal electrodes covering the quartz surface. A conducting layer is deposited on quartz surface which is connected to oscillator. Although the commercial QCM devices used in most applications operate at frequencies between 5 and 10 MHz, the f range for the fundamental mode of quartz is between 5 and 20 MHz (Fig. 14.5). Mass of the crystal increases after adsorption of template molecules on MIP films resulting decrease in crystal's oscillating frequency. For the situation of pure elastic mass added to the surface, the well-known linear Sauerbray equation was first observed (Sauerbray 1959) and used to precisely quantify, with ng sensitivity,

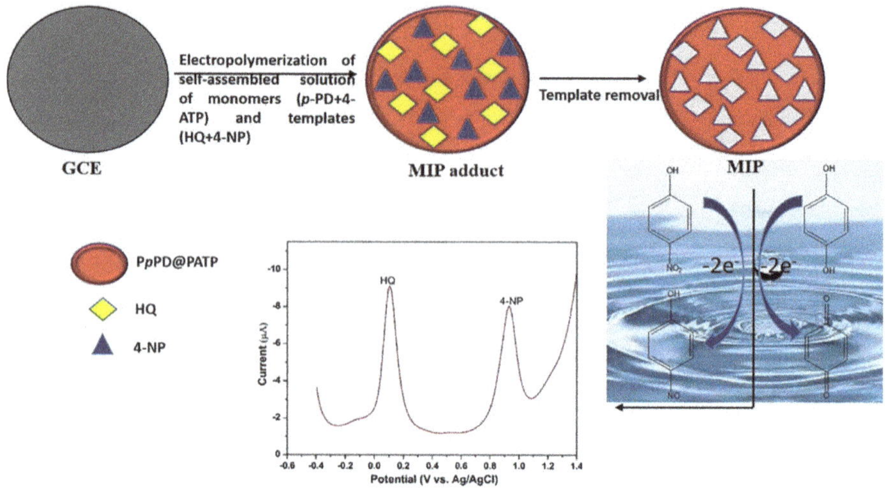

Fig. 14.4 Schematic representation for stepwise fabrication of the MIP sensor for the detection of HQ and 4-NP. *Courtesy* Singh and Singh (2022)

the quantity of elastic mass added to the surface

$$\Delta f = -2\Delta m f^2 / A(\mu \rho_q)^{0.5} = -C_f \Delta m \tag{14.1}$$

where Δf is the measured resonant frequency decrease (Hz), f is the intrinsic crystal frequency, Δm is the elastic mass change (g), A is the electrode area (0.196 cm^2 in many applications), ρ_q is the density of quartz (2.65 g/cm^3), and μ is the shear modulus (2.95 × 1011 dyn/cm^2).

The viscoelastic behaviour is also simultaneously monitored through the resistance measurement which enables real-time monitoring of the viscoelasticity of

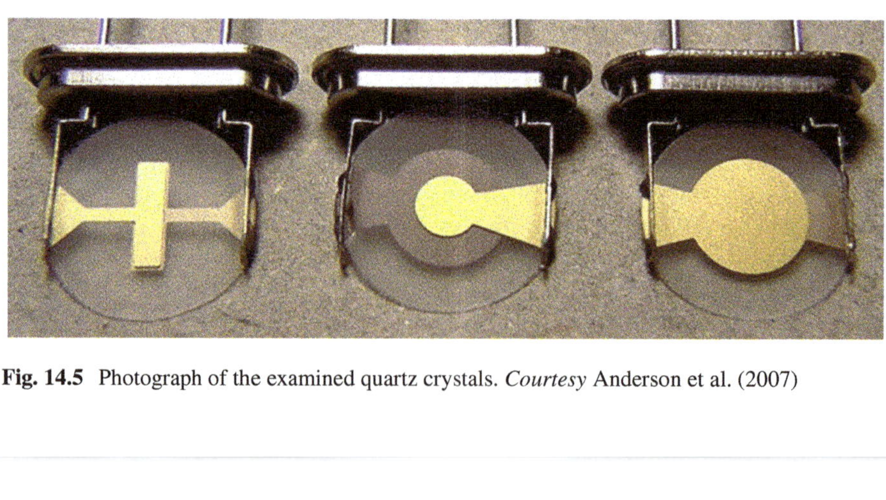

Fig. 14.5 Photograph of the examined quartz crystals. *Courtesy* Anderson et al. (2007)

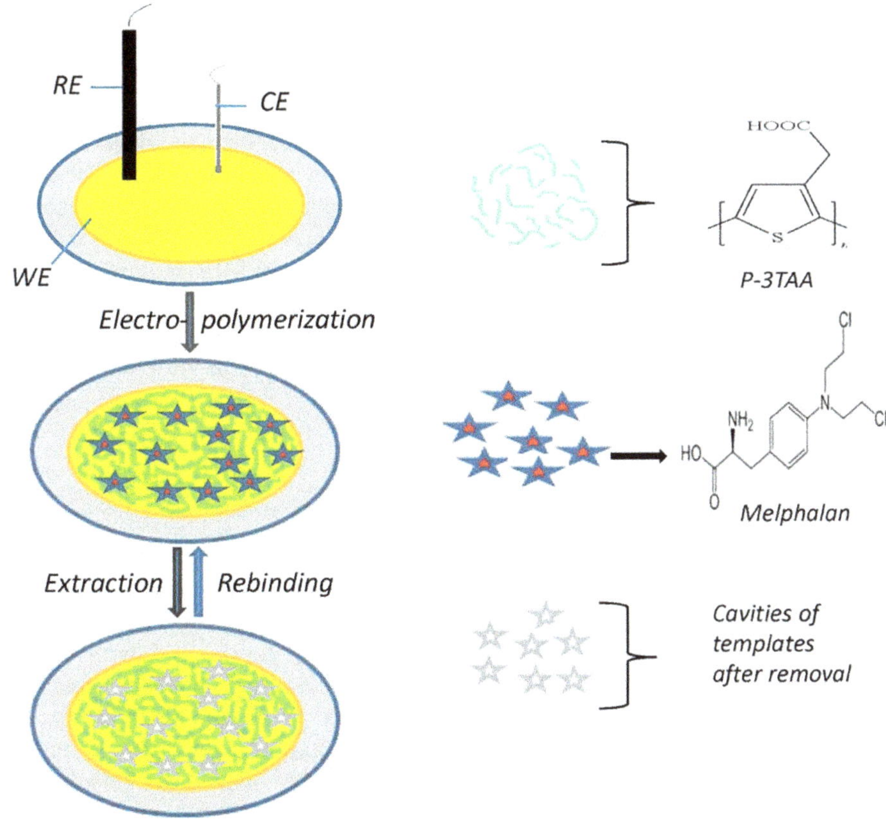

Fig. 14.6 Schematic representation for the preparation of electrochemical MIP sensor on EQCM. *Courtesy* Singh and Singh (2015)

molecular events occurring on the surface. Such measurements and real-time monitoring has enabled the detection of various (bio)molecules including proteins, nucleic acids, peptides, oligonucleotides, hormones, etc. (Fig. 14.6).

14.3 Challenges in MIP Commercialization

Three particular characteristics of MIP sensors that promises their bright prospects are: polymers are exceedingly stable; compatible with microfabrication technology; and are much less expensive than any biological recognition agents such as enzymes, antibodies, etc., making them commercially attractive. Though there are several examples of successful MIP-based electrochemical sensor research, the extent of commercial activity related to their development and marketing is not very pleasant.

Commercial MIP sensing devices might not be promoted due to integration issues with these artificial recognition components and transducers. The scope of marketing and commercial activity associated to MIP sensor technology is still limited. The situations where a sizable market definitely demands MIP technology, i.e. where traditional procedures and standard assays are constrained, including field applications, PoC testing, real-time analyte determination, and wearable sensors. These are the scenarios where MIP sensors are most likely to be used in the field. In situations when biological receptors are absent or insufficiently stable, MIP sensors may have a specialized use (Ahmad et al. 2019). Commercialization of MIPs require a holistic and synergistic approach for acceptance with the probable vendors in the various field where all the risk-assessment should be evaluated and must be addressed appropriately with a focus on transferring technology from laboratory (in-vitro to in-vivo) to the industries and clinic.

14.3.1 Nonspecific Adsorption

In 1995, Wulff reported that only 15% of cavities generated during imprinting will be reoccupied after binding of template molecules to a polymer based on non-covalent interactions (Wulff 1995). The success of the imprint is determined by several structural factors, including a good balance between the polymer network's rigidity, which preserves the contours of the imprinted cavity, and flexibility, which makes substrate access and rebinding easier. According to studies carried out by Andersson and Nicholls (1997), there are not many stable interactions between the monomer and template in the pre-polymerization solution. The template to functional monomer ratios employed for non-covalent imprinting in previous investigations equated to 0.3–0.6% of complex formation. It is important because according to literature figures, only 0.5–1% of the predicted yields of binding sites produced by an imprinting technique fall into the category of 'medium to high affinity' binding sites. It follows that the quantity of highly selective recognition sites present in the relevant MIPs corresponds with the fraction of complex formation prior to polymerization (Mahony et al. 2005). The most straightforward approach to the problem of the charged polymer-derived non-specific binding of MIPs could be to use surfactants in conjunction with artificial antibodies, or MIPs, in applications like immunoassay or separation where complex biological and culture matrices (Chou et al. 2020). In the synthesized MIP systems, Rusling et al. demonstrated how these detergents may successfully avoid non-specific binding (Weerasuriya et al. 2021; Bhakta et al. 2017). As shown by Qiang and colleagues biomimetic immunoassay, adding surfactants during the synthesis process can also increase the general specificity of the MIPs (Guoning et al. 2020). The MIP's overall capabilities were considerably enhanced by the decrease in non-specific interactions, leading to higher performing tests.

14.3.2 Macromolecule Imprinting

Targeting macromolecules has always proven to be challenging, even though imprinting small molecules is comparatively easy. Most instances of preparing MIPs for small molecules include non-covalent interactions between a single type of functional monomer, or a limited number of types of functional monomers, and a template. Hydrogen bonding has been found to be comparatively favorable energy of contact in non-polar solvents, hence often used to associate the template with monomers. Small molecules mostly contain a small number of the heteroatoms such as oxygen and nitrogen, which are typically the focus of molecular imprinting. As a result, at least in non-polar organic solvents, very few and powerful interactions account for the majority of the binding free energy of small molecules to their imprints. Numerous spectroscopic studies via NMR (Sellergren et al. 1988; Whitcombe et al. 1998), FT-IR (Dhal and Arnold 1996), ESR (EPR) (Osmani et al. 2008), and UV (Svenson et al. 1998) spectroscopy as well as approaches to the in-silico modeling (Subrahmanyam et al. 2013) of the pre-polymerization mixture to aid in monomer selection (Subrahmanyam et al. 2001; Chianella et al. 2002; Wyszomirski and Prus 2012; Li et al. 2014) have been conducted in order to better understand the imprinting process due to the significance of these relatively few but significant interactions that determine the recognition properties of the imprinted polymer. The same principles that apply to small molecules are also applicable to the synthesis of polymers imprinted with biological species, although the situation is significantly altered by the macromolecular template's needs for size, structure, functionality, and solubility. In order to maintain the binding sites produced during polymerization, traditional imprinted polymers frequently have very dense networks (small pore sizes for diffusion of the template into and out of the matrix). Large templates, such as proteins, can stay stranded in the network after polymerization and find it difficult to diffuse back into the network to find binding sites afterward. There are both forward and backward network diffusion constraints, which result in poor recognition characteristics. Additionally, delayed template leakage is occasionally seen. This can have a considerable influence on identification outcomes, particularly when testing for trace substances. Macromolecular imprinting find uses in production of sensors, media for affinity chromatography, detoxifying agents, or biomaterials with molecularly tailored surfaces, etc. (Peppas and Clegg 2016).

14.3.3 Portable Molecularly Imprinted Sensors with Multiplexing Capabilities and High Integration

Despite significant developments in imprinting technology and micro and nano electronics, reports on commercially viable MIP-based sensor systems have mostly been published in research publications only. In addition to the commercial bottlenecks, the process of valuing lab-based technology is complicated, and to make MIP sensors a true competitor for commercial biosensors, molecular diagnostics, and conventional lab analysis and other such technical obstacles must be overcome. From the perspective of the device, technological challenges might be found in the development of disposable, functionalized electrodes or in the design of handheld or user-friendly portable systems. The former is demonstrated by the fact that portable glucose meters for use on the go are often built on a basic conductometric reading platform that makes miniaturizing and calibration easy. Physiological glucose concentrations, on the other hand, are often in the millimolar range, and developing sensors for other substances may demand for higher sensitivity, which would need the use of different readout principles that may have lower detection limits.

A significant portion of the R&D budget will go into developing repeatable functionalized electrodes in addition to downscaling device manufacture and miniaturizing readout design. Research focuses on an approach of microfabrication like electropolymerization, surface grafting, and/or imprint lithography, for particle immobilization onto electrodes. The repeatable production and scale-up of MIP particles or layers is another bottleneck. It is difficult to eliminate inter-batch variability due to the difficult challenge of achieving uniformity in both morphological traits and binding affinity, which is obviously unfavorable when implementing commercial sensor applications (Fig. 14.7).

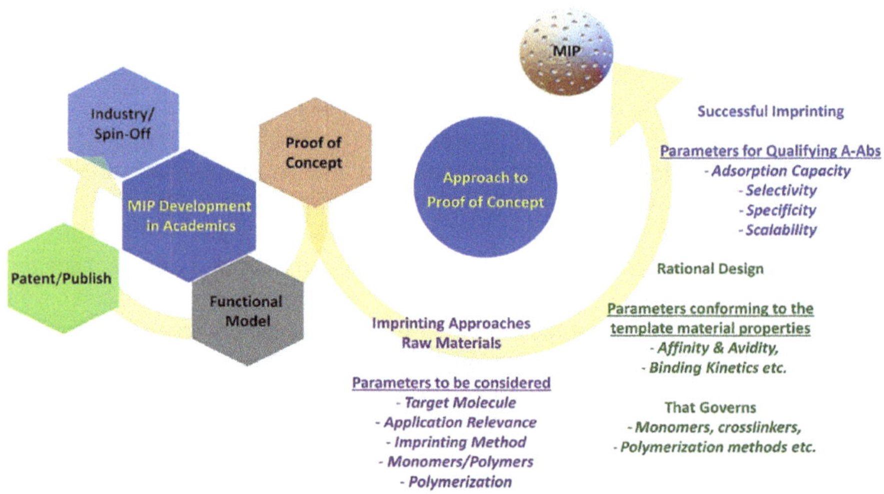

Fig. 14.7 The overview of the critical factors for the successful development of MIPs and a workflow of such development in a typical research setting. *Courtesy* Dixit et al. (2022)

14.4 Sensitivity and Selectivity

Sensitivity and selectivity are the two critical parameters for any sensor. The chemical functionalities of binding sites are significant for selectivity of MIPs as their structural contours induce the development of imprinted binding sites. Despite the success in the development and application of MIPs, relatively little has been discovered about the chemical structure of the specific binding sites located in each MIP. In this regard, computer modeling of MIPs has a lot of promise (Suryana et al. 2021), but there does not seem to be any direct experimental data to back up the predicted binding site configurations. Many studies have come to the opinion that most MIPs, and notably the non-covalently imprinted ones (which constitute the bulk of MIPs), contain heterogeneous binding sites as a result of adsorption tests using MIPs. These sites differ in the extent to which they bind the template. Additionally, the stronger sites are considered to be more selective about the template than the weaker ones. One may approach selectivity phenomenologically if there is not a satisfactory mechanistic foundation for MIP selectivity to be explained.

Sensitivity issues can indeed be a significant challenge in the commercialization of MIPs. The sensitivity of MIP-based products, however, can be impacted by a number of complications, which could prevent their effective commercialization. The sensitivity of such product depends on the binding affinity between the polymer and the target molecule. If the binding affinity is weak, the MIP may not effectively capture and release the target molecule, leading to low sensitivity. Incomplete removal of analyte and complex structure of template molecules can result in reduced sensitivity also.

To address these sensitivity issues in the commercialization of MIPs, extensive research and development efforts are often required. This includes optimizing the MIP synthesis process, improving the choice of template molecules, enhancing binding affinity and validating the sensitivity of MIP-based products through rigorous testing. Collaboration between scientists, engineers, and industry experts is crucial to overcome these challenges and bring sensitive MIP-based products to market successfully.

However, the incorporation of nanomaterials in MIP matrix offers an interesting approach to add some novel properties and can overcome the problems related to sensitivity and selectivity (Roy et al. 2017). The remarkable qualities of the nanostructured modified molecularly imprinting polymer, including its high surface to volume ratio, accessible cost, and ease of production and handling, varied from bulk MIP have garnered significant interest in the research field recently. The majority of reported studies use carbonaceous nanomaterials, such as graphene and carbon nanotubes, as a core and a polymer matrix as a shell, throughout the manufacturing process (Thakur et al. 2023). Li et al. have reported the development of a simple yet efficient molecularly imprinted sensor for the selective detection of bovine serum albumin on the surface of GO-modified GCE using pyrrole electropolymerization. Using the simple DPV method, the modified sensor is utilized to selectively determine the amount of bovine serum albumin in the range of 1.0–100 $\mu M\ L^{-1}$ with excellent selectivity and recovery (Li et al. 2013). Additionally, by using the combination of magnetic graphene oxide (MGO) and metal nanoparticles, Li et al., developed an effective MIP using sunset yellow (dye) as template by modifying the GCE (glassy carbon electrode) with functionalized MGO, β-cyclodextrin (β-CD), ionic liquid (IL), and AuNPs. The proposed sensor possesses excellent selectivity, high sensitivity and immediate response towards sunset yellow (Li et al. 2016).

14.5 Commercially Available MIPs for Analytical Applications

Sensors are important tools during sample analysis owing to their excellent stability, sensitivity, selectivity, and reproducibility. Hence the molecular imprinting techniques could provide an excellent platform for sensors. However, the number of MIPs in commercial manufacturing still lags behind what was anticipated based on laboratory experience. How well this technique competes in the market against traditional solid-phase extraction sorbents, LC stationary phase materials, and sensor recognition components will be a major factor in future research into the field of molecular imprinting. Piletsky et al. outline the potential of MIP technology in a succinct summary from the end of 2000 (Piletsky et al. 2001). The estimated US market for separation methods will thus be around $1.19 billion. Particularly in the $500 million per year market for chromatography columns, MIP materials might gain a portion of 1–3% of this industry. Since up to 68% of medications now under research are chiral, the capacity of imprinted polymeric materials to resolve racemic mixtures might be extremely beneficial to the pharmaceutical sector. This translates to a $120 billion industry for chiral chemicals.

Additionally, others have drawn attention to the $4 billion market for specialized sensing devices that assess the quality of consumer goods as being potentially lucrative and a key area of use for MIPs (Anon 1998). The development of an enzyme-linked MIP sorbent test by Haupt and colleagues is a demonstration of the adaptability of MIPs as well as the viability of employing them in place of costly and frequently less stable biological recognition components like antibodies in such assay applications (Surugiu et al. 2000).

Nevertheless, a number of companies have released MIP-based devices on a commercial scale for a variety of applications, including sensing (van Nostrum 2005). For instance, 'MIP4SPE®' for solid phase extraction, 'MIP4LC®' for HPLC, and 'MIP4Discovery®' for drug screening, all of which were marketed by MIP Technologies, which is now a part of Biotage (http://www.biotage.com). For solid phase extraction, they also market a product called 'SupelMIP™ SPE' in conjunction with Supelco (http://www.sigmaaldrich.com.). Additionally, Biotage has made a variety of MIP-based products known as 'AFFINILUTE MIP products' available for sale. These products are utilized for solid phase extraction (SPE) analysis of substances such chloramphenicol, tobacco-specific nitrosamine derivatives, NNAL, triazines, clenbuterol, and b-agonists and b-blockers. MIPs for solid phase extraction (MIP[4] SPF) of specific analytes are made available by Chrysalis Scientific Technologies Inc. There are MIP[4] SPF products that target amphetamines (class specific), β-agonist drugs, β -blocker drugs, β -receptor (β-agonists + β-blockers), chloramphenicol, clenbuterol, fluoroquinolones, nitroimidazoles, NNAL (carcinogenic tobacco-specific nitrosamine), nonsteroidal anti-inflammatory drugs (NSAIDs), polycyclic aromatic hydrocarbon (PAHs), riboflavin (Vitamin B2), tobacco-specific nitrosamines (NNN, NNK, NAB, and NAT), and triazine herbicides (Mustafa and Lieberzeit 2013).

A few companies have recently been established that focus on mass-producing MIPs for sensing. MIPs are synthesized by Semorex for protein identification, with the potential to be expanded to cancer and bowel disease therapies. The bulk manufacture of protein imprinted polymers, however, needs a significant volume of the template, which entails expenses, if conventional methods are used. An alternative approach has been developed by MIP diagnostics, which might completely revolutionize the industry. The Sharnbrook, UK-based firm was established in 2015 with a focus on MIP development utilizing the solid-phase imprinting method to create high affinity nanoparticles, which was invented by the Piletsky group at the University of Leicester (Poma et al. 2013). The benefits of this strategy include the ability to recycle the template to reduce costs as well as the ability to produce products in a (semi)automated manner by eluting the polymers rather than the template (Canfarotta et al. 2016). Their produced nanoMIPs have been used to remove micropollutants and microorganisms from drinking water, as an alternative to ELISA and newer sensor formats. For scaling-up the operations, more investment is needed (Lowdon et al. 2020).

The company MIP diagnostics recently announced an exciting advancement in the collaboration with Stream Bio, an enterprise that is specialized in bioimaging to combine nanoMIPs and conjugated magnetic nanoparticles in a lateral flow assay for the development of a fast (10 min) diagnostic test for SARS-COV-2. By providing PoC diagnostics, the test might enhance the ability of healthcare systems throughout the world to diagnose diseases, but it can also be used to check effluent water for the SARS COV-2 virus. The latter is thought to be a possible route for disease propagation (Yeo et al. 2020). Futuristic MIP test are anticipated, and due to the probable intrinsic biocompatibility of nanoMIPs, potential for use in oral treatments and drug administration also can be explored (Cecchini et al. 2017).

In the future, developing reliable and repeatable sensors that can be used on actual samples may pose problems. This implies that for clinical samples, manufacturing in bulk of sensors that will enable comparable outcomes to be consistently attained. The detector would have to be either fully disposed of or reused, and the majority of the solid electrodes, like glassy carbon electrodes (GCEs), are used in current designs that do not meet to these rules. Future studies, on the other hand, may concentrate on produced carbon nanostructures or screen-printed electrodes that could be inexpensive enough to be disposed off. A molecularly imprinted polymer electrochemical sensor was developed by Lopes et al. (2017) using a commercially available screen-printed carbon electrode (SPCE) and modified to enable quick and easy detection of NLX.

The reason for MIPs holding a little market share may be related to the 'unsustainability' of traditional MIPs, which has hindered their widespread practical use, large-scale manufacturing, and global availability. Consequently, greenified MIPs (GMIPs) emerged as a novel class of synthetic receptors that have adopted one or more greenification principles in their approach to sustainability. This has mitigated or reduced the harmful effects of molecularly imprinted technology (MIT) on the environment and human health, and it has opened up new avenues for MIP uses (Ostovan et al. 2022). In comparison to the same polymers prepared using common synthetic routes, the introduction of biomass waste, such as chitosan, cellulose, or cyclodextrins, by replacing typical highly toxic chemicals in MIP preparation can determine an increase of adsorption and selectivity performance, high availability as well as greater sensitivity, morphology control and low cost of production (Sole et al. 2021). Another approach towards green chemistry is the use of modern technology such as computational modelling that allow the rational design of MIP by screening the probable monomers before the synthesis of desired polymers (Tabandeh et al. 2012). For instance, density functional theory is investigated in the rational design of MIPs, where a significant deal of work has been accomplished in the careful selection of pre-polymerization reagents that reduce experimental costs and minimize waste formation Furthermore, the application of MIPs, particularly in the food and biological sectors, is being investigated using miniaturized techniques that fall neatly into the purview of green analytical chemistry (Madikizela et al. 2018).

Instrumentation presents another difficulty for field implementation. Since its recent invention, microsystems have been employed as a potent tool for combining several analytical processes into one. Due to MIPs' complete compatibility with nanotechnology and lab-on-a-chip, it is feasible to use them for both sample preparation and detection on the same microfluidic platform. Furthermore, as many electrochemical methods or equilibration durations for carbon nanomaterial sensors might be sluggish, it would be appropriate to concentrate on quick detection. Potential sensors have a wider variety of uses than only clinical research. Promising applications include those in industry, the environment, and pharmaceuticals, which may make use of a variety of sensors for quick event monitoring.

Another objective is to create implanted sensors for biosensors that might track proteins linked to human illness continually. The potential toxicity of the materials is a significant problem with carbon nanomaterial-based instruments for these kinds of applications. Many of the current research on the toxicity of carbon nanotubes have focused on the consequences of inhaling the particles, and the results indicate that aggregates of CNTs may behave similarly to asbestos. Longer CNTs are more difficult to remove than short CNTs and are more resistant to phagocytosis. Beyond CNTs, there have been other novel kinds and geometries of carbon nanomaterials that need to be studied; some of these materials could be less harmful than others. Implantable sensors have a sizable potential market if they can be manufactured using nontoxic materials or with as little leaching as possible (Ansari 2017).

Vlatakis et al. (1993) and Andersson et al. (1995) indicated sensitivities in the μmol/l range for hydrophobic analytes, whereas more recent studies reveal detection limits in the nmol/l range, enabling MIP-based afinity tests at clinically meaningful doses (Vlatakis et al. 1993; Andersson et al. 1995; Bengtsson et al. 1997; Andersson 1996). These detection limits, however, are larger than what is typically attained with radioimmunoassays. The leaking of leftover template molecules is a constraint on decreasing detection limits in MIP-based affinity tests. Better cleaning procedures of the MIP before use in the assay are necessary to remedy this circumstance.

The application of NPs in conjunction with MIPs for colorimetric tea residue detection of the pesticide Cartap was described by Wu et al. (2018). Using methacrylic acid as a functional monomer, magnetic molecular imprinted microspheres (Fe_3O_4@mSiO_2@MIPs) were used to extract cartap from both spiked and non-spiked tea samples. AgNPs were used to elicit a color shift, converting the elution from yellow to grey depending on the presence of the target when cartap was eluted from the MIPs. This led to the development of a semi-quantitative test that can detect Cartap at concentrations as low as ± 5 mg L^{-1}. A UV–vis spectrophotometer allows for an additional quantification of the reaction with a limit of detection of 0.01 mg L^{-1}. This method's simplicity makes it highly fascinating, but it's not particularly sensitive and requires a multistep technique, which takes time.

By utilizing AuNPs for signal amplification, Zhao et al. presented a technique that can solve the latter issue. They were able to build a colorimetric sensor with a LOD of 1.2 μg L^{-1} and a LOQ of 4.0 μg L^{-1} for the detection of atrazine (Zhao et al. 2019). Because of this, the assay can now compete with the majority of sensor applications that can identify targets in the lower nanomolar concentration range. Nonetheless, sample pretreatment is required for sample analysis, which restricts the method's commercial use. There are other reports of comparable applications for nanoparticles, but they are all plagued by the same problems that make them unfeasible for commercial usage (Feng et al. 2017).

The commercialization of the assays generated was impeded by the use of radiolabeling, despite significant improvements in the binding of target analytes and environmental analysis. The feasibility of radiolabeled experiments was deemed impossible due to concerns raised about the availability, handling, cost, and coupling of isotopic labels. Piletsky et al. were among the first to develop fluorescently tagged analogues of triazine for the detection of unlabeled triazine (Piletsky et al. 1997). Other approaches were subsequently developed that showed more promise. Fluorescent monomers, crosslinking agents, and initiators would soon be used to include fluorescent qualities into the physical characteristics of the imprinted polymers (Turkewitsch et al. 1998). This made it possible to examine fluorescence enhancement and quenching as viable methods for analyte direct detection.

Pardieu et al. reported an electrochemical sensor for the detection of atrazine by creating poly (3,4-ethylenedioxythiophene-co-thiophene-acetic acid), a molecularly imprinted conducting polymer (MICP) (Pardieu et al. 2009). The polymer was electrochemically synthesized on a platinum electrode while atrazine was present as a template. With a low detection limit of 10^{-7} mol L^{-1}, this MICP-based sensor demonstrated selectivity for the triazine family throughout a wide dynamic range of

10^{-7} mol L^{-1} to 1.5×10^{-2} mol L^{-1} atrazine. The 2,4-dichlorophenoxyacetic acid (2,4-D) was detected in this example using cyclic voltammetry, a method that was previously successful but did not involve evaluation of the sensor response in real-world samples (Kroger et al. 1999). However Udayan reported that by the utilization of MnO_2-NiO as the catalytic probe, a novel label free electrochemical sensor has been fabricated for selective and sensitive identification of atrazine with detection limit of 400×10^{-9} M; the linear range for atrazine was from 0.8×10^{-6} to 120×10^{-6} M in pulse amperometric mode (Udayan 2023). In the electrochemical impedance spectroscopic investigation, MnO_2-NiO modified electrode exhibited a decrease in the electron transfer resistance in response to increase in the concentration of atrazine indicating improved electron-transfer kinetics for atrazine reduction. The sensor was successfully illustrated for detecting atrazine in tap and waste samples.

Martin et al. in 2001 reported that despite the clear differences in the molecular structures of the template molecules, it was discovered that, following imprinting a polymer against the drug tamoxifen for solid-phase extraction purposes, a control polymer imprinted against propanolol displayed greater affinity for tamoxifen than the tamoxifen imprinted polymer (Martin et al. 2001). Such irregularities need an in-depth and rigorous investigation and analysis of MIP systems before they are widely accepted as a simple analytical method.

Jia et al. investigated the process of creating imprinted polymer nanoparticles (MIPNPs) using glucose (Glu-FH) and bleomycin (BLM) as the dual templates for the 71–80 peptide of human fibroblast growth-factor-inducible14 (Jia et al. 2019). In order to create FH-MIPNPs/BLM, the prepared MIPNPs may load BLM and Glu-FH concurrently. The in vitro findings demonstrated that FH-MIPNPs may specifically target human fibroblast growth-factor-inducible 14 (FN14)-overexpressing BxPC-3 cells. On cells lacking FN14 expression, however, the targeting effect was not discernible. However, further study is required on stability, biocompatibility, and toxicity, particularly long-term toxicity, especially when MIP is used in intravenous drug delivery systems. Second, as organic solvent is used to synthesis the majority of MIP, organic solvent residue needs to be closely monitored.

In order to address the needs of clinical applications, more perfect ingredients like functional monomers, need to be produced for the preparation of molecularly imprinted polymers. It is possible to requisition of computer-based high-throughput screening methods. Finally, the two most important issues that need to be resolved for MIP to be successfully applied in a clinical setting are the transition from bench to clinical application and large-scale manufacturing (He et al. 2021).

According to a 2011 study by Puoci, Cirillo et al., the primary drawback of using conventional MIPs in the biomedical area is that their optimal performance is found in hydrophobic organic solvents (Puoci et al. 2011). This might be explained by the fact that the contacts involved in the chemical recognition mechanism have the optimal environment when created in apolar solvents, which remove nonspecific hydrophobic interactions (Piletska et al. 2008). There are some areas for MIPs application, for example, SPE of natural products, where the use of hydrophobic solvents could give an advantage; but when considering some of the others, full compatibility with water is required (Shi et al. 2007). For the use of MIPs in biomedical applications, and medication delivery in particular, the ability to apply molecular imprinting to construct synthetic receptors for water-soluble compounds would be highly promising (Piletsky et al. 1999; Oral and Peppas 2006; Bures et al. 2001).

Venkatesh et al. (2021) reported that in order to have a handheld and fully quantitative device with easy operation further research and product development is required in the enhancement of sensitivity and reduction of device variance via electrode design refinement, additional transduction layers and automation of testing procedures.

14.6 Concluding Remarks and Future Prospects

Molecular imprinting has potential to grow synergistically with technological developments in the field of separation, catalysis, diagnostic tools and many more in industry in a cost-effective manner. The society can be benefited by its multipronged capability. Molecular imprinting has a potential for diagnostics, chromatography, and purification columns, and many more. Although few companies have started setting up smaller companies but still the potential is not tapped or rather one can say still in infancy stage. Attempts are being made up to scale-up the laboratory scale experiments, still many more steps need to be taken up. This chapter was conceptualized thoughtfully for looking out the limitations/barriers/hurdles being faced by imprinting fraternity preventing its commercialization/industrialization, as expected looking at its capabilities and efficacy. The factors affecting the key steps in fabrication of MIPs which are crucial in their successful application to "real" samples and affecting the potency for commercial production are identified and critically discussed to improve upon them. One of the major barriers that hinders the commercial viability of MIPs is the ability to scale up synthesis with high homogeneity and reproducibility for subsequent incorporation into sensors (Lowdon et al. 2020). The technology of nanoMIPs based on a solid-phase approach by Canfarotta et al. shows promises for overcoming the batch-to-batch variation and exploring the scale up potential for MIPs (Canfarotta et al. 2016). The generic nature of MIP technology as opposed to its natural counterpart allows for development of detection kits for molecules of any type and size and thus holds the potential to change the dynamics of the diagnostics industry in the coming years. The success of molecular imprinting at laboratory scale shows better prospects for the better future.

Acknowledgements Financial support from IoE, BHU (6031) is acknowledged. AS acknowledges the financial assistance from UGC [Scheme no 31(A)].

References

Abass AM, Rzaij JM (2020) A review on: molecularly imprinting polymers by ion selective electrodes for determination drugs. J Chem Rev 2:148–156
Ahmad OS, Bedwell TS, Esen C et al (2019) Molecularly imprinted polymers in electrochemical and optical sensors. Trend Biotechnol 37(3):294–309
Akgönüllü S, Özgür E, Denizli A (2022) Recent advances in quartz crystal microbalance biosensors based on the molecular imprinting technique for disease-related biomarkers. Chemosensors 10(3):106
Altintas Z, Guerreiro A, Piletsky SA et al (2015) NanoMIP based optical sensor for pharmaceuticals monitoring. Sens Acutator B Chem 213:305–313
Anderson H, Jönsson M, Vestling L et al (2007) Quartz crystal microbalance sensor design: I. Experimental study of sensor response and performance. Sens Actuators B Chem 123(1):27–34
Andersson LI (1996) Application of molecular imprinting to the development of aqueous buffer and organic solvent based radioligand binding assays for (S)-propranolol. Anal Chem 68(1):111–117
Andersson HS, Nicholls IA (1997) Spectroscopic evaluation of molecular imprinting polymerization systems. Bioorg Chem 25(3):203–211
Andersson LI, Müller R, Vlatakis G et al (1995) Mimics of the binding sites of opioid receptors obtained by molecular imprinting of enkephalin and morphine. Proc Natl Acad Sci 92(11):4788–4792
Anon (1998) The Economist (Lond.) 348:78
Ansari S (2017) Combination of molecularly imprinted polymers and carbon nanomaterials as a versatile biosensing tool in sample analysis: recent applications and challenges. TrAC 93:34–151
BelBruno JJ (2018) Molecularly imprinted polymers. Chem Rev 119(1):94–119
Bengtsson H, Roos U, Andersson LI (1997) Molecular imprint based radioassay for direct determination of S-propranolol in human plasma. Anal Commun 34(9):233–235
Bhakta S, Dixit CK, Bist I et al (2017) Albumin removal from human serum using surface nanopockets on silica-coated magnetic nanoparticles. Chem Comm 53(66):9254–9257
Bures P, Huang Y, Oral E et al (2001) Surface modifications and molecular imprinting of polymers in medical and pharmaceutical applications. J Control Release 72:25–33
Canfarotta F, Poma A, Guerreiro A et al (2016) Solid-phase synthesis of molecularly imprinted nanoparticles. Nat Protoc 11(3):443–445
Cecchini A, Raffa V, Canfarotta F et al (2017) In vivo recognition of human vascular endothelial growth factor by molecularly imprinted polymers. Nano Lett 17(4):2307–2312
Chen L, Xu S, Li J (2011) Recent advances in molecular imprinting technology: current status, challenges and highlighted applications. Chem Soc Rev 40(5):2922–2942
Chen W, Meng Z, Xue M et al (2016) Molecular imprinted photonic crystal for sensing of biomolecules. Mol Imprinting 4(1):1–12
Chianella I, Lotierzo M, Piletsky SA et al (2002) Rational design of a polymer specific for microcystin-LR using a computational approach. Anal Chem 74:1288–1293
Chou CY, Lin CY, Wu CH et al (2020) Sensing HIV protease and its inhibitor using "helical epitope"—imprinted polymers. Sensors 20(12):3592
Das TR, Bhattacherjee P, Patra S et al (2021) Metal−organic framework-based artificial antibodies for sensing applications. In: Metal−organic frameworks for environmental sensing, J Am Chem Soc, pp 155–175
Del Sole R, Mele G, Bloise E et al (2021) Green aspects in molecularly imprinted polymers by biomass waste utilization. Polym 13(15):2430
Dhal PK, Arnold FH (1996) Substrate selectivity of molecularly imprinted polymers incorporating a rigid chelating monomer bis-methacrylato (4-methyl 4-vinyl)2,2-bipyridine Cu(II). New J Chem 20:695–698
Dixit CK, Bhakta S, Reza KK et al (2022) Exploring molecularly imprinted polymers as artificial antibodies for efficient diagnostics and commercialization: a critical overview. Hybrid Adv 100001

El-Akaad S, Mohamed MA, Abdelwahab NS et al (2020) Capacitive sensor based on molecularly imprinted polymers for detection of the insecticide imidacloprid in water. Sci Rep 10(1):14479
Feng S, Hu Y, Ma L et al (2017) Development of molecularly imprinted polymers-surface-enhanced Raman spectroscopy/colorimetric dual sensor for determination of chlorpyrifos in apple juice. Sens Actuators B Chem 241:750–757
Guoning C, Hua S, Wang L et al (2020) A surfactant-mediated sol-gel method for the preparation of molecularly imprinted polymers and its application in a biomimetic immunoassay for the detection of protein. J Pharm Biomed Anal 190:113511
Hart BR, Shea KJ (2014) Molecularly imprinted polymers. Wiley, New York, pp 684–704
He S, Zhang L, Bai S et al (2021) Advances of molecularly imprinted polymers (MIP) and the application in drug delivery. Eur Polym J 143:110179
Hoshino Y, Shea KJ (2011) The evolution of plastic antibodies. J Mater Chem 21(11):3517–3521
Janata J (2010) Principles of chemical sensors. Springer Science & Business Media
Jia C, Zhang M, Zhang Y et al (2019) Preparation of dual-template epitope imprinted polymers for targeted fluorescence imaging and targeted drug delivery to pancreatic cancer BxPC-3 cells. ACS Appl Mater Interfaces 11(35):32431–32440
Kroger S, Turner AP, Mosbach K et al (1999) Imprinted polymer based sensor system for herbicides using differential-pulse voltammetry on screen printed electrodes. Anal Chem 71:3698–3702
Li H, Wang Z, Wu B et al (2012) Rapid and sensitive detection of methyl-parathion pesticide with an electropolymerized, molecularly imprinted polymer capacitive sensor. Electrochim Acta 62:319–326
Li F, Xia JF, Wang ZH et al (2013) Graphene modified molecular imprinted electrochemical sensor for specific recognition of bovine serum albumin. Adv Mat Res 709:891–894
Li S, Cao S, Whitcombe MJ et al (2014) Size matters: challenges in imprinting macromolecules. Prog Polym Sci 39(1):145–163
Li J, Wang X, Duan H et al (2016) Based on magnetic graphene oxide highly sensitive and selective imprinted sensor for determination of sunset yellow. Talanta 147:169–176
Liu JQ, Wulff G (2004) Functional mimicry of the active site of carboxypeptidase A by a molecular imprinting strategy: cooperativity of an amidinium and a copper ion in a transition-state imprinted cavity giving rise to high catalytic activity. J Am Chem Soc 126(24):7452–7453
Lopes F, Pacheco JG, Rebelo P et al (2017) Molecularly imprinted electrochemical sensor prepared on a screen printed carbon electrode for naloxone detection. Sens Actuators B Chem 243:745–752
Lowdon JW, Diliën H, Singla P et al (2020) MIPs for commercial application in low-cost sensors and assays—an overview of the current status quo. Sens Acutator B Chem 325:128973
Madikizela LM, Tavengwa NT, Tutu H et al (2018) Green aspects in molecular imprinting technology: from design to environmental applications. Trends Environ Anal Chem 17:14–22
Mahony JO, Nolan K, Smyth MR et al (2005) Molecularly imprinted polymers—potential and challenges in analytical chemistry. Anal Chim Acta 534(1):31–39
Martin PD, Wilson TD, Wilson ID et al (2001) An unexpected selectivity of a propranolol-derived molecular imprint for tamoxifen. Analyst 126(6):757–759
Marx KA (2003) Quartz crystal microbalance: a useful tool for studying thin polymer films and complex biomolecular systems at the solution—surface interface. Biomacromol 4(5):1009–1120
Minoura N, Idei K, Rachkov A et al (2004) Preparation of azobenzene-containing polymer membranes that function in photoregulated molecular recognition. Macromol 37(25):9571–9576
Montagut Y, Garcia JV, Jimenez Y et al (2011) Validation of a phase-mass characterization concept and interface for acoustic biosensors. Sensors 11:4702–4720
Morsi SM, Abd El-Aziz ME, Mohamed HA (2023) Smart polymers as molecular imprinted polymers for recognition of target molecules. Int J Polym Mater 72(8):612–635
Mosbach K, Haupt K (1998) Some new developments and challenges in non-covalent molecular imprinting technology. J Mol Recognit 11:62–68
Mustafa G, Lieberzeit PA (2013) MIP sensors on the way to real-world applications. Designing receptors for the next generation of biosensors, pp 167–187

Nash ME, Fan X, Carroll WM et al (2013) Thermoresponsive substrates used for the expansion of human mesenchymal stem cells and the preservation of immunophenotype. Stem Cell Rev 9:148–157

Oral E, Peppas NA (2006) Hydrophilic molecularly imprinted poly(hydroxyethylmethacrylate) polymers. J Biomed Mater Res 78A:205–210

Osmani Q, Hughes H, Flavin K et al (2008) The use of FTIR and NMR spectroscopies to study prepolymerization interactions in nitrogen heterocycles. Anal Bioanal Chem 391:1229–1236

Ostovan A, Arabi M, Wang Y, Li J, Li B, Wang X, Chen L (2022) Greenificated molecularly imprinted materials for advanced applications. Abstr Adv Mater 34(42). https://doi.org/10.1002/adma.202203154

Ozcelikay G, Kaya SI, Ozkan E et al (2022) Sensor-based MIP technologies for targeted metabolomics analysis. TrAC 146:116487

Özgür E, Parlak O, Beni V et al (2017) Bioinspired design of a polymer-based biohybrid sensor interface. Sens Actuators B Chem 251:674–682

Pardieu E, Cheap H, Vedrine C et al (2009) Molecularly imprinted conducting polymer based electrochemical sensor for detection of atrazine. Anal Chim Acta 649:236–245

Peppas NA, Clegg JR (2016) The challenge to improve the response of biomaterials to the physiological environment. Regen Biomater 3(2):67–71

Piletska EV, Guerreiro AR, Romero-Guerra M et al (2008) Design of molecular imprinted polymers compatible with aqueous environment. Anal Chim Acta 607:54–60

Piletsky SA, Piletskaya EV, El'Skaya AV et al (1997) Optical detection system for triazine based on molecularly-imprinted polymers. Synth React Inorg Met 30(3):445–455

Piletsky SA, Andersson HS, Nicholls IA (1999) Combined hydrophobic and electrostatic interaction-based recognition in molecularly imprinted polymers. Macromolecules 32:633–636

Piletsky SA, Alcock S, Turner AP (2001) Molecular imprinting: at the edge of the third millennium. Trends Biotechnol 19(1):9–12

Poma A, Guerreiro A, Whitcombe MJ et al (2013) Solid-phase synthesis of molecularly imprinted polymer nanoparticles with a reusable template- "plastic antibodies." Adv Funct Mater 23(22):2821–2827

Puoci F, Cirillo G, Curcio M et al (2011) Molecularly imprinted polymers in drug delivery: state of art and future perspectives. Expert Opin Drug Deliv 8(10):1379–1393

Qin L, He XW, Yuan X et al (2011) Molecularly imprinted beads with double thermosensitive gates for selective recognition of proteins. Anal Bioanal Chem 399:3375–3385

Roy E, Patra S, Tiwari A et al (2017) Introduction of selectivity and specificity to graphene using an inimitable combination of molecular imprinting and nanotechnology. Biosens Bioelectron 89:234–248

Sadana A, Sadana N, Sadana R (2018) A fractal analysis of chemical kinetics with applications to biological and biosensor interfaces, 1st edn. Elsevier Science, p 337

Sauerbray G (1959) Z Phys 155:206–222

Saylan Y, Erdem Ö, Inci F et al (2020) Advances in biomimetic systems for molecular recognition and biosensing. Biomimetics 5(2):20

Sellergren B, Lepistö M, Mosbach K (1988) Highly enantioselective and substrate-selective polymers obtained by molecular imprinting utilizing noncovalent interactions—NMR and chromatographic studies on the nature of recognition. J Am Chem Soc 110:5853–5860

Shea KJ (1994) Molecular imprinting of synthetic network polymers: The de novo synthesis of macromolecular binding and catalytic sites. Trends Polym Sci 2:166–173

Shi X, Wu A, Qu G et al (2007) Development and characterization of molecularly imprinted polymers based on methacrylic acid for selective recognition of drugs. Biomaterials 28:3741–3749

Shiraki Y, Tsuruta K, Morimoto J et al (2015) Preparation of molecule-responsive microsized hydrogels via photopolymerization for smart microchannel microvalves. Macromol Rapid Commun 36(6):515–519

Singh AK, Singh M (2015) QCM sensing of melphalan via electropolymerized molecularly imprinted polythiophene films. Biosens Bioelectron 74:711–717

Singh R, Singh M (2022) Highly selective and specific monitoring of pollutants using dual template imprinted MIP sensor. J Electroanal Chem 926:116939

Subrahmanyam S, Piletsky SA, Piletska EV et al (2001) "Bite-and-switch" approach using computationally designed molecularly imprinted polymers for sensing of creatinine. Biosens Bioelectron 16:631–637

Subrahmanyam S, Karim K, Piletsky SA (2013) Computational approaches in the design of synthetic receptors. In: Piletsky SA, Whitcombe MJ (ed) Designing receptors for the next generation of biosensors. Springer series on chemical sensors and biosensors, pp 131–165

Surugiu I, Ye L, Yilmaz E et al (2000) An enzyme-linked molecularly imprinted sorbent assay. Analyst 125(1):13–16

Suryana S, Mutakin RY et al (2021) An update on molecularly imprinted polymer design through a computational approach to produce molecular recognition material with enhanced analytical performance. Molecules 26(7):1891

Svenson J, Andersson HS, Piletsky SA et al (1998) Spectroscopic studies of the molecular imprinting self-assembly process. J Mol Recognit 11:83–86

Tabandeh M, Ghassamipour S, Aqababa H et al (2012) Computational design and synthesis of molecular imprinted polymers for selective extraction of allopurinol from human plasma. J Chromatogr B Anal Technol Biomed Life Sci B 898:24–31

Thakur N, Das TR, Gupta S et al (2023) Material selection for the fabrication of MIP-based devices: an overview. Molecularly imprinted polymers (MIPs), pp 101–114

Turkewitsch P, Wandelt B, Darling G D et al (1998) Fluorescent functional recognition sites through molecular imprinting. A polymer-based fluorescent chemosensor for aqueous cAMP. Anal Chem 70(10):2025–2030

Udayan APM (2023) Electrochemical detection of herbicide atrazine using porous MnO2-NiO nanocatalyst. Mater Sci Eng B 290:116302

Ugalmugale S, Swain R (2018) Global market insights Inc.

van Nostrum CF (2005) Molecular imprinting: a new tool for drug innovation. Drug Discov Today Technol 2(1):119–124

Venkatesh S, Yeung CC, Li T et al (2021) Portable molecularly imprinted polymer-based platform for detection of histamine in aqueous solutions. J Hazard Mater 410:124609

Vlatakis G, Andersson LI, Müller R et al (1993) Drug assay using antibody mimics made by molecular imprinting. Nature 361(6413):645–647

Wang H, Yi J, Velado D et al (2015) Immobilization of carbon dots in molecularly imprinted microgels for optical sensing of glucose at physiological pH. ACS Appl Mater Interfaces 7(29):15735–15745

Weerasuriya DRK, Bhakta S, Hiniduma K et al (2021) Magnetic nanoparticles with surface nanopockets for highly selective antibody isolation. ACS Appl Bio Mater 4(8):6157–6166

Whitcombe MJ, Martin L, Vulfson EN (1998) Predicting the selectivity of imprinted polymers. Chromatographia 47:457–464

Wu HG, Ju XJ, Xie R et al (2011) A novel ion-imprinted hydrogel for recognition of potassium ions with rapid response. Polym Adv Technol 22(9):1389–1394

Wu M, Deng H, Fan Y et al (2018) Rapid colorimetric detection of cartap residues by AgNP sensor with magnetic molecularly imprinted microspheres as recognition elements. Molecules 23(6):1443

Wulff G (1995) Molecular imprinting in cross-linked materials with the aid of molecular templates—a way towards artificial antibodies. Angew Chem Int Ed Engl 34(17):1812–1832

Wulff G (2013) Fourty years of molecular imprinting in synthetic polymers: Origin, features and perspectives. Microchim Acta 180:1359–1370

Wulff G, Knorr K (2001) Stoichiometric noncovalent interaction in molecular imprinting. Bioseparation 10:257–276

Wulff G, Liu J (2012) Design of biomimetic catalysts by molecular imprinting in synthetic polymers: the role of transition state stabilization. Acc Chem Res 45:239–247

Wulff G, Chong BO, Kolb U (2006) Soluble single-molecule nanogels of controlled structure as a matrix for efficient artificial enzymes. Angew Chem Int Ed 45(18):2955–2958
Wyszomirski M, Prus W (2012) Molecular modelling of a template substitute and monomers used in molecular imprinting for aflatoxin B1 micro-HPLC analysis. Mol Simulat 38:892–895
Yang B, Fu C, Li J et al (2018) Frontiers in highly sensitive molecularly imprinted electrochemical sensors: challenges and strategies. TrAC 105:52–67
Yarman A, Kurbanoglu S, Zebger I et al (2021) Simple and robust: the claims of protein sensing by molecularly imprinted polymers. Sens Acutators B Chem 330:129369
Ye L, Mosbach K (2008) Molecular imprinting: Synthetic materials as substitutes for biological antibodies and receptors. Chem Mater 20:859–868
Yeo C, Kaushal S, Yeo D (2020) Enteric involvement of coronaviruses: is faecal–oral transmission of SARS-CoV-2 possible? Lancet Gastroenterol Hepatol 5(4):335–337
Yilmaz E, Schmidt RH, Mosbach K (2005) The non-covalent approach. Marcel Dekker, New York, pp 25–57
Zhang H, Ye L, Mosbach K (2006) Non-covalent molecular imprinting with emphasis on its application in separation and drug development. J Mol Recognit 19:248–259
Zhao WF, Fang BH, Li N et al (2009) Fabrication of pH-responsive molecularly imprinted polyethersulfone particles for bisphenol-A uptake. J Appl Polym Sci 113(2):916–921
Zhao B, Feng S, Hu Y et al (2019) Rapid determination of atrazine in apple juice using molecularly imprinted polymers coupled with gold nanoparticlescolorimetric/SERS dual chemosensor. Food Chem 276:366–375

GPSR Compliance
The European Union's (EU) General Product Safety Regulation (GPSR) is a set of rules that requires consumer products to be safe and our obligations to ensure this.

If you have any concerns about our products, you can contact us on

ProductSafety@springernature.com

In case Publisher is established outside the EU, the EU authorized representative is:

Springer Nature Customer Service Center GmbH
Europaplatz 3
69115 Heidelberg, Germany

www.ingramcontent.com/pod-product-compliance
Ingram Content Group UK Ltd.
Pitfield, Milton Keynes, MK11 3LW, UK
UKHW022216051026
15624UKWH00001B/184
* 9 7 8 3 0 3 1 5 8 9 9 4 2 *